SYNERGY

SYNERGY

MARK L. LATASH

OXFORD
UNIVERSITY PRESS

2008

OXFORD
UNIVERSITY PRESS

Oxford University Press, Inc., publishes works that further
Oxford University's objective of excellence
in research, scholarship, and education.

Oxford New York
Auckland Cape Town Dar es Salaam Hong Kong Karachi
Kuala Lumpur Madrid Melbourne Mexico City Nairobi
New Delhi Shanghai Taipei Toronto

With offices in
Argentina Austria Brazil Chile Czech Republic France Greece
Guatemala Hungary Italy Japan Poland Portugal Singapore
South Korea Switzerland Thailand Turkey Ukraine Vietnam

Published by Oxford University Press, Inc.
198 Madison Avenue, New York, New York 10016

www.oup.com

Library of Congress Cataloging-in-Publication Data

Latash, Mark L., 1953-
Synergy / Mark L. Latash.
p. ; cm.
Includes bibliographical references.
ISBN: 978-0-19-533316-9
1. Human mechanics. 2. Biomechanics. I. Title.
[DNLM: 1. Movement—physiology. 2. Biomechanics. 3. Motor
Activity—physiology. 4. Neurophysiology—methods. WE 103 L351s 2008]
QP303.L33 2008
612.7'6—dc22
2007031088

PREFACE

We live in very interesting times. A few previous centuries were those of physics. The great scientists of the past 300 years have developed an adequate set of notions and computational apparatus to deal with the laws of inanimate nature, and have reached such enormous success that it has become possible to exterminate the whole population of our planet in many different ways. If human beings want to survive as a species, the twenty-first century has to be the century of biology. Our ignorance in the very basic issues of life is so profound that we do not even have proper words to express it. Biology is still to develop an adequate language that would allow researchers to formulate problems before trying to solve them. One of the main goals of this book is to sketch directions of research that may lead to discovery of such adequate words in an area of biology that deals with movements—certainly in the very subjective opinion of the author. I will try to persuade the reader that *synergy* may well be such a word if it is properly and operationally defined.

Why movements and why synergy? Being a physicist by training, I have always believed that any theory should be grounded in an experiment and should make predictions that can be observed and measured. Biological movements (human movements being one example) obey the well-established laws of classical mechanics, and there are experimental and computational methods to quantify them. On the other hand, human movements are products of the brain and can be used to peek into this small crack at the functions of the arguably most enticing and mysterious organ of the human body.

The word *synergy* has been around for millennia, and it meant a lot of different things, from its direct meaning of "work together" (Greek) to something esoteric and mystical. I believe that both extremes are unproductive, as most extremes usually are. To claim that any material object is a synergy, for example, that an atom is a synergy of elementary particles does not sound very exciting. If non-synergies do not exist, why bother using a new word? Let us call these things simply material objects. On the other hand, to claim that synergy is highly mystical, nondefinable, and nonmeasurable may make the word attractive for some people but not for scientists, definitely not for physicists. So, both "every material object is a synergy" and "synergy cannot be measured" are equally unattractive. A word starts making scientific sense and can be used in research only after it has been operationally defined. In other words, one should know how to tell things that this word describes from those that the word is inapplicable to. One should have in one's mind at least a hypothetical procedure that could be used to measure things this word means. To me, *synergy* starts making scientific sense if I know procedures that allow me to tell a synergy from a non-synergy and that can be used to measure synergies.

The book is composed following a certain logic. In Part 1, I start with an attempt to build a definition for *synergy* based on a few axiomatic statements and an intuitive feeling of what this word means. Then, since the book is mostly about movements, Part 2 presents a brief historical account of movement science with a rather large section on the history of movement science in the Soviet Union in the twentieth century.

The next major section (Part 3) discusses issues of motor control: This is a very subjective section that will not be taken benevolently by many of my respected colleagues. This is likely to happen because in Part 3, I claim that the only theory of motor control that is compatible with physics, common sense, and the existing knowledge about the human body is the equilibrium-point hypothesis. All the currently available alternatives are so obviously wrong that I wonder how they have managed to survive and crawl from one textbook into another for years and decades. In this part, there are quite a few *Digressions*; their purpose is to introduce basic facts about the human system for movement production and methods of its analysis without interrupting the main story about motor control and coordination. There are a few more *Digressions* in further sections.

Further, the book develops in two dimensions. First, it moves deeper into physiology and physics (with a bit of math—not too much, just enough) to develop a computational apparatus to identify and quantify synergies (Part 4). Second, it expands "sideways" using more and more examples illustrating how the suggested definition and computational methods can be used to analyze a wide variety of motor synergies. Part 5 describes a zoo of motor synergies. In Part 6, more synergies are discussed, including those that can be seen in atypical persons (e.g. in persons with Down syndrome), in elderly, and in patients with neurological disorders. This part also addresses issues related to effects of practice on synergies.

The next few sections (Part 7) discuss the role of various neurophysiological structures and mechanisms in synergies. And then, in Part 8, I describe several models of very different nature that try to show how salient features of synergies could be based on various computational and neural principles. I am a bit ashamed of the last couple of sections of Part 8 that are about issues that I simply do not know well enough, the sensory function and the language. But these are such exciting topics, and their relevance to the notion of synergy is so obvious to me that it was impossible to fight the temptation. I would like to beg forgiveness of those of my colleagues who work in those areas and who will likely be appalled by the depth of my ignorance reflected in those sections.

This book is very personal. It reflects what I truly think about a variety of things, scientific and nonscientific. Is it a reference book? Yes, it presents a review on a large body of knowledge in the area of motor control and coordination. Is it a textbook? Yes, it can be used as a textbook, although building a course exclusively on this book may be a challenge. Is this book about movements? Yes, but not that alone it is also about many other things.

The book was designed and written having in mind a variety of potential readers, from inquisitive high school kids to graduate students and professionals such as researchers, university professors, and clinicians. On the one hand, I have tried to present briefly all the necessary background information. On the other hand, I have assumed that the reader is familiar with simple equations and graphs, is not afraid of vectors and matrices, and knows what a derivative is. There is a certain progression in the complexity of the material from the first sections to the last ones: It starts with simple examples and historical essays, and then the material develops and covers problems that are complex, controversial, and yet unsolved.

How did it happen that this book got conceived and written? This is not an easy question that begs for a lengthy answer. In brief, it has been written because I have been very lucky with parents, friends, mentors, colleagues, and students.

My first and most influential mentor was my father, Lev Latash. He was to me an embodiment of a true and uncompromising scientist. His scientific career was very far from smooth. After getting a medical degree, he was fired from the graduate school for "supporting the anti-scientific views" of his advisor, academician Lina S. Shtern, the first female member of the Academy of Sciences of the Soviet Union. Then, he spent several years in exile in Kazakhstan and, after Stalin's death, returned to Moscow where he worked as a physician for a few more years before finally getting a chance to return to research in neurophysiology. I am sure that, if he lived in middle ages, he would be a tzadik. However, in the Soviet Union, he became a scientist. He taught me by example how to see exciting and enigmatic things in everyday life—potential topics for scientific research. He insisted that in a nonfree country, the only chance to have a rich and uncompromising life was to become a scientist. And then I got into the hands of Victor Gurfinkel, a world-famous scientist in the area of postural control and movement disorders.

Over my scientific life, four senior colleagues have been my formal and informal mentors. My first advisor, Victor Gurfinkel helped me, an ignorant and anything but humble undergraduate student, during the very first steps in a research environment. His sense of humor combined with the most serious attitude to science has formed a truly unique cocktail. He taught me that performing high-quality research did not mean being boringly serious and high-nosed. Victor Gurfinkel has been famous for great one-liners that were frequently very instructive. I still remember that "there are no boring topics for research, there are boring ways of doing any research." Unfortunately, many of his great stories and pseudo-scientific illustrations are too graphic to be put on paper. But I will try to recollect one of the softer stories that he told me when I showed him my first recording of muscle activity that was relatively free of noise and erratic jumps of the signal. He said: "OK, this is already an intermediate result." And then explained: "One of the Soviet research institutes got a very important task to develop a chemical process that turns human excrement into butter. After the first 5 years of research, they were asked for a report. The director of the group declared that they had achieved an impressive intermediate result: It could already be spread over a slice of bread!"

My next mentor–friend was (and still is) Anatol Feldman. By the time I was fired from my first job because of my application for emigration, Feldman did not have younger colleagues to work with. So, he suggested me to come to his laboratory and work together. This was a very noble and brave move: During those times, working with an applicant for emigration—an outcast of the society—presented a clear danger to Feldman's career. Fortunately, that now-famous laboratory in the Institute for Problems of Information Transmission was headed by another very noble and independent person, Levon Chailakhian. Anatol came to him in the hallway and asked whether Chailakhian would mind if I came to the laboratory meetings. And Anatol started a phrase about my application for emigration. But Chailakhian interrupted: "Do you know anything about his application? I do not. Until somebody tells me officially, I see no reason for Mark not to come to the meetings." I would like to take this opportunity and thank Levon and all other members of that laboratory who knew everything about my status but pretended that my visits to the laboratory were routine and normal.

Anatol Feldman went quite a few steps further. When our first paper was written (mostly by Anatol), a question emerged about its publication. At those times, to publish anything in the USSR or abroad, the authors had to file a lot of nonsensical forms that were then signed by the officials of the institute. One of those forms stated explicitly that the manuscript contained no element of novelty and thus could be published in an open-access journal! Since all the bosses of the institute knew my father's name and were aware of the fact that he and his family had applied for emigration, Anatol invented a clever strategy: He showed the bosses the manuscript with only his name on it and, after getting all the signatures, added my name and mailed it to the Editor. He correctly figured out that the party officials and KGB informants did not read foreign journals; so, we published three papers in a row in 1982 in *Biological Cybernetics*, *Journal of Motor Behavior*, and *Neuroscience Letters*, using the same trick, and nobody suspected dirty play! This was truly brave on the part of Anatol, and I am forever indebted to him.

My third advisor was Gerry Gottlieb. When I came to Chicago, Gerry and his group were among the most vocal opponents of the equilibrium-point hypothesis (described in detail in Part 3), while I was (and still am) in love with the hypothesis. Gerry and his group were looking for a young colleague, and I happened to be in the right place at the right time. Gerry listened to my emotional and probably not very well-articulated arguments expressed in broken English and was amused sufficiently to offer me to stay in his laboratory and use his equipment to try to persuade him that the equilibrium-point hypothesis was worth something. This was a formidable task, and at some point, after having published half a dozen of papers with Gerry on the equilibrium-point hypothesis, I thought that the task was close to being successfully accomplished. Alas, Gerry happened to be at least as strong-willed as Anatol Feldman is and continued the quest against the equilibrium-point hypothesis since the times our paths parted.

It is probably too arrogant on my part to name Israel Gelfand as a mentor, but this is how I sincerely feel. There is quite a bit said about Israel Gelfand in the book. Here, I would only like to say that without the discussions with Gelfand, this book would have never been conceived. I feel that his influence on my thinking has been comparable only to that of my father.

By the time this book is published, I hope to get close to 100 journal papers co-authored with Vladimir Zatsiorsky. I first met Vladimir in 1976 when he was the very young Chairman of the Department of Biomechanics in the State Institute of Physical Culture in Moscow, and I was a fresh graduate of the Moscow Physico-Technical Institute, a junior researcher in the Department of Physiology of the Institute of Physical Culture. At those times, our contacts were limited to exchanging greetings in the hallway. A quarter of a century later, I accepted an offer from Penn State University, and the presence of Vladimir Zatsiorsky on the faculty was a major factor in making this decision. Over the past 12 years or so, we have been working side by side, overcoming differences in the characters and views, listening to each other, and being patient. At least Vladimir has been exceptionally patient with many of my poorly formulated ideas and projects. I sincerely believe that we have worked in a synergy (read the book if you want to understand what I mean).

So many colleagues to thank! And it is probably impossible not to miss some of the very important names of those with whom I have collaborated on exciting scientific projects. My dear Gyan Agarwal, Gil Lucio Almeida, Greg Anson, Alexander Aruin, Mireille Bonnard, Richard Burgess, Daniel Corcos, Mats Djupsobacka, Dmitri Domkin, Marcos Duarte, Simon Goodman, Dusko Ilic, Slobodan Jaric, Tatsuya Kasai, Jeffrey Kroin, Jozsef Laczko, Mindy Levin, Mark Lipshitz, Onno Mejier, Jeff Nicholas, Ida Neyman, David Patterson, Om Paul, Richard Penn, Konstantin Popov, David Rosenbaum, John Rothwell, Robert Sainburg, John Scholz, Gregor Schöner, Mark Shapiro, Pelle Sjolander, Dagmar Sternad, and Michael Turvey—Thank you all very much!

In any laboratory, the hard work is done by graduate and postdoctoral students while professors are having fun inventing poorly conceived projects that are impossible to carry out. The Motor Control Laboratory and the Biomechanics Laboratory at Penn State have not been an exception. Many of my (and Zatsiorsky's) former and current students have become personal friends, and I would like to thank them for the hard work, patience, and optimism, without which they would have probably never graduated. Among them are Tomoko Aoki, Frederic Danion, Alessander Danna-Dos-Santos, Adriana Degani, Sander DeWolf, Kazuhisa Domen, Fan Gao, Stacey Gorniak, Ning Kang, Sun Wook Kim, Vijaya Krishnamoorthy, Brendan Lay, Sheng Li, Zong-Ming Li, Halla Olafsdottir, Thomas Robert, Alexandra Shapkova, Elena Shapkova, Jae Kun Shim, Minoru Shinohara, Takako Shiratori, Siripan Siwasakunrat, Harmen Slijper, Kajetan Slomka, Alan Walmsley, and Wei Zhang.

Now it is time to thank those who helped me a lot with this project without realizing the importance of their help. For them, these were social occasions, chats at conferences, and informal conversations while sharing a bottle of wine or two, but these conversations made me think about issues that I otherwise would have missed. I am particularly grateful to Josef Feigenberg, Fr. Michael Meerson, Olga Meerson, Andrey Smilga, Voldemar Smilga, and Prince Andrey Volkonsky for such accidental insights.

Several of my friends–colleagues volunteered to read earlier versions of different parts of this book. Their comments were exceptionally helpful, and the author's stubbornness is the only factor to blame for the fact that the book remains imperfect. I would like to thank François Clarac, Anatol Feldman, James Houk, Slobodan Jaric, Mindy Levin, Richard Nichols, David Ostry, John Scholz, Doug Stuart, and Vladimir Zatsiorsky for sharing with me their wisdom and critique softened with words of encouragement.

And—for dessert—I would not be where I am without the love of the three most important women of my life, my mother Sara, my wife Irina, and my daughter Liza.

CONTENTS

Part 1: Building a Definition for Synergy 1
1.1 Synergies and Non-Synergies: A Few Examples 1
1.2 Palama's Concept of Synergy 5
1.3 Inanimate "Synergies": The Table and the Rusty Bucket 7
1.4 Examples of Biological Synergies 12
1.5 The Definition: Three Components of a Synergy 13

Part 2: A Brief History of Movement Studies 17
2.1 Ancient Greece and Rome 19
2.2 Renaissance 20
2.3 The Century of Frogs, Photography, and Amazing Guesses 23
2.4 The Twentieth Century: Wars of Ideas 25
2.5 Nikolai Alexandrovich Bernstein and Movement Science in the Soviet Union 30
2.6 History of Synergies and the Problem of Motor Redundancy 35
2.7 Problems with Studying Biological Movement 45

Part 3: Motor Control and Coordination 51
3.1 Israel Gelfand and Michael Tsetlin 51
3.2 Structural Units and the Principle of Minimal Interaction 56
3.3 Motor Control: Programs and Internal Models 63

Digression #1. The Muscle: Slow and Visco-Elastic 65
Digression #2. Neural Pathways: Long and Slow 70
Digression #3. Sensors: Confusing and Unreliable 72
Digression #4. Adaptation to Force Fields and After-Effects 79
Digression #5. Brain Imaging Techniques: What Do They Image? 81
3.4 The Equilibrium-Point Hypothesis 88
3.4.1 Experimental Foundations of the Equilibrium-Point Hypothesis 89
Digression #6. Reflexes and Nonreflexes 91
3.4.2 Equilibrium-Point Control of Simple Systems 100
3.4.3 Three Basic Trajectories within the Equilibrium-Point Hypothesis 104
3.4.4 Equilibrium-Point Control of Multi-Muscle Systems 105
3.4.5 The Mass–Spring Analogy and Other Misconceptions 109

Part 4: Motor Variability: A Window into Synergies 119
4.1 The Uncontrolled Manifold Hypothesis 120
4.2 Modes as Elemental Variables 131
4.2.1 Force Modes 132
Digression #7: Digit Interaction and Its Indices 134
4.2.2 Muscle Modes 139
Digression #8: Electromyography 140
4.2.3 Experimental Identification of the Jacobian 147
4.3 Stability, Variability, and Within-a-Trial Analysis of Synergies 148
4.4 Other Computational Tools to Study Synergies 155
4.4.1 Principal Component Analysis and Uncontrolled Manifold 155
4.4.2 Analysis of Surrogate Data Sets 159
4.5 Timing Synergies: Do They Exist? 162

Part 5: Zoo of Motor Synergies 167
5.1 Kinematic Synergies 167
5.1.1 Postural Synergies in Standing 170
5.1.2 Sit-to-Stand Task 174
5.1.3 Reaching 176
Digression # 9: Optimization 176
5.1.4 Reaching in a Changing Force Field 180
5.1.5 Multi-Joint Pointing 182
5.1.6 Quick-Draw Pistol Shooting 184
5.2 Kinetic Synergies 188
5.3 Multi-Digit Synergies 192
5.3.1 Force and Moment Stabilization during Multi-Finger Pressing 192
5.3.2 The Role of Timing Errors 195
5.3.3 Emergence and Disappearance of Synergies 197
5.3.4 Anticipatory Synergy Adjustments and Purposeful Destabilization of Performance 199
5.4 Prehensile Synergies 204

5.4.1 Hierarchical Control of Prehension 207
5.4.2 Principle of Superposition 209
5.4.3 Adjustments of Synergies: Chain Effects 212
5.4.4 Hierarchies of Synergies 213
5.5 Multi-Muscle Synergies 217
5.5.1 Anticipatory Postural Adjustments 220
5.5.2 Making a Step 222
5.5.3 Multi-Muscle Synergies in Hand Force Production 224

Part 6: Atypical, Suboptimal, and Changing Synergies 227
6.1 Is There a "Normal Synergy"? 227
6.2 Principle of Indeterminicity in Movement Studies 232
6.3 Plasticity in the Central Nervous System 233
Digression #10: Transcranial Magnetic Stimulation 235
6.4 Changes in Synergies with Age 238
6.4.1 Effects of Age on Muscles and Neurons 238
6.4.2 Effects of Age on Motor Coordination 242
6.5 Synergies in Persons with Down Syndrome 248
6.5.1 Movements in Persons with Down Syndrome 249
6.5.2 Multi-Finger Coordination in Down Syndrome 254
6.5.3 Effects of Practice on Movements in Down Syndrome 257
6.5.4 Relation of Atypical Synergies to Changes in the Cerebellum 261
6.6 Synergies After Stroke 263
6.7 Learning Movement Synergies 266
6.7.1 Traditional Views on Motor Learning 266
6.7.2 What Can Happen with a Synergy with Practice? 268
6.7.3 Practicing Kinematic Tasks 269
6.7.4 Practicing Kinetic Tasks 273
6.7.5 Plastic Neural Changes with Learning a Synergy 280

Part 7: Neurophysiological Mechanisms 285
7.1 Neurophysiological Structures and the Motor Function 285
Digression #11: What Is Localized in Neural Structures? 285
7.2 Synergies in the Spinal Cord 290
7.3 Synergies and the Cerebellum 302
7.4 Synergies and the Basal Ganglia 307
7.5 Synergies and the Cortex of the Large Hemispheres 309
7.5.1 TMS and the Equilibrium-Point Hypothesis 310
7.5.2 Studies of Neuronal Populations 314

Part 8: Models and Beyond Motor Synergies 321
8.1 Synergies and the Control Theory 323
8.1.1 Control: Basic Notions 323
8.1.2 Open-Loop and Closed-Loop (Feed-Forward and Feedback) Control 325
8.1.3 A Simple Feedback Scheme of Synergic Control of a Multi-Joint Movement 328

8.1.4 Optimal Control and Synergies 329
8.2 Synergies and Neural Networks 331
8.3 Synergies without Feedback 334
8.3.1 Do Synergies Improve Accuracy? 334
8.3.2 A Feed-Forward Model with Separate Specification of Good and Bad Variability 337
8.4 Synergies and the Equilibrium-Point Hypothesis 339
8.5 Sensory Synergies 344
8.5.1 Sensory Synergies in Neurological Disorders 345
Digression #12: Sensory and Motor Effects of Muscle Vibration 346
8.5.2 Sensory–Motor Interactions 349
8.5.3 Sensory Synergies in Vertical Posture 352
8.5.4 Multi-Sensory Mechanisms 354
8.6 Language as a Synergy 357
8.7 Concluding Comments: What Next? 360

References 363

Index 405

SYNERGY

Part One

Building a Definition for Synergy

1.1 SYNERGIES AND NON-SYNERGIES: A FEW EXAMPLES

The word "synergy" has recently become very common in both scientific and nonscientific fields. This word is used in the names of companies, cereals, methods of education, interactions among humans and animals, and certainly in basic and applied studies of movements. In order to use a notion in scientific research, it has to be defined with sufficient exactness, at least to make it possible to distinguish objects and processes, to which this notion can be applied, from those, to which it cannot. When a word has been used for many years without a definition, the situation becomes very complicated. A definition has to be introduced, but it should not contradict the intuitive feeling of what the word has meant; it should be compatible with the earlier usage of the word. So, a definition needs to be built or discovered that would create a reasonable blend of the intuitively accepted meaning of the word and exactness typical of scientific research.

We can learn a lesson from one of the greatest minds of the twentieth century, Nikolai Alexandrovich Bernstein (1896–1966), whose name will be featured in many pages of this book. In the mid-1940s, Bernstein wrote a book, *On Dexterity and Its Development*, which was later translated into English and published in 1996. In that book, he attempted to introduce a rigorous definition for the notion of dexterity—a notion that has been used for centuries as a lay term. Through much of the book, Bernstein uses examples from such diverse areas as evolution of life, Chinese fairy tales, contemporary neurophysiology, athletics, design of cars and sewing machines, and everyday activities to convince the reader that dexterity does have a specific meaning that can and should be defined and studied rigorously. Bernstein also appeals to the linguistic sensibilities of the readers asking them to imagine certain situations and decide for themselves whether particular examples describe situations of behaviors that carry an element of dexterity.

In the first few sections, I will try to follow Bernstein's example and build a definition for synergy that would fit its intuitively accepted everyday meaning. This definition should, at the very least, be able to separate "synergy" from "non-synergy." When you think of synergy, many examples come to mind—from the motion of the fingers of a musician to circus acrobats to workers carrying a heavy piano upstairs. But try to think of non-synergy. It is much harder to come up with similarly unambiguous examples. So, let me start with a few examples from my life in the Soviet Union to illustrate what I mean by "non-synergy" or, at least, truly bad synergy.

Imagine two workers who get an assignment to dig a pit 2 m × 2 m and 1 m deep. This example is particularly close to my heart because of the many cubic meters of soil I dug in archaeological expeditions during the years of being one of the army of Soviet "refuseniks." Let me remind those who do not remember refuseniks or were born too late to hear about them. Soviet Union was a closed society, and its citizens were not allowed to travel abroad [with a few exceptions, mostly limited to party members and Komitet Gosudarstvennoj Bezopasnosti (KGB) informants]. For decades, emigrations were absolutely out of the question. However, in the late 1960s and early 1970s, the system developed a crack in its iron curtain and started to allow Jews to emigrate to Israel. This was a miracle! Jews, who had been second-rate citizens of Russia and the Soviet Union for centuries, suddenly got a privilege: They were able to leave the country while non-Jews could not! Ridiculous things started to happen: Non-Jews started to invent a Jewish grandmother to become eligible for application to emigrate, Jewish and non-Jewish couples divorced, and cross-married and then remarried in the right combinations again after the emigration, etc. The Soviet authorities did not want to make emigration too attractive, and so they imposed a number of regulations to make sure that future emigrants did not enjoy life too much, and others were discouraged from following in their footsteps. In particular, most applicants for emigration were fired from their jobs. Some of the applications, particularly from those who had worked in areas considered sensitive by the communist regime, were refused.

Our family applied to emigrate at the least opportune moment, exactly 2 months before the Soviets invaded Afghanistan. The relations with the West had turned very sour—remember the two Olympic Games, in Moscow and in Los Angeles, each boycotted by quite a few countries from the opposite camp—and the emigration all but stopped. Applications of those trapped inside the Soviet Union were refused without any explanation, and a whole army of refuseniks emerged. The "explanation" our family got was "Your emigration would be against the interests of the Soviet Union." Nobody cared to explain what those interests were, and why our emigration would hurt those interests. So, we were stuck from 1979 up to 1987 when Mikhail Gorbachev re-allowed emigration. (I am still very grateful to him and to Ronald Reagan who pressured the Soviet system, until it developed cracks sufficiently large for our family to squeeze through!)

Archaeological expeditions at that time were a kind of internal immigration where outcasts of the Soviet society could work, live, and enjoy life. The work itself was exciting, combing elements of intellectual challenge, treasure hunting, and physical workout. The evenings and nights were filled with drinking local wines, playing cards, singing songs, and enjoying each other's company. In a somewhat ironic way, I am even grateful to the Soviets who turned me into a refusenik, did not allow me to work in my own area, and forced me to become a semi-professional archaeologist.

I am sorry for the lengthy digression! Let us get back to the two workers who have nothing to do with the emigration of Jews and archaeology. They "work together," so by the literal meaning of the original Greek word, they are a synergy. The workers agree to share the task evenly and to finish it in 8 hours. One of them happens to be much stronger and more experienced, so, in 8 hours he digs out not only his fair share (2 m^3) but 50% more. What can be expected from the other worker?

Let us consider two scenarios. First, the other worker pays no attention to what his comrade is doing or how the overall work is progressing. So, he continues to dig at his preferred pace. He proudly digs out 2 m^3 of soil in 8 hours. As a result, the pit becomes deeper than planned. Then, the supervisor appears and tells the workers that they were idiots and should now start filling the pit with the freshly dug soil to make it exactly 1 m deep. The two workers scratch their heads, blame each other for the mistake, and work for 30 more minutes. In this scenario, each of the two workers paid no attention to what the other was doing, and the result was disappointing. It required the interference of the supervisor and resulted in extra work.

Alternatively, the second worker could see that his buddy was digging faster than expected. So, he could have either simply slowed down such that they got to the depth of 1 m in exactly 8 hours, or he could have also speeded up but monitored the actual depth of the pit and at some point, for example 6 hours later, told his overzealous friend: "Hey, it looks like we are at 1 m deep already. Shall we take a measure and go to the pub?" When the supervisor appeared, he would have been happy to see the pit exactly 1 m deep.

In both scenarios, the workers worked together. However, the results were much better in scenario #2: They avoided doing extra work as well as being reprimanded by the supervisor. Shall we call this team a synergy in both scenarios? Probably not. Or, at the very least, a team that works according to the first scenario deserves to be called very bad synergy.

In the old Soviet times, there was a great comedian, Arkady Raikin. Somehow, the communist party officials allowed him much more freedom in his sketches than was typical of the Soviet times. In one of the sketches, Raikin pretended being a construction worker, who explained how his brigade had got a lot of prizes for exceeding the official plans. His monologue went something like this: "OK, guys, the local authorities came up with a plan to build 10 high-rise buildings. And so, we are asked to excavate 10 sites for those buildings. So, we offer the officials our own super-plan to dig fifteen sites. And we work our butts off, without smoking breaks and, lo and behold, we do it. So, we dig fifteen sites, but there are only 10 buildings planned to be built. Not a problem: The bosses order us to fill the five extra sites with soil, and we bust our butts even more to do this. So, you see, we not only fulfilled our super-plan, but we exceeded it. And this is how we got all the fame and prizes." This ridiculous monologue was always greeted with laughter since the audience immediately recognized the crooked logic of the planned socialist economy.

Collective farms and other socialist enterprises were excellent examples of non-synergies or very bad synergies abundant in the Soviet Union. I want to apologize for these examples to all those honest collective farmers and workers who suffered through the surrealistic Soviet system and should not be ridiculed. But the temptation is too strong: The poorly coordinated actions at different steps involved in growing, harvesting, and storing food present a great example of very poor synergy. Even in a situation where each element might have been performing at its best (despite the complete lack of incentives), the result was an overall pitiful performance.

Let us consider another, less politically flavored example. Imagine that you are conducting a choir. You want the choir to sing at a certain level of sound, but you have 50 singers to deal with. One strategy would be to tell each singer how loudly to sing. Potentially, this could solve the problem. If each singer performs exactly as instructed, this strategy will be very successful and lead to a perfectly correct level of sound. However, if one of the singers sings at a wrong volume, or gets sick and decides to quit altogether, the overall level of sound would be wrong. What would the alternative be? To make use of the fact that singers can not only sing but also hear. Then, the instruction could be "Listen to the level of sound. If it is lower than necessary, sing louder; if it is louder than necessary, sing softer." Now, even if one or more of the singers decide to quit, others will hear an "error" in the overall level of sound and correct it without any additional instruction. In both cases, the singers sang together. But probably everybody would agree that in the first case, there was no synergy among the singers because each one did his or her own job without paying attention to what the others were doing. In the second case, there was a synergy because actions of individual singers at all times depended on how loudly the other members of the choir were singing.

We can already come up with a few preliminary conclusions. If several persons "work together," and they care only about their own activity, then this group of people does not deserve to be called a synergy. In contrast, if several persons adjust their efforts, depending on the actions of others, or on how well the overall goal is achieved, they do form a synergy.

1.2 PALAMA'S CONCEPT OF SYNERGY

Synergy in Greek means "work together." These two words imply that a synergy always does something (*work*) and that more than one participant share the activity (*together*). This word has been used for centuries in an area that is rather far away from contemporary science. The *doctrine of synergy* was used by the Greek Fathers of the Christianity to imply the collaborative effort of man and God to overcome man's corruption, help man surpass himself, and to reveal God to him. This doctrine was developed by one of the outstanding philosophers of Christianity, St. Gregory Palamas. His development of the doctrine of synergy comes amazingly close to the contemporary meaning of this word in biology. However, first, let me say a few words about Gregory Palamas, his biography, and his philosophical views (Meyendorff 1964, 1974).

Gregory Palamas was born in 1296 into a noble family. He grew up at the court of Emperor Andronicus II Paleologus, an intellectual known for his deep religious convictions and patronage of writers and scholars. Until he was about 20 years old, Gregory was involved mostly in secular studies and then joined a monastery on Mount Atos, where he spent about 20 years. At the age of 30, Palamas was ordained a priest in Thessalonika. Ultimately, Gregory Palamas became Metropolitan of Thessalonika in 1347 and kept this position until his death in 1359. Soon thereafter, he was canonized and has been one of the most venerated saints of Thessalonika. However, Palama's life had not been smooth. He had participated in politics, had been arrested, tried, and even excommunicated. Fortunately, all these misfortunes were transient and reversible.

Much of Palamas' teachings represent arguments with another religious philosopher of the same time, Barlaam (1290–1348) (St. Gregory Palamas 1983). Barlaam based his position on two postulates—first, the Aristotelean postulate that all knowledge, including knowledge of God, comes from perceptual experience and second, following Plato, that God is beyond experience by senses and, therefore, is unknowable.

In his teachings, Palamas argues with Plato who was one of the first philosophers to turn the human mind into an object of study. Plato—a religious man himself—thought that the human mind was naturally able to surpass itself and "study itself" introspectively, as it would study any other object. However, he also thought that this facility was not enough to reach God (it does not matter whether one writes God or Gods in this context). For Gregory Palamas (and other Christian philosophers) such a limitation was unacceptable; it went against their own beliefs and religious experiences.

Through their lives, Gregory Palamas and others observed that some people turned into true believers, while others did not. Two explanations could be offered. First, some people may be inherently unable to transcend the mind and reach God. This view would be politically incorrect in our times (just imagine a politician claiming that some people are born inferior to others in an important aspect of life!), and it was not acceptable to Palamas either. Second, the mind itself has limitations and requires external help. This help cannot come from other minds, which are similarly limited, and, hence, it can only come from God.

Palamas came to the conclusion that the mind must be transformed by grace and not only the mind but also "all the facilities and powers of the soul and of the body." Man then rises to what Palamas calls a "divine state," the result of collaboration (σινεργια in Greek) between grace and human effort. Further development of this concept led Palamas to his famous teachings about *natural energies*—a notion rather different from the contemporary usage of energy in physics, engineering, and everyday life. For Palamas, natural energies were nonquantifiable and synonymous with *grace*. They reflected the help a person can get from God. In his most famous book, *Triads*, Palamas wrote: "God in His overflowing goodness to us, being transcendent to all things, incomprehensible and ineffable, consents to allow our intelligence to participate in Him…" (St. Gregory Palamas 1983; *Triads* I, 3 $10, p. 129).

According to Palamas, true self-study, self-introspection, and reaching the ultimate religious feeling required free collaboration of man with the redeeming action of God. It required a synergy of human effort and grace. In that synergy, weaker men need and get more of God's collaborative effort. Those with strong faith do not require too much help to reach God, and God's contribution to this process is accordingly decreased. As we will see further in this book, one of the distinguishing features of synergies is a cooperation among its elements such that if one element does too little, another element does more.

To visualize this example, let us consider it a combined effort of two participants, Man and God, who try to reach toward each other over a distance of say ten units (Figure 1.1). If a particular man can only reach over three units, God will reach over seven (point A). For another man who can reach over nine units, God will reach only over a single unit (point B), and so on. All the points on this graph showing the relative efforts of Man and God form a straight line with

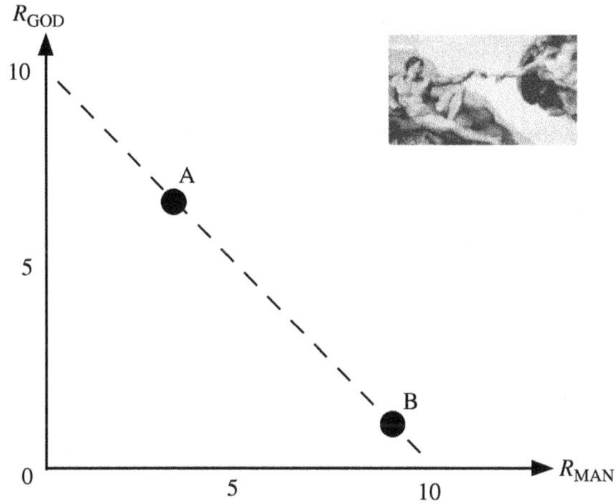

Figure 1.1. An illustration of the Palama concept of a synergy between Man and God. Imagine that salvation is reached when the total distance of 10 units is covered by the combined effort of Man (R_{MAN}) and God (R_{GOD}). God offers more help (larger R_{GOD}) to a weaker Man (point A) than to a stronger Man (point B).

a negative slope. We will return to this figure later, when similar graphs will be viewed as examples of either synergies or non-synergies.

1.3 INANIMATE "SYNERGIES": THE TABLE AND THE RUSTY BUCKET

Let us consider a couple of examples of simple inanimate objects. Probably, it makes little sense to use the word *synergy* as a synonym for "any material object." If a stone placed on scales is a synergy of its molecules that all "work together" pressing down on the scales, then anything is a synergy, and this book should not have been written in the first place (for an alternative opinion, see Corning 2003, 2005). But I am not ready to stop yet, so I would like to use *synergy* only with respect to multi-element systems that can, at least theoretically, change interactions among the elements and make them "work together" toward different goals. So, to identify salient features of synergies—those that make them different from non-synergies—it would be very helpful to see whether certain features of synergies can be seen in inanimate objects. Then, if we observe similar features in behaviors of biological objects, we would be rightfully suspicious about these being produced by certain nonspecific factors such as structural design or material properties of the objects and will search for more convincing proofs to claim that a synergy is responsible for the behaviors.

First, consider a table standing on the flat horizontal floor (Figure 1.2). The table has four legs, and they share the weight of any object that may be placed on the table. If a heavier object is placed on top of the table, all four legs show an increase in the absolute forces they produce on the supporting surface. Shall we call this a synergy? I would rather not. The main reason is that if a table qualifies as a synergy because its four legs apparently share the load, then any object does as well. Look at any large object, for example a large, irregular stone on the ground. It makes contact with the ground with parts of its surface. Nothing prevents one from considering different parts of the contacting surface of the object as its "legs," as many as one cares to define. If a person sits on the stone, all the apparent "legs" will show an increase in their forces exerted on the ground. This is not different from the table example. So, if a table is a synergy, then any object that can in principle share a load among its components, whether these are real or apparent, qualifies as well. Such a broad definition would make the term *synergy* meaningless.

So, let us agree that a table is not a synergy. This is an important step, since in future, if particular examples and analyses can be equally applied to a table,

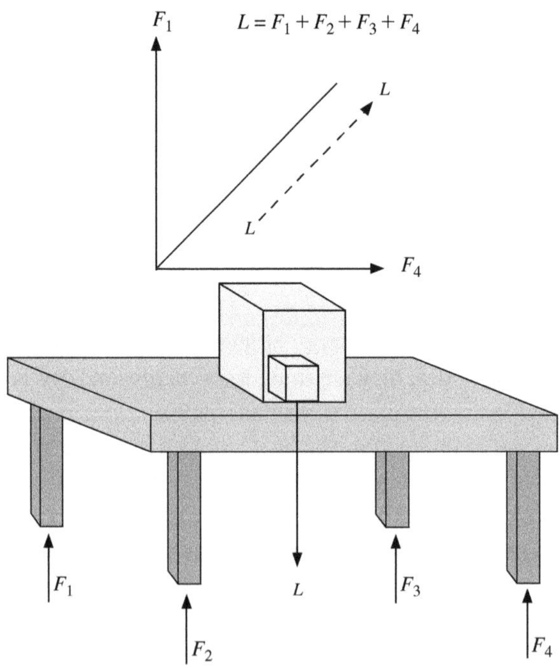

Figure 1.2. The total weight (L) of an object placed on the top of a table is shared among the four legs. The reactions forces (F_i) under the legs scale together with changes in the weight of the object (as in the graph).

I would immediately claim that what is being studied may or may not be a synergy. The most obvious example is measuring how much individual elements contribute to a task when the task magnitude varies. Imagine that force sensors are placed under each leg of the table. Now, let us put objects of different mass on the top of the table. The forces measured under each leg will scale very closely with the mass of the load (as in the graph over the table in Figure 1.2). In other words, all the elements will show parallel changes in their apparent contributions to the overall task of supporting the load, with changes in the magnitude of the load. The legs of the table share the "work," and this sharing pattern is preserved over variations in the load. But we have already agreed that the table is not a synergy. So, observations of parallel changes in the contributions of elements to a task fall short of suggesting that one is dealing with a synergy. We are going to continue exploring the table example a few paragraphs later. Now let me turn to another simple object that can show more sophisticated behavior, while also being part of the inanimate material world.

Imagine you have an old rusty bucket full of holes (Figure 1.3). The holes are irregular, with uneven edges, and with crumbs of metal falling off occasionally. There are also pieces of garbage on the bottom of the bucket that sometimes block and unblock some of the holes. Now let us pour water into the bucket at a constant rate. The water will pass out through the holes such that each hole will show a flow related to its time-dependent area, $s_i(t)$, and the level of water accumulated in the bucket, $H(t)$. At some level of water, the pressure will be such that the amount of water poured into the bucket, $Q(t)$, will be equal to the combined

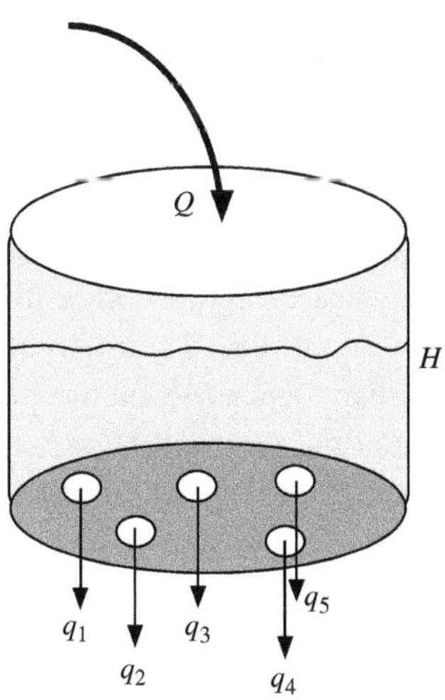

Figure 1.3. An illustration of the old rusty bucket example. The water is poured into the bucket at a rate Q. It is flowing out of the holes at rates q_i (i = 1, 2, 3, 4, 5). At a certain level of water H, an equilibrium is reached: $Q = \Sigma q_i$.

amount of water flowing out of the holes, $\Sigma q_i(t)$, and the system will be in an equilibrium. If we increase the flow of water into the bucket, a new equilibrium will be reached at a new water level corresponding to a higher pressure and a proportionally higher outflow. Changes in the rate of water inflow will lead to proportional changes in the rates of water outflow from each of the holes. This is similar to how changes in the weight of an object placed on the table lead to proportional changes in the forces under each of the legs. So, the bucket can show a preserved sharing pattern among its holes with changes in the water inflow. But it can do more than that.

Let us now imagine that a crumb of metal falls off the edge of one of the holes. The hole becomes bigger, and the amount of water flowing out of the hole increases. As a result, the water equilibrium is violated, and the level of water in the bucket starts to fall down. This leads to a decrease in the hydrostatic pressure and in the amount of water flowing out of all the holes. Ultimately, the water in the bucket reaches a new, lower level, which corresponds to a new level of pressure. If one ignores the object and looks only at the behavior of its elements, one can see quite an amazing picture emerging. One element (the crumbling hole) introduces a change in its behavior, namely, it becomes bigger and starts to let more water through. Within a short time, all other holes start to let less water through such that the total amount of water flowing out of the bucket is the same as before, exactly equal to the amount of water that is being poured in. We will return to a more quantitative analysis of the rusty bucket later. For now, let us use more general words to describe what happened. One element introduced an error into the functioning of the system. Other elements corrected the error (we will call such phenomena *error compensation* among elements). This looks like the result of the action of a smart controller. However, we know that there is only an old rusty bucket without a brain or even a simplest calculator built in. I absolutely refuse to call this bucket a synergy of its holes!

Actually, a similar apparent error compensation can be observed in the former example with the table. Imagine placing an object of a certain weight on top of the table many times, and measuring the forces under each of the four legs (Figure 1.4). Each time, the object is placed at different spots—sometimes closer to one leg, sometimes to another. Depending on the exact location of the object, the distribution of forces under the legs will be different, while their sum will always be equal to the weight of the object. In other words, if in a particular trial, the object is placed closer to leg #1 than to any other leg, that leg will show more than its fair 25% share of the total weight of the object, while legs #2, 3, and 4 will, on average, show smaller forces. If we ignore what is going on and look only at the outputs of the elements, the forces under the four legs, we can come up with an absolutely wrong conclusion that there is a smart controller that measures forces under the legs of the table and adjusts them to make sure that

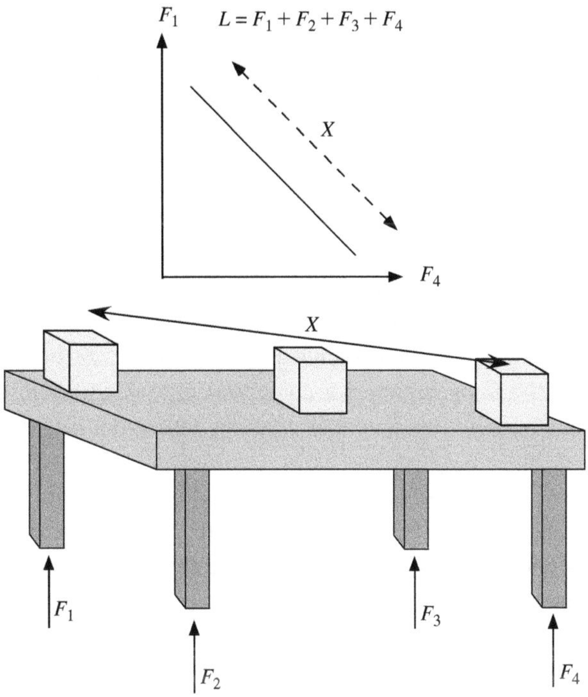

Figure 1.4. If an object of the same weight (*L*) is placed in different locations on the table, the sharing of the load among the legs will change. Compare the graph of the top of the figure to the one on the top of Figure 1.2.

their sum is constant. (As was said many years ago at one of the concerts in my *alma mater*—the Moscow Physico-Technical Institute—"Physicists are people who seriously believe that electrons know the Coulomb Law and measure distances between each other.")

If we agree that the table is not a synergy of its legs, and the rusty bucket is not a synergy of its holes, a few conclusions may be drawn. First, looking only at the outputs of the elements of an alleged synergy may lead to a misleading impression of the existence of a smart controller, while none may actually exist. Second, sharing a task among a set of elements, with changing magnitude of the task, can happen in objects that we do not want to call synergies. So, such sharing cannot be used as a convincing proof of a synergy. Third, even an apparent error compensation among elements may be a result of a certain rather unsophisticated mechanical design, not a controller. In the example with the table, the mechanical laws of statics defined the distribution of forces under the four legs. In the example with the bucket, the laws of hydrostatics plus the law of conservation of matter organized the water outflow from individual holes.

1.4 EXAMPLES OF BIOLOGICAL SYNERGIES

Now, let us consider several biological examples trying to identify which ones are synergies and which are not. Since we would like to use the "work together" definition, we have to define *work* and *elements* in each case. In some of these examples, it is very hard to state unambiguously that we are dealing with a synergy. So, I am going to use a measure for synergy from 0 to 10, where 0 is an obvious non-synergy, while 10 is an undisputable synergy. Despite not being a native speaker of English, I am going to use my imperfect sense of language to make the following judgments. The table and the old bucket get zero on this scale, despite their ability to show such features as a stable sharing pattern among apparent elements and apparent error compensation, while the action of human fingers playing a musical instrument or working with a tool gets 10.

1. *A swarm of mosquitoes attacking a hiker.* The elements are obvious—they are individual mosquitoes. However, is there a common work in this example? It is not likely, unless we view eating the hiker alive as the ultimate goal of the swarm—a goal that is not achieved. In fact, each mosquito tries to get its portion of the nutritious food and is happy if it manages to do so without being squashed. The mosquitoes fly in a cloud that seems to move as a whole, keeping close to the hiker, drawn by the temperature of his or her body. We do not know the rules that define movements of mosquitoes in the cloud, but these may not be absolutely random. The mosquitoes somehow coordinate their actions, probably using their sensory systems as well as the neural elements in their bodies (for a more serious discussion of behaviors of large groups of insects see Kugler and Turvey 1987). Maybe, they are a synergy, and the task is keeping a certain optimal distance from the source of food rather than eating it. I feel like scoring them at about 3.
2. *A school of piranhas attacking an animal.* This seems to be a very similar example, although with evolutionary higher animals as elements. Each piranha tries to get as much food as it can, caring only about getting access to a bare portion of the body. Interestingly, unlike the mosquito example, a school of piranhas shows a feature that looks like error compensation, since the whole animal is usually eaten. This means that if one piranha gets more meat than its fair share, others will get less. Nevertheless, I am not impressed with this example any more than I am with the swarm of mosquitoes and give it also a score of 3.
3. *Termites (elements) building a mound (work).* This example is borrowed from a wonderful book by Peter Kugler and Michael Turvey (1987). When termites build a mound, apparently they act individually, but their acts are modified with the actual state of the mound. Since the actual state of the mound reflects accumulated effects of actions of all the termites, each termite

in fact coordinates its actions with those that have already contributed to the building process. This is similar to how the overall pressure of water in the rusty bucket defines the outflow from each of the holes. At the same time, it reflects the balance between the water inflow and the amount of water flowing out of all the holes. But the termites are smarter. They use their sensory organs to judge the shape of the building and define where to drop the next piece of the building material. For that, they deserve a score of 5.

4. *A pride of lions hunting an antelope.* I am sure that those of us who like to watch the Animal Channel have been fascinated by the perfectly coordinated efforts of a pride of lions trying to hunt down an unfortunate animal. Action of each lion (each element) at each moment of time is defined by at least three factors: what this lion does at that particular moment, what other lions do, and what the prey does. And the goal of the work together is unambiguous. So, this example gets a 10 on the introduced scale.
5. *The human hand.* This is a proverbial synergy. So, let us not even waste time but give it 10.
6. *Getting back to collective farms.* On the 10-point scale, a Soviet collective farm does not deserve more than 5 for its synergic activity.
7. *The world economy.* The elements may be viewed as corporations or even countries. However, is there a universally accepted definition of a goal? It is unlikely that there is one; at least, there is none as yet. Actions by the corporations are driven primarily by these corporations' interests, while the countries (or, rather, their governments) are frequently driven by ideological, religious, or political factors that have complex relations with and are not reduced to world economy. This is a poor synergy but maybe a bit better than the Soviet collective farms. The score is 5.5. If at any time in the future the world economy starts to function with a common goal of assuring a decent living standard for all people, while avoiding irreversible negative changes in the environment, the score could be adjusted upwards.

1.5 THE DEFINITION: THREE COMPONENTS OF A SYNERGY

I feel ready to introduce a qualitative definition for a synergy. This definition is going to be built on three pillars—sharing, error compensation, and task-dependence. As argued later, *error compensation* may be a misleading term in certain situations; so, I am going to use another term, *flexibility/stability,* to address the feature of co-variation among elemental variables that sometimes leads to error compensation. *Flexibility/stability* implies that a task is accomplished by *flexible* (variable) solutions and that a common feature of these solutions is providing *stability* of an important performance characteristic.

These three pillars correspond to three conditions that should all be met for a group of elements to be considered a synergy. Mathematicians like to formulate necessary and sufficient conditions; so, I am going to suggest that the following three conditions have to be met for a group of elements to be considered a synergy.

First, the elements should all contribute to a particular task. When the human hand operates a tool, the digits produce forces and moments of force (further, I am going to call them simply *moments,* for brevity) on the handle of the tool, and these forces and moments should sum up to produce a required mechanical action of the tool on an external object. The production of an adequate mechanical action may be considered a task of this multi-digit synergy. The digits share the task in some way. Patterns of such sharing may be characterized quantitatively in different ways. For example, one can compute the percentage of the required grip force produced by individual digits.

When four people carry a heavy object, they share the task of counteracting the force of gravity by producing forces directed upwards. The sum of these forces should equal the weight of the object. A pattern of sharing the load may be quantified for this example in a similar way, that is, as the percentages of the total required upward force produced by each person. I am going to use the term *sharing pattern* to address such distributions of a task over a set of elements (persons, digits, muscles, etc.).

Second, if one element produces more or less than its expected share, other elements should show changes in their contributions such that the task is performed properly or at least better, compared to what could be expected if all elements acted independently. In the previous example of carrying a heavy object, imagine that one person stumbles and temporarily stops contributing to the task. To deserve being called a synergy, the other three persons of the team should increase their upward forces applied to the object such that the total force is quickly restored to its required value. Grasp a mug with the thumb opposing the four fingers. Now lift one of the fingers trying not to move the mug. This is a very easy task. The lifted finger stopped contributing to the task of holding the mug. Despite the finger action, the mug did not slip out of the hand and did not rotate. This means that other digits redistributed their efforts such that their total mechanical effect on the mug remained virtually unchanged.

In earlier sections, we discussed examples from very different areas that demonstrated these two features, sharing and error compensation. Even such unsophisticated objects as a table and a rusty bucket could show apparent sharing patterns and error compensation. So, if we want the definition of synergies to be specific for biological objects, something else needs to be added. Something that tables and buckets cannot do, while lion prides, human hands, and other "10-point synergies" can.

The missing third component is *task-dependence* or the ability of a synergy to change its functioning in a task-specific way or, in other words, to form

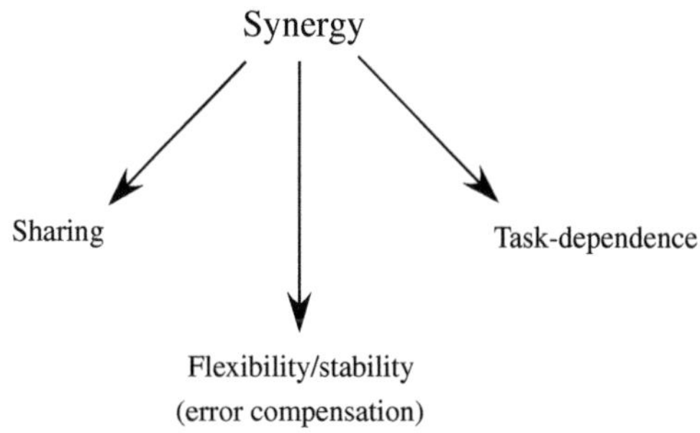

Figure 1.5. Three main components of a synergy.

a different synergy for a different purpose based on the same set of elements. The hand can open can lids, turn a screwdriver, write with the pen, play music, wave good-bye, and even speak using sign language. A team of construction workers can build many different things. An orchestra can play different tunes and even carry the piano upstairs. A pride of lions can hunt an antelope, a giraffe, or a buffalo, or defend their meal from a pack of hyenas. The table and the bucket cannot do such things; their design allows apparently synergic behaviors with respect to a limited set of tasks such as supporting the weight of an object placed on the table or making sure that all the water that gets into the bucket gets out.

The third condition means, in particular, that there is no such a thing as an abstract synergy. Synergies always do something, their elements "work together" toward a particular goal. Hence, analysis of any alleged synergy should always involve a hypothesis on what the synergy is supposed to do. Figure 1.5 summarizes the three components of a synergy.

Before moving to methods of analysis and features of synergies, I would like to present a brief history of movement science. Although the idea of synergy is applicable not just to movements (and at the end of the book I will try to speculate about synergies in perception and language), most examples considered in this book are movements.

Part Two

A Brief History of Movement Studies

The first thing we learn on entering this world is that things move. The mother approaches and leaves, the toys roll and fall, our arms and legs flap and swing, and the whole world seems to rotate around us. We do not know yet how this universe came about, but we feel that whatever started it gave it a lot of movement. Soon we start to notice regularities of movements. If the source of a delicious food approaches the mouth, it is smart to open the mouth and get ready. If an object increases in size, it is likely to come closer and may be touched or eaten. If a ball rolls toward us, it is likely to continue rolling and may be caught or hit.

When we grow, we start to perceive movement as being equivalent to change. Music is movement of sounds, whereas many beautiful scenes of nature are movement of colors. Movement may be fascinating, particularly if it repeats itself but not exactly in the same way. Watching the fall of snowflakes, the roll of ocean waves, or the flight of fire sparks can be mesmerizing.

After entering school, we witness with dismay how movement turns into an object of study of one of the school subjects, mechanics. The magic is partly gone. However, even after learning that $F = ma$, we preserve the ability to experience awe, while watching movement of objects, shapes, colors, and sounds.

The history of movement studies has two distinct branches. Physics studies movement of inanimate objects. Within classical mechanics, motion of these objects can be predicted, if one knows their original state and all the external forces acting on them. This motion can be very complex and hard to predict, particularly if many objects move and exert forces on each other. For example, it is hopeless to try to predict motion of a molecule of the ideal gas for a long period of time. But then laws of statistics come into play and allow to describe motion of such large ensembles of objects with sufficient accuracy. The history of physics is a very encouraging example of the development of a science—the history of discovering the origins of forces and refining the laws that define their characteristics.

Movements of animate objects present a very different story. These objects do not violate any rules of physics, but they themselves become origins of forces that are produced in a purposeful way. These objects are active. They try to act on the world and fit it to their needs. The needs may be relatively simple such as avoiding being eaten or getting closer to food. They may also be very complex such as playing a Mozart sonata exactly the way the pianist wants it to sound, or giving a speech persuading the jury to set a murderer free.

There are two opinions with respect to movements of living beings and to biology in general. One view is that biology, including movement science, is nothing more than a complex physics. The problem is to measure things properly and find the sources of all the forces. As one of my best friends and colleagues once said: "Motor control is done by people who cannot measure properly." To be fair, this opinion later softened, and the person who said it became one of the most productive researchers in the area of motor control. True, measuring properly is no small problem in itself. However, I would like to suggest that biology is more than a complex physics. It is a different area of science that needs a different set of notions and maybe a different mathematical apparatus. We are going to return to this issue a little later. For now, let me emphasize that in this book we are interested in movements of living beings. This certainly does not mean that laws of mechanics are to be ignored. But we should always keep in mind that analysis based exclusively on these laws is only the first step toward understanding how movements are controlled.

Motor control in higher animals, including humans, can also be viewed as an area of natural science, the purpose of which is to understand the functioning of the central nervous system (CNS). Purposeful, natural movements are products of the functioning of the CNS. They are relatively readily observed, and their properties can be relatively easily and objectively quantified. In this sense, movements have clear advantage over such products of the CNS as thoughts, dreams, or emotions. Besides, movements are ultimate outcomes of the functioning of the CNS. Playing music, reciting poetry, and even expressing love and hatred are all done through movements (my sincere apologies to Tolstoy and Dostoevsky).

So, motor control and studies of movement synergies can be viewed as a quest for two Saint Graals—first, to develop a scientific description of processes involved in the production of purposeful movements by living beings, a description that would be comparable in its rigor to that offered by physics for inanimate objects and second, to get insights into the functioning of the CNS, and particularly of the brain.

There are several volumes containing historical reviews of the studies of biological movement (e.g. Cappozzo et al. 1992; Latash and Zatsiorsky 2001). Hence, only a brief overview is presented here with the focus on scientists of the past, whose ideas and works were particularly influential for the development of the contemporary understanding of motor control and coordination. I apologize for omitting many great names and inventions. Also, the following subjective representation of some of the ideas may not necessarily be shared by many of my colleagues. I respectfully ask for permission to reserve the right to be subjective and make my own errors, rather than repeat views that may be more commonly accepted but possibly not less erroneous. Let me refer to the very old and highly respected opinion that humans can know for certain only that they cannot know anything for certain. Well, to be consistent, one has to be uncertain in that highly respected opinion too.

2.1 ANCIENT GREECE AND ROME

The origins of movement and the intimate relations between human movements and their controller have been fascinating scientists, at least since the times of the great Greek philosophers of the past. At that time, the problem of movement–controller relation was more commonly formulated as that of the relation between the body (that moves) and the soul (that controls). Three questions related to movement were commonly discussed. First, what are the origins of movement in the world? Second, what makes the soul command the body to move? And third, how does the soul induce body movement?

Pythagoras (571?–497? BC) viewed movement as the evolution of metaphysical numbers. Movement of heavenly spheres was supposed to produce movement of the soul, which in itself was a self-moving number. It remained unclear how movement of the soul induced body movement. Approximately at the same time, Heracleitus of Ephesus (540–??? BC) and his school came to be known as "those who believe in movement." According to these philosophers, one could not know objects but only their trajectories. In the twentieth century, the traditions of the Heracleitus school were rejuvenated by theories of motor coordination, based on analysis only of movement kinematics.

A century later, Democritus (460–370? BC) developed his atomic theory of the universe. In that theory, he came up with a conclusion that movement of the soul was transmitted to the body by the movement of atoms. If one substitutes "atoms" with "ions" and "soul" with "intention" ("soul" and "intention" mean similarly undefined and unmeasurable things, but the latter is, however, more palatable to the contemporary scientific community), this statement sounds very modern.

According to Plato (428–347 BC), the Demiurge (Creator) ordered time and movement. Self-motion was viewed as a sign of the immortal soul, which was supposed to possess a unique ability to be moved by itself. This view of Plato is close to some of the religions of the Orient, with their respect for all living beings, including worms and insects that possess the feature of self-movement. The soul was a gift to human beings from the Creator, as illustrated in the great Michelangelo's fresco, "Creation of Adam," in the Sistine Chapel. Plato compared human voluntary movements to a chariot moved by horses controlled by the ultimate charioteer, the Soul. Movements of parts of the body were prescribed by the charioteer as those of marionettes. In contemporary language, one could say that the Soul produced motor programs that were later implemented by the body parts.

Aristotle (384–322 BC) defined movement as any quantitative or qualitative change in bodies. He argued that, for a movement to be, there had to be a mover and a moved body. Moved bodies are usually readily observable, while the movers are not. Hence, the soul was reconfirmed as the prime mover of the body. Aristotle was arguably the first to pay particular attention to a distinguishing feature of biological movement—that is, its coordination. For him, coordination was synonymous with harmony, which occurred naturally, by the design of creation. In contemporary parlance, compared to Plato, Aristotle made a step from the motor program theory to dynamic systems.

Despite the differences in opinions, all these philosophers would probably agree that purposeful movement is a characteristic that distinguishes living beings from the rest of the natural objects. But how are such movements produced? It took a few hundred years to take the next step.

In the second century AD, the great Roman physician Galen (129–201) got important insights into the origins of human movement. He realized that voluntary movements of body segments were produced by pairs of muscles (now called *agonist* and *antagonist* muscles) that receive their ability to produce force via nerves conveying "animal spirits." The origin of movement was still the soul, while body parts, hands in particular, were instruments that were controlled by the soul pretty much according to Plato's traditions.

The Greek–Roman understanding of the relation between the Soul and the Body was summarized by St. Augustine (354–430): "The way in which souls are cling to bodies is completely wonderful, and cannot be understood by man; and this is man himself" (cited after de Montaigne 2003, p. 489).

2.2 RENAISSANCE

The fifteenth and sixteenth centuries presented humanity with an unmatched constellation of great artists, writers, and philosophers. For many, the embodiment of the early Renaissance was Leonardo da Vinci (1452–1519). No words can adequately describe his contribution to art, science, and philosophy. In this book, however, I would like to emphasize his studies of the human anatomy and, particularly, the functional anatomy of muscles that allow us to regard him as the first scientist who made movement in humans and animals the focus of his research.

Michel Eyguem de Montaigne (1533–1592) was quite a character. A striking representative of Renaissance skepticism, he was famous for arguing that humans were in no way superior to beasts, and in many respects, quite the contrary. He used an example of the human inability to comprehend the origin of movements of animals to emphasize how presumptuous it was for humans to view themselves as ultimate, perfect creations endowed with an omnipotent mind. For de Montaigne, movements were exciting as means of expression and communication, understandable to all humans, irrespective of their native language, and as a reflection of the Soul. He wrote in the famous *Apology for Raymond Sebond* (de Montaigne 2003): "There is no movement that does not speak both a language intelligible without instruction and a public language" (p. 403).

Half a century later, René Descartes (1596–1650) inaugurated dualism, which became a dominant branch of philosophy. Descartes taught that every human being was composed of two independent entities, the Soul (tentatively placed by Descartes in the pineal gland) and the Body. The Soul was responsible for thinking and other things, many of which would now be probably categorized as "cognition," while the Body obeyed the Soul and the laws of nature. To me, this sounds like a very reasonable starting point for scientific research directed at understanding mechanisms of motor control. In contemporary scientific community, being called a dualist is not complimentary; nevertheless, the author of this book admits to being one, albeit maybe in a form modified from the original formulation by Descartes. Descartes had a good understanding of neuroanatomy, as it was then known. Following traditions set by Galen, Descartes thought that movements were produced by the Soul via animal spirits. Some movements were independent of the Soul, for example, the beating of the heart. Other movements were induced by senses and mediated by the brain, for example, protective arm movements during a fall. In his analysis of eye movements, Descartes mentioned that action of a muscle was commonly accompanied by relaxation of a muscle with an opposing action, an antagonist. Descartes and a younger British anatomist, Thomas Willis (1628–1678), viewed reflexes as fundamental neural mechanisms and tried to link them to different brain structures. The noun "reflex" was, however, introduced later by a French scientist, Jean Astruc (1684–1766). For more on reflexes, see Digression #6.

Another younger contemporary of Descartes, Giovanni Alfonso Borelli (1608–1679) was a disciple of Galileo Galilei (1564–1642). Borelli is now viewed as the father of biomechanics, for example, the highest award of the American

Society of Biomechanics is the Borelli Award. Borelli tried to integrate physiology and physics and performed numerous studies of muscles and their actions in a variety of static and dynamic tasks. In particular, the analysis of static tasks allowed Borelli to estimate the maximal forces of major muscle groups. He described muscles as being composed of thin fibers that possessed the property of passive elasticity. His analysis of the mechanics of jumping led him to a perfectly correct conclusion on the importance of release of the accumulated energy. However, when it came to issues of control, he assumed that movements were still produced by the Soul. To prove that "nerve spirits" were not gaseous, Borelli submerged a struggling animal in water and then slit its muscles to demonstrate that no bubbles appeared (Brazier 1959). According to Borelli, to produce movement, nerves were shaken by the Soul leading to the secretion of droplets of nerve juice. This was quite a remarkable insight: Just substitute "nerve juice" with "acetylcholine"! Let us not be too choosy about words in our assessment of the views of Descartes and Borelli. If one substitutes "soul" with "mind" or "intention," their insights sound rather modern and attractive. (By the way, if anybody knows a profound difference among these three terms that makes one of them more scientific than others, please contact me.)

The great Isaac Newton (1642–1727) also contributed to the discussion on how the biological movement came about. Newton was a religious man, and his theory of motor control could not start with anything but the Soul. He was quite aware of the problem of communication between the Soul and the Body, and as a hard-core physicist, he solved it by introducing a medium, ether. This medium was supposed to obey the laws of mechanics; however, the fact that it was apparently nonobservable made it a poor candidate for experimental studies, and, unfortunately, Newton did not spend much time developing this notion. If he had, quite possibly, the electrical nature of information transmission between the brain and the muscles would have been discovered much earlier.

The world waited until the end of the eighteenth century to learn about the importance of electricity for intrinsic processes in animals. In 1791, a great Italian scientist, Luigi Galvani (1737–1798), wrote in his "Commentary" that there was a link between neural tissue and electricity. And in the next century, new tools were invented that were able to record electrical currents associated with neural and muscular activity. Among those were galvanometers, named after Galvani, which became indispensable in physiological laboratories for the next 100 years. The introduction of these tools led to the rapid progress in the neuromuscular physiology in the nineteenth century. In 1838, Carlo Matteucci (1811–1868) showed that electrical currents could originate in muscles. A great French scientist, Etienne DuBois-Reymond (1818–1896), was the first to detect an electrical signal of the muscle with a galvanometer. In 1849, he published a book with illustrations of electrical signals recorded in human muscles, and this led to a new exciting area in movement studies, electromyography (EMG).

2.3 THE CENTURY OF FROGS, PHOTOGRAPHY, AND AMAZING GUESSES

In the nineteenth century, movement science advanced by progress in two seemingly unrelated areas. The first one was physiology, which at that time was to a large degree a "frog science." The second one was related to the development of new methods of observation and measurement, such as the fascinating technique that allowed to "stop time"—photography.

Let us start with the frogs. These defenseless creatures were among the favorite specimens for physiological studies. In the middle of the nineteenth century, a German scientist, Eduard Friedrich Wilhelm Pflüger (1829–1910), performed a study of the wiping reflex in decapitated frogs. To be fair, let me mention that before Pflüger, similar studies had been performed by another German physiologist, J.A. Unzer (1727–1799). In Pflüger's experiments, a headless frog was suspended from a frame, and a small piece of paper soaked in a weak acid solution was placed on its back. After a short delay, the frog produced a beautifully coordinated motion of the ipsilateral (same side of the body as the one to which the stimulus is applied) hindlimb that wiped the piece of paper off its back. When the hindlimb was amputated and another piece of paper was placed on the same side of the back, a longer delay was followed by a wiping motion of the remaining, contralateral hindlimb. Obviously, these movements could be controlled only by the spinal cord. So, Pflüger opined that the spinal cord was able to control targeted movements, and that spinal reflexes could switch and lead to activation of different muscle groups. These remarkable experiments and conclusions were all but forgotten for about a century. In the second half of the twentieth century, interest in the wiping reflex in frogs was revived by studies in several laboratories, leading to exciting insights into the organization of the control of multi-joint limb movements (Fukson et al. 1980; Berkinblit et al. 1986a; Schotland et al. 1989; Bizzi et al. 1991).

Ivan Mikhailovich Sechenov (1829–1905) was not only a great scientist but also the founder of the Russian school of classical physiology, which culminated in the early twentieth century with works by Sechenov's student, Ivan Petrovich Pavlov (1849–1936). In 1863, Sechenov was the first to discover that electrical processes within the nervous system could lead not only to excitation but also to inhibition. One of his students, Spiro, described in 1876 that stimulation of a "flexion center" of a hindlimb of a frog was accompanied by inhibition of the flexion center of the other, contralateral hindlimb suggesting that these phenomena could be related to the spinal coordination of locomotion, an insight developed later by Sir Charles Sherrington and his school.

Everybody would probably agree that in order to understand the nature of a phenomenon, one first needs to learn how to measure it properly. Brothers Emst Heinrich (1795–1878), Wilhelm Eduard (1804–1891), and Eduard Friedrick

Wilhelm Weber (1806–1871) were among the first who realized the importance of mechanical observation and analysis of biological movement. The Weber brothers worked in Leipzig and Göttingen during the second half of the nineteenth century. Ernst was the author of the famous Weber's Law that describes how perception scales with the magnitude of a perceived stimulus (for more details on Weber's Law see any contemporary textbook of psychology). In 1894, Wilhelm and Eduard published a book, *The Mechanics of Human Walking Apparatus.* In it, they emphasized the importance of the mechanical (pendulum-like) motion of the leg during locomotion and suggested that this motion occurred under the action of gravity, without active participation of the muscles.

The development of photography allowed researchers to stop the time, record movements, and analyze their characteristics. Etienne-Jules Marey (1830–1904) and Eadweard J. Muybridge (1830–1904) are credited with the development of photographic methods, specifically for analysis of natural movements. Making sequences of photographs at a high speed allowed them to analyze changes in the configuration of the body and its segments in different phases of the movements. Marey invented a photographic gun and the first version of the photographic camera with moving film. Ultimately, he reached a very respectable speed of 60 frames per second in his records. Marey also developed a pneumatic system, with an ingenious recording technique to study the gait of horses and humans, as well as the flight of birds and insects. Muybridge was famous for his photographs of locomoting animals, particularly horses. He also photographed naked humans locomoting on their hands and knees, for reasons that probably combined sense of humor and scientific inquiry. These developments allowed other researchers to perform analysis of movement kinematics and make inferences about the role of different forces in biological movements.

At the end of the nineteenth century, Christian Wilhelm Braune (1831–1892) and Otto Fischer (1861–1917) photographed human subjects with Geissler tubes taped to the body. The tubes flashed at 26 Hz, and the subjects were filmed in complete darkness. Further, based on stick figures of the human body at different phases of the locomotor cycle, Braune and Fischer performed mechanical analysis of the movement kinematics and computed the muscle contribution to forces acting at each segment of the human leg during the swing phase. They found an important contribution of muscles to leg motion and thus concluded that the pendulum theory of the Weber brothers was wrong.

Scientists of the nineteenth century reformulated the question of how the Soul controls the Body in a form that appeared better suited for research, namely, how the brain controls movements. Physicists, physicians, and physiologists addressed this issue and came up with important and very deep insights.

Hermann Ludwig Ferdinand von Helmholtz (1821–1894) was a universal scientist with major contributions to physics, in particular optics, biology, mathematics, engineering, medicine (he invented the ophthalmoscope), and philosophy.

In particular, von Helmholtz paid attention to the following observation known to any inquisitive elementary school student. If you close one eye and move the other eye or the head in a natural way, changes in the image of external objects on the retina are correctly interpreted as motion by the observer in the motionless environment. However, if you press on the eyeball with a finger, the motion of the eyeball leading to motion of the images of external objects over the retina is associated with a feeling that the external world moves. This suggests that perception depends not only on patterns of sensory signals but also on accompanying motor action. Von Helmholtz was among the first to recognize the importance of relations between perception and action. At about the same time, Sechenov wrote: "We do not hear but listen, we do not see but look," also with an emphasis on the role of activity in perception. Von Helmholtz also analyzed the mechanics of the human eye and described that the eyes did not use all their mechanical degrees of freedom during natural movements, an insight currently known as Donder's law.

Until the very end of the nineteenth century, little attention was paid to such a ubiquitous feature of human movements as their inaccuracy. Although the ability of humans to err had been well acknowledged, in particular by Michel de Montaigne, nobody had performed scientific studies of movement errors. Robert Sessions Woodworth (1869–1962) filled that gap. In 1899, he wrote his Ph.D. dissertation under the direction of James McKenn Cattell (1860–1944). In the dissertation, Woodworth focused on errors and variability in motor performance. He studied the relations between speed and accuracy, effects of vision on motor variability, and emphasized the importance of action–perception relations. Based on those studies, he came up with arguably one of the first inferences on motor control. He concluded that control of a fast movement consisted of an initial pulse and later corrections, an idea that can be now reformulated as a combination of feed-forward and feedback control processes.

On the threshold of the nineteenth and twentieth centuries, the progress in the functional anatomy of the nervous system formed the background for studies of neural mechanisms involved in the control of movements. Santiago Ramon Y. Cajal (1852–1934) provided essential support for the idea of the nervous system consisting of independent interacting neurons, while Sir Michael Foster (1826–1907) and Sir Charles Sherrington (1852–1952) coined the term "synapsis" (later transformed into "synapse") for hypothetical sites of interaction between pairs of neurons.

2.4 THE TWENTIETH CENTURY: WARS OF IDEAS

The twentieth century will probably be known in future history books as the century of World Wars. If any other century in future has more than two World Wars, history books will likely become redundant because of the lack of readers. The twentieth century also set the stage for wars of ideas in many sciences.

Some of these wars crossed the border between science and politics, had a strong ideological flavor, and resembled the witch-hunts of the Middle Ages much more than scientific arguments among civilized people. The most obvious examples were the "Arian science" of the Nazi Germany and the "Communist science" of the Soviet Union. The author is particularly aware of the latter because of personal experiences. In the late 1940s, his father, a bright young neurophysiologist Lev Pavlovich Latash (1924–2002) was expelled from graduate school in Moscow and sent into exile to Kazakhstan for "supporting the false, anti-scientific views of academician Shtern." His advisor, Lina Solomonovna Shtern (1878–1968), the first female academician of the Soviet Union, spent several years in prison and was freed only after the death of Stalin in 1953. Most great scientists who took part in the wars of ideas of the twentieth century did not use political means to prove their theories—the totalitarian political systems accepted theoretical views of one of the arguing sides as the only correct ones, usually without consulting the authors of the theories, and repressions followed.

One of the best known examples having a direct relation to the topic of this book is the deification of Pavlov and his theory of conditioned reflexes by the official Soviet ideology. Ivan Petrovich Pavlov (1849–1936) was a great Russian physiologist. He was awarded the Nobel prize in 1904 for his work on the physiology of digestion and later was officially crowned "the king of the world physiology" (*Princeps Physiologorum Mundi*) at the International Physiological Congress in 1935. Pavlov's theory of the functioning of the nervous system assumes that all animals are born with a set of inborn reflexes (stereotypical actions induced by external stimuli) that define the behavior during the first stages of life. Later, the animals learn to associate other stimuli with either positive or negative consequences, for example, food or pain, and enrich their repertoire of motor actions in response to the new, conditional stimuli. Pavlov performed series of fascinating experiments on the elaboration of conditioned salivation reflexes in dogs in support of his theory. After the revolution, he even built "towers of silence" in a suburb of St. Petersburg to optimize the conditions for the elaboration of conditioned reflexes by removing any extraneous, distracting stimuli. To his frustration, however, dogs confined to those towers of silence lost interest in all stimuli, old or novel, with disastrous consequences for the elaboration of new conditioned reflexes.

Until his death, Pavlov was revered as the highest, undisputed authority among Russian physiologists. His theory was eagerly accepted by the ruling communist party because of its strongly materialistic foundation: The system of reflexes did not leave room for any soul or intention, since actions of an animal or a person were predefined by a set of inborn reflexes (assumed to be common in all representatives of a particular species) and external stimuli that the animal or person experienced in life. This view gave the communist leaders a scientific foundation for their desire to breed a perfectly obedient people by manipulating external stimuli (information) available to individuals during their development, education, and throughout their life. Amazingly, the author of this super-materialistic theory remained highly religious until his death.

Pavlov probably did not pay much attention to criticisms of his theory in papers published by a young physician, who had started to study the mechanics of labor movements, soon after the 1917 revolution in Russia. This young physician, Nikolai Alexandrovich Bernstein (1896–1966), was, however, to become Pavlov's most formidable adversary (Meijer 2002). Bernstein's analysis of the mechanics of labor movements, described later in the book, led him to conclude that even the most stereotypical movements, such as hitting the chisel with the hammer by a professional blacksmith, could not be built on reflexes.

After World War II, the communist leaders headed by Stalin inspired a series of pseudoscientific sessions whose purpose was to proclaim their interpretation of Marxism (sometimes dubbed "dialectic materialism") as the only correct ideology in science. The purpose of one such session in 1951 was to eradicate all deviations from Pavlov's teachings in physiology. Even at that session, however, the witch-hunters missed the main opponent, Bernstein, and instead focused on the most talented and independent thinking students of Pavlov such as Leon Orbeli (1882–1958) and Nicolas Beritashvili/Beritov (1885–1974). However, later this "mistake" was corrected, Bernstein was fired and remained officially unemployed until his death in 1966. During that time, he started creating the physiology of activity, which assumed that most actions performed by animals and humans did not represent reactions to external stimuli but were initiated from within the organism.

The Pavlov–Bernstein controversy was only one example of several other disputes centered on the same question: What is the relative role of neural patterns generated within an organism versus those that represent reactions to external stimuli in the motor repertoire of an animal or a human? Simply put, are movements controlled by a soul (mind, intention, cognition, etc.) or driven by the environment through both direct mechanical interactions and sensory signals? For an update review on the physiology of free will, I recommend a recent paper by Mark Hallett (2007).

Sir Charles Sherrington (1852–1952) is considered by many as the father of contemporary neurophysiology. His contributions to the field are too many to be enumerated (for a review, see Stuart et al. 2001). In particular, Sherrington incorporated the notion of synapse into neurophysiology, introduced the idea of active inhibition within the CNS as a method of coordination of movements, described and emphasized the importance of reflex connections among muscles, etc. He was the first to view muscle reflexes not as hardwired stereotypical responses to stimuli but rather as tunable mechanisms that formed the basis of motor behavior. Control of movements, according to Sherrington, was performed by changing parameters of reflexes, an idea that later formed the foundation of the equilibrium-point hypothesis (Feldman 1966).

One of the trainees and colleagues of Sherrington, Thomas Graham Brown (1882–1965), had a different opinion. His studies of locomotion led him to conclude that the spinal cord contained neural structures able to produce rhythmic patterns without a sensory input. Hence, he viewed movements, such as locomotion, as predominantly originating from within the body, controlled by *central pattern generators,* rather than based on modulation of reflex loops originating from sensory receptors. The difference in scientific opinions soured the relations between Graham Brown and Sherrington, and the pioneering works of Graham Brown were all but forgotten for about 40 years. The controversy between sensory feedback-based control and central pattern generation continued throughout the century, taking different forms and contributing to such discussions as motor programming versus perception–action coupling and internal models versus equilibrium-point control.

Contemporary motor control owes much of its success to the progress in the mechanical analysis of movements over the twentieth century. This analysis was built on classical Newtonian mechanics. Despite the high level of the computational apparatus of classical mechanics, analysis of biological movement progressed slowly, because of the complexity of the biological effectors and changes in their mechanical properties during movements (in particular, because of changes in the shape of the muscles, muscle visco-elastic properties, blood flow that changed mass–inertial characteristics of the moving body parts, etc.).

Several brilliant scientists contributed to our current understanding of the muscle—the motor that drives all movements. Sir Archibald Vivian Hill (1886–1977) studied muscles as both mechanical objects and thermodynamic machines. Among his major contributions to muscle physiology and mechanics are the discovery of the mechanisms of heat production in the muscle, the introduction of the notions of *active state* and of *series elastic element,* which are commonly used in modern models of muscle behavior, and the introduction of the famous *Hill equation* describing the relation between muscle force and velocity of its shortening. Hill also studied physiology of the hemoglobin molecules in the blood and introduced the notion of *oxygen debt,* one of the central notions in exercise physiology. It is quite ironic that Hill's name is cited in many contemporary works on biomechanics and motor control in association with the notions of muscle stiffness and viscosity. Indeed, Hill focused on quantifying mechanical muscle properties but concluded that the classical physical notions of stiffness and viscosity were inapplicable to muscles. He came to this conclusion partly because the observed relationships between force and length (force and velocity) depended on active mechanisms in the muscle.

Approximately at the same time, Wallace Osgood Fenn (1893–1971) performed classical studies of heat production during muscle work and showed that muscles did not behave as prestretched passive springs. This loss of energy during muscle contractions was to be known as the *Fenn effect.* Fenn also performed classical works on the storage and release of energy in tendons and muscles and energy transfer among body parts.

Studies of Warren Plimpton Lombard (1855–1939) covered many topics such as the knee jerk, muscular fatigue, blood pressure, and metabolism. In our time, he is best known in association with *Lombard's paradox*, the coactivation of quadriceps and hamstring muscles during the transition from sitting to standing. The paradox is due to the bi-articular muscle action and fits well with classical mechanics. Lombard's paradox implies, in particular, that a bi-articular muscle can accelerate a segment in a direction opposite to the moment of force the muscle produces at that segment.

Among the numerous contributions to motor control by Nikolai Alexandrovich Bernstein (1896–1966) is his development of a novel method for mechanical analysis of movements, kimocyclography. This method was based on earlier works by Marey and Muybridge, already mentioned. It used a camera with a film that moved slowly at a continuous speed and a rotating shutter that opened the lens for a short time for every revolution. Light bulbs were placed on the subject and filmed, leading to a sequence of stick-figure images. The number of stick figures that such a system could record within a second was defined by the speed of the rotating shutter. In his analysis of the mechanics of arm motion during piano playing by professional pianists, Bernstein reached the amazing frequency of about 500 frames per second (see the next section).

Studies of patterns of muscle activation during fast voluntary movements became a common tool in motor control studies, in the middle of the twentieth century. One of the pioneers of these studies was Kurt Wachholder (1893–1961). He was the first to describe the tri-phasic EMG pattern during a fast single-joint movement performed by a human subject. Wachholder and his colleague, Hans Altenburger, published a series of 11 papers between 1923 and 1928, in which they analyzed changes in the tri-phasic EMG pattern with changes in movement characteristics. They were particularly interested in interactions between a pair of muscles crossing a joint, an agonist and an antagonist. In one of those papers, Wachholder and Altenburger asked a seemingly naïve question: How can muscles be relaxed at different joint positions in the absence of an external load, taking into account their elastic properties? Indeed, if two muscles acting at a joint are relaxed at a certain position, motion of the joint would lead to the stretching of one muscle and the shortening of the other. Muscle elastic properties should produce forces that have to move the joint back to its initial position. So, if a joint has to stay at the new position, muscle activation levels must change to counteract those elastic forces. This logic led Wachholder and Altenburger to perform meticulous studies of muscle activation levels at different joint positions. As a result, they concluded that muscles could indeed be relaxed at different joint positions, and that, consequently, elastic muscle properties were regulated by the CNS—an amazing insight for the 1920s!

Within this brief review, it is impossible to pay proper tribute to all the important discoveries made in the area of neurophysiology in the twentieth century. Classical neurophysiological and behavioral works by Magnus, von Holst, Renshaw, Granit, Eccles, Lundberg, Henneman, Matthews, Evarts, and their colleagues formed the body of knowledge that allowed to formulate and test hypotheses on how the CNS controls the motor function. In the early 1950s, progress in neurophysiology had led to the creation of arguably the first model of motor control, the *servo-hypothesis* of Merton (Merton 1953). The servo-hypothesis was the first to specify hypothetical physiological variables that could be used by the CNS to control voluntary movements. The model was later proven to be wrong, but its emergence signified an important stage in the development of the area of motor control.

2.5 NIKOLAI ALEXANDROVICH BERNSTEIN AND MOVEMENT SCIENCE IN THE SOVIET UNION

I would like to end this historical section with a brief description of the development of movement science in my native Soviet Union. This is an example of how a handful of dedicated, bright researchers inspired by an outstanding scientist moved ahead in an area of science under the constant pressure from the ideologically biased society and virtually without any financial or other support. Soviet science was a strange, artificial creation of the communist system, which tried to control the thoughts of all its citizens but simultaneously encouraged exploration in areas that were given priority, mainly for military reasons. As a result, mathematicians and physicists were allowed relative freedom of expression, and their respective areas were relatively well funded. On the other hand, virtually all areas of biology were in a pitiful state, maybe with the exception of biological warfare. [The author guesses. He has no information, but if Komitet Gosudarstvennoj Bezopasnosti (KGB) archives prove otherwise, he would welcome their publication and is ready to apologize, if he is proven wrong.]

Genetics and physiology suffered particularly badly, following two infamous sessions of Soviet Academies, inspired and staged by the communist party leaders. The first session of the Academy of Agricultural Sciences in 1948 was led by an ignorant "academician" Trofim Lysenko, who claimed that species could turn one into another under appropriate external conditions. According to Lysenko, genes and chromosomes were fictitious objects invented by bourgeois scientists to cheat the workers and peasants. This session threw the young Soviet genetics into the Stone Age. The second session has been known as the Joint Session of the Academy of Sciences and the Academy of Medical Sciences. It occurred in 1951 and attacked all "anti-Pavlovian" directions in physiology.

As mentioned earlier, the attack was largely misguided, but nevertheless, movement science seemed to be doomed, together with other areas of physiology.

Undoubtedly, the Soviet school of movement science would not have flourished but for the personality and scientific contributions of Nikolai Alexandrovich Bernstein (1896–1966), who is considered by many as the founder of contemporary motor control.

Bernstein's parents were broadly educated persons. His father, Alexander Nikolaevich Bernstein was a famous psychiatrist, a student of Sergei Korsakov, who was the creator and leader of the Moscow school of psychiatry. After the revolution, Alexander Bernstein became a professor and was for some time a major figure in the Ministry of Science and Education. Bernstein's mother, Alexandra Karlovna Bernstein (Johansson) was a nurse. She played a central role in the early education of Nikolai Bernstein and his younger brother Sergei, particularly in their language and music lessons. Throughout his life, Bernstein remained deeply attached to music. To complete the family picture, it is necessary to mention Bernstein's uncle, a famous mathematician, Sergei Natanovich Bernstein (1880–1968), whose early career was marked by his solving two of the famous Hilbert problems. After the 1917 revolution, Sergei Natanovich Bernstein became a professor and a member of the USSR Academy of Sciences.

During World War I, Nikolai Alexandrovich Bernstein studied medicine in Moscow University and volunteered as a physician's assistant in a military hospital. He graduated in 1919 with an M.D. degree and was enlisted in the Red Army as a military physician. In 1921, Bernstein returned to Moscow and worked as a psychiatrist for about a year. Then, he joined the newly formed Institute of Labor, whose purpose was to develop new, better forms of labor movements for the workers of the communist state. The Institute was organized by Alexei Gastev (1882–1939), an avant-garde poet and a professional revolutionary, who also served as the head of the Ministry for Scientific Labor Organization. During his work at the Institute of Labor, Bernstein developed a number of methods for quantitative movement analysis and performed his now classical studies of the kinematics and kinetics of a variety of movements. His analysis of the motor variability led him to the conclusion that skilled movements could not emerge as a result of "treading a path" through a chain of neurons, in contradiction to the dominant views of Pavlov's school.

In particular, in the 1920s, Bernstein performed a study of the kinematics of hitting movements, when professional blacksmiths strike the chisel with the hammer (Bernstein 1930). His subjects were perfectly trained: They had performed the same movement hundreds of times a day for years. For this analysis, Bernstein used his newly developed method to record movement kinematics, *kimocyclography*. In the precomputer age, data processing created the bottleneck; it was not unusual to spend months to measure the kinematics of movements recorded in a single subject. Using this system, Bernstein observed that the variability of the trajectory of the tip of the hammer across a series of strikes by a blacksmith was smaller than that of the trajectories of individual joints of the subject's arm holding the hammer (Figure 2.1). Note that at any time during the striking movement, a deviation of any joint from its average angular trajectory was expected to produce a larger spatial deviation of the tip of the hammer, compared to deviations of markers placed over individual arm joints from their respective spatial trajectories. Since, apparently, the brain could not send signals directly to the hammer, Bernstein concluded that the joints were not acting independently but correcting each other's errors. This observation suggested that the CNS did not follow a unique solution for the problem over repetitive strikes but rather used a whole variety of joint trajectories to ensure more accurate (less variable) performance of the task.

During the same period, Bernstein performed, from my subjective point of view, one of his most amazing studies (Bernstein and Popova 1930; Kay et al. 2003). He placed electric bulbs over major joints of the arms and hands of professional pianists and recorded their motion during the task of octave strike at

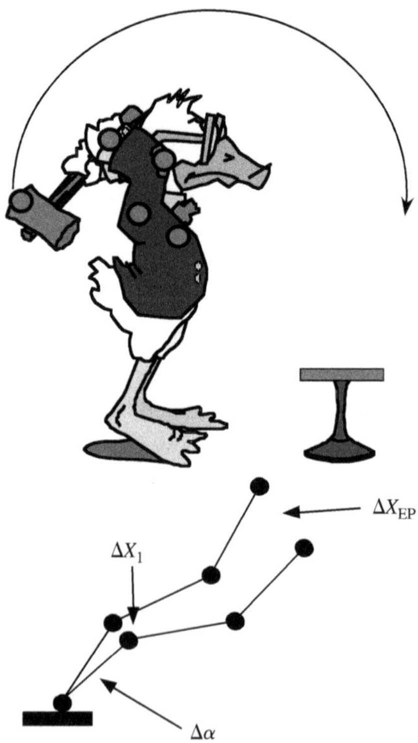

Figure 2.1. When a blacksmith hits an object with the tip of the hammer, small errors in proximal joint angles ($\Delta\alpha$) are expected to lead to larger deviations of the endpoint (ΔX_{EP}, the tip of the hammer) than of more proximal joints (ΔX_1).

a fixed tempo but different loudness and at a fixed loudness but varying tempi. The selection of such an amazingly complex object of study could be explained only by Bernstein's youthful optimism. After reconstructing the movement kinematics, Bernstein performed several analyses that became commonplace in movement studies only half a century later. In particular, he performed analysis of joint torques and computed torque components that originated in joints due to the motion of other joints of the limb (so-called *interaction torques*).

While working in an area of science dominated by intuitive rather than precisely formulated notions and hypotheses, Bernstein realized that before solving problems, one needed to focus on their exact formulation. Later in his life, he liked to recount the following story (as recollected by Prof. V.M. Zatsiorsky): "You probably do not know that God has a cousin who has never been very famous. So, the cousin asked God to help him achieve fame and glory in science. To please the cousin, God gave him an ability to get any information about physical systems in no time and to travel anywhere within a microsecond. First, the cousin decided to check whether there was life on other planets. No problems; he traveled to all the planets simultaneously and got an answer. Then he decided to find out what the foundation of matter was. Again, this was easy: He became extremely small, crawled inside the elementary particles, looked around, and got an answer. Then, he decided to learn how the human brain controls movements. He acquired the information about all the neurons and their connections, sat at his desk and looked at the blueprint. If the story has it right, he is still sitting there and staring at the map of neuronal connections."

In contemporary literature, Bernstein is best known for his formulation of the problem of motor redundancy, commonly known as the Bernstein problem (Turvey 1990). This problem, and attempts at solving it, will be discussed in detail further in this book. We should not underestimate the contribution of Bernstein to other areas of movement science and physiology in general. His most comprehensive book, *On the Construction of Movements* (1947), has never been translated into English. This book presents a theory of movement control and coordination and discusses, such issues as the evolution of movements, their development, effects of practice, and many others.

In the late 1940s, Bernstein was a well-known figure in the Soviet biomechanics, physiology, and sport science fields. In 1947, "On the Construction of Movements" was awarded the highest USSR prize, the Stalin Prize. However, the mentioned Session of the Two Academies changed the situation dramatically. A few months after the Session, Bernstein was fired. Since that time, and until his death, Bernstein worked mostly alone, with only a few informal students, among them being a young psychologist Josif Feigenberg and a young neurophysiologist Lev Latash.

In the late 1950s, the progress in cybernetics reached the Soviet Union. The father of cybernetics, Norbert Wiener (1894–1964) visited Moscow and presented a lecture at the Moscow State University. The interpreter struggled with the technical aspects of the lecture, and then an elderly gentleman from the audience offered his help. This was Bernstein. Wiener probably never learned who helped translate his lecture.

Approximately at the same time, in the late 1950s, two outstanding scientists, a mathematician Israel Gelfand (born in 1913) and a physicist Michael Tsetlin (1924–1966), met Bernstein and decided to start a seminar series to address the problem of an exact description of biological objects and their functions. This seminar produced what is currently known as the Russian (Soviet) school of motor control. It represented a very fortunate blend of ideas of one of the greatest physiologists of the century (Bernstein), the clarity of critical thinking of one of the greatest mathematicians (Gelfand), and the ability to synthesize information of a great physicist (Tsetlin).

I was too young at the time, but my senior friends/colleagues shared their recollections with me. Apparently, this was the most speaker-unfriendly seminar in the history of science. Some speakers were never allowed to move beyond the title slide because of discussions over terms used in the title. Gelfand was most active during the seminar; he frequently interrupted the speakers in the least friendly manner. Being a famous mathematician, he was particularly annoyed when a speaker tried to write mathematical formulas, while obviously being rather ignorant about math. In such situations, Gelfand used to say: "Simply by watching how you grasp the chalk, I already know that you are not a mathematician. Stop pretending, and tell us something you do know." A former speaker recollected: "When you presented at that seminar, you felt like a cow in the process of being milked. But just like the cow, you had no idea what would be done with the milk." The general feeling among the young scientists who attended the seminar was a mixture of awe and optimism. This was largely because of the talented Michael Tsetlin, who was typically quiet during the talks but, after the talk was over, he would rise and say: "I think that I know what the talk was about." Then, he would summarize the essence of the talk in 3 min. The "Gelfand seminar" was arguably the only place in the Soviet Union, where people were told the truth to their face. It was the place where real things happened, in contrast to the rest of the country deeply bogged down in the swamp of ideological, nonsensical rhetoric. The seminar produced a generation of outstanding researchers, such as Arshavsky, Berkinblit, Feldman, Fookson, Gurfinkel, Orlovsky, Severin, Shik, Smolyaninov, and many others, whose works will be featured prominently in this book.

It is getting dangerous to continue with this historical review of studies performed over the last 50 years, their significance being still debated. So, to be safe, I would rather stop this section at the early 1960s.

This brief and biased historical review teaches a lesson. I will try to summarize it in a very subjective way. One should not try to create a general theory of everything. Rather, one should always start with the identification of an object of research and assume that inputs to this object, which can be "physical" (such as mechanical interactions with the environment) or "intelligent" (in other words, purposeful, coming from the mind, intention, or soul), are beyond current analysis, although there have been attempts to approach the physiology of free will (reviewed in Hallett 2007). If we want to understand how the CNS controls movements, we should not confound our analysis with a question of how it comes up with an idea to perform a movement in the first place. Some other researchers will address this very important question. At least for now, this issue seems to be a subject of philosophy, not of natural science, and our understanding of it has not changed much since the times of Plato, Aristotle, and Galen.

2.6 HISTORY OF SYNERGIES AND THE PROBLEM OF MOTOR REDUNDANCY

A great British neurologist, J. Hughlings Jackson (1835–1911), did not use the term *synergy* when he described his theory of a three-level representation of movements in the CNS. He wrote in 1889: "the central nervous system knows nothing about muscles, it only knows movements" (p. 358), and continued on the same page: "In the highest motor centres (prae-frontal lobes)... the same muscles are represented (re-re-represented) in innumerable different combinations, as most complex and most special movements."

The notion of muscle synergies was developed by a great French neurologist, Joseph Felix Francois Babinski (1857–1932), whose main works were published at the end of the nineteenth century and the beginning of the twentieth century (Smith 1993). Babinski studied the coordinated activity of muscles in patients with neurological disorders and in healthy persons. In 1899, he linked impaired muscle coordination to a pathology in the cerebellum and called such discoordinated movements "cerebellar asynergies." This insight was developed later in the twentieth century, leading to several models of motor synergies based on the neuronal structure of the cerebellum (see section 7.3).

For most researchers in the field of movement studies, the word *synergy* is associated with the name of Bernstein and is inseparable from the famous problem of motor redundancy (the Bernstein problem, Turvey 1990; Latash 1996). So, let me first introduce the problem of motor redundancy using Bernstein's example.

Touch your nose with the tip of your right index finger. Now try to move the arm without losing the contact between the fingertip and the nose. This is easy to do. This means that one can touch the nose with very many combinations of arm joint angles. Nevertheless, when the task was presented, you did it with a particular joint combination. How did your brain select it?

In other words, to place the tip of the finger at any point in space, one needs to specify three coordinates corresponding to the three-dimensional space we happen to live in. At a kinematic level of description, the human arm has at least seven axes of joint rotation, even if one does not count the joints of the hand and fingers. There are three axes of shoulder rotation (flexion–extension, abduction–adduction, and internal–external rotation), one axis of elbow joint rotation (flexion–extension), two axes of wrist rotation (flexion–extension and ulnar–radial deviation), and one axis of rotation shared between the wrist and the elbow (pronation–supination). To define unambiguously seven angles about the seven axes, one needs to have seven constraints. In other words, to solve a system of equations with seven unknowns, one needs seven equations in the system. But the task supplies only three equations, which means that there are an infinite number of solutions. Figure 2.2 illustrates this problem with two configurations that are equally successful at pointing with a four-joint arm in a two-dimensional space.

Similar problems emerge at other levels of description of the neuromotor system. For example, consider the task of producing exactly 10 Nm of flexion torque in the elbow joint. The joint is crossed by six muscles, three flexors (biceps, brachialis, and brachioradialis) and three extensors (three heads of the triceps) (Figure 2.3). One may ask: "How much torque should each muscle produce?" Here, we are dealing with one equation and six unknowns. Obviously, it also has an infinite number of solutions.

We can take another step and look at a single muscle. Consider, for example, the biceps muscle of the arm. It contains many fibers and is innervated by many neural cells in the spinal cord, called alpha-motoneurons (Figure 2.4; see Digression #8). There are fewer motoneurons than muscle fibers such that when

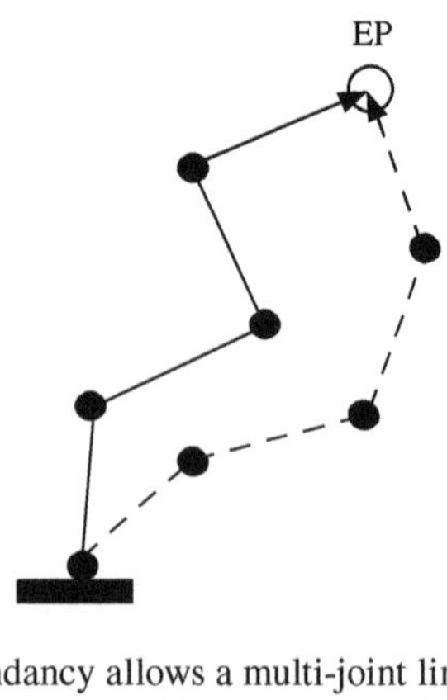

Figure 2.2. Kinematic redundancy allows a multi-joint limb to reach the same endpoint (EP) location with an infinite number of joint configurations (two are illustrated).

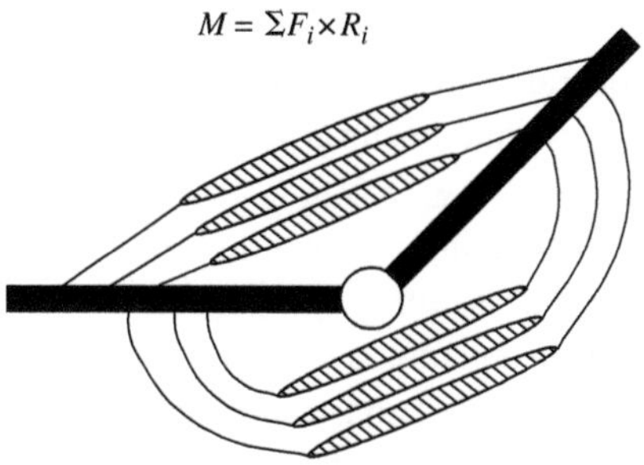

Figure 2.3. A joint crossed by three flexor and three extensor muscles. There is an infinite number of ways muscle forces may be combined to produce a required magnitude of joint moment of force: $M = \Sigma F_i R_i$, where i corresponds to different muscles ($i = 1, 2, \ldots, 6$), F is muscle force, and R stands for the lever arms.

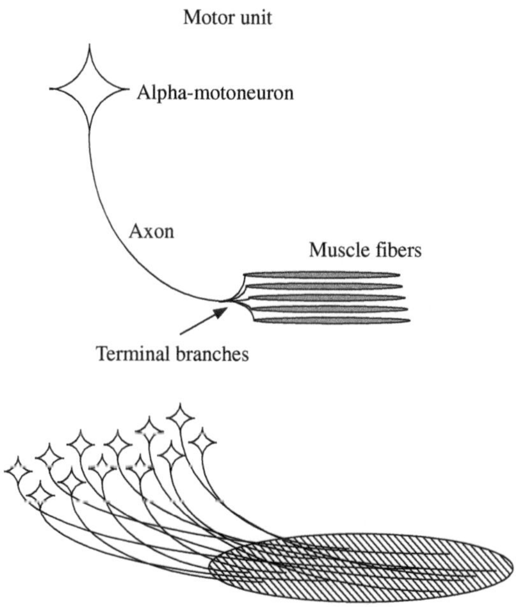

Figure 2.4. A muscle is composed of motor units. Each motor unit (the top drawing) consists of neural cells (alpha-motoneuron) and a group of muscle fibers that receive signals from the terminal branches of the axon of that neural cell. A muscle is innervated by many alpha-motoneurons (bottom).

a motoneuron sends a signal (called an action potential) to the muscle, the signal is received by a group of muscle fibers. Such groups serve as units of muscle activation, since all the fibers within a group always respond together when their alpha-motoneuron generates an action potential. Quite appropriately, these groups are called motor units. The CNS can modify muscle force by changing the number of recruited alpha-motoneurons and/or by changing the frequency of action potentials generated by individual motoneurons. An increase in the frequency of the firing of a motoneuron increases its contribution to the total muscle force.

So, the question is: "How many motoneurons and at what frequencies should the CNS recruit to achieve a certain level of muscle activation?" Obviously, here we are dealing with a single equation and many, possibly hundreds of unknowns (Figure 2.5). There are established constraints on patterns of motor unit recruitment, in particular, the size principle (the Henneman principle, Henneman et al. 1965) discussed in Part 3 of the book. These constraints reduce the number of alternative recruitment patterns, but they still fail to produce a unique solution to the problem.

How does the CNS handle all these apparent problems of choice when more elements than necessary contribute to each task? No agreement has been reached on this issue. A lot of research has been motivated by Bernstein's suggestion that the CNS unites elements into groups such that each group is controlled by a single variable. Bernstein called such groups synergies (as discussed later, this understanding of synergies differs from the one advocated in this book). The presence of such synergies decreases the number of variables the controller needs to manipulate and (partly) solves the problem of redundancy. In a sense, the physiological organization of muscle fibers into motor units may be viewed as a low-level synergy, which allows the controller not to bother about individual muscle fibers but to deal with a fewer number of motor units.

So, according to Bernstein, the CNS organizes elements (joints of a limb, muscles acting at a joint, fingers of the hand, etc.) into synergies in a task-specific, flexible way. Ultimately the problem of motor redundancy is solved and a single optimal solution emerges.

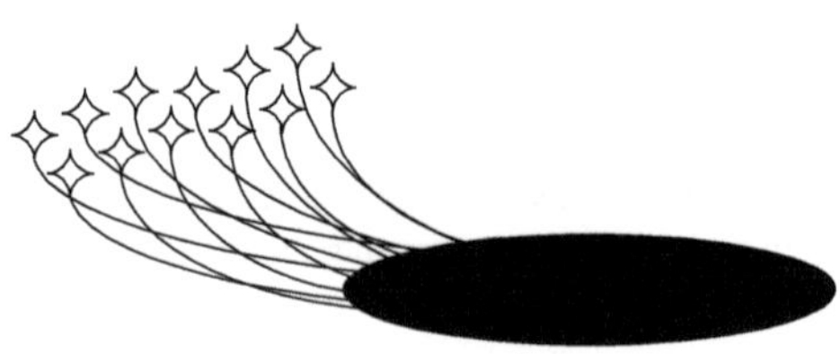

$$F_M = \Sigma f(F_i; \phi_i)$$

Figure 2.5. Muscle force (F_M) results from activation of many motor units. The contribution of each motor unit to the muscle force is a function f of its size (which may be represented by the peak force of its single contraction, F_i) and the frequency at which this motor unit is activated (ϕ_i).

In line with Bernstein's thinking, many studies of different motor systems and tasks explored different characteristics of motor synergies. Most studies, however, focused on only one feature of synergies, typically on sharing. In particular, the term *synergy* has been used to describe correlated outputs of muscles/joints/effectors in voluntary multi-joint limb movements, force production tasks, quiet standing, locomotion, postural adjustments, quick reactions to perturbations, and other motor actions and reactions (Nashner and Cordo 1981; Keshner et al. 1988; Gottlieb et al. 1996; Alexandrov et al. 1998; Li et al. 1998; Vernazza-Martin et al. 1999; Saltiel et al. 2001; Ivanenko et al. 2004). A number of clinical studies reported atypical sharing patterns among kinematic (joint angles) or kinetic (joint torques or muscle forces) variables in different patient groups and interpreted them as atypical synergies (Levin et al. 2002; Cirstea et al. 2003; Beer et al. 2004).

When several variables contribute in an additive way to a common output, sharing can be easily quantified, for example, as the percentage of the total output produced by each variable. Invariant sharing patterns have been described for many tasks, including vertical posture, locomotion, reaching, finger force production, etc. (Smith et al. 1985; MacPherson et al. 1986; Desmurget et al. 1995; Li et al. 1998; Wang and Stelmach 1998; Santello and Soechting 2000; Pelz et al. 2001). However, stable sharing and the associated positive pairwise correlations between pairs of individual variables do not require a controller and may result from the structural design of the system: As in an earlier example, the forces under the four legs of a table scale together with the weight of an object placed on the top of the table.

Imagine the following example. When a person presses with the four fingers of a hand such that the total force increases with time, for example, linearly as in Figure 2.6, the forces of the individual fingers can vary within a wide range as long as they sum up to the required total force. These variations show certain regularities that apparently reflect a particular control strategy selected by the CNS. Typically, forces of individual fingers show close to linear profiles such that the percentage of the total force produced by each finger remains nearly constant within a wide range of the total force, that is, they show a nearly constant sharing pattern (Li et al. 1998).

Now let us ask the subject to perform the same task several times and compute indices of variability of the finger forces and of the total force across the trials for each moment of time. Figure 2.7 shows force levels reached by individual fingers (symbols correspond to finger names, I—index, M—middle, R—ring, and L—little) and by all of them together (total) in successive trials (this is a cartoon, not real data). Comparison of variability can be done using a statistical index termed variance. This index has certain advantages compared to other common indices of variability such as standard error and standard deviation. In particular, if several independent noisy signals are summed up, the variance of their sum

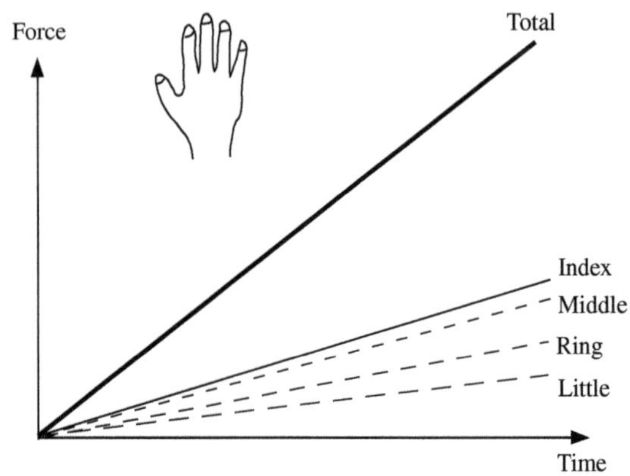

Figure 2.6. The total force produced by a set of fingers is typically shared among the fingers in a relatively stereotypical way such that a linear increase in the total force leads to similar linear increases in the individual finger forces.

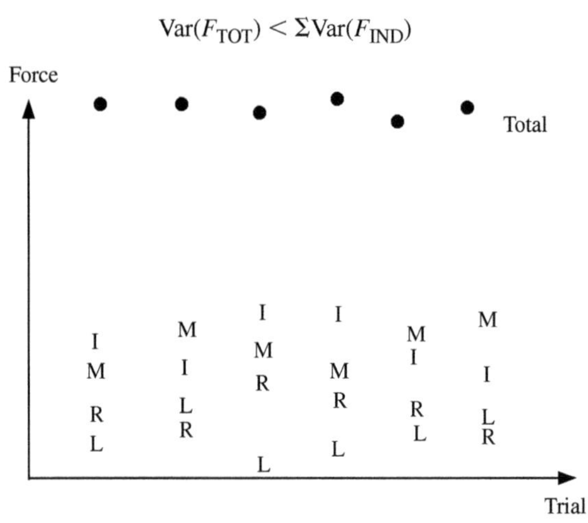

Figure 2.7. If a person is asked to produce a certain value (e.g. the maximal value) of the total force by pressing with the four fingers of a hand, the trial-to-trial variability of the total force is smaller than the trial-to-trial variability of the individual finger forces (shown by symbols I, M, R, and L corresponding to the index, middle, ring, and little fingers, respectively). This may be formalized as an inequality between the variance of the total force and the sum of the variances of the individual finger forces: $\mathrm{Var}(F_{\mathrm{TOT}}) < \Sigma\mathrm{Var}(F_{\mathrm{IND}})$.

is expected to be equal to the sum of the variances of individual signals. For example, let us compute the variance of the total peak force [$\mathrm{Var}(F_{\mathrm{TOT}})$] across a set of trials and compare it to the sum of the variances of individual finger peak forces [$\Sigma\mathrm{Var}(F_{\mathrm{IND}})$]. The cited paper by Li et al. (1988) has shown that the former index is smaller than the latter, that is, $\mathrm{Var}(F_{\mathrm{TOT}}) < \Sigma\mathrm{Var}(F_{\mathrm{IND}})$. This inequality means that some of the variance in the individual finger forces is missing. This is likely to result from a negative co-variation among individual finger forces from trial to trial, that is, if one finger in one trial produces more force than its average contribution, other fingers will produce less force, and the variability of the total force is decreased. In other words, this is an indication of flexible solutions used by the fingers that are organized in such a way that they lead to smaller variability (higher stability) of the total force compared to what one could have expected if finger forces varied independently of each other.

Such apparent error compensation among elements of a synergy has been reported in many studies. For instance, in a study in which subjects pointed at targets in two dimensions with a three-joint limb moving in a plane, the variability of the end point final position was smaller than the variability of a marker placed on the wrist (Jaric and Latash 1999). The variability in the final positions of the elbow and the shoulder joint could be expected to lead to higher variability in the end point marker location, since it was farther from the joints than the wrist marker, and equal joint deviations were expected to lead to larger spatial displacements. The opposite result has been interpreted as suggesting that the wrist rotation partly compensated for the errors in the location of the end point marker introduced by imprecise rotation of the proximal joints.

Other studies used perturbation techniques to change the natural pattern of individual elements in a well-practiced multi-element motor task. For instance, in studies of the coordination of articulators during speech production, an unexpected mechanical perturbation applied to one of the articulators induced rapid changes in the performance of unperturbed articulators leading to the relatively undisturbed production of required sounds (Kelso et al. 1982, 1984; Abbs and Gracco 1984; also see Saltzman and Kelso 1987).

Several researchers studied the coordination of the elbow and wrist muscles during motor tasks that required movement in one of the two joints (Gielen et al. 1988; Koshland et al. 1991; Latash et al. 1995). Imagine that you place the upper arm on the table and keep the forearm and the hand vertical (Figure 2.8). If you now move the elbow joint very fast into flexion or extension, the wrist will not show any visible motion. However, motion of the elbow produces inertial torques at the wrist joint that could be expected to move the joint. The mechanical coupling between the joints requires coordinated changes in commands to muscles crossing the two joints to avoid flapping of the wrist. Indeed, a fast elbow movement is accompanied by tightly coupled patterns of activation seen in the muscle pairs crossing both joints (Figure 2.9). Similarly, a fast wrist movement could be expected to lead to elbow joint motion because of the joint coupling. This does not happen because of changes in the activation of muscles acting at the elbow during a voluntary wrist movement.

One may conclude that there exists a wrist–elbow synergy that coordinates signals to muscles crossing the two joints. This synergy may be characterized by

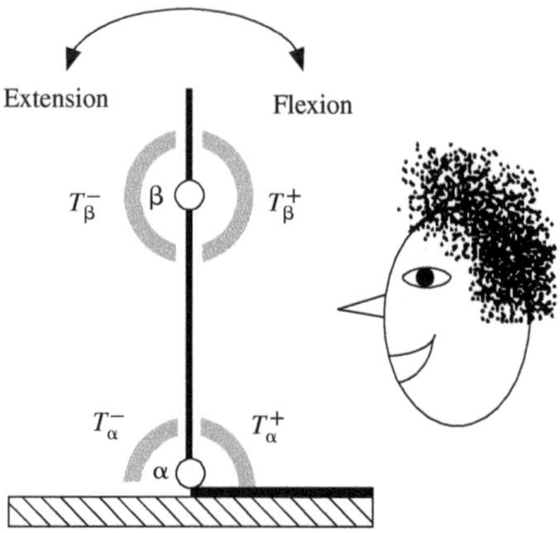

Figure 2.8. When a sitting person places the upper arm on a table and tries to move one joint at a time, the elbow (α) or the wrist joint (β), the mechanical joint coupling requires simultaneous changes in the joint torques (T_α and T_β). Reproduced by permission from Latash ML, Aruin AS, Shapiro MB (1995) The relation between posture and movement: a study of a simple synergy in a two-joint task. *Human Movement Science* 14: 79–107, © Elsevier.

a sharing pattern (defined by the mechanical properties of the arm segments that may differ in different persons). However, it shows limited flexibility: If movement in one of the joints is blocked with a splint, the coupling of the muscle-activation patterns is preserved despite its apparent inefficacy (Koshland et al. 1991). These observations make the alleged synergy questionable. On the other hand, several experiments have shown that the wrist–elbow synergy may possess a feature of flexibility/stability if we assume that the purpose of this synergy is to stabilize the trajectory of the endpoint, for example, of the tip of the index finger. Studies with unexpected joint blocking and release during elbow–wrist motor tasks have shown quick responses (at the delay between 50 and 90 ms) in the muscles crossing both the explicitly perturbed joint and the other joint (Gielen et al. 1988; Koshland et al. 1991). These responses could be interpreted as directed at preserving the planned trajectory of the end point of the arm (Latash 2000).

Let us now consider a very different object, a frog in which the spinal cord has been surgically separated from the brain, the so-called spinal frog. As described earlier, spinal frogs have been favorite objects of study in the nineteenth century. However, about a 100 years later, researchers returned to the wiping reflex in the spinal frogs, described and studied in detail by Pflüger, at a different level of analysis. To recall, if an experimenter places a stimulus (a small piece of paper soaked in a weak acid solution) on the back of a sitting spinal frog, the frog, after a certain latent period, performs a series of finely coordinated movements wiping the stimulus from the back and, sometimes, throwing it away from the body

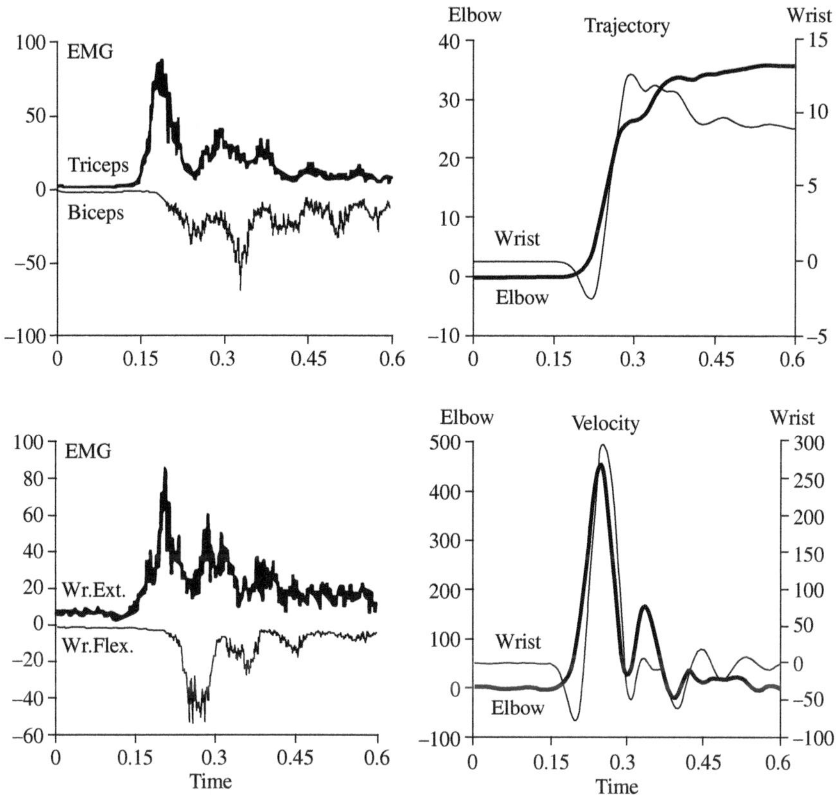

Figure 2.9. The mechanical coupling illustrated in Figure 2.8 requires parallel changes in the levels of activation of the muscle pairs acting at the elbow and wrist joints. This figure illustrates the electromyographic (EMG) patterns for the elbow extension movements. Note that the elbow moved over about 30°, while the wrist flapped over less than 10°. However, the EMG bursts are comparable in the elbow flexor–extensor pair (biceps and triceps) and in the wrist flexor–extensor pair (Wr.Flex. and Wr.Ext.). Reproduced by permission from Latash ML, Aruin AS, Shapiro B (1995) The relation between posture and movement: a study of a simple synergy in a two-joint task. *Human Movement Science* 14: 79–107, © Elsevier.

(Figure 2.10). The joints of the hindlimb share the task and show reproducible patterns of their trajectories. There are a number of unique attractive features in this type of reflex behavior: First, since the animal is spinal, there is no influence of a poorly controlled supraspinal neural inflow (intention or soul). Second, the wiping movement is stimulus-directed (it has a goal), reproducible, and can easily be evoked. Third, it is a multi-joint movement performed by a kinematically redundant effector: There are four major joints in the frog's hindlimb, even if one ignores joints of the toes and whole-body movements.

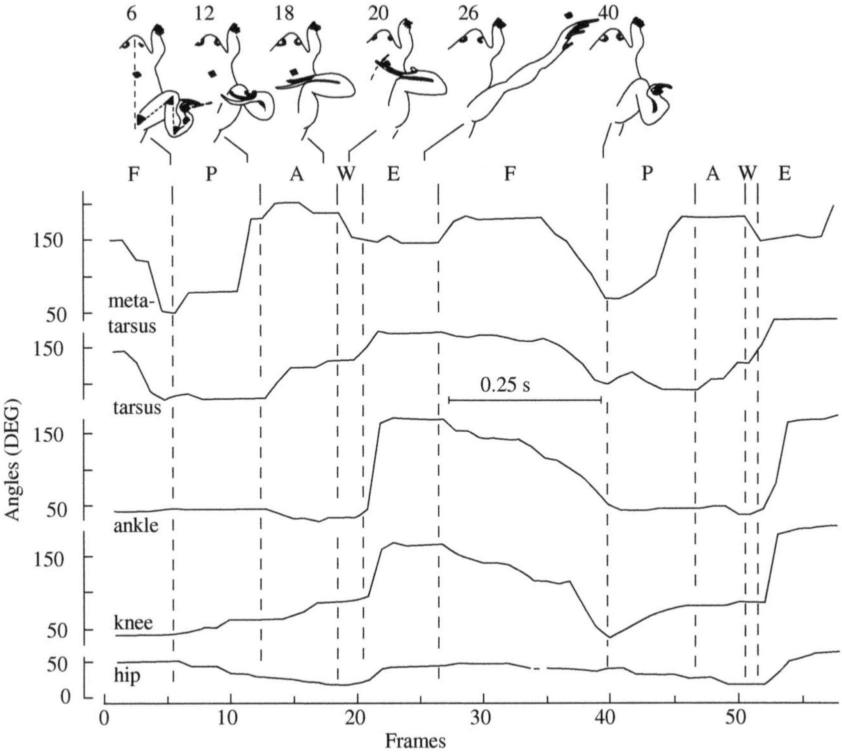

Figure 2.10. An illustration of typical postures during the wiping reflex by the spinalized frog. F—flexion, P—placing, A—aiming, W—wiping, and E—extension. The lower graphs show changes in the five major joints of the hind limb. Reproduced by permission from Berkinblit MB, Feldman AG, Fukson OI (1986a). Adaptability of innate motor patterns and motor control mechanisms. *Behavioral and Brain Science* 9: 585–638, © Cambridge University Press.

A series of experiments studied the effects of unexpected perturbations on the wiping patterns (Figure 2.11). In one of these experiments, a loose thread loop was placed on the hindlimb preventing movements in the knee joint beyond a certain limit so that the maximal knee joint motion was about 5°. When the wiping was performed by an unconstrained limb, changes in the knee joint angle during certain phases of wiping were tenfold bigger. The frog was able to remove the stimulus from its back during the first attempt. Then, the knee was released, and a cast was placed preventing movements in the next (more distal) joint. The frog once again wiped the stimulus at the first trial. Then, a lead bracelet was placed on the distal part of the hindlimb; the weight of the bracelet was similar to the weight of the hindlimb itself. The frog still was able to wipe the stimulus accurately. Obviously, in experiments with joint fixation, the frog used the multi-joint design of the limb to compensate the unexpected errors introduced by the restrained joint

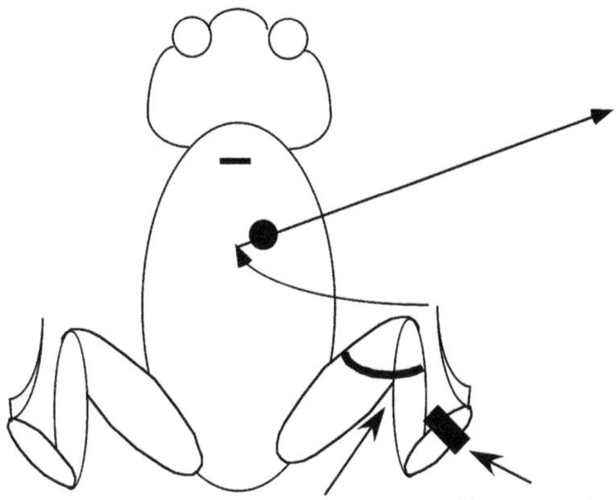

Figure 2.11. The spinalized frog (the level of the surgery is shown by a black bar) was able to perform accurate wiping in conditions of loading of a distal leg segment with a heavy bracelet and in conditions of blocking motion in one of the major joints (knee joint).

using modifications of motion of other joints of the limb. Its spinal cord showed error compensation. It also showed task dependence of wiping patterns in experiments, when the stimulus was placed on the forelimb of the frog and the forelimb could be moved to different positions with respect to the body (Fukson et al. 1980). Moreover, the pattern of joint coordination could undergo qualitative changes, when the stimulus was moved closer to the bottom part of the frog's back, to spots that were hardly reachable by the toes of the hindlimb: In such conditions, the frog wiped the spot with more proximal parts of the hindlimb. We may conclude, therefore, that the wiping reflex of the spinal frog meets all three criteria, rendering it fit to be called a synergy and, therefore, deserves a score of 10 on our scale.

2.7 PROBLEMS WITH STUDYING BIOLOGICAL MOVEMENT

As a graduate of the Moscow Physics-Technical Institute (MPTI or FizTech), I ought to believe that there is only one science, and that is physics. One of the most popular college songs of our times had a line: "Only physics is salt, the rest is nil." This statement reflects the rather outrageous arrogance typical of those who went through the tough selection during the entrance exams and made it into FizTech. Certainly, not everybody would immediately agree with this statement, particularly those who work in the humanities and different areas of biology. Is psychology a science? Is linguistics a science? Is neuroscience a science? Or maybe it is better to ask: Are psychology, linguistics, and neuroscience (and motor control) subfields of physics?

For a long time, the MPTI has had a major called *Physics of Living Systems*. This name sounds highly scientific and implies that there is such an area of physics, while actually it has never existed. My friends and I were taught the full range of physics and mathematics, as well as bits and pieces of biology, including general biology, biochemistry, biophysics, physiology, and even such outlandish disciplines as *Aviation and Space Medicine*. This major produced only a few graduates a year, and a surprisingly high percentage of those ended up studying human movements. Or maybe this should not be that surprising. Many of the students in that major were interested in knowing how the brain worked. Much of the brain functioning is reflected in movements, even such "high-level" actions as language and music may be viewed as combinations of appropriately timed movements of the articulators, fingers, breathing organs, etc. Movements are objectively observable and measurable, unlike such products of the brain as thoughts and emotions. So, for a physicist interested in the brain, movement is a very attractive object of study. Let me quote Richard Feynman (1918–1988), one of the great physicists of the twentieth century: "The key to modern science…is to look at the thing, to record the details, and to hope that in the information thus obtained might lie a clue to one or another theoretical interpretation" (Feynman 1994, p. 5).

The very first attempts at using the well-established tools and principles of physics to "record the details" of biological movement show young physicists how challenging even the seemingly simplest measurements could be. Even if one were interested only in the mechanical analysis of a natural movement, the difficulties are mind-boggling. At first glance, these difficulties seem to be technical, not conceptual. In fact, biological movement obeys the same Newton laws as motion of inanimate objects. So, if body mechanics is the object of study, one "only" needs to know mechanical characteristics of the moving objects and have accurate methods of motion measurement.

All serious researchers in the field of biomechanics would appreciate the quotation marks in the previous sentence. Let me use the simplest example. To write equations of motion, one needs to know inertial characteristics of the moving objects such as their mass and the location of the center of mass. For many human movements, these objects are limbs and their segments. The problem is that the process of movement production is associated with changes in the inertial properties of effectors, in particular, because of such factors as changes in muscle shape (muscle bulging) and in the blood flow. Muscle elasticity (the dependence of muscle force on muscle length) and damping (the dependence of muscle force on the rate of change in muscle length) also change with muscle activation, length, and velocity. This means that parameters in the equations of motion are changing continuously, even if one accepts the doubtful assumptions that the equations are written correctly, they describe the body adequately, and their general form does not change with time. (By the way, all these assumptions are commonly wrong.)

However, this is only the tip of the iceberg of problems in movement analysis. If one wants to use movement studies to get an insight into the functioning of the CNS, more serious problems emerge: The object of study (the CNS) reacts to the process of measurement and even to its own activity. These two points deserve a more detailed explanation.

One of the common methods in the area of motor control is to perturb a motor action by an external force and to observe reactions of the system to such a perturbation. Assuming that the system can be characterized by certain properties that do not change during its reaction, a comparison of the system's state before perturbation and immediately after may allow to compute these properties (parameters of the system). But the problem is that the system does indeed react to virtually any perturbation and changes its own parameters. Signals produced by sensory neural cells in response to a perturbation travel along a variety of neural pathways and induce changes in the activity of muscles in the vicinity of the perturbation and also in distant muscles. These reactions occur at different time delays; they are not very reproducible and are stereotypical, making it difficult to decipher results of experiments with perturbations. Just imagine that you are trying to measure the mass of an object, which reacts to application of external forces by changing its mass in a poorly predictable fashion.

This is not all. The CNS itself undergoes changes in response to both external forces and its own activity. One of the most exciting and amazing properties of the CNS is its plasticity (reviewed in more detail in section 6.3). In a way, the system is rewiring itself all the time. For many years, it had been assumed that plastic changes within the CNS happened mostly in childhood and that adult brains were rigid and limited in their ability to change connections among neurons. However, over the last 20 years or so, researchers have witnessed many examples of dramatic changes in neuronal projections onto each other (so-called neuronal maps) in adult animals and humans. Plastic changes in neural structures have been reported, following relatively dramatic injuries to the system such as a stroke or an amputation (Cohen et al. 1991a,b; Fuhr et al. 1992; Nudo 2003; Cauraugh 2004; Celnik and Cohen 2004; Ward 2005). But more recently, such changes have been documented in healthy persons proficient in a particular skill such as playing a musical instrument or reading Braille (Pascual-Leone et al. 1995; Cohen et al. 1997; Sterr et al. 1998; Pascual-Leone 2001), and even following relatively brief practice sessions limited to 1 or 2 hours (Classen et al. 1998; Latash et al. 2003).

This means that repeating a task by itself may cause changes in the neural structures involved in the production of motor actions associated with the task. Now, recall that natural human movements are characterized by variability, that is, natural changes in motor patterns. As we will see in further sections, characteristics of motor variability provide important windows into the organization of movements. To study motor variability, one needs to have a person perform several trials at "the same task." Even if task characteristics remain perfectly unchanged across trials (which is in itself impossible, for example, due to small variations in the initial conditions), the body of the subject changes and, as a result of these changes, the CNS may use different strategies while solving "the same task."

In a sense, this general problem is similar to problems of physics of very small particles: To observe a particle (e.g. to make its photographic image), one needs to illuminate it with light. But if the particle is very small, photons of the light beam perturb it and change its very properties that one may be interested in, such as the coordinates and the momentum of the particle. This problem was formalized in physics as the principle of indeterminicity: There is a limit to how accurately one can measure both location and momentum of a very small particle. In physics, this problem disappears (or becomes practically irrelevant) when one deals with large objects, which may be assumed not to react to such methods of observation as light beams. Biological objects are large by physical standards, but there is something like the principle of indeterminicity in movement studies: Biological systems change with experience and react to perceivable external stimuli. Hence, one has to be very careful in trying to measure their properties and in making conclusions from experimental observations that involve application of external forces and repetitive observations.

Sometimes, I tell students that motor control is the *physics of unobservable objects*. This expression sounds like an oxymoron. Physics rests on such pillars as a possibility of experimental testing of hypotheses through observation and of reproducing results of an experiment. Yes, studies of biological movement are physics, but they seem to be a very weird physics. Another factor that makes them not exactly physics is the apparent dearth of good theories.

A naïve physicist or mathematician would think that there is no basic difference between animate and inanimate systems. The former are more complex, but they consist of a large number of "usual" elements such as elementary particles and molecules. So, theories of behavior of large ensembles of elements may be expected to help in biology. The problem, however, is that elements of any biological system do not seem more simple than the system itself. Hence, taking a biological system apart into smaller parts may not help in the understanding of its functioning, unlike, for example, such man-made systems as a television set or a computer. Even the most complex systems made of inanimate elements by contemporary engineers are also inanimate, while every sensible element of a living system is also a living system (e.g. individual neurons forming the CNS). In a sense, each element is a biological system of its own with its own goals. Its behavior is, therefore, defined by a complex interaction between its own goals and the goals of the larger system to which it contributes.

Nevertheless, it is so tempting to take an already available, well-developed theory from another area and apply it to biological objects! Through the twentieth century, there have been numerous examples of attempts at importing theories from such diverse fields as information theory, classical mechanics, control theory, and theory of nonlinear differential equations. These studies have led to a number of important findings and interesting conclusions. The problem, however, is that these studies try to squeeze all phenomena into the Procrustean bed of available mathematical theories, rather than take a step backward to the origins of biologically specific problems and then move forward. There is certainly nothing wrong with all these theories, but one should realize that they were developed to deal with particular classes of objects and particular groups of problems. They were not developed to deal with weird biological objects that are not computers, or ballistic missiles, or rigid bodies, and so on. This does not mean that the human brain cannot perform formal operations or that human limbs violate the laws of classical mechanics. But this is not what makes biological movement an exciting object of study. It is much harder to formulate problems in biology in an exact way that allows for their unambiguous experimental testing than to solve them. The heart of the problem seems to be the absence of an adequate language for studies of biological objects (see van Hemmen 2007).

Having an adequate language (adequate set of notions and ideas) is a necessary prerequisite for the development of any area of science. One of the greatest mathematicians of our times, Israel Gelfand, used to say that mathematics is a tool for finding adequate languages (Gelfand 1991). Let me use a couple of examples to illustrate this statement.

The early period of the development of geometry led to the formulation of an adequate language for an analysis of spatial relations among objects. Such a language allowed to discuss problems of geometry in terms of reasoning. An example is the Eucledian definition of a point as an object that does not have width, length, and height. This definition has no logical meaning—we know from everyday experience that such objects do not exist. Besides, the definition has been reviewed in the twentieth century. However, this illogical definition was a major element of the adequate language for geometry. This definition of a point allowed to introduce a definition for a line, for a surface, etc.

The successful scientific analysis of movements of inanimate objects has been based on an adequate language reflected in the apparatus of differential equations, which provided the basis for classical physics. However, for biological objects, the main problem is not in the description of their mechanics or other physico-chemical properties. Even the simplest movement of lifting the arm that can be well described with forces and coordinates, turns into a whole science fiction novel, if the process is considered within its biological context, taking into account the existence of an ever-changing neural controller. Thus, for a biological system, an adequate language seems to be a much more transcendental notion.

It has become obvious that adequate language/languages is/are urgently needed to make biology a science of its own. I am optimistic that presently we are at an early stage of the development of an adequate language for biological sciences and that *synergy* is a basic word for such a language (see Gelfand and Latash 1998, 2002).

Part Three

Motor Control and Coordination

3.1 ISRAEL GELFAND AND MICHAEL TSETLIN

This book would have never been written if it were not for a series of lucky coincidences that led in 1996 to my first meeting with Israel Moiseevich Gelfand. That was the first in a series of meetings and discussions with Gelfand—discussions that had the most profound effect on my thinking and the style of research I have been conducting since. Certainly, I had heard about Israel Gelfand before meeting him personally. In Moscow Physics-Technical Institute, we were taught linear algebra using Gelfand's classical textbook. During the early stages of work in the Laboratories of Victor Gurfinkel and Anatol Feldman, I had heard many stories about Gelfand and Tsetlin, some of which sounded more like legends about fairy tale heroes. So, when I received an invitation to visit Gelfand in his house in New Bruinswick (New Jersey) and to tell him about my work, it felt like getting an invitation from Albert Einstein or Nikolai Bernstein—a great honor, a great responsibility, and also like entering a fairy tale.

Israel Gelfand was born in a small Jewish town (*mestechko* or *shtetl*) Okna in Ukraine in 1913. When Gelfand was 16, his father was deprived of citizen's rights because, prior to the Revolution, he had worked on a mill and had had a couple of employees. So, by the crooked Soviet standards, he was a former "exploiter." As a result, Israel was fired from the only school in Okna that was a professional training school, with an emphasis on agricultural chemistry. He had no chance to continue education in Okna, and the family decided to send him to Moscow, where they had distant relatives.

It took Israel several days to find the relatives in Moscow. During that time, he spent nights sleeping on a bench across the Bolshoy Theater in the center of the city. His relatives managed to find a position for Israel in the Central Library (currently, the Lenin Library, if it has not been renamed again)—to check the library cards at the entrance. There he had ample time and access to all the books. So, he read a lot and was once caught by a man who noticed the young Jewish boy reading a book on advanced mathematics. This man happened to be a graduate student in the Department of Mathematics of the Moscow State University. He asked Gelfand whether he could understand the book and, after getting an affirmative answer, offered him a job—teaching late evening classes in the Moscow Institute of Chemical Technology. This is how Gelfand became a University Instructor without even graduating from high school.

Moscow of the 1920s and early 1930s was a surrealistic and dangerous place to live in but also a rather exciting one. Appointment of a high school dropout as an Instructor of Mathematics was not atypical of those times. Another example directly related to the topic of this book was the creation of the Laboratory of Biomechanics, arguably the first in the world, in an experimental theater directed by a famous actor and director Vsevolod Meyerhold (1874–1940). Meyerhold established this Laboratory in the belief that actors needed to know the mechanics of their bodies to be successful on stage. The actors participating in Meyerhold's productions acted according to the principles of theatrical biomechanics, the system of actor training and approach to theater, which Meyerhold developed. This approach taught acting based on such physical capabilities as balance, strength, coordination, agility, and flexibility. The actors learned a broad range of skills, including tumbling, acrobatics, partner work, and work with objects. Unfortunately, this Laboratory and its creator did not survive the terror of the late 1930s. Meyerhold was executed, and his theater was closed.

While lecturing at the Moscow Institute of Chemical Technology, Gelfand started to attend graduate seminars in the Moscow State University led by a famous mathematician, Leontjev. Leontjev noticed the quiet, very young man who sometimes asked very inquisitive and relevant questions and, after some time, offered Gelfand the opportunity to join the graduate program in mathematics. Nobody was worried by the fact that Gelfand had never studied formally beyond the seventh grade. In a few years, Gelfand defended Candidate of Science dissertation (close to Ph.D. in the Western countries) and became a Member of the Moscow Mathematical Society. By 1940, he became one of the youngest Doctors of Science in mathematics.

His career was truly stellar, and by the end of the 1940s, Gelfand had turned into a recognized leader of the Moscow School of Mathematics. His contributions to mathematics of the twentieth century are too many to be enumerated. They are reflected in a multi-volume series, *Collected Papers by Israel Gelfand,* published by Springer-Verlag in 1987–1989. His classical textbook, *Lectures on Linear Algebra,* has been used to educate several generations of mathematicians. Gelfand was awarded honorary degrees by the most prestigious Universities in the World and memberships by a dozen national academies. Among the many prizes awarded to Gelfand for his contributions to mathematics are the Wolf Prize (1978), the Wigner Prize (1979), the Kyoto Prize (1989), and the MacArthur Award (1994).

By the end of the 1950s, Gelfand felt that mathematics alone was not sufficiently challenging and wanted to study new problems. He turned to biology, joined by his former graduate student, Michael Tsetlin (1924–1966), a brilliant physicist and a decorated World War II veteran. Tsetlin possessed a lovable personality with a great sense of humor. Anatol Feldman recounted the following story to me. After many discussions on the cerebellum and its possible functions, a special mailbox was hung on the wall of Gelfand's laboratory in the Institute for Problems of Information Transmission, and all laboratory members were encouraged to deposit there pieces of paper with ideas about the cerebellar function. After several months, the mailbox was opened. There was only one piece of paper written by Tsetlin: "Kiss me in the . . . cerebellum!"

Tsetlin's University education had started just before Germany attacked the Soviet Union in 1941. After the War ended, Tsetlin was forced to serve in the Army for two more years. In 1947, he resumed studies in the Moscow State University, where he met Israel Gelfand who became his advisor. Tsetlin managed to avoid persecution during the anti-Semitic campaign of the early 1950s and graduated in early 1953, with several published works both in mathematics and applied physics. In the mid-1950s, Tsetlin started a collaboration with Victor Gurfinkel, one of the leading figures in the areas of neurophysiology and movement studies in Moscow. In particular, they worked on the development of an arm prosthesis with bioelectrical control. This was a very successful project, which had an enormous, long-lasting impact in the field of prosthetics.

In 1957, Tsetlin was recruited by A.A. Ljapunov, a great Russian mathematician, to join the newly formed section of cybernetics in the Institute of Mathematics. Since 1961, Tsetlin worked on problems of purposeful collective actions by automatons starting from the simplest models of behavior of "a small animal in the big world" (Tsetlin's favorite expression). Later, Tsetlin used these studies as models of the social and economic structure of the human society. His interests were very broad and included pedagogy, linguistics, and poetry.

In 1958, Gelfand, Gurfinkel, and Tsetlin organized a seminar, initially in Gelfand's apartment with only a handful of participants. I have heard from many participants of the seminar that, without Tsetlin, the seminar would have probably not survived. Michael Tsetlin had a rare talent of explaining complex mathematical ideas with examples from everyday life, funny stories, jokes, etc. He was a great intermediary between mathematicians and biologists who took part in the seminar. The Gelfand–Tsetlin team was a true synergy. Unfortunately, the synergy was short-lived: Michael Tsetlin died in 1966 at the age of 42 because of liver failure, soon after the death of Nikolai Bernstein. This was a major loss for the Moscow School of Neurophysiology and movement studies.

The papers by Gelfand and Tsetlin on biological systems were very influential (Gelfand and Tsetlin 1961, 1962, 1966). In particular, they were the first to classify variables that describe biological systems as essential and nonessential. For example, topological variables that distinguish maple leaves from oak leaves are essential, while size of a leaf is nonessential. These insights were later developed by Michael Bongard in his now classical studies on recognition of objects (Bongard 1970). Rather quickly, Gelfand and Tsetlin realized that the well-developed methods of theoretical physics and the tools of mathematics had obvious limitations in dealing with biological problems. They decided to conduct a seminar and get as much reliable information about biological systems as they could.

I have already mentioned this seminar in the earlier historical section. Its impact on the development of biology in Moscow was enormous and not limited to neurophysiology and movement studies. In particular, Gelfand's school has also contributed significantly to such diverse areas as cell biology and epidemiology. In the early 1960s, Gelfand and Tsetlin invited Bernstein to present at the seminar (as recollected by Josif Feigenberg, a prominent psychologist and an informal student of Bernstein). Bernstein presented his ideas about the control of voluntary movements. During the presentation, Gelfand paced impatiently between the rows with a dissatisfied expression on his face and, after the talk was over, he summarized: "But this is all rubbish, of course!" Bernstein answered very politely: "No, this is not rubbish, and I can prove it." "Misha," Gelfand appealed to Tsetlin, "would you not agree that this is all rubbish?" "No," answered Tsetlin, "it is not that obvious to me." Gelfand made a pause, thought for a couple of seconds, and then concluded the discussion: "OK, now I can see that all this makes sense." Since that seminar, development of Bernstein's ideas was at the center of attention of the seminar.

Gelfand's laboratory, established in the Institute of the Problems of Information Transmission, later led to the setting up of several laboratories. Many well-known scientists worked there, including Yuri Arshavsky, Mikhail Berkinblit, Levon Chailakhian, Vitali Dunin-Barkovsky, Anatol Feldman, Olga Fookson, Victor Gurfinkel, Efim Lieberman, Grigory Orlovsky, Raisa Person, Feodor Severin, Mark Shik, Vladimir Smolyaninov, and many others.

Let me skip 30 years and arrive in the 1990s. After settling down in New Jersey and accepting a position at Rutgers University, Gelfand continued to keep contacts with some of his Moscow students. One of them, Olga Fookson recommended me to Gelfand as someone who could tell him what had happened in neurophysiology and movement science over the past 30 years. During that time, Gelfand had been more involved in other areas of biology (and certainly, mathematics). But by the mid-1990s he had decided to rekindle the old interest in neurophysiology and looked for an update.

During our first meeting, Gelfand asked unexpected and very direct questions, sometimes seemingly very far from the field of neurophysiology and frequently very personal. I was particularly impressed when he asked me: "I know that you were too young to attend our seminar with Misha Tsetlin. But what would you say was our biggest weakness?" And I knew that he was not expecting a typical polite answer, like: "Come on, Israel Moiseevich, there was no weakness. You were the two smartest persons in the field," even if this was in fact true. So, I answered: "Your biggest weakness was the inability to see ideas developed and brought to fruition." Gelfand chuckled but apparently liked the answer and did not kick me out of the house.

Further meetings were as unpredictable as the first one. I could bring results of new experiments and get ready to discuss them, and then we would spend hours listening to Shostakovich music performed by Shostakovich himself and by other famous pianists. These were rather exhausting meetings for me. It was clear that one could not say imprecise things in front of Gelfand. The typical everyday babbling, full of jargon and irresponsible statements that forms the core of functioning of any laboratory was clearly unacceptable. Gelfand's mere presence required absolute concentration and precise selection of words that was not easy to maintain over several hours. I viewed those meetings as mind-cleansing exercises, and their effects were obvious: After each meeting, for several days, there was a strong illusion that the brain was working much better.

Throughout his lifetime, Gelfand has been fascinated by the cerebellum. He has viewed it as one of the most enigmatic and exciting structures of the brain. On different occasions, he suggested that the cerebellum was related to such brain functions as intuition, creation of general complex percepts, establishing trade-offs among different factors, and judging situations. He said during one of our meetings, "Freud was a cortical person, unlike Jung who was probably a cerebellar person, although I have never been able to understand his writings."

After presenting this quotation, it is impossible not to say a few words about Gelfand's ability to speak directly, exactly, succinctly, and frequently unexpectedly. Many scientists were offended by Gelfand's direct comments, not appreciating that in scientific discussions, he gave very low priority to politeness and very high to exactness. If he heard something foolish, he could call the speaker a fool, publicly and directly. Not all fools appreciated this.

Let me present a few quotations that I wrote down immediately after meetings with Gelfand in the late 1990s and tried my best to give an adequate translation. My sincere apologies to those who may be offended by some of these statements. These comments are authentic, and rather than taking offense, I would recommend thinking them over. After all, they were made by Israel Gelfand!

> The worst possible way of discussion is talking about complex things with hints.
>
> I think in synergies.
>
> Mathematics is not a science. It is a tool provider to help find adequate languages for natural sciences.

Application of contemporary mathematics and physics to biology is a dead-end. We should not waste our time. These areas can only supply tools to accelerate data acquisition and processing. Adequate formulation of essential problems should come from within biology. Do not waste time on mathematics—think.

Dynamic systems approach to movements is as foolish as Bernstein's analysis of locomotion as pendular motion of the leg.

Science in the eighteenth and nineteenth centuries was successful in developing the language of differential equations and partial derivatives driven by the technological progress. The twenty-first century will be the century of biology. I would love to develop mathematical biology.

The equilibrium-point hypothesis with its lambda is probably the brightest example of identifying essential variables in neurophysiology. (More on the equilibrium-point (EP) hypothesis is in section 3.4.)

I dream of recreating the Moscow Laboratory, a small institute, as small as that.

(And he showed with his hands how small the Institute would be.) This phrase contains a very direct allusion to Leon Feichtwanger's "The Judean War." After Roman legions crushed the revolt in Judea and burned down the Jerusalem Temple in the first century, a tiny elderly rabbi, Johanaan ben Zakkai asked Titus, the commander of the Roman legions and future Emperor, to allow the rabbi and his remaining colleagues to establish a small university in Yamnia to continue studies of Judaism, and the rabbi showed with his small hands how tiny the University would be.

Menukhin wrote somewhere that arm movement during playing violin consists of three basic oscillations. We should find those.

Listen to Schostakovich performing his concerts. Nobody, even better pianists, could play like him.

Here is the principle of Wiegner–Gelfand: Reasonable inefficacy (!!!) in biology.

We are the only two who can develop a new adequate language for biology. Why? Because we are Jews, and this means that we are not afraid of saying with respect to some questions: "These are beyond our comprehension."

Let us write a book with the title: *Beyond Our Comprehension.*

Let us revive the former glory of the Georgian tea! It is not clear whether there was glory in the first place, whether there will ultimately be any tea, but we should try anyway.

To appreciate the sarcasm of this statement, one should know that Georgian tea sold in stores was rather bad, particularly during the Brezhnev years in the Soviet Union.

3.2 STRUCTURAL UNITS AND THE PRINCIPLE OF MINIMAL INTERACTION

Nikolai Bernstein was probably one of the first to realize that specific features of behavior of biological objects cannot be reduced to specific features of elements forming these objects. Bernstein further suggested that the development of the central nervous system had been driven by the design of the peripheral motor apparatus, that is, by its numerous mechanical degrees of freedom that needed to be controlled. In his view, the problem of motor redundancy had to be tightly related to the principles of neural control used by the brain. He came to the conclusion that the central nervous system was built on a hierarchical principle, with a higher level producing *engrams* of planned movements based on task parameters. These *engrams* served as inputs for lower levels of the hierarchy, which ultimately led to the generation of commands to individual muscles (Bernstein 1935; Figure 3.1). Bernstein's *engrams* were assumed to reflect mechanical requirements and constraints associated with different tasks, but these *engrams* were never supposed to represent muscle forces, joint torques, or their direct neural precursors. Unlike further developments of the idea of *engrams* into *motor programs* and *internal models* (reviewed in Schmidt 1975, 1980; Kawato and Gomi 1992; Wolpert et al. 1998; Kawato 1999), Bernstein never assumed that the brain precomputed forces and torques required to perform particular motor tasks. We will get to this important distinction later in section 3.3.

Gelfand and Tsetlin took these general ideas of Bernstein as a starting point and developed them into a scheme based on the principle of nonindividualized control. This principle accepts the idea of hierarchical control but rejects prescriptive, authoritarian control by hierarchically higher levels. According to the principle of nonindividualized control, elements of a system are not controlled individually, like segments of a marionette's body by attached strings, but united into task-specific or intention-specific *structural units*. Structural units were assumed to be organized in a flexible, task-specific way. Gelfand and Tsetlin used the word *synergy* for purposes

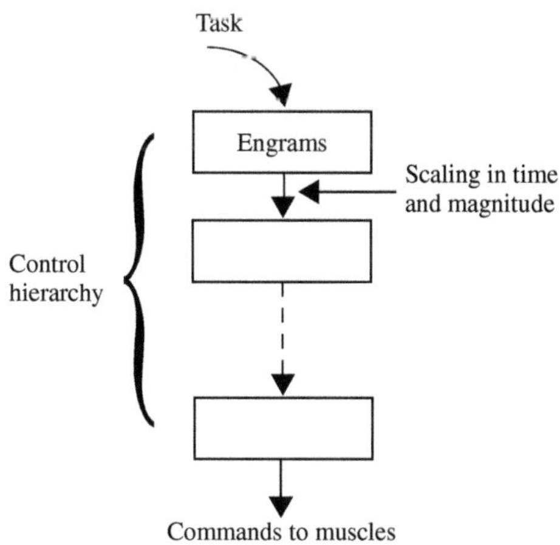

Figure 3.1. Bernstein's scheme assumes that neural precursors of motor commands (engrams) are selected at a higher level of the control hierarchy, can be scaled in time and magnitude, and ultimately lead to commands to muscles.

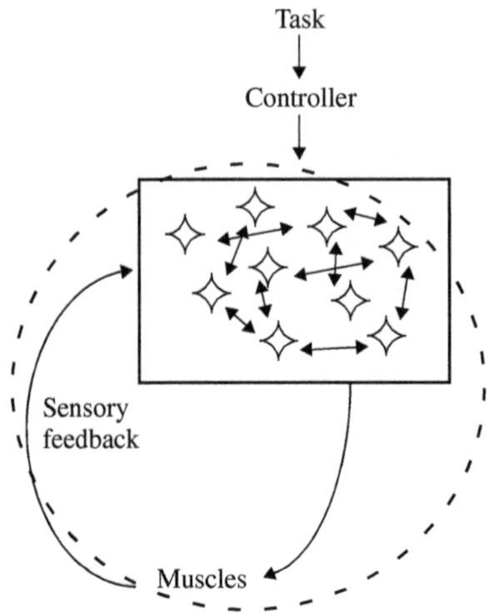

Figure 3.2. An illustration of a structural unit. The relations among elements and the effects of the sensory feedback are tuned by the controller in the process of learning to make the structural unit specific for a group of tasks. A structural unit is able to execute a task in an autonomous regime without continuous supervision by a hierarchically higher controller.

of structural units. External behavior generated by a structural unit is defined by its purpose (i.e. by the corresponding synergy) and by current external conditions. Note that Gelfand and Tsetlin (probably, purposefully) did not introduce clear definitions for the central notions of their scheme, such as *elements*, *structural units*, and *synergies*. They appealed to the common sense and intuitive understanding by the readers of the main idea on nonindividualized control. The terminology has changed over time, and, for example, most contemporary researchers would probably use *synergy* instead of *structural unit* in this intuitive scheme.

For example, consider locomotion—a behavior, which had been viewed as a proverbial synergy by Bernstein and many other researchers. Gelfand and Tsetlin would call the organization of neurons and effectors (including all the central interneuronal connections and feedback loops) that produce locomotion a structural unit (Figure 3.2). Actual locomotor patterns, such as walking and running, are defined by the purpose of the structural unit (synergy) and external conditions. They will be different for walking uphill and downhill, for walking on high-heel shoes and in sneakers, etc.

In an attempt to build an exact scheme, acceptable to mathematicians and physicists, Gelfand and Tsetlin (1966) introduced a set of three axioms to describe essential properties of structural units:

Axiom-1. The internal structure of a structural unit is always more complex than its interaction with the environment (which may include other structural units).

Axiom-2. Part of a structural unit cannot itself be a structural unit with respect to the same group of tasks.

Axiom-3. Elements of a structural unit that do not work with respect to a task.

Axiom-3a are eliminated and a new structural unit is formed or

Axiom-3b find their own places within the task.

The first axiom reflects a feature of biological objects that has already been mentioned: Elements of biological systems are not less complex than the systems themselves. Each element of a structural unit is also a structural unit. Hence, a biological system cannot be reduced to simpler elements interacting with each other. A structural unit is characterized by the fact that the number of internal connections within a structural unit is at least by an order of magnitude higher than the number of its external connections. A structural unit is not simply an ensemble of elements, for example, neurons, but a system with a particular function. Changes of connections between the elements may lead to the creation of different structural units, based on the same set of elements.

The second axiom implies that one cannot cut a structural unit into subunits that perform particular variations or components of a task. A structural unit either works as a whole with respect to a task, or it does not. This does not mean, certainly, that structural units cannot evolve and optimize their structure and functioning with practice of particular tasks. Let me mention here that a number of recent studies have suggested that at least some structural units responsible for particular behaviors may represent a superposition of several structural units responsible for components of the behaviors—the principle of superposition (Arimoto et al. 2001; Zatsiorsky and Latash 2002; Latash et al. 2006).

Axiom 3a illustrates the principle of economy, when a minimal number of elements carry out each given task. Hence, the set of axioms 1, 2, and 3a may be more applicable to the description of established, stereotyped reactions. Axiom 3b illustrates the very important principle of abundance developed in later works (Gelfand and Latash 2002), when many more elements than necessary participate in the activity of a structural unit with respect to each task. The set of axioms 1, 2, and 3b describes systems that may be expected to evolve and form structural units able to solve tasks, for which they had not been originally designed. The principle of abundance is considered typical of the organization of everyday voluntary movements in the unpredictable and continuously changing environment.

Structural units can be introduced for systems of different complexity. For example, a cell, a subsystem within the human body, an organism, or a group of organisms can each be viewed as a structural unit.

Let me illustrate the notion of structural unit and the three major axioms using the organization of a scientific laboratory as an example. The laboratory can be viewed as a structural unit, while individual researchers can be seen as its elements. Let us assume that the laboratory has been organized to solve a specific problem, for example, to design a bridge. Note that the brains of individual researchers (elements) are not less complex than the laboratory (system). The ultimate result of the functioning of the laboratory, that is the bridge, can be described in a much simpler way than the way researchers interact with each other (axiom 1). Let us imagine that the task takes 6 months to be completed. If after 3 months half of the personnel are fired, the rest will likely be unable to solve the same problem (axiom 2) without a major reorganization, which would lead to the formation of a new structural unit based on a new set of available elements.

The laboratory may be organized based on axiom 3a or on axiom 3b. In the first case, based on the principle of economy, a minimal number of researchers are hired, and each researcher is assigned a unique, specific function. In this case, the laboratory will be able to fulfill its purpose successfully, but it will likely have problems performing other tasks and will be very sensitive to the functioning of each researcher. In particular, it will be unable to complete the original task if one of the members of the laboratory suddenly falls ill or retires.

Alternatively, based on the principle of abundance, a large group of talented researchers may be assembled and asked to deal with the problem. Each researcher is expected to find his or her own place in the team and contribute to the process. Such a team may take time to form a structural unit adequate for the task. However, after doing this, it will be able to solve the original task and can also be expected to reorganize successfully if the task is modified, for example, if it involves designing a skyscraper. Such a laboratory may also be expected to continue its successful functioning, if one of the members falls ill and takes a few days off.

In a way, elements of a structural unit may be compared to a class of lazy students, whose main purpose is to keep the teacher from giving them new assignments (an example used by Michael Tsetlin). When a task is given by the teacher, the students interact in such a way that their overall output (e.g. the level of hum in the classroom) keeps the teacher happy, or even better—asleep. If the class becomes too quiet or too noisy, the teacher is likely to wake up and give a new task.

One may conclude that the principle of abundance renders the problem of redundancy irrelevant. Numerous elements are not a source of computational problems for the nervous system but a useful, flexible apparatus that requires proper organization.

As already mentioned, within the introduced hierarchical organization, a functional goal is formulated by an upper level of the hierarchy, but this formulation does not suppress the freedom of elements at the lower level of the hierarchy. But how do these elements behave? Can one formulate a simple, universal rule that would explain patterns of their interaction?

In the 1960s, Gelfand and Tsetlin (1966) suggested a principle of minimal interaction, which stated that the interaction among elements at the lower level of the hierarchy was organized so as to minimize the external input to each of the elements. If the output of an element is a simple function of its input, this formulation means that each element tries to keep its output constant at the lowest possible level compatible with the task. Using the terminology introduced earlier for synergies, this formulation of the principle of minimal interaction implies a constant sharing pattern among the elements.

However, as mentioned earlier, there is another feature of biological synergies, which is flexibility/stability (error compensation). Actually, the earlier example of the teacher and the class suggests that elements should change their outputs if they want to avoid waking up the teacher, which sounds more like error compensation than a constant sharing pattern. So, here is another facet to the principle of minimal interaction: The elements behave so as to minimize changes in signals the whole structural unit receives from the hierarchically higher level (the controller).

These two formulations seem to be in competition. Indeed, consider a very simple structural unit that tries to keep the sum of the outputs of individual elements constant, despite possible external and internal noise (Figure 3.3). Keeping a constant sharing pattern, for example, 50:50 illustrated by the slanted dashed line in Figure 3.3, means that if one element introduces an error (a deviation from its expected output), other elements will likely amplify the error by scaling their outputs in proportion in the same direction.

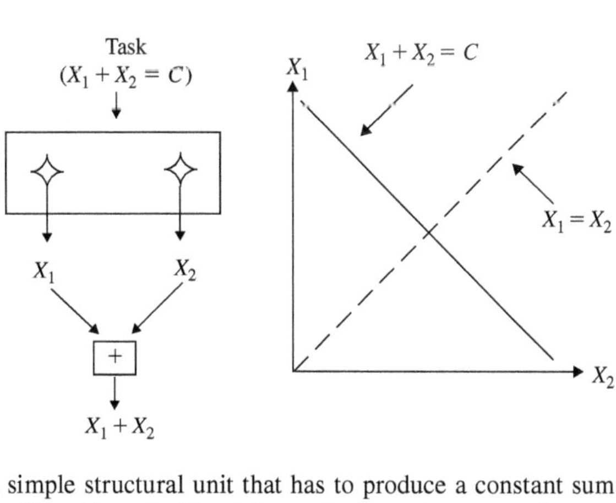

Figure 3.3. A simple structural unit that has to produce a constant sum of the outputs of two elements ($X_1 + X_2 = C$). The right graph shows that preserving a constant sharing between the two elements (the dashed line corresponds to a 50:50 sharing) potentially leads to deviations of their sum from C. To keep the sum of the elements at C, the sharing pattern has to be violated (the slanted solid line).

Then, the summed output will be violated, and the controller will have to act. This is exactly what the lazy elements want to avoid. To keep the controller happy, the elements should adjust their outputs in such a way that the total, functionally important output of the whole system is unchanged. To ensure this, if one element changes its output, other elements should change theirs in the opposite direction (the solid slanted line in the graph in Figure 3.3). This latter mode of functioning should probably be called the principle of maximal interaction.

The principle of minimal/maximal interaction implies, in particular, that if a perturbation is applied to one of the elements of a structural unit, it is expected to lead to changes in the contributions, not only of the perturbed one but also of other elements. The purpose of these changes is to correct errors in the common functional output of the structural units that were introduced by the changed contribution of the perturbed element. Quick corrections to unexpected perturbations applied to an element of a multi-element system during natural, complex movements have been reported for many different activities such as speech production (Abbs and Gracco 1984; Kelso et al. 1984), multi-joint arm movement (Gielen et al. 1988; Latash 2000), vertical posture maintenance (Nashner 1976; Nashner and Woollacott 1979; Cordo and Nashner 1982), and locomotion (Forssberg 1979; Grillner 1979; Dietz et al. 1984).

Recently, a number of studies have tried to revisit the principle of minimal interaction and link it to specific variables that can be minimized within the central nervous system. Aron Gutman (1994) suggested a direct link of this principle to neurophysiological mechanisms forming the foundation of the EP hypothesis (see section 3.4). Along similar lines, studies of invariant kinematic features of the trajectories of pointing movements involving different sets of kinematic degrees-of-freedom have also suggested links between the principle of minimal interaction and the EP hypothesis (Feldman et al. 2007). The principle has also been invoked in studies of multi-finger action to interpret the natural variations in sharing the force within a set of fingers (Dumont et al. 2006). Application of the principle of minimal interaction to the EP hypothesis will be discussed in one of the following sections.

A particular computational method to deal with abundant systems, the methods of ravines, was introduced by Gelfand and Tsetlin in the early 1960s (1961, 1966). This method was later modified to explain certain observations of the wiping reflex in the decerebrated frog, in particular its ability to wipe the stimulus off its back when one of the joints is blocked (Berkinblit et al. 1986b). The application of this method to the frog wiping reflex is illustrated in Figure 3.4. Imagine that rotation of each joint is defined by a simple rule: The vector of its angular velocity equals the cross product of two vectors, from the joint to the endpoint and from the joint to the target. No single joint is "particularly important" in this scheme. Each joint moves under the influence of two factors. First, the discrepancy between the locations of the endpoint and of the target and, second, effectiveness

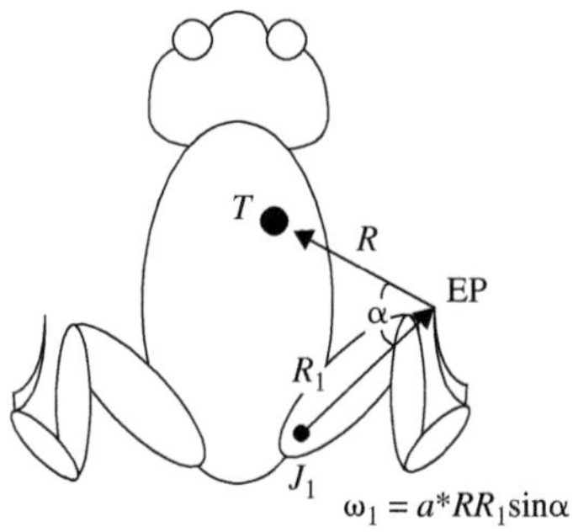

Figure 3.4. A simple computational model at the level of joint kinematics that can account for accurate wiping movements of the spinal frog in conditions of loading and joint blocking. Reproduced by permission from Gelfand IM, Latash ML (1998) On the problem of adequate language in motor control. *Motor Control* 2: 306–313, © Human Kinetics.

of each joint's motion in bringing this discrepancy down. This rule ensures that joint motion continues until the endpoint reaches the target. Obviously, when the endpoint reaches the target, all joint motion stops.

Is *structural unit* (or *synergy*) a word for a future adequate language for motor control (and maybe biology in general)? I would like to believe so, but currently this is only a belief.

3.3 MOTOR CONTROL: PROGRAMS AND INTERNAL MODELS

To move ahead, let me consider a simple motor task. Imagine that you need to move the tip of the index finger from some initial position to some final position (Figure 3.5). To make analysis of this problem easier, let us consider movement of only the arm, while the trunk remains motionless. First, I am going to approach this problem as an engineer trained to deal with inanimate objects, for example ballistic missiles or robots. As the reader has probably already guessed, later I am going to discard this approach and offer an alternative that would be specific to the actual design of the human body.

Step-1. Both positions of the fingertip, initial and final, can be characterized by three coordinates in the external space. To place the endpoint of a multi-link chain into a certain point, one needs to define joint angles that satisfy the task. As mentioned earlier, the human arm is kinematically redundant: The number of major axes of joint rotation is more than three. So, this step already represents a problem of motor redundancy. How does the controller select a particular joint configuration from an infinite number of possibilities? This problem is known as the problem of inverse kinematics (Mussa-Ivaldi et al. 1989; Zatsiorsky 1998).

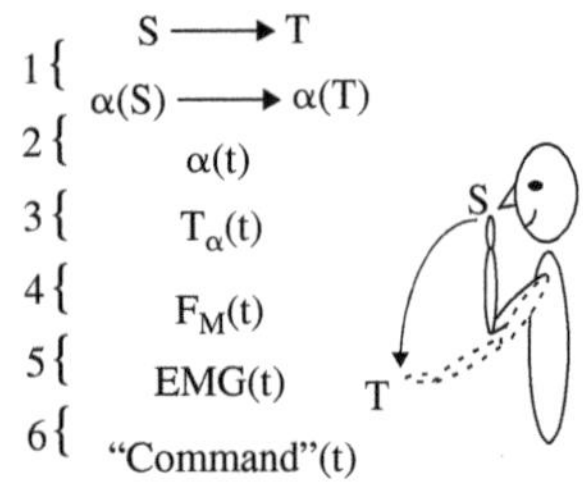

Figure 3.5. An illustration of the six steps involved in the generation of a "motor command" (in quotation marks because it is undefined) for a simple movement of a limb from a starting position (S) to a target (T). Step-1: From positions in the external space to joint configurations. Step-2: From joint configurations to joint trajectories. Step-3: From joint trajectories to joint torque profiles. Step-4: From torque profiles to time profile of muscle forces. Step-5: From muscle forces to muscle activation patterns. Step-6: From muscle activation patterns to "commands".

Let us, however, skip this problem and assume that the controller somehow solves it.

Step-2. To implement the selected set of joint angles, the controller needs to define trajectories from the initial joint configuration to the final joint configuration. This problem also has many (an infinite number of) solutions. For example, you can start movement by bringing one joint to its final position and then move another joint, or you can move all the joints simultaneously. Joint motion can be done smoothly or not so smoothly. And so on. Let us assume that joint trajectories have somehow been computed and selected.

Step-3. To move the joints, one has to apply appropriately timed joint torques. So, the next step is to define a pattern of joint torques that would implement the required movement kinematics. This problem is known as the problem of inverse dynamics (Hollerbach and Atkeson 1987; An et al. 1988; Atkeson 1989; Zatsiorsky 2002). Let us once again assume that it has been somehow solved.

Step-4. Each joint is crossed by several muscles. Skeletal muscles are unidirectional actuators, and they can only pull but not push. Therefore, at least two muscles are needed to control a joint. Actually, all arm joints are crossed by more than two muscles. Some of these muscles cross more than one joint. For example, the well-known biceps and triceps of the human arm have muscle heads that cross both the elbow and the shoulder joints. So, we now face a new problem: What forces should be produced by individual muscles to ensure the required pattern of joint torques? This is another problem of motor redundancy. As at all previous steps, let us assume that this problem is also somehow solved.

Step-5. To produce a muscle contraction, action potentials need to be sent to the muscle by neural cells in the spinal cord called alpha-motoneurons. Muscle force depends on the level of excitation it receives from its motoneuronal pool and also on the actual muscle length and rate of its change (velocity) (reviewed in Zatsiorsky 2002). Therefore, to define signals that should be generated by alpha-motoneurons to a muscle to produce a required profile of muscle force, one has to take into account expected changes in muscle length. Imagine that this problem is also solved.

Step-6. Alpha-motoneurons produce output signals (sequences of action potential) that depend on the total input to the motoneurons. For simplicity, let us consider this input as the sum of two components (this is a very crude and even misleading simplification). First, there are signals that come from the controller (from the brain or from other neural structures within the spinal cord). Second, there are signals that come from peripheral sensory endings (receptors). The former component can be called central and the second peripheral. The peripheral component is a complex function of signals related to muscle length, velocity, force, and may also depend on such signals from other muscles (Nichols 2002). To ensure an adequate total input into a motoneuronal pool, the controller needs to take into account predicted peripheral contribution and generate a central component that would make the total input to the motoneuronal pool adequate to produce the output computed at the previous step.

These six steps are sometimes addressed as an inverse model of the system. It is called inverse, because the sequence of the six steps follows an order opposite to the natural order of events, when a movement is generated. Naturally, a signal from the controller is summed up (better, interact) with peripheral inputs and forms the total input to motoneuronal pools; the pools excite the muscles, the muscles produce forces, the forces lead to joint torques, the joint torques lead to joint displacements, the joint displacements lead to motion of the endpoint, and ultimately to "the house that Jack built." But our computation started from endpoint motion and ended up with signals from the controller—inverse to the natural sequence.

Solving problems at each of the six mentioned steps is not easy. Actually, there are no universally accepted methods of solving any of them. But the six steps represent only the tip of the iceberg. From the engineering point of view, the human body was designed by an undergraduate student who had probably failed courses on principles of engineering design, control theory, theory of stability, and many other useful disciplines. Among its apparent shortcomings are the sluggish actuators (the muscles), the confusing sensors, and the long-time delays in all the pathways that are used to deliver control signals to the muscles and sensory information to the controller. So, let me digress and describe in more detail some of the major "flaws" in the design of the human body.

Digression #1. The Muscle: Slow and Visco-Elastic

Properties of skeletal muscles are likely to appear, to put it mildly, suboptimal, to engineers. Given a task of designing a human-looking moving system (a robot), an engineer would probably start with selecting powerful torque motors that are able to produce joint torques in a predictable and reliable fashion independent of possible changes in the mechanical properties of the moved object and external forces. Then, the problem will be reduced to computing requisite torque patterns and sending control signals that bring about those patterns.

The properties of human muscles are far removed from such ideal torque-motors. Let me review briefly the physiological processes leading to force production by skeletal muscles, as well as properties of muscles and tendons that make the human motors very different from torque-motors.

The functioning of excitable tissues in the human body, in particular, neurons and skeletal muscles, uses as a unit of information transmission a certain short-lasting change in the potential across the membrane that separates the cell from the environment. These so-called *action potentials* show a time profile and peak amplitude that are highly standardized for a given tissue and external conditions (in particular, temperature). As a result, all action potentials generated by a cell are identical to each other—this is referred to as the *law all-or-none*, that is, an excitable cell either generates a standard action potential or it does not. The only way a cell can send information to another cell is by modulating the frequency of action potentials.

Muscle contraction is a complex process that converts an electrical control signal to contraction force. A contraction starts with a neural signal, an action potential arriving along a neural fiber from a neuron in the spinal cord to a target muscle fiber. When an action potential arrives at the junction of the neural fiber and a muscle cell, it triggers a sequence of physico-chemical effects that ultimately lead to changes in the membrane potential in the muscle cell. Such junctions are called neuromuscular synapses. Neuromuscular synapses are obligatory: This means that an action potential arriving along a neural fiber always leads to the generation of an action potential in the muscle cell the fiber innervates.

An action potential running along the membrane of a muscle fiber triggers a sequence of physico-chemical events, ultimately leading to a brief episode of force production by molecular elastic links called cross-bridges. As mentioned earlier, muscles are unidirectional force generators—they can only pull but not push. There is, however, a notable exception to this rule. The ciliary muscle that controls the shape of the lens in the human eye is important for focusing the eye on objects at different distances. When this muscle is activated, it bulges and pushes on the lens thus increasing its curvature and allowing the eye to focus on a closer object. An age-related weakening of this muscle may be responsible for the well-known development of far-sightedness in the elderly.

If no other stimulus arrives, the cross-bridges disengage quickly and the muscle fiber relaxes. A typical time profile of force produced during such a single contraction (it is called *twitch contraction*) is shown in Figure 3.6. Typical times of force rising and falling during a twitch contraction vary depending on muscle fiber properties; typical twitch contractions take about 50–100 ms to reach peak force. The phase of force decline is typically longer. This means that it is even harder to stop a muscle contraction abruptly than to initiate it quickly.

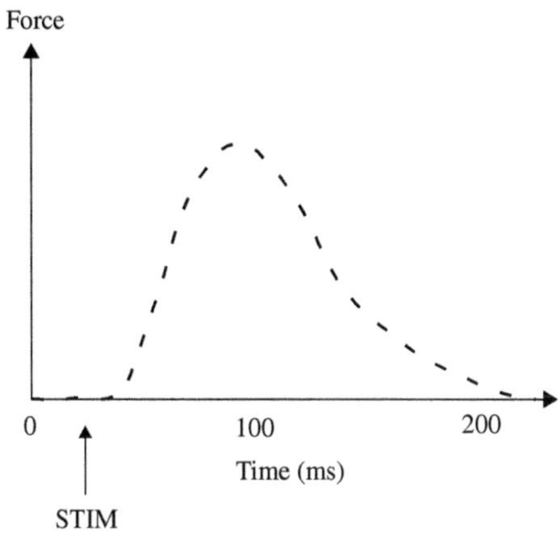

Figure 3.6. A typical muscle contraction to a single stimulus (twitch contraction). Muscle fibers typically take longer to relax than to generate force.

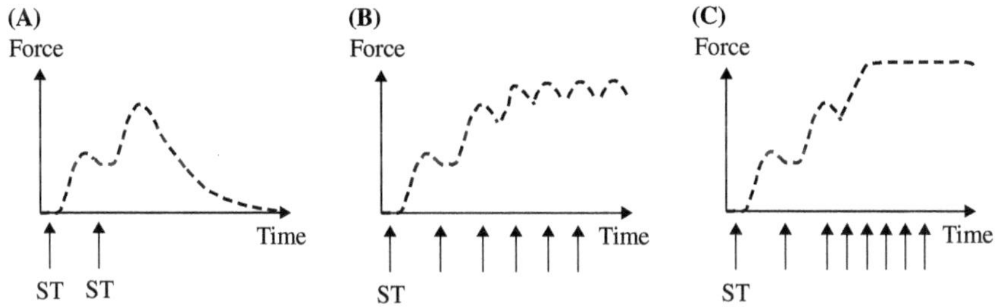

Figure 3.7. (A) When two stimuli come to a muscle at a short time interval, two twitch contractions superimpose and reach a higher peak force compared to that of a single twitch. (B) If many stimuli come at a sufficiently high frequency, a saw-tooth tetanus is observed. (C) After the frequency of the stimulation reaches a certain level, individual twitch contractions merge and produce a smooth tetanus.

If two stimuli come to a muscle fiber at a small time interval, the first twitch contraction may still continue when the second stimulus arrives. In this case, the second twitch contraction starts from an already elevated force level and reaches a higher peak force (Figure 3.7A). If a sequence of stimuli come to a muscle fiber at a high enough frequency, it may reach the state of *smooth tetanus* (Figure 3.7C), that is a state of high constant force production. Both twitch contractions and smooth tetanus are rare during everyday muscle function. More typically, muscle fibers receive action potentials that come at frequencies below those that could have produced smooth tetanus. Each muscle fiber produces a contraction consisting of peaks and valleys that neither reach the smooth tetanus force, nor drop to zero (Figure 3.7B); such contractions are called *saw-tooth tetanus*.

The magnitude of force produced by a muscle fiber in both twitch and tetanic contractions is not fixed but depends on the fiber length and its rate of change.

These visco-elastic properties of muscle force depend partly on the mechanical properties of cross-bridges and partly on properties of passive connective tissue including those of tendons. In static conditions, longer muscle fibers produce larger forces in response to a standard stimulus up to a certain length after which their force shows a drop (Figure 3.8). Note, however, that this drop is typically observed at fiber length values that are beyond the muscle anatomical range. So, during natural movements, muscle fibers behave like nonlinear springs with a damping element.

When muscle fiber length changes, the force produced by the fiber shows a dependence on the fiber length velocity. When a muscle is stretched during a contraction, its force is higher compared to the force produced in static conditions at the same length; when a muscle is shortened, the force is lower. This general relation is illustrated in Figure 3.9. It was formalized by Sir A.V. Hill in the form of the famous Hill equation:

$$(F + a)V = b(F - F_0),$$

where a and b are constants, F stands for muscle force, V is muscle fiber velocity, and F_0 is the force generated by the muscle at zero velocity, that is, in static

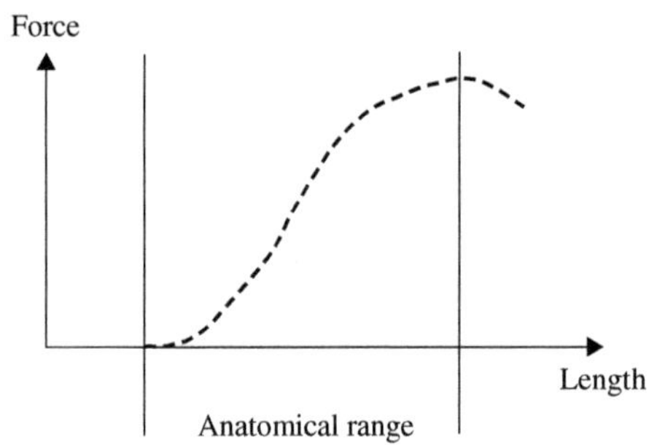

Figure 3.8. Active muscle force depends on muscle length (even in the absence of muscle reflexes). It increases with muscle length over the typical anatomical range but can fall at higher length values.

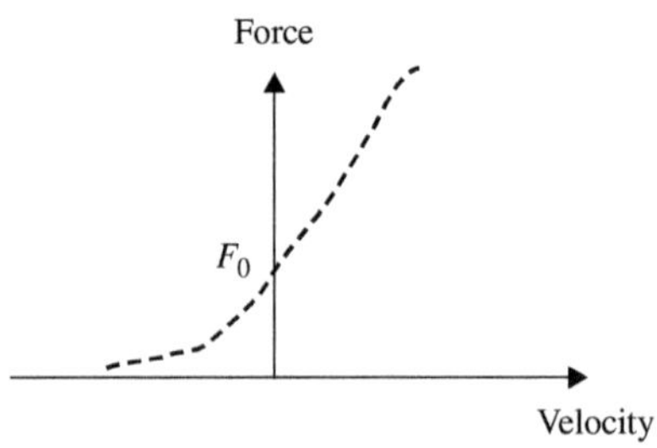

Figure 3.9. Active muscle force depends on the velocity of contraction. When the muscle is shortening (negative velocity, concentric contraction), its force is lower than in static conditions (F_0). When the muscle is stretching (positive velocity, eccentric contraction), the force is higher than F_0.

conditions. Although muscles are unidirectional active force generators, both shortening and lengthening contractions happen all the time during natural movements. They are sometimes addressed as concentric and eccentric, respectively. This is due to the fact that all joints are crossed by at least one pair of muscles acting in opposite directions (flexors and extensors, adductors and abductors, etc.). Typically, both muscles show nonzero levels of activation. As a result, one of them performs a concentric contraction, while the other one contracts eccentrically.

Now, we have to remind ourselves that muscles represent large groups of muscle fibers innervated by neurons in the spinal cord that send their long branches (axons) to muscles, so-called alpha-motoneurons. Each alpha-motoneuron innervates a group of muscle fibers (Figure 3.10). The axon of an alpha-motoneuron ends with a "brush" of short fibers called terminal branching. Each branch normally makes connection with only one muscle fiber. As a result, when a motoneuron sends a signal (an action potential), all muscle fibers innervated by this alpha-motoneuron receive excitatory signals virtually simultaneously. Hence, an alpha-motoneuron and all the muscle fibers it innervates always work together and may be viewed as a unit of muscle action called a motor unit. Note that, although the muscle fibers "work together," a motor unit can hardly be considered a synergy, since an increase in the contribution of one of its muscle fibers is always accompanied by an increase in the contribution of other fibers. This built-in relation makes the fibers act more like legs of a table, which, as we have agreed, is not a synergy.

Not all motor units are created equal. There are small and large motor units (Figure 3.10), fast and slow, fatigable and fatigue-resistant. These features correlate such that smaller alpha-motoneurons have thinner axons, innervate

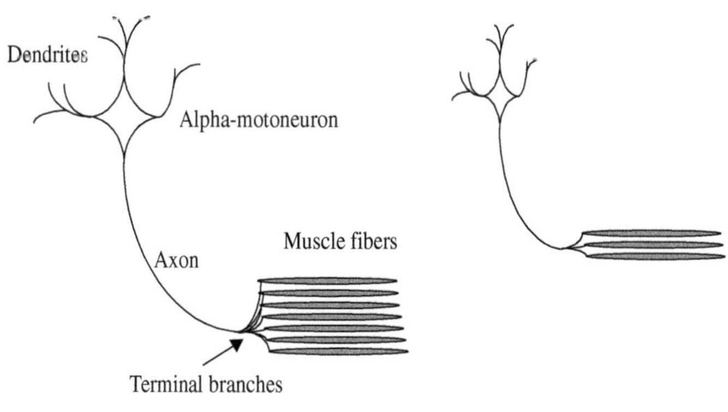

Figure 3.10. Muscles are innervated by many neural cells (alpha-motoneurons). There are large and small motoneurons that innervate correspondingly larger and smaller number of muscle fibers. Larger motor units (the motoneuron and the fibers it innervates) are stronger and faster; smaller motor units are more fatigue resistant. During natural movements, motor units are recruited from the smallest to the largest.

fewer muscle fibers, and produce less force (in both twitch and tetanus contractions). These motor units are typically slower and fatigue-resistant. In contrast, the largest motoneurons have the thickest axons, innervate larger groups of muscle fibers, and produce more force. These motor units are typically faster and fatigue quickly.

During natural contractions, motor units are typically recruited in an orderly manner, starting from the smallest ones and gradually involving larger and larger motor units. This rule is known as the size principle or the Henneman principle (Henneman et al. 1965). So, overall, muscle force is a variable produced by many motor units that may change their contribution to the total force by changing the frequency of the action potentials generated by their respective alpha-motoneurons. This gives the controller two methods of force modulation, recruitment or de-recruitment of motor units and changes in the frequency of action potentials to the already recruited motor units.

To summarize, human muscles are slow in force generation and in relaxation. The force they produce cannot be predicted in advance unless one knows all the details of movement kinematics, that is, changes in muscle length during the movement. Not surprisingly, Bernstein wrote in one of his early papers (Bernstein 1935) that the central nervous system cannot in principle predict forces that would be produced as a result of its own central commands.

End of Digression #1

Digression #2. Neural Pathways: Long and Slow

Although the unit of information transmission (the action potential) is electrical, its speed of conduction within the human body is relatively slow, very far from the speed of electric current. This is due to the fact that action potentials do not truly travel along the membrane of excitable cells but rather emerge and disappear, giving rise to new action potentials on adjacent membrane segments. The process of action potential generation involves ion diffusion, which happens over short distances and, as a result, is fast but not as fast as electric current. It takes just under 1 ms for the action potential to be generated after the membrane receives adequate excitation. This may not seem like a lot of time. However, consider that typical distances within the human body are of the order of 1 m, for example, the distance from the brain cortex to the lumbar enlargement of the spinal cord where a lot of neurons that control the lower part of the body are packed, or the distance from the spinal cord, where the bodies of the alpha-motoneurons reside, to a muscle in the foot. Conducting an action potential generated over the membrane in the body of a neuron along its axon over a distance of about 1 m may take tens or hundreds of episodes of action potential generation along the axon. This may be expected to add up to several tens or several hundreds of milliseconds of time delay.

There are two types of neural fibers in the human body. Some of the fibers are covered with a sheath made of a fat-like substance, myelin (Figure 3.11).

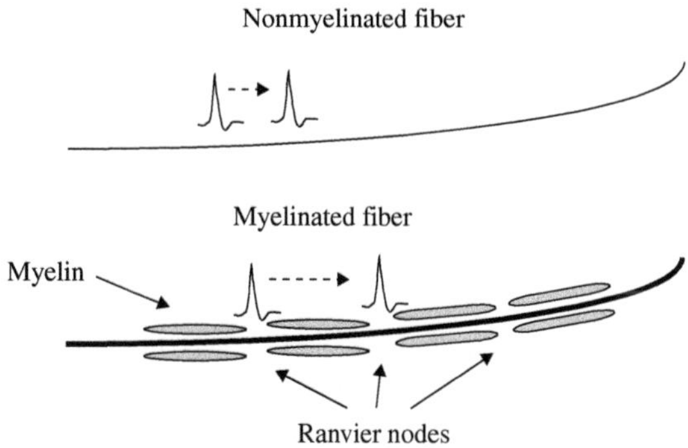

Figure 3.11. Neural fibers can be covered with a special substance, myelin. Nonmyelinated fibers are typically thinner; they conduct action potentials at considerably lower speeds. Myelinated fibers generate action potentials only at special sites—breaks in the myelin sheath that are called Ranvier nodes. The action potentials jump from one node to the next one.

This sheath helps the action potentials to jump over larger distances and leads to a substantial increase in the conduction speed. For example, action potentials travel along the fastest conducting myelinated axons with the speed of up to 120 m/s. In contrast, axons that are not covered with myelin conduct action potentials at speeds of about a few meters per second or even slower.

The obvious advantage of the myelinated fibers led Nikolai Bernstein to one of his less well-known theories that the dinosaurs became extinct because there were eaten alive by rats. Well, not exactly by rats but by the first rat-like mammals that apparently shared the Earth with dinosaurs for some time. Bernstein assumed that dinosaurs were equipped only with nonmyelinated fibers, while the first mammals already had myelinated fibers. If one considers a 3 m long dinosaur that has just been bitten by a "rat," it would take the dinosaur at least 2 s to feel the bite and react to it. This is more than enough time for the small mammal to get a morsel of food and run away using its much quicker action potentials that have to travel over much shorter distances. Dinosaurs were helpless against their tiny rivals equipped with sharp teeth and much faster conducting axons.

However, even the fast-conducting myelinated fibers lead to time delays of the order of several tens of milliseconds over typical distances in the human body. For very fast actions and reactions, such delays are not insignificant. For example, the tennis ball served at 150 km/h travels 1 m over 25 ms. Besides, the advantages of having myelinated fibers are accompanied by disadvantages: Any new high-tech invention makes the system prone to malfunctioning. The price of having myelinated fibers is multiple sclerosis, a progressive disorder characterized by loss of myelin sheath over major conducting pathways within the central nervous system. This disorder leads not only to the slowing down of the speed of action potential conduction but also to the stopping of it altogether: The myelinated fibers lose the ability to conduct action potentials in the good old way, as nonmyelinated fibers do, and for the fiber, losing part of the myelin sheath means an inability to conduct action potentials over the damaged zone.

End of Digression #2

Digression #3. Sensors: Confusing and Unreliable

The human body is equipped with a variety of sensors that allow adequate knowledge of one's own body, the external world, planning of human actions, and their corrections in cases of unexpected changes in external forces, motor goals, etc. Some sensors (*receptors*) inform the central nervous system about the functioning of internal organs; this information (*interoceptive information*) is typically not consciously perceived. Other sensors, such as those located in the human eye, deliver information on such important aspects of the world as the location of external objects and visible parts of one's own body. Still other receptors inform the brain on properties of objects that are in direct contact with the body (*haptic perception*), as well as on relative position of body parts with respect to each other (*proprioception*). Proprioceptive information has been viewed as crucial for the control of voluntary movements. It is hard to imagine how the aforementioned six-step program would be accomplished in the absence of reliable information on the body configuration and its interaction with external forces. So, let us consider features of the sensors that supply this crucial information.

Proprioceptive information about the limbs and the trunk is provided by specialized neural cells that reside in *ganglia* (many cells united by an anatomical and sometimes physiological principle), located close to the spinal cord. These cells have axons with two very long branches. One of the branches reaches a point somewhere in the body with a sensitive ending, a structure that can generate action potentials in response to a particular physical stimulus. The other branch of the axon enters the spinal cord, where it can make connections with neurons at the level of the entry point; these connections can lead to quick responses to peripheral stimuli, commonly known as *reflexes*. The central branch of the axon can also travel along the spinal cord to other neural structures, even all the way to the cortex of the large hemispheres of the brain.

Sensory endings of proprioceptors are typically sensitive to deformation, although most endings can produce action potentials in response to other stimuli, for example, electrical stimulation. There are several groups of proprioceptors that are of particular importance for motor control. At first glance, they seem to provide necessary information about all the potentially important physical variables such as joint position, muscle length and velocity, muscle force. However, a closer analysis reveals, that deciphering this information may not be so easy and straightforward.

Arguably, the most famous sensors in the human body are muscle spindles. Actually, spindles are structures that house many sensory endings sensitive to both muscle length and velocity. A pictorial illustration of a muscle spindle is shown in Figure 3.12. Spindles are scattered in skeletal muscles, oriented parallel to the

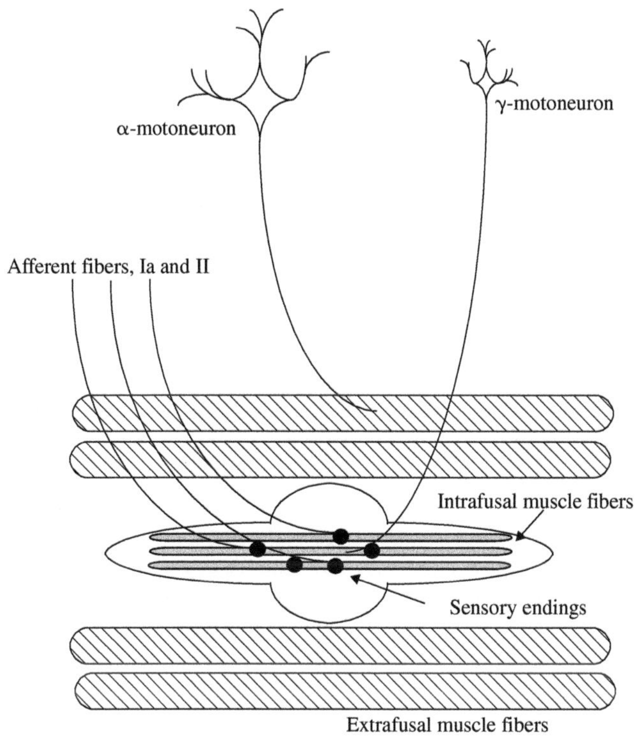

Figure 3.12. Muscles contain sophisticated structures called muscle spindles. The spindles are located in parallel to the power-producing (extrafusal) fibers. They contain muscle fibers (intrafusal fibers) with sensory endings that are sensitive to muscle length and velocity. These endings send information into the central nervous system via relatively thick and fast-conducting afferent fibers (groups Ia and II). The intrafusal fibers are innervated by small motoneurons (gamma-motoneurons) that can change the sensitivity of the sensory endings to both muscle length and velocity.

power-producing muscle fibers. These fibers are sometimes called *extrafusal* in contrast to *intrafusal* fibers, which are muscle fibers located inside the spindles. The shell of a spindle has internal elastic connections to the intrafusal fibers and external connections to extrafusal fibers. As a result, when extrafusal fibers are stretched, the spindle and the intrafusal fibers are stretched too.

There are two types of sensory endings on the intrafusal fibers, primary and secondary. The primary endings generate action potentials at a higher frequency, when the muscle is longer and stretched (its velocity is positive). They may be completely silent when the muscle is shortened. The secondary endings are sensitive only to muscle length but not to its velocity.

Why would the Ultimate Designer use muscle fibers—very complex structures—as the place to put sensory endings inside the spindle? Could one use something cheaper, like passive, connective tissue that would be equally able to transfer changes in the muscle length to the sensory endings? Moreover, the axons of the proprioceptive neurons that innervate primary spindle endings are among the fastest-conducting in the human body. They are thick, myelinated, and can transmit action potentials at the speed of up to 120 m/s, faster than neurons in the cortex that send signals to the spinal cord, when a person wants to initiate a very quick action. This complex design looks purposeful and suggests that receiving adequate information on muscle length and its changes is important for the central nervous system.

To add to the complexity of the design, there is a specialized group of small neurons in the spinal cord that send signals to intrafusal fibers, the so-called *gamma-motoneurons.* These motoneurons produce contractions of the intrafusal fibers—contractions that have no direct effect on the mechanical functioning of the muscle. This design appears wasteful. However, it seems to be an opportune moment to recall a basic philosophical postulate: If something in the design of the human body looks suboptimal, it is likely we have missed something important.

The role of the gamma-motoneurons is to change the sensitivity of the spindle endings to muscle length and its changes. This mechanism allows the spindle endings to continue generating information for the central nervous system, even when a muscle is being actively contracted. If it were not for the system of gamma-motoneurons that tend to be activated together with alpha-motoneurons (the so-called *alpha–gamma coactivation*), an active muscle contraction could have led to a temporary interruption of information about muscle length and velocity.

Proprioceptive sensory endings of another type are located at the junction between the extrafusal muscle fibers and tendons (Figure 3.13). They are sensitive to muscle force, which means that they generate action potentials at a frequency roughly proportional to the force produced by the muscle fibers on the tendon at that particular point. These endings are called *Golgi tendon*

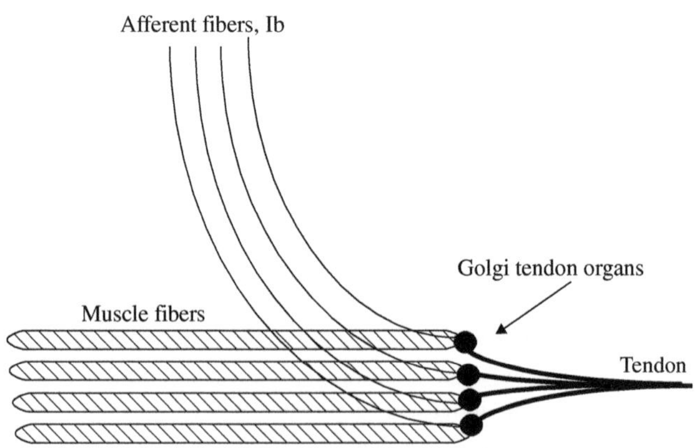

Figure 3.13. Force-sensitive receptors are located at the junction of muscle fibers and tendons. They are called Golgi tendon organs. They send signals to the central nervous system via fast-conducting afferent fibers of group Ib.

organs. They have no active mechanism of changing sensitivity that would be analogous to the system of gamma-motoneurons. They are also innervated by very fast-conducting axons, just a little slower than those innervating the primary spindle endings.

Let me mention one more group of proprioceptors that used to be considered the main source of joint position information. These are articular receptors located inside the articular capsule. These receptors seem to be in the best possible location to signal joint angle and its changes. However, studies of properties of these receptors produced disappointing results (Burgess and Clark 1969; Clark and Burgess 1975). First, most articular receptors are sensitive to joint position, only when the joint is close to one of the limits of its anatomical rotation; this does not look like a good design to signal joint angle during natural movements. Second, articular receptors change their activity when the joint capsule tension changes. For example, if you co-contract muscles acting at the elbow joint, the articular receptors are likely to show an increase in their activity, while the joint does not move. Articular receptors also change their activity with joint inflammation. Taken together, these results suggest that articular receptors may play the role of safety sensors informing the central nervous system that the joint may be in danger of being damaged. However, they do not look like reliable sources of joint position information.

Nevertheless, muscle spindles and Golgi tendon organs seem more than adequate to supply the central nervous system with information on position and force. However, there are problems. Let me start with the most obvious one. Natural limb movements involve joint rotations. While forces are adequate variables to describe changes in linear motion of material objects (their linear acceleration), moments of force are adequate variables to describe changes during rotations. Moment of force is the product of the force and its lever arm. At different joint positions, lever arms of muscle action may differ (Figure 3.14), leading to differences in the magnitude of the moment produced by the same muscle force. So, even if Golgi tendon organs are perfect force sensors, they are not perfect sensors for moment of force, which is a more adequate variable to describe mechanical consequences of muscle action.

$$M_1 = F \times d < F \times D = M_2$$

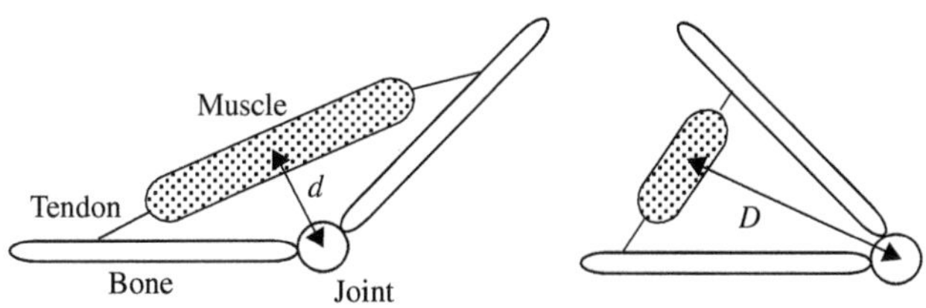

Figure 3.14. Information about muscle force does not allow estimation of the moment of force (*M*), which depends on the lever arms (*d* and *D*). The same force (*F*) may correspond to different moments of force (M_1 and M_2) at two different joint positions. Note that this is a cartoon, which distorts the actual geometry of muscle attachment.

The location of muscle spindles parallel to the extrafusal fibers allows them to measure the length of the muscle fibers and velocity. What matters for movements, however, is the length of the "muscle + tendon" complex, which does not necessarily have a simple relation to muscle fiber length. When a muscle is relaxed, it is typically very compliant—more compliant than the tendon. When the same muscle is activated, it is typically more stiff than the tendon. So, under activation, muscles can bulge and shorten causing tendon stretch even if there is no joint motion. How can the controller decide whether a drop in signals from a spindle corresponds to a joint motion causing shortening of the muscle fibers or to force production against a stop (the so-called, *isometric contraction*) without any joint motion? Figure 3.15 shows two conditions, where the length of the muscle fibers is the same, while joint positions are quite different.

To complicate matters, there is the system of gamma-motoneurons that can lead to changes in the activity of muscle spindle endings unrelated to joint motion. There are also bi-articular muscles crossing pairs of joints; signals related to their length changes cannot be easily deciphered in terms of joint rotations. One seems to need a powerful computer to decipher this messy stream of sensory signals, in terms of joint angles and joint moments of force—the two variables that seem to be truly important for movements.

The solution to this problem lies in the mentioned classical observations of the famous scientist Hermann von Helmholtz: If a person closes one eye and moves the other one in a natural way, he or she has an adequate perception that the eye moves while the external world does not, although eye movement leads to a shift in the image of the external world over the retina. If the same person moves the eye in an unusual way, for example, by pressing slightly on the eyeball with a finger, there is a strong illusion that the external world moves. Von Helmholtz suggested that this observation points to an important role played by motor command in sensory perception. At about the same time, a similar idea was expressed by Sechenov who wrote that humans do not see but

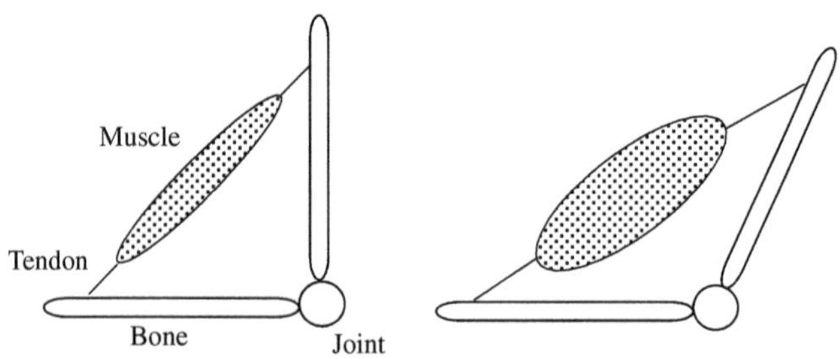

Figure 3.15. Information about muscle length does not allow estimation of joint angle. One reason is the relative difference of length of the muscle fibers in the "muscle + tendon" complex at different levels of muscle activation. The right drawing shows an activated muscle that stretched the tendons. Muscle length is the same in the two drawings, while the joint angles are different. Note that this is a cartoon, which distorts the actual geometry of muscle attachment.

look, they do not hear but listen. The term *efferent copy* (or sometimes *efference copy*) was coined to address hypothetical motor command signals that take part in perception (Von Holst 1954). Obviously, to use this idea, one has to at least assume what these motor command signals might be. We will return to this issue later after addressing controversies in the area of motor control. At this point, let me only state, somewhat vaguely, that motor command plays the role of a reference frame within which sensory signals are measured.

This general view has been used to interpret not only adequate kinesthetic perception but also kinesthetic illusions. Within this general scheme, there are two mechanisms that can bring about illusions. First, there may be an inadequate stream of sensory information produced by an artificial stimulus (inadequate afferent signals). Second, muscles may receive signals from the central nervous system that are different from what could be expected, given a current motor command (inadequate efferent copy).

Kinesthetic illusions of the first group have been observed in many experiments with high-frequency, low-amplitude muscle vibration (Goodwin et al. 1972; Feldman and Latash 1982; Roll and Vedel 1982). Such vibration is a very powerful stimulus for the primary spindle endings because of their high sensitivity to velocity of muscle fibers. It can drive virtually all the endings within a muscle, that is, force them to produce action potentials in response to each cycle of the vibration, up to frequencies of 100 Hz. This is a very high level of spindle activity, much higher than the levels observed during natural movements. The central nervous system bombarded by action potentials from the spindles tends to misinterpret them as corresponding to the lengthening of the muscle. As a result, illusions of joint position may be perceived leading sometimes to the perception of an anatomically impossible joint position (Craske 1977) and also deviations of vertical posture, if vibration is applied to a major postural muscle such as, the ankle plantar flexors (Lackner and Levine 1979; Hayashi et al. 1981). Illusions of the second group can be seen when a movement in a joint is produced by muscle activation that is artificially altered compared to a natural movement. This can be done, for example, by applying external electrical stimulation directly to the muscle nerve.

End of Digression #3

To summarize the three digressions, our body is equipped with weak, sluggish force generators. The force generators are controlled using signals that take significant time to travel from the controller. Sensory information on such vital variables as joint position and torque is not readily available; hence, it has to be deciphered.

These apparently suboptimal features of the design of the human body require additional procedures that would allow such a "poorly designed" structure to behave with acceptable accuracy and reliability. In other words, in addition to solving the six-step problems one has to add "crutches" to help the body behave like a perfect robot equipped with powerful and quick motors and electrical wires that conduct signals from the controller to the motors and from the perfect sensors to the controller, with negligible delays.

Since the six-step problems obviously have to be solved before a command is generated to initiate a movement, most of the required information is unavailable and has to be predicted. For example, muscle force depends on the neural input for the muscle and muscle kinematics. Peripheral inputs into motoneuronal pools also depend on actual time changes in such variables as muscle length, velocity, and force. All these variables have to be known (estimated) before they actually emerge. Sensory signals inform the central nervous system about these variables, but this information is delayed (and may not be easy to decipher). Signals from the controller also take time to reach alpha-motoneurons and muscles. So, the controller always deals with outdated information and sends commands that will be outdated when they reach the effectors.

To deal with this problem, computational systems of another type need to be invoked, that is, predictors. There are different types of predictors that can, in principle, be used to compute changes in the peripheral apparatus expected to happen, given current feedback on its state and command signals. For example, one of the popular predictors is the so-called *Kalman filter* (Grush 2004). An important component of any predictor is the *direct model* of the controlled system. Unlike the inverse model, it computes the effects of a particular command on the state of all the "lower" structures, including the motoneurons, the muscles, the joints, and the end-point. So, its computations follow the natural course of events, justifying the term "direct." A combination of inverse and direct models is potentially able to alleviate many of the apparent problems associated with the nonrobotic design of the human body (Schweighofer et al. 1998; Wolpert et al. 1998; Imamizu et al. 2003).

Many studies have been performed to support the idea that somewhere in the human brain, there are structures that compute direct and inverse models associated with particular motor tasks (reviewed in Kawato 1999; Grush 2004; Shadmehr and Wise 2005). Most studies of this group follow a similar logic. They present a person with a relatively simple motor task, for example, to move the dominant hand along a straight line from a starting position to a certain final position. Little practice is required to perform such a task well. Then the external conditions are changed. These changes can be mechanical, for example, an external force starts to push the moving arm thus changing its mechanical behavior in response to neural signals in an unusual way (Shadmehr and Mussa-Ivaldi 1994; Bhushan and Shadmehr 1999). They can also be sensory. For example, when a person starts to move in a certain direction, the feedback informs him or her that the hand actually moves in a different direction (Imamizu et al. 1995; Tong and Flanagan 2003; Wang and Sainburg 2006).

The person tries to perform the task under the new, unusual conditions, and, not surprisingly, starts making mistakes. After some practice, however, the person learns to perform the required movements accurately. Has the person learned a new set of internal models associated with the new conditions? The authors of many of these studies claim that this is so and cite in support of this view an observation that when the unusual conditions are unexpectedly turned off, the person starts making mistakes in the opposite direction to those observed when the novel conditions were first presented. Let me make another digression and describe a typical study in greater detail.

Digression #4. Adaptation to Force Fields and After-Effects

Imagine that a person sits in front of a table and moves the arm parallel to the ground. There is a starting position, and several targets are located at the same distance but in different directions from the starting position (Figure 3.16). This is sometimes called the *center-out motor task.* Imagine also that there is no friction and gravity does not play a role (this can be achieved with a few simple mechanical methods, for example, see Bagesteiro and Sainburg 2005). Each trial starts with the hand in the starting position, then one of the targets appears, and the subject is required to move into the target as quickly and accurately as he or she can.

This is an easy task, and after a few practice trials, subjects show relatively straight trajectories directly into the target (Figure 3.17A). Imagine now that an external force field is applied (this can be done using a programmable robot), for example, external force that always acts perpendicular to the movement velocity and is proportional to the magnitude of the velocity (as in the top drawing in Figure 3.17). In the starting position, the subject does not feel the force since velocity is zero. However, as soon as a movement is initiated, it becomes perturbed by the force. This results in curved trajectories (Figure 3.17B).

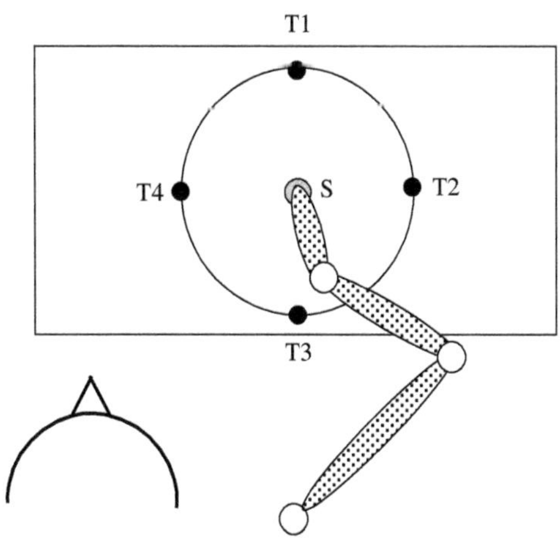

Figure 3.16. An illustration of a "center-out task." A person places the tip of the index finger of a hand into the starting position (S) and then, after a signal, performs a very fast movement into one of the targets (four are shown, T1, T2, T3, and T4).

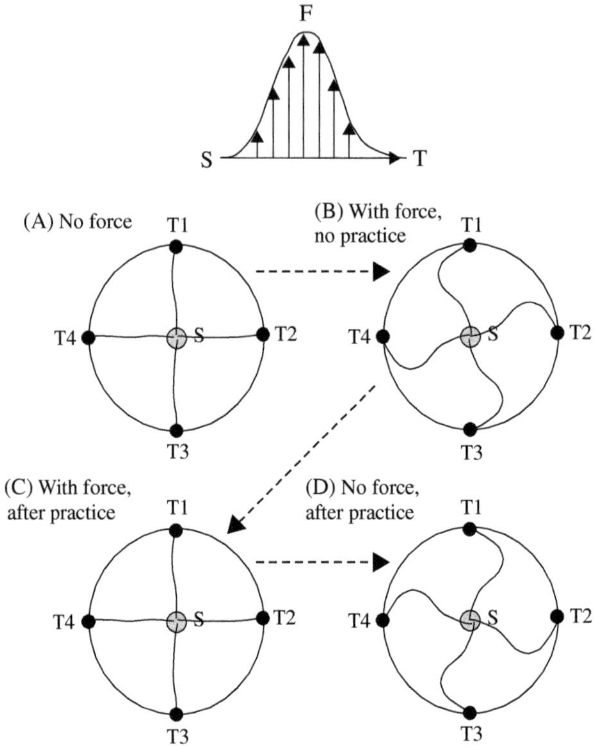

Figure 3.17. (A) The endpoint trajectories in the center-out task are typically straight. (B) If an external force field is turned on (e.g. producing a force acting perpendicular to the trajectory with the magnitude proportional to velocity—see the top drawing), the trajectories become curved. (C) After some practice, the subjects learn to perform straight trajectories with the external force field. (D) If then, unexpectedly, the force field is turned off, the trajectories again become curved, but in the opposite direction.

After the subject performs many movements in such novel conditions, his/her movements adapt to the new force field, and the trajectories become straight again (Figure 3.17C). Now, imagine that the force field is turned off. The first few trials will show curved trajectories that look like mirror images of those that were observed immediately after the force field was turned on (Figure 3.17D).

A common interpretation of these findings has been that the subjects learned new internal models during practice under the unusual force field conditions. When the conditions were changed back to "normal," the application by the subjects of inappropriate internal models resulted in errors in the opposite direction. This interpretation sounds very reasonable, if one accepts as an axiom that movements are controlled with sets of direct and inverse internal models. There is, however, an alternative view on the organization of motor control, which leads to a different interpretation of these findings (see Ostry and Feldman 2003; Malfait et al. 2005).

In particular, a recent series of studies by David Ostry and his group (Malfait and Ostry 2004; Malfait et al. 2005) have suggested that learning to move in a novel force field has only local effects, that is, it does not show generalization when the task is modified. These observations cast doubt on the appropriateness of the term "model" to describe results of such studies (see also Gupta and Ashe 2007).

Another study by the group of Ostry suggested a simple algorithm that could produce effects of adaptation to novel force fields using the framework of the equilibrium-point hypothesis (see the next major section). In that study (Gribble and Ostry 2000), the authors postulated that in each trial, time-shifts of the hypothetical control signals (related to equilibrium trajectory) were adjusted directly in proportion to the positional error between desired and actual movements in the previous trial. The authors simulated the behavior of a two-joint, six-muscle system in a velocity-dependent force field and showed that they closely corresponded to experimental observations.

End of Digression #4

If internal models do exist, where are they? In personal conversations, some of the most eminent champions of this idea typically agree that hypothetical internal models are probably broadly distributed across the structures of the central nervous system. In publications, however, much attention has been attracted by the cerebellum as the likely site of the internal models (Wolpert et al. 1998; Kawato 1999; Grush 2004). This is partly due to the enigmatic nature and function of the cerebellum that has been hypothesized to play many diverse roles from being an internal clock of the body (Braitenberg 1967; Llinas 1985; Ivry et al. 1988; Ivry and Spencer 2004) to organizing sets of elements into functional units (structural units or synergies, Bloedel 1992; Thach et al. 1992a; Houk et al. 1996). Besides, the structure of the cerebellum looks very attractive for neural models. There are only five types of cells in the cerebellum, it receives only two types of inputs, and neurons of only one type (Purkinje cells) generate its output (see more on the cerebellum in section 7.3). Results of studies with brain imaging have been cited in support of the idea that internal models are being created and/or stored in the cerebellum (Imamizu et al. 2000, 2004; Bursztyn et al. 2006).

Digression #5. Brain Imaging Techniques: What Do They Image?

Brain imaging has become popular these days. Publications ranging from respected scientific Journals such as *Science* and *Nature* to more journalistic outlets such as *Scientific American* and *New York Times* are filled with reports on changes in the brain activity with action, learning, aging, development, neurological disorder, treatment, etc. Questions like "What does happen in such-and-such brain area when a person performs such-and-such action?" and "What are the relations between changes in the activity of area #1 and area #2 in the brain?"

are being asked in numerous studies. Before addressing particular methods of brain imaging, I would first like to ask a more basic question: "What are we going to gain in understanding the brain function, if we suddenly get answers to all these questions?" This basic question, which is sometimes addressed as the problem of inter relationship between structure and function has been debated for a long time (reviewed in Bernstein 1935, 2003). Views have been oscillating between "any neuron can do anything" and "there are places in the brain that do specific functions." Bernstein contributed significantly to this discussion.

In the very first book published by Bernstein, about 40 years after his death (this is not a typo, Bernstein 2003), he develops his ideas on the relative roles of morphology and what he calls "dynamics"—that is activity-dependent plastic changes in neural projections: "If the features of neural dynamics are tightly linked to the anatomical structure of the elements and to the system construction (organization) of these elements, the relations between the dynamics and morphology could be qualitatively different at different levels of the organization. At the lowest elementary levels of the construction of the neural process, the role of morphology may dominate compared to the role of dynamics. As the organization becomes more complex, an increase in the new, dynamic features of the neural process…may overcome the constraints imposed by the morphology and dominate them" (p. 316). And then he makes a wonderful, truly Bernsteinian logical leap: "As such, an increase in the morphological, localizational separation of brain structures leads to a more intensive development of the foundations for the development in the brain of nonlocalizable processes" (p. 317).

How can brain imaging, even if working perfectly, help us understand the nonlocalizable processes in the brain? Imagine that an experiment shows an increase in the neuronal activity in a certain area of the cerebellum or the basal ganglia, when a person makes a certain functionally relevant action. What can be concluded about the role of those neurons in the action? I leave it to the reader to answer this question in a single word.

There are two types of brain imaging techniques. First, there are methods based on recording electrical phenomena in the brain. Second, there are methods of recording signals related to brain metabolic processes. Apparently, methods of the first group provide information more directly related to neural activity. Unfortunately, these methods have certain limitations and pitfalls.

The oldest and best known method is electroencephalography (EEG). It involves placing electrodes on the scalp and recording electrical events that happen in the neural tissue. This is the most direct method of measuring electrical events in the brain, and it provides excellent time resolution: One can record millisecond-to-millisecond changes in the brain activity online; however, there are problems. First, the electrodes are rather far away from the sources of neural activity. As a result, they record very small signals and their spatial resolution is very low. In addition, this method may be adequate to record neural activity in structures that are close to the surface of the brain but not in deep brain structures that are very far from the recording site. To complicate matters, there are other sources of electrical activity in the vicinity of the recording sites such as head muscles. If a person moves the eyes or clenches the teeth, virtually no useful signal can be recorded from brain neurons. Nevertheless, EEG has proven a very useful clinical and research tool.

A fancier and more sensitive method of measuring changes in the electromagnetic field produced by brain neuronal activity is magnetoencephalography (MEG). This method provides much better spatial resolution compared to EEG while maintaining the same level of temporal resolution. It can also record signals that originate from deeper brain structures. It is, however, extremely expensive and requires continuous, highly skilled technical support, which keeps it beyond the reach of most research laboratories and hospitals.

The other group of methods addresses neural activity only indirectly. This group includes magnetic resonance imaging (MRI) and positron emission tomography (PET). Despite the difference in the physical mechanisms underlying the two methods, they both produce signals reflecting local metabolic processes. For example, the most commonly used signal in MRI is the so-called BOLD (blood-oxygenation-level-dependent) signal. It reflects the proportion of hemoglobin molecules that are oxygenated in the area of interest. The logic of these methods is as follows. The rate of metabolic processes is known to correlate with neural (in particular, synaptic) activity. The proportion of hemoglobin molecules that lose oxygen correlates with the rate of metabolic activity. Hence, one can expect the BOLD signal to reflect intensity of neural processes in the area. These methods typically have a very good spatial resolution (much better than EEG); but despite the recent technological progress their temporal resolution is still marginal. Temporal resolution of these methods is naturally limited by the relatively slow processes of blood flow. Both MRI and PET are relatively expensive and put severe constraints on the participants of such a study. In particular, there are constraints on having metallic objects (including implanted objects) in the area of analysis and on movements of the head. There is another potentially serious confound: Changes in the hemoglobin concentration in a brain area may reflect processes that occur "upstream" of the blood vessels happening to run through this area.

However, even if all the technological problems were solved, would we move closer to understanding how important brain functions are organized? What do all these fancy figures with bright red spots in certain brain areas mean? Can one say that if a lot of neurons are active in a particular brain area during a particular activity, this brain area plays a central role in controlling that activity? Are most important decisions in a country made during football games or rock concerts when a lot of people shout together?

End of Digression #5

Most researchers working with the idea of internal inverse and direct models assume that computational complexity is not a problem. We have been brought up with a strong belief that the brain is more powerful than any computer. So, unlike many more trivial steps, such as delivering sensory information to the controller and command information to the muscles, it has been assumed that computations happen nearly instantaneously and perfectly.

By itself, this is quite a leap of faith. Most computers can now beat most people in chess, backgammon, versions of tic-tac-toe, and other games that require quick computations. But even if we consider a much simpler task, limitations of the brain in making quick, correct decisions become obvious.

Imagine that a person is asked to perform the most simple, well-practiced movement, for example, to press a button, as quickly as possible after hearing a beep or seeing a flash of light. In such conditions, frequently called a *simple reaction time* paradigm, the delays between the signal and the action are substantial. The shortest delay to the first detectable changes in the activity of muscles that initiate the required action is typically over 100 ms. One tenth of a second is a very long time for at least some of the human actions. Over this time, a good sprinter runs over 1 m, the baseball pitched by a professional may be expected to move over 4 m, the tennis ball after a powerful serve travels about 7 m, etc. For quick everyday movements of people without special skills, 100 ms is also not a negligible time, which may be about half of the total movement time.

Less than half of the delay in a simple reaction time task can be accounted for by transmission delays from the sensory organs sensing the stimulus (the eyes or the ears) to brain structures and from brain structures to the muscles. Why does the omnipotent controller take so long to initiate the simplest action in the most reproducible conditions? There is no answer. If the task involves making a choice, for example, of pressing the left button when the left light turns on and pressing the right button when the right light turns on, the time delays become substantially longer—they nearly double. The task of selecting which button to press seems so trivial compared to the computation of muscle forces required to move quite a complex multi-joint mechanical structure to a predefined position. So, maybe brain computations are not that perfect and instantaneous.

Consider the following mental experiment. Imagine that a person is asked to flex the wrist joint from one position to another as quickly and accurately as possible. After a few trials, the subject of this experiment becomes very fast and very accurate; it is not uncommon to see movements completed in about 200 ms. According to the idea of movement control with internal models, during practice trials, the subject learned to precompute a neural command that took into account all the important mechanical and physiological factors, including the typical external forces and produced a necessary torque profile in the joint. Let us imagine now that unexpectedly, in one of the trials, the hand is quickly pushed by an external force into flexion. The push is strong and brief, and it happens after the movement has been initiated, for example, it starts 30 ms after the movement begins and lasts for 50 ms. What can be expected from the final position reached by the joint? If the subject does not introduce adjustments into the precomputed torque profile, there should be an overshoot of the final position.

However, typically the final position is achieved accurately in the perturbed trial without a major increase in the movement time. This particular feature of natural fast movements is commonly referred to as *equifinality* (reviewed in Feldman 1986; Feldman and Levin 1995). We will discuss it in more detail later.

What can the origin of equifinality in such an experiment be? If one continues to use the idea of internal models, the controller has to perceive the consequences of the unexpected push, quickly update all the inverse and direct models, and introduce corrections into the ongoing neural signals. This means that the controller gets the sensory information, performs all the computations, and implements them within about 100 ms. This happens despite the fact that the controller did not expect the perturbation, did not know its magnitude, when it would begin, and end. This explanation does not look too attractive, particularly taking into account the mentioned long-time delays in much simpler tasks, such as a two-choice reaction time task.

Equifinality in conditions of unexpected perturbations can be displayed by much less sophisticated creatures. Consider the already-mentioned wiping reflex in a frog whose spinal cord has been surgically separated from the brain. The frog performs a targeted multi-joint movement by a hindlimb toward an object such as a piece of paper soaked in a weak acid solution placed on its back or on a forelimb. The six-step problems are apparently solved in the spinal frog by the spinal cord, since no signal is received by the brain from the periphery and no signal from the brain can reach hindlimb muscles. According to the idea of control with internal models, the spinal cord should be able to handle both inverse and direct modeling.

The spinal frog can also show adaptation of the wiping movements to changes in the external conditions. In particular, if a lead bracelet is placed on a distal leg segment, the frog is able to wipe the stimulus off its back, despite the dramatically changed requisite joint torques. It is also able to wipe the stimulus off, if motion in one of the joints of the limb is blocked mechanically. Amazingly, accurate wiping under such manipulations is commonly observed during the first movement, meaning that all adjustments of the movement are performed over its course. Does this mean that the spinal cord detects that a movement does not proceed as planned, updates all the internal models, performs all the computations of new required muscle forces and joint torques, and implements a new control strategy in the course of a single fast movement? If the spinal cord of a frog can do such amazing circus tricks within a single trial, why does it take dozens or even hundreds of trials for a healthy, young graduate student to learn to move accurately in a novel force field (e.g. in Hinder and Milner 2003)?

These observations plant serious doubts with respect to the whole approach of controlling movements with computations of patterns of muscle forces and joint torques. There should be an alternative.

Bernstein was probably the first to realize the importance of all the apparently complicating factors for the control of human movements. In his very early writings, he emphasized that the presence of long-time delays and the relatively slow, elastic muscles made it impossible for the brain to control movements using pre-computed performance variables, such as forces and displacements, or their functions. Among such functions are derivatives of these variables, for example, the rate of force production, velocity, acceleration, as well as their ratios such as "stiffness" and "damping." I use quotation marks to emphasize the frequent abuse of these terms in movement studies. If you are interested in this aspect, please read the reviews by Latash and Zatsiorsky (1993) and Zatsiorsky (2002).

For years, these very important conclusions of Bernstein were all but forgotten, and many hypotheses in the area of motor control emerged assuming direct computation of requisite forces and torques by neural structures. Frequently, the authors of these hypotheses referred to Bernstein's papers in support of their views. One development of the Bernstein idea of *engrams* (Bernstein 1935, 1967) gained particular prominence under the name of a generalized motor program (Schmidt 1975).

The idea of a generalized motor program was based on an everyday observation that people can perform "the same movement" faster or slower, weaker or stronger. When a child learns how to ride a bicycle, he or she can then easily ride under various conditions and at different speeds. These everyday observations have been interpreted as elaboration of a particular function of control variables stored in the central nervous system. Such a function can be scaled in magnitude and also used at different time scales, faster or slower. Many studies have been performed in support of this postulate. In particular, a study of professional typists showed that when the typists typed a standard phrase at different speeds, they preserved the relative timing of pressing individual keys (Viviani and Terzuolo 1980). Other classical observations include the apparently preserved ability of a person to show individual features of handwriting, when writing on a piece of paper and on the blackboard. Note that different joints and muscle groups are used during writing under these two conditions. Moreover, individual handwriting features were reported for writing with the nondominant hand and even with the pen attached to the elbow or to a foot, or even held between the teeth (Bernstein 1967; Raibert 1977; Latash 1993; Figure 3.18). All these observations suggest that learning a movement (a skill) is associated not with learning a set of commands to muscles but a pattern of a more abstract variable—an engram.

So far, so good. It is hard to argue with a conclusion that something is learned during the practice of a novel motor task, related to particular features of the task, and that this something can later be applied to similar tasks and maybe even to tasks performed with different effectors. However, further developments of the idea of generalized motor program drew direct links between the hypothetical programs and variables such as forces, torques, and patterns of muscle activation. As such, the generalized motor program was a precursor of the internal models; both ideas share a view that peripheral performance variables of the

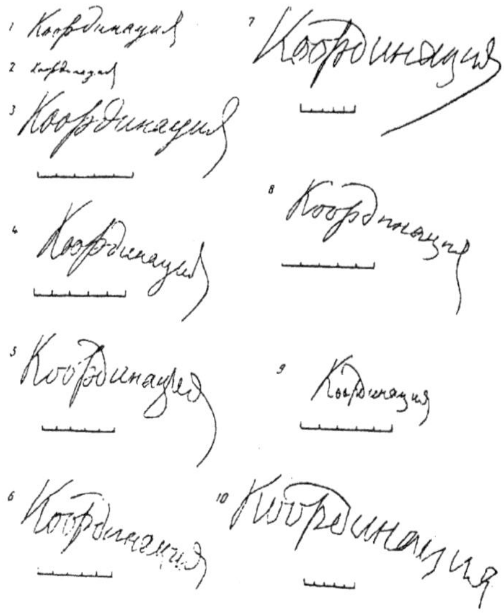

Figure 3.18. The word "Coordination" in Russian written by the pencil in the dominant (right) hand (1,2), produced by the wrist motion (fingers not moving, 3), attached to the dominant forearm (4), attached above the dominant elbow (5), attached above the dominant shoulder joint (6), attached to the right foot (7), gripped in the teeth (8), gripped by the fingers of the nondominant (left) hand (9), and attached to the left foot (10). Reproduced by permission from Bernstein NA (1947) *On the Construction of Movements.* Medgiz: Moscow (in Russian).

neuromotor system, such as forces and torques, are somehow precomputed by neural structures before a movement starts. These developments were certainly not in line with Bernstein's concept of engrams.

In the mid-1960s, a hypothesis on the control of movements was introduced by a young scientist, Anatol Feldman (Asatryan and Feldman 1965; Feldman 1966). This hypothesis, known as the equilibrium-point (EP) hypothesis, has offered an alternative to the view that the human body is a poorly designed robot with a powerful computer on its shoulders. The fate of this hypothesis is truly unique: It has neither been disproved nor accepted by most researchers for about half a century. When Anatol Feldman described this hypothesis to Michael Tsetlin for the first time, Tsetlin reacted: "This is a very arrogant hypothesis! Everybody is talking about how complex the system is, with all those DNAs, RNAs, numerous neurons and projections, and you suggest that the whole system is just a spring with regulated parameters!" Note that Tsetlin had immediately realized that the spring analogy referred not to the muscles but to the whole neuromuscular system. He was thus able to avoid the widespread misrepresentations of the hypothesis, based on the view that that the spring analogy referred to the muscle itself.

Personally, I am very fond of the EP-hypothesis. Most people whom I know, and who have actually tried to work with this hypothesis (very few indeed!), also like it. So, why are most colleagues so critical of it and reluctant to accept it? Maybe, the problem is in the contrast between the seemingly simple formulation of the hypothesis (see Tsetlin's quotation in the previous paragraph) and its depth and complexity when one actually tries to understand it at a level that allows to turn it into a research tool. This hypothesis definitely deserves a special subsection.

3.4 THE EQUILIBRIUM-POINT HYPOTHESIS

Let me start with a very general description of the main idea of the EP-hypothesis. Then, I will describe its historical background and development. At the conclusion of this section, I will try to address some of the main controversies that have surrounded the hypothesis over the 40 years or so of its existence. In this section, analysis will be limited to single-muscle and single-joint control. Relations between the EP-hypothesis and motor synergies will be discussed in more detail in section 8.4.

The EP-hypothesis is based on a major principle of the design of the neuromuscular system: its *threshold nature*. As described in more detail later, changes in descending signals to the spinal segmental apparatus may be described as setting threshold values of muscle length. If the length of a muscle is below this threshold, the muscle is silent. If it is over the threshold, the muscle is activated, and the level of activation grows with the difference between the actual muscle length and the threshold value. This activation tends to produce muscle contraction (shortening) thus bringing muscle length closer to the threshold value. In the absence of external resistance, an active muscle always tries to reach a state corresponding to its activation threshold. When dealing with pairs of muscles that oppose each other's action at a joint, assuming zero net external torque, the descending signals can define a position, at which the joint would reach an equilibrium (the torques produced by the muscles would exactly balance each other), and a joint angle range within which both muscles are active. These have been referred to as *r*-command and *c*-command. Control of a limb movement or a whole-body movement can be described as the process of defining a reference configuration for important elements of the limb (body) and its stability about this reference configuration.

In other words, the difference between the threshold positions defined mainly by descending signals and the actual positions (sensed by proprioceptors) leads to activation of motoneurons and muscle fibers they innervate. These neural and muscular elements interact with each other and with the environment through both mechanics and neural loops. These interactions tend to reduce the activity of motoneurons and minimize the difference between the actual position and the threshold position. This may be viewed as a *principle of minimal end-state action* (or *the principle of ultimate slothfulness*), related to the principle of minimal interaction described in section 3.2 (see also Feldman et al. 2007). In an ideal world, the body would come to rest at a configuration when all the muscles are silent. In the real world, muscle length changes are constrained by the body anatomy and its interaction with the environment. As a result, actual equilibrium states may correspond to certain patterns of nonzero muscle activations.

Note that the natural confluence of the principle of minimal end-state action and the EP-hypothesis offers a physiological mechanism to solve the problem of motor redundancy: Setting a new reference configuration leads to an interaction among the neuromuscular elements and between them and the environment producing an action. Repetitive actions can be different from each other due to many factors such as unavoidable differences in the initial state of all the elements and variations in the environment. They may also differ because of other constraints imposed by the controller such as performing other task components and optimizing certain characteristics of performance. More on this topic can be found in the section on the EP-hypothesis and synergies (section 8.4).

There are several well established properties of muscles, neurons, and reflexes that have been crucial for the emergence and development of the EP-hypothesis. The purpose of the next several pages is to describe these properties and, by doing so, help the reader understand the main message of the previous several paragraphs.

3.4.1 Experimental Foundations of the Equilibrium-Point Hypothesis

Elastic properties of human muscles have been known for over a century. Weber brothers were arguably the first to pay attention to muscle elasticity as a possible important contributor to the movement mechanics (see the earlier section on history of movement studies). In the first half of the nineteenth century, classical studies on muscle mechanical properties were performed by Sir A.V. Hill (Hill 1938, 1953). Hill analyzed the mechanical behavior of muscles under changes in the external load and also energy production by muscles. He came up with a few conclusions on applicability of classical mechanical notions to muscles. In particular, Hill concluded that muscles could not be viewed as ideal springs and that the notion of viscosity was inapplicable to muscles. Still, numerous publications continue to cite early Hill's works in support of applying these very notions to muscles!

A number of scientists of the first half of the nineteenth century pondered the importance of muscle elastic properties for the control of voluntary movements. Bernstein considered muscle elasticity as a major problem for the controller. He used the following illustration (Bernstein 1996). Imagine that you have a heavy ball attached by a rigid horizontal rod to the wide belt strapped around the waist.

There are also two long elastic ropes attached to the ball. Imagine now that you grab the free ends of the elastic ropes and try to move the ball quickly and accurately by pulling on the ropes. This would be a very hard task indeed, particularly if the ball is heavy and the ropes are long and compliant. The problem is that forces and movements of the hands will be different from the net forces and displacements of the ball.

In the 1920s and 1930s, a prominent German scientist, Kurt Wachholder and his younger colleague Altenburger published a very influential series of papers on the patterns of muscle activity associated with very simple movements performed predominantly about one joint. Wachholder was well aware of the muscle elasticity. He asked a seemingly naïve question: How can a person relax muscles acting at a joint at different joint positions? Indeed, imagine that two opposing muscles acting at a joint (let us call them a flexor and an extensor) are relaxed at an angular position α_1 corresponding to the length of the flexor f_1, and the length of the extensor e_1 (Figure 3.19). If there are no external forces acting at the joint, and it is at an equilibrium, the moments of force produced by the two muscles should be equal in magnitude to each other. Let us, for simplicity, consider that the moment arms (r_f and r_e) of the two muscles do not change by much during small joint rotations and that the two muscles are ideal linear springs (despite Hill's conclusions!). Then, we get:

$$k_f r_f(f_1 - f_0) = k_e r_e(e_1 - e_0), \qquad \text{Equation (3.1)}$$

where f_0 and e_0 are resting lengths of the muscle springs, and k_f and k_e are their stiffness coefficients.

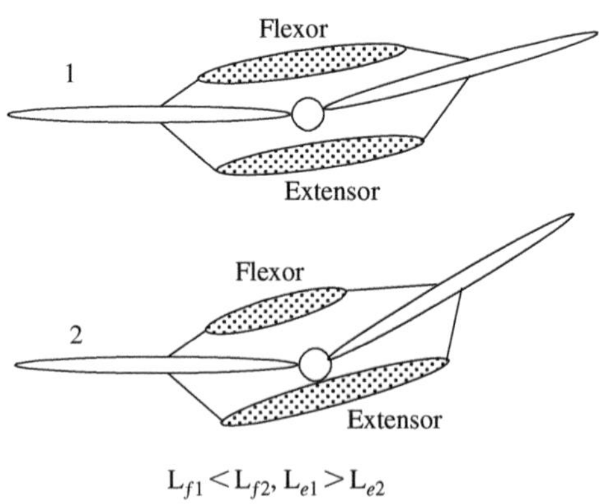

Figure 3.19. If a pair of muscles produce a net zero moment of force at the joint they cross, any motion of the joint (compare drawings 1 and 2) leads to shortening of one of the muscles and stretching of the other muscle. In drawing 2, the flexor muscle is relatively shorter, while the extensor one is relatively longer. Because of the muscle spring properties, this change is expected to generate a nonzero net moment of force that would try to move the joint back to position 1.

If the joint is moved to a new position, for example, into flexion, the length of the flexor decreases, while the length of the extensor increases. For simplicity, let us consider both length changes equal in magnitude to Δx. The moment of the flexor force will become $k_f r_f(f_1 - \Delta x - f_0)$, while the moment of the extensor force will be $k_e r_e(e_1 + \Delta x - e_0)$. Given Equation (3.1), these two moments cannot be equal to each other. There should be a nonzero net moment of force acting at both segments with respect to the joint equal to $(k_f r_f + k_e r_e)\Delta x$. This moment should act to move the joint to its initial position.

Wachholder recorded muscle activity and claimed that people could indeed relax both muscles at two different positions. Let us not be picky about the sensitivity of his equipment and its ability to detect small changes in the muscle activity. This study was an excellent example of a general rule: If an experiment is well designed, even bad equipment cannot make it fail. To reconcile the mechanical analysis with the results of his experiments, he came up with a revolutionary conclusion that was at that time (and for many many years thereafter) completely ignored by other researchers. He concluded that during voluntary movements the central nervous system had to change the spring properties of muscles, in particular their resting length values, f_0 and e_0 in Equation (3.1). Indeed, this is one of the very few ways to make the results of his study and Equation (3.1) compatible.

The early progress in the analysis of the mechanical muscle properties was paralleled by studies of muscle involuntary reactions to peripheral stimuli, so-called *muscle reflexes*. Attitude to reflexes has undergone substantial changes over the past 100 years or so. It ranged from viewing all behaviors as combinations of reflexes and voluntary movements as results of modulation of basic muscle reflexes to considering reflexes as useful only for quick reactions in unexpected situations, while they are supposed to be turned off or at least be negligibly weak during natural everyday movements. This discrepancy has partly originated from the lack of a clear definition for a reflex. This situation asks for another digression.

Digression #6. Reflexes and Nonreflexes

What is a reflex? Some of my colleagues view reflex as an established useful notion reflecting important physiological mechanisms within the body. They see the main problem in defining *reflex* properly. Other colleagues, however, question the usefulness of this notion and suggest that the apparent dichotomy between reflexes and voluntary actions is artificial and that all movements are actions irrespective of complexity of the involved neural pathways and methods of their initiation. The two views are very clearly summarized in a collective paper, written by several prominent researchers (Prochazka et al. 2000).

I do not think that there is a dichotomy, particularly if one views voluntary movements as consequences of changes in parameters of reflexes (as described later for the EP-hypothesis). On the other hand, there are motor reactions to sensory stimuli that come at short time delays, shorter than the fastest voluntary (instructed) motor reaction. For some (very few) of these reactions, neural pathways that bring them about have been defined. So, it seems reasonable to keep a special term for these reactions, and *reflex* sounds quite appropriate.

There are several classifications of reflexes based on their particular characteristics. For example, the great physiologist Ivan Pavlov classified reflexes as *inborn* and *conditioned* and tried to build a theory of all actions, based on combinations of those reflexes. Another classification of reflexes into *phasic* and *tonic* is based on their time course: Phasic reflexes are transient, short-lasting, while tonic reflexes are steady-state. It is also possible to say that phasic reflexes are produced by a change in the magnitude of a sensory stimulus, while tonic reflexes are produced by the magnitude of the stimulus itself. Reflexes can lead to muscle contractions in the area of stimulus application; then, they are commonly called *homonymous* or *autogenic*. Alternatively, they can be seen in remote muscles; in such cases, they are referred to as *heteronymous* or *heterogenic*.

One of the most frequently used classifications of muscle reflexes is based on the number of synaptic connections between neural cells in the loop that brings about the reflex. Figure 3.20 illustrates the most simple reflex that is produced by stimulation of a sensory ending, which leads to excitation of motoneurons innervating a muscle. Such a reflex is called *monosynaptic*, although apparently it involves two synapses, one between the sensory neuron and the motoneuron and the other between the axonal terminals of the motoneuron and muscle fibers. However, since all muscle responses to a neural stimulus have to involve the neuromuscular synapse, it is not taken into account.

There is only one well-known monosynaptic reflex in adult humans. This reflex originates from the primary sensory endings in the muscle spindles. These endings are innervated by very fast-conducting afferent fibers (group Ia),

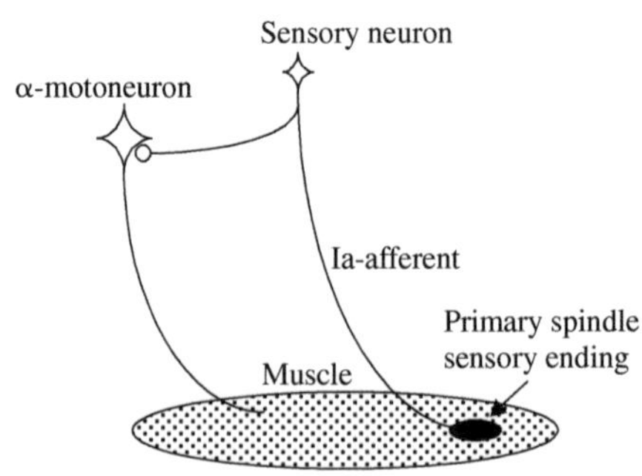

Figure 3.20. A scheme of the monosynaptic reflex. A quick muscle stretch produces a burst of activation of the primary spindle sensory endings. These action potentials travel along the Ia-afferent fibers to the sensory neurons in the spinal ganglia and then enter the spinal cord through the dorsal roots. They make direct excitatory projections (synapses, the open circle) on alpha-motoneurons that send their axons to the muscle containing the spindles.

making direct projections on alpha-motoneurons that send their axons to the same muscle. The spindle endings can be stimulated naturally, for example, by a very quick muscle stretch (it can be produced by a tendon tap). Their afferent fibers can also be stimulated artificially by an external electrical stimulator. The two incarnations of this reflex are called the tendon tap reflex (or T-reflex) and the H-reflex (after a German scientist, Hoffman who described this reflex in early nineteenth century). These reflexes come at the shortest possible delay that is mostly due to the conduction time along the afferent (sensory) and efferent (motor) axons. Typical delays (latencies) of monosynaptic reflexes in the arm and leg muscles are within the range of 20–40 ms.

Monosynaptic reflexes can be seen in newborn babies, so they are inborn. They are phasic, which means that they produce quick twitch muscle contractions. Typically, they are homonymous, although heteronymous monosynaptic reflexes (in particular, in antagonist muscles) have been described in babies and in some pathological states (Myklebust et al. 1982; Myklebust and Gottlieb 1993). Since monosynaptic reflexes are seen only in response to a very quick muscle stretch, their role in most everyday movements is questionable. Besides, there is a report (Latash 1993; Latash and Penn 1996) of using a powerful reflex suppressing drug in a patient who had strong uncontrolled muscle contractions (spasticity) on one side of the body, while he perceived the other side of the body as normal. The drug acted on both sides of the body and completely eliminated the monosynaptic reflexes in the apparently healthy side. The patient, however, did not notice any changes in the way he moved the limbs on that side. This semi-anecdotal observation suggests that maybe monosynaptic reflexes are not that necessary for the normal motor function.

Another group of reflexes, for which the neural pathways have been described with sufficient confidence involve *oligosynaptic* reflexes. This group unites reflexes with two or three central synapses. Among the best known reflexes in this group are those originating from the primary endings of the muscle spindles (Figure 3.21) and from Golgi tendon organs (Figure 3.22). The Ia-afferents from muscle spindles located in a certain muscle make projections on a group of small neurons (Ia-interneurons) in the spinal cord, which in turn make inhibitory projections on alpha-motoneurons innervating a muscle with action opposing that of the parent muscle of the spindles, the so-called *antagonist muscle*. This disynaptic reflex is called *reciprocal inhibition*. Golgi tendon organs have oligosynaptic reflex projections on alpha-motoneurons that innervate both the parent muscle and its antagonist muscle. The homonymous projections are inhibitory; they involve one interneuron (Ib-interneuron). The projections to the antagonist muscle are facilitatory; they involve two interneurons. Muscle responses produced by oligosynaptic reflexes are also typically phasic. Their latencies are similar to those of monosynaptic reflexes, just a bit longer, corresponding to more time spent on synaptic transmission (on average, about 0.5 ms is lost at each synapse).

All the reflexes with undefined neural pathways are assumed to involve more than two to three central synapses and are called *polysynaptic*. These reflexes

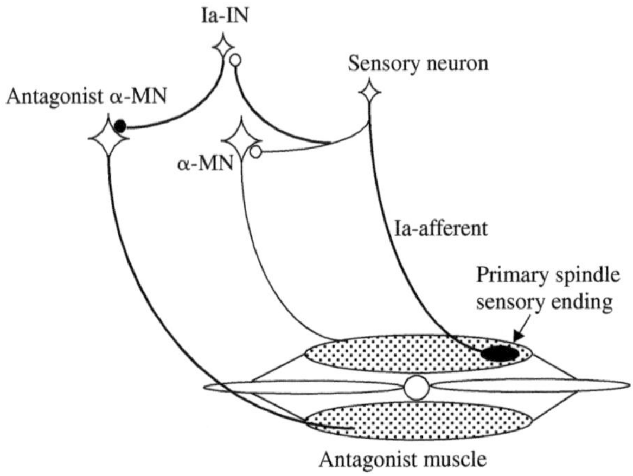

Figure 3.21. A scheme of the disynaptic reciprocal inhibition loop. Activation of the primary spindle sensory endings in a muscle results in action potentials traveling along the Ia-afferent fibers. They make excitatory projections (the open circle) on small interneurons (Ia-interneurons), making inhibitory projections (the filled circle) on alpha-motoneurons, which send their axons to a muscle that opposes the one containing the spindles. Symmetrical projections extend from the spindles within the antagonist muscle to the first one.

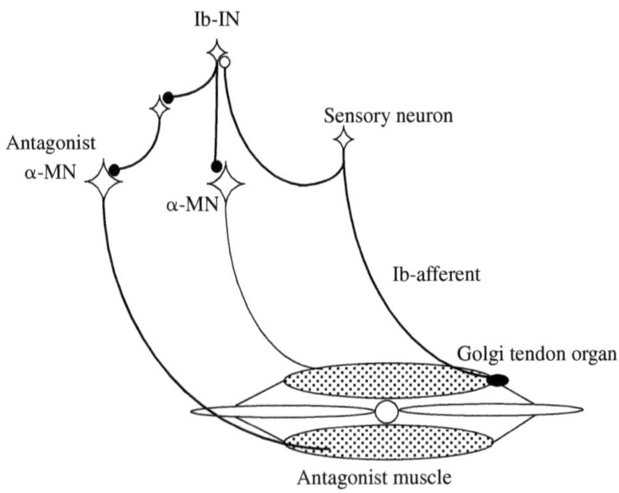

Figure 3.22. Activation of Golgi tendon organs at the junction between the fibers of a muscle and its tendon leads to two effects. The signals travel along Ib-afferents and project in the spinal cord on small Ib-interneurons. These interneurons have direct inhibitory projections on alpha-motoneurons innervating the original muscle. In addition, Ib-interneurons inhibit another group of interneurons that make inhibitory projections of alpha-motoneurons innervating an antagonist muscle. Inhibition of inhibition results in a net excitatory effect of the antagonist alpha-motoneurons. Excitatory synapses are shown with open circles; inhibitory synapses are shown with filled circles.

typically show longer delays between the stimulus and the response, and they can be both phasic and tonic. Among the phasic polysynaptic reflexes let me mention the flexor reflex and the crossed extensor reflex. They originate from various peripheral receptors united under the name *flexor reflex afferents*. If these receptors or their afferent fibers are stimulated within an area of an extremity, a reflex response is seen in most flexor muscles of the extremity. It is commonly accompanied by a reflex response seen in extensor muscles of the contralateral extremity. Another polysynaptic reflex that we are going to consider in more detail a little later is the tonic stretch reflex; as the name suggests, this reflex is tonic.

There is one more group of semi-automatic actions in response to sensory stimuli that are sometimes viewed as reflexes and sometimes as voluntary actions. Imagine that you hold a position in a joint against an external load by activating a muscle. You are given an instruction: "If a change in the external force happens, try to bring the joint to its original position as quickly as possible." Then, without any additional warning, such a perturbation comes, for example, a load increase stretching the activated muscle. Typical muscle responses are illustrated in Figure 3.23. If the perturbation is strong, it can lead to a monosynaptic response in the muscle (M_1). This response is followed by two, not always well-differentiated, responses that come at the latencies of about 50–60 and 70–90 ms, respectively. They are referred to as M_2–M_3. Later, a voluntary reaction can be seen, typically after a delay of about 150 ms.

Imagine now that the instruction has been changed into "let the joint move, do not react." A similar perturbation would produce a similar M_1 but much smaller M_2–M_3. Are M_2–M_3 responses reflexes? On one hand, they come at a rather short time delay. On the other hand, they can be modulated by the instruction,

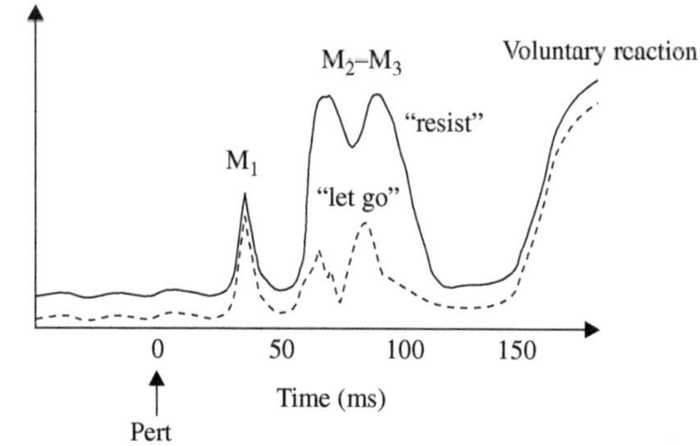

Figure 3.23. A quick stretch of a muscle leads to a sequence of effects. The earliest response (M_1) is likely to be of a monosynaptic origin (see Figure 3.20). Then, two (sometimes poorly differentiated) responses come at latencies under 100 ms (M_2 and M_3). Voluntary reaction is seen after about 150 ms. A change in the instruction from "resist" to "let go" does not affect M_1 but leads to a dramatic drop in M_2–M_3.

which makes them look much more like a voluntary action. Differences in the understanding of these reactions have led to many different terms being used to address them, including the mentioned M_2–M_3, long-loop reflexes, trans-cortical reflexes, functional stretch reflexes, triggered reactions, and pre-programmed reactions.

Without going into too much detail, let me suggest my personal understanding of the nature of these reactions, which is shared by some but not all of my colleagues. These reactions (let me address them as pre-programmed reactions) represent voluntary corrective actions to expected perturbations that are prepared by the controller (the central nervous system) in advance and triggered by adequate sensory stimuli. Such reactions have been described in association with a variety of functional motor actions such as standing, walking, reaching, grasping, and speaking (reviewed in Forssberg et al. 1976; Chan and Kearney 1982; Latash 1993). Shall we call them *reflexes* or not? This does not seem to be an important question.

End of Digression #6

Sir Charles Sherrington viewed muscle reflexes as a set of built-in mechanisms that were modulated to produce natural movements (Sherrington 1910). In particular, he viewed locomotion as a consequence of alternating flexor and crossed extensor reflexes in the limbs. He was also the first to describe the tonic stretch reflex (together with Liddell; Liddell and Sherrington 1924), that is, a change in the muscle activation resulting from slow changes in the muscle length. If a muscle is slowly stretched, its level of activation increases leading to higher muscle forces compared to those expected from the same muscle under unchanged activation level. In other words, the tonic stretch reflex modifies the natural relation between muscle force and length. It modifies muscle elastic properties described in Digression #1.

Figure 3.24 illustrates a dependence between muscle force and length during slow muscle stretch in an experiment, when there are no changes in signals from the brain. This was achieved in animal experiments by transecting all the pathways leading from the brain to the spinal cord and placing a stimulator at the "spinal end" of the cut fibers (Matthews 1959). In human experiments, similar results were obtained, when a person was asked "not to interfere voluntarily" (Feldman 1966). This is a less convincing method since "not interfering" proves to be a hard instruction to follow: Humans typically tend to interfere and oppose external forces. So, to avoid controversy, let us consider an animal experiment, similar to those performed by Peter Matthews in the late 1950s.

Let us imagine that a muscle at a relatively short length is completely relaxed, that is, it does not show signs of electrical activity. Then, the muscle would resist an externally imposed stretch by generating opposing forces due to its (and also the tendon's) peripheral elasticity. At some value of the muscle length, the muscle will show signs of increasing electrical activity. This length value

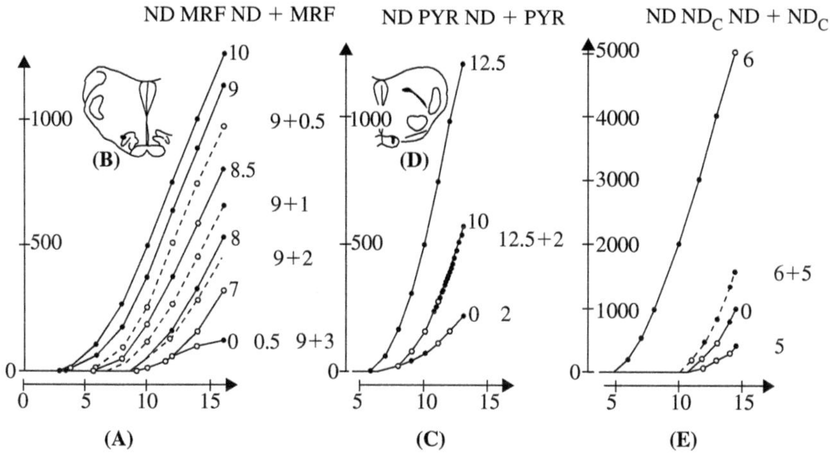

Figure 3.24. Dependences of muscle force (plantaris for A and C; gastrocnemius for E) on muscle length in experiments on decerebrate cats. Note that a change in the level of stimulation of different supraspinal structures could lead to a shift of the dependence nearly parallel to itself. Stimulation was applied to ipsilater (ND) and contralateral (NDc) Deiters' nuclei, pyramidal tract (PYR), and mesencephalic reticular formation (MRF). Reproduced by permission from Feldman AG, Orlovsky GN (1972) The influence of different descending systems on the tonic stretch reflex in the cat. *Experimental Neurology* 37: 481–494.

(λ) was termed the threshold of the tonic stretch reflex. If we continue to stretch the muscle beyond λ, it will oppose the stretch more vigorously, as a much stiffer spring. In Figure 3.25, this is reflected in the much steeper force–length curve after λ compared to the curve before λ. Let us stop stretching the muscle at a certain length $L_{1,}$ at which it generates a force F_1. This combination of length and force values can be referred to as an equilibrium-point (EP) of the system: "the muscle with its reflex connections and the external force." If a small, transient change in the external force moves the muscle away from the $\{F_1, L_1\}$ point, the muscle will return as soon as the transient force change is over. Note that for a given external force and fixed λ, there is only one muscle length value, at which the muscle can be in an equilibrium. We can now define the tonic stretch reflex as a neural mechanism that ensures a particular relation between muscle length and active muscle force (the curve to the right of λ in Figure 3.25). This definition makes this reflex a major posture-stabilizing mechanism.

Now, let us consider the following question: What will happen if some external with respect to the muscle factor moves the muscle from its EP to a new length value? If the force–length curve remains unchanged, the new length value, for example, L_2 in Figure 3.26, would correspond to a different force produced by the muscle F_2. The difference between this force and the original external force F_1 will tend to move the muscle back to its equilibrium position

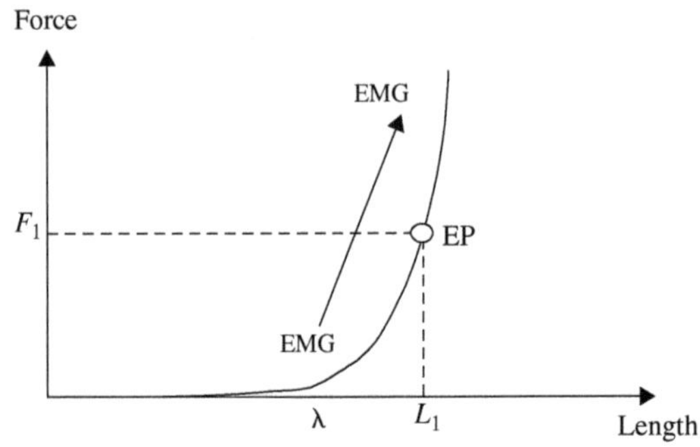

Figure 3.25. A slow stretch of a muscle leads to an increase in its force due to passive properties of the "muscle + tendon" complex. At a certain length, the muscle shows signs of activation (electromyogram, EMG). This length is called the threshold of the tonic stretch reflex (λ). With further stretch, the level of force (and EMG) increases. If a muscle acts against a constant external load (F_1), it has only one equilibrium length (L_1) that corresponds to its active force exactly equal to F_1. The combination of muscle force and length, at which it is at an equilibrium, is called the equilibrium point (EP).

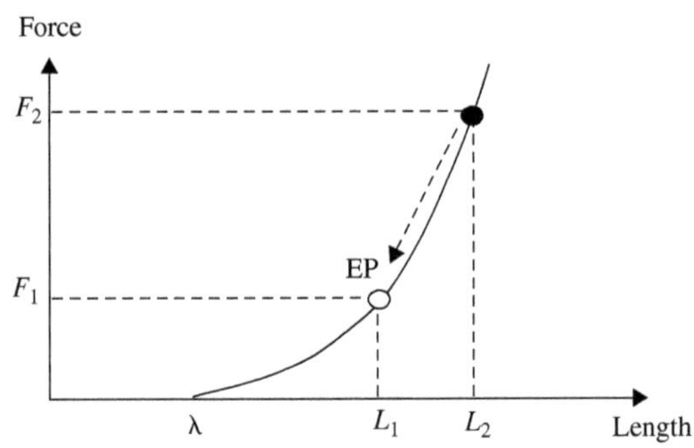

Figure 3.26. If a muscle is at an equilibrium point (EP), an external force can stretch it to a new length (L_2) corresponding to a new force level (F_2). However, when the force is removed, the muscle will return to EP (posture-stabilizing mechanism). If the controller wants the muscle to stay at this new length, its has to deal with the posture-stabilizing mechanism.

$\{F_1, L_1\}$. If the system wants to stay at the new position after the external force is removed, it has to deal with this posture-stabilizing mechanism. There seem to be two ways of dealing with the problem. First, to produce forces (e.g. by other muscles) that would counteract the posture-stabilizing forces produced by the tonic stretch reflex. Second, to do something with parameters of the tonic stretch reflex such that it stops trying to move the muscle back to its old equilibrium position.

Von Holst and Mittelstaedt (1950/1973) considered this problem and preferred the second solution. This is not surprising. The first solution is rather wasteful—one always needs to apply excessive force to prevent the joints from snapping back to their original equilibrium positions. Besides, the mentioned observations of Wachholder and Altenburger suggest that muscles can indeed be relaxed at different joint positions. So, von Holst and Mittelstaedt introduced what they called the principle of re-afference. They suggested that afferent signals from proprioceptors could be referred by the controller to different joint positions (different muscle length values), rather than inhibited when an intentional movement is produced. Then, different positions can be stabilized by the tonic stretch reflex mechanism, depending on where the controller wants the joint to be in an equilibrium. Actually, Von Holst and Mittelstaedt formulated the re-afference principle in a more general way, implying under posture-stabilizing mechanisms not only the tonic stretch reflex but all the homonymous and heteronymous reflexes of agonist and antagonist muscles involved in the task.

Let us consider this in greater detail. In Figure 3.27, deviations of muscle length from L_1 induce afferent signals that produce changes in the muscle force that tend to move the muscle back to L_1. According to the principle of re-afference, to perform a voluntary movement to a new muscle length value L_2, the controller needs to make that value an equilibrium. Then, deviations from L_2 start to produce afferent signals leading to changes in the muscle force. In particular, the original length L_1 becomes a deviation from the new equilibrium length value and leads to force production that moves the muscle to L_2. The very same tonic stretch reflex mechanism that used to stabilize L_1 is now producing a movement from L_1 to L_2. However, there is one unanswered question: How does the controller (the brain) change the EP, that is, the combination of muscle length and force values

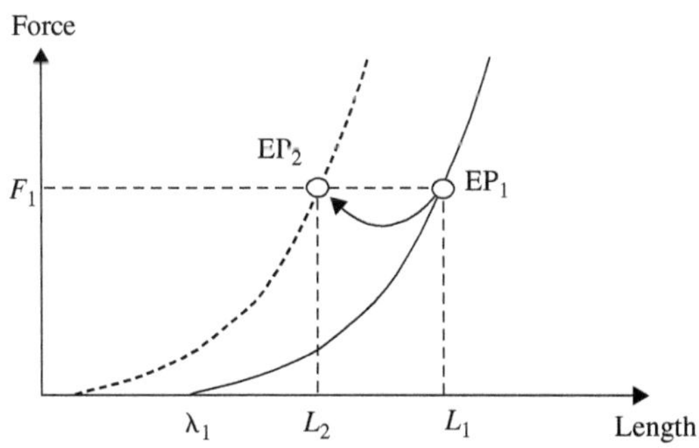

Figure 3.27. Originally, a muscle is in an equilibrium point (EP$_1$) at a certain length L_1, acting against a constant force F_1. Point EP$_2$ corresponds to another position (muscle length L_2). It is a deviation from EP$_1$, and posture-stabilizing mechanisms would try to bring the muscle back to EP$_1$ if it somehow happened to be in EP$_2$. However, if the neural controller can change a "command" to the muscle and shift its force–length characteristic (the dashed curve) such that EP$_2$ becomes the new equilibrium point, EP$_1$ will become a deviation from that point, and the same posture-stabilizing mechanisms will produce a motion to EP$_2$.

at which the muscle reaches an equilibrium? An answer to this question forms the core of the EP-hypothesis.

3.4.2 Equilibrium-Point Control of Simple Systems

The EP-hypothesis was formulated in the mid-1960s based on a series of experiments on human subjects, who were given a rather vague instruction (Asatryan and Feldman 1965; Feldman 1966; reviewed in Feldman 1986). The subjects were asked to occupy a position in the elbow joint against a certain load acting against the joint flexor muscle group. Then, the load was smoothly removed. The subjects were asked "not to interfere voluntarily," or in other words, "to allow the arm to move as it naturally moves." This was not an easy instruction: It is much more natural for humans to react to a change in the load, either to resist the induced motion of the arm or to relax completely. However, after a few trials, subjects learned "not to think about the arm and the load." Then, a reproducible behavior of the arm was observed.

Figure 3.28 illustrates typical results of such an experiment. Imagine that initially the subject held the elbow joint at an angle corresponding to a certain length of the flexor muscles, L_1. The external load was F_1. The combination of the two defined the initial EP of the system (the top open circle in Figure 3.28). When the load was decreased to F_2, the joint moved to a new position corresponding to a new muscle length, L_2. These two values defined a new EP. If a different change in the load was applied, for example to F_3, the muscle moved to L_3, and so forth. According to the assumption, all the EPs correspond to the same unchanged voluntary "command" but different external loads. I use "command" in quotation marks since this notion has not been defined yet. These points can be connected by a smooth line, which has been termed an invariant characteristic (IC) of the muscle.

The same experiment can be performed starting from another elbow joint position and using the same set of external loads. Apparently, to stay at a different position against the same initial load, a different effort is required corresponding to a different "command." Unloading procedures can be repeated starting from the new EP; they will result in a new IC (IC_2, filled circles in Figure 3.28). A major experimental finding was that ICs did not intersect but shifted more or less parallel to each other for experiments starting from different initial EPs. This result led to a very important conclusion that "command" might be associated with selecting a particular IC, while actual values of muscle force and length, as well as the level of muscle activation, depend on both the command and the external load.

When muscle activity is recorded in a "do-not-interfere-voluntarily" experiment with smooth unloadings, it shows a decrease with a decrease in the muscle length, and at some value of muscle length, muscle activity disappears. This value of muscle length was termed the threshold of the tonic stretch reflex, and Feldman introduced the Greek letter λ to refer to this value; hence, the hypothesis has

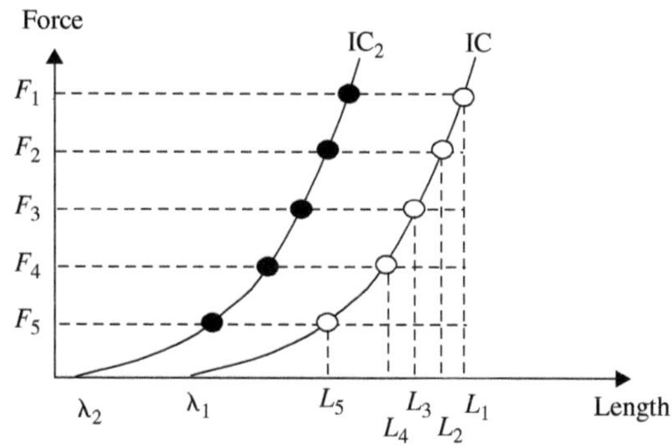

Figure 3.28. An illustration of an experiment with reconstruction of muscle invariant characteristics (ICs). A person is asked to occupy a position in a joint and act against a constant load. These correspond to muscle length L_1 and load F_1. Unloading the muscle leads to a sequence of equilibrium points (the open circles). Their interpolation is an IC. If the same experiment is repeated starting from a different length–load combination, a new IC can be reconstructed (the filled circles).

also been commonly called the λ-model. A value of λ defines a particular IC and, therefore, may be viewed as a measure of neural command. For a fixed λ, the muscle does not show activity when its length is shorter than λ; then, its spring-like behavior is defined by the peripheral elasticity of the muscle and its tendon. When the muscle is longer than λ, it shows electrical activity, which increases with the muscle length.

These experiments, as well as earlier experiments on animals by Peter Matthews, allow to draw a few conclusions. First, an unchanged voluntary command to a muscle does not encode a value of muscle force, or of muscle length, or a level of muscle activation. All these variables change depending on the command and external forces acting on the muscle. Second, a command may be associated with a value of λ. In other words, voluntary control of a muscle may be described with only one parameter! This was a revolutionary hypothesis given the apparent complexity of the muscle and its reflex connections with neuronal structures in the spinal cord. This apparent oversimplification raised the first of a series of objections that the EP-hypothesis has faced from the research community.

Note that selecting a command to a muscle is equivalent to selecting its IC, that is, a line consisting of an infinite number of possible combinations of muscle force and length. Which one of those points will be realized? This problem may be viewed as another example of the problem of redundancy, but choice in this case is made by the external force field.

Within the EP-hypothesis, movements may result from two different sources. On the one hand, a movement may be a consequence of a change in the external load, while the person does not change the voluntary command to the muscle or, in the experiments described, “does not interfere voluntarily.” A change in the

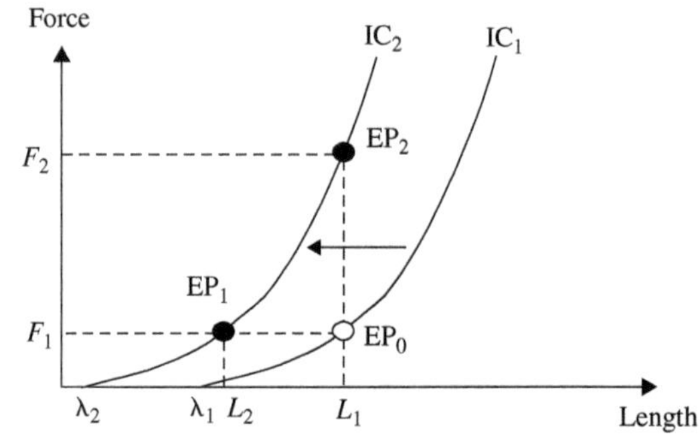

Figure 3.29. According to the equilibrium-point hypothesis, movement and force generation are different peripheral consequences of the same control process. A shift of the control variable λ leads to a shift of the invariant characteristic (IC). If the muscle acts against a constant external force, a movement will occur from the original length L_1 to a new length L_2. If a muscle presses against a stop (isometric conditions), the same shift of λ will lead to a change in the muscle force from F_1 to F_2.

load while keeping the command constant results in a new combination of muscle force and length along the same IC, that is, a movement. One may address such a movement as "involuntary."

On the other hand, a voluntary shift in λ can also lead to a movement. For example, in Figure 3.29, a shift from λ_1 to λ_2 is expected to lead to a movement from L_1 to $L_{2,}$ if the external load stays constant at F_1. On the other hand, if a movement is blocked by an external stop, the same change in λ will lead to a change in the muscle force without a movement. So, within the λ-model, voluntary movements and voluntary force changes are different peripheral consequences of basically the same central control processes. Commonly, voluntary motor actions are associated with a shift in λ, while the external load changes as well, such that both factors influence the movement.

Until now, the description has been limited to voluntary control of a single muscle. It is rather easy to generalize it for a system of two muscles that act at a joint and produce joint torques in opposite directions, for example a flexor and an extensor muscle. Figure 3.30 illustrates ICs for two such muscles with opposing actions. It uses a different pair of mechanical variables, torque and angle (rather than force and length), that are more appropriate to describe rotational actions. Each muscle is controlled with its own command variable, λ_f for the flexor and λ_e for the extensor. These commands define the positions of the ICs for each muscle. A pair of such characteristics define an overall joint characteristic shown by the bold line in Figure 3.30. Equilibrium state of the joint and its mechanical behavior will also depend on the external torque. For example, if there is a constant external torque acting on the joint in Figure 3.30, the system will be at an equilibrium at a combination of torque and angle values $\{T_0, \alpha_0\}$.

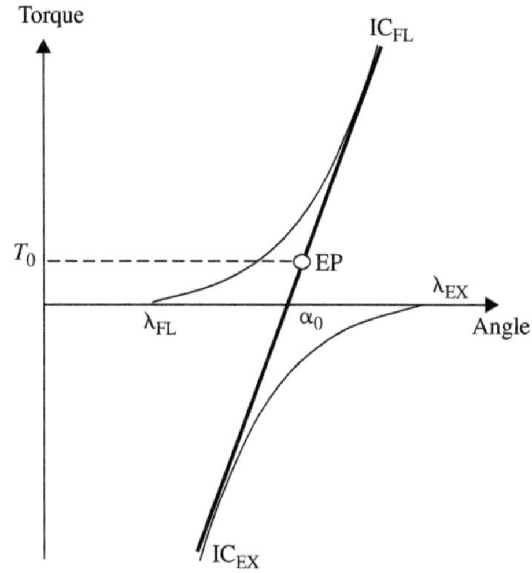

Figure 3.30. Imagine a joint spanned by two muscles with opposing actions (a flexor and an extensor). Their invariant characteristics (IC_{FL} and IC_{EX}) can be drawn on a torque-angle plane (extensor torques are assumed negative). The behavior of the joint will be defined by the algebraic sum of the two characteristics (the thick, straight line). Its equilibrium point (EP) will be defined by both the position of this line and the external torque (T_0).

Voluntary joint motion and/or torque production result from shifts of the two λs. It is easy to observe that if both λ_f and λ_e shift in the same direction along the angle axis the mechanical characteristic of the whole joint, which is the algebraic sum of the two muscle characteristics, shifts parallel to itself along the angle axis without a change in its shape (Figure 3.31). On the other hand, if λ_f and λ_e shift in opposite directions, there is little change in the location of the joint characteristic but a major change in its slope. These shifts of the joint characteristic are also illustrated in Figure 3.31.

The availability of two opposing muscles allows to change joint behavior in two ways. First, one can try to activate one muscle and relax the opposing muscle (illustrated by unidirectional shifts of the two λs). This will lead to the motion of the joint in the direction corresponding to the shortening of the activated muscle. Second, one can try to activate both muscles simultaneously. This will not move the joint by much but will "stiffen" it. To reflect these two modes of joint control explicitly, another pair of variables can be used that is equivalent to the $\{\lambda_f; \lambda_e\}$ pair. These variables have been referred to as reciprocal command, *r*, and coactivation command, *c* (Feldman 1980, 1986). They can be defined as $r = (\lambda_f + \lambda_e)/2$; $c = (\lambda_f - \lambda_e)/2$.

A common misconception about the EP-hypothesis is that the *r* command to a joint specifies its equilibrium position, while the *c* command specifies its stiffness. It is more appropriate to say that, because of the threshold nature of the

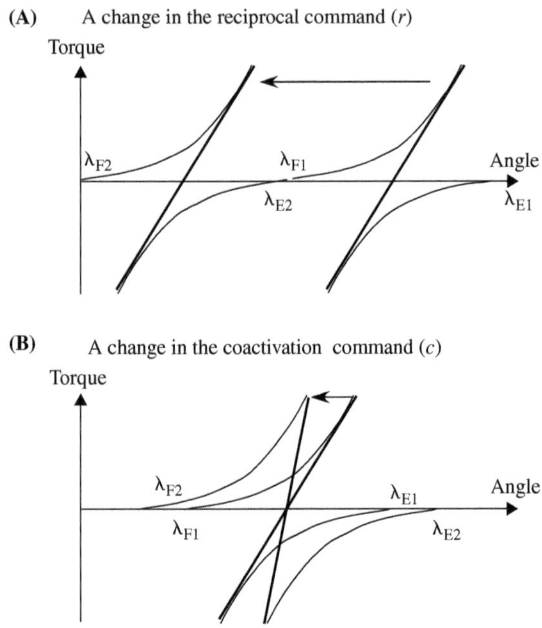

Figure 3.31. Control of a joint can be described with two parameters λ for the flexor and extensor muscles (λ_F and λ_E) or with an equivalent pair {*r*, *c*}. (A) A change in the reciprocal command *r* corresponds to unidirectional shifts of λ_F and λ_E, leading to a shift of the joint characteristic without a major change in its slope. (B) A change in the coactivation command *c* corresponds to shifts of λ_F and λ_E in opposite directions, leading to a change in the slope of the joint characteristic without a major change in its location.

tonic stretch reflex, the *r* command influences location of an angular range, in which muscles can be activated. The *c* command defines the size of the angular range where both muscles are active; in more mechanical terms, it defines the effectiveness of external load in moving the joint away from position *r*. As a result, the equilibrium position of a joint acting against a nonzero load is always influenced by both *r* and *c*.

3.4.3 Three Basic Trajectories within the Equilibrium-Point Hypothesis

For now, let us assume that a hypothetical central controller can set values of λs for each muscle, and correspondingly, values of {*r*, *c*} for each joint independently of other factors. As we will see further, this is generally not true, but let us accept this assumption for the purpose of this subsection. Under this assumption, control of a muscle can be adequately described with a time function λ(*t*). This time function may be viewed as a control variable supplied to the segmental apparatus controlling the muscle. We can call the time profile of λ a *control trajectory.*

If at any moment of time, t_0, the current value of λ is frozen at $\lambda(t_0)$, and the system is allowed to reach an equilibrium, it will come to rest at a combination of muscle length and force (l_{EP}, F_{EP}), corresponding to the current value of the external load. Such a combination is as an equilibrium point EP_0. In other words, a control trajectory λ(*t*) may be associated with a time sequence of EPs, EP(*t*) or an *equilibrium trajectory*. If one is interested only in movement kinematics, EP(*t*) may be represented as a time sequence of only the length coordinate, $l_{EP}(t)$.

Note that while the control trajectory is assumed to be specified centrally the equilibrium trajectory emerges with an equally important role played by the external force field. For example, if a movement is practiced against a constant external load, repeating the same control pattern (same control trajectory) can be expected to lead to the same equilibrium trajectory but only if the load does not change. Generating the same control trajectory against a changing load would result in a different equilibrium trajectory. This feature was used in a series of studies with the reconstruction of time patterns *r*(*t*) and *c*(*t*) using external loads that could change smoothly and unexpectedly (Latash and Gottlieb 1991a,b).

For a given equilibrium trajectory, actual behavior of the system, its *actual trajectory*, *l*(*t*), will depend on many factors that may be united under a not-very-precise notion of *dynamics*. These factors include, in particular, the external force field, the mechanical properties of the moving segments, the time delays in the reflex arcs that bring about changes in muscle activation via the tonic stretch reflex loop, the properties of the transformation from muscle activation to force generation, etc.

Unfortunately, parameters that describe the mechanics of the system and all the mentioned processes can be estimated only very crudely. Since mechanical behavior is the only reliable observable in human experiments, the complexity of the transformation from the equilibrium trajectory to an actual trajectory has been a major stumbling point in quantitative analysis of control patterns associated with voluntary movements, particularly with fast voluntary movements. Existing estimates of control trajectories are few (Latash and Gottlieb 1991a,b; Latash 1992a; Gomi and Kawato 1996) and have been rightfully criticized as being based on crude and inadequate models of the last transformation, between EP(*t*) and *l*(*t*) (Gribble et al. 1998).

3.4.4 Equilibrium-Point Control of Multi-Muscle Systems

Generalization of the EP-hypothesis to multi-muscle systems, particularly to redundant systems is not so trivial. It requires another important step. One needs to realize that the previous description of the EP-hypothesis is a major simplification in a couple of important aspects. One of them is that EP control can be introduced for systems of different complexity, not built using the simpler building bricks of single muscle control schemes. Recall a major point made by Gelfand and Tsetlin that biological structural units are built hierarchically on sets of elements that are also biological structural units, which are not simpler than the "bigger" structural units.

Imagine that a person puts a fingertip into a certain point of the accessible space and balances an external force (pushes against the palm of another person, the experimenter). The subject of this mental experiment is once again instructed "not to intervene voluntarily," that is, not to change his/her voluntary motor command (whatever it is) when there are changes in the external load. We assume that the subject is able to follow this instruction. The experimenter may now push the fingertip in different directions by changing the original force. Apparently, the location of the fingertip, its force, as well as joint angles and torques will all change while the central command stays presumably the same. Muscle activation patterns will change as well because of the changes in the length of the muscles and the action of the tonic stretch reflex.

The pattern of the dependence between external force and fingertip coordinates is defined by an involuntary mechanism that may be considered reflex. Let me call it a generalized displacement reflex (GDR). GDR is a multi-joint, three-dimensional expansion of the notion of the tonic stretch reflex. Its action can be characterized by a set of parameters. Fixed voluntary command corresponds to unchanged parameters of the GDR, while a change in a voluntary command may be viewed as a change in some of these parameters. Is defining parameters of GDR equivalent to defining parameters of the tonic stretch reflexes for all the muscles of the limb? In other words, can a command to a multi-joint limb be adequately and unambiguously described in the "lambda-space"?

Currently, there is no unambiguous answer to this question. A reference body configuration hypothesis introduced by Anatol Feldman and Mindy Levin (1995) suggests that this is the case: For any movement, there is a reference body configuration that defines equilibrium states in all joints. This view is compatible with several other hypotheses on motor control including, in particular, the posture-based *knowledge hypothesis* advanced by the group of David Rosenbaum (Rosenbaum et al. 1993, 2001). This latter hypothesis assumes that performing any movement starts with selecting a target posture, and then a trajectory to this posture is computed based on certain optimization criteria. However, it is also possible that selecting a position of the endpoint does not imply selecting a posture of the limb. Because of the kinematic redundancy, the same endpoint position may correspond to many joint postures. Moreover, many experiments, starting from the mentioned classical Bernstein's study of blacksmiths, have shown that trying to repeat the same motion (position) of the endpoint leads to substantial variability in the trajectories of the joints (postures). So, it may be that controlling motion of the endpoint uses the same principles as controlling motion of a single joint but different control variables (see section 8.4).

Another simplification was in the introduction of λ as a purely central control variable. A closer look at the neurophysiological pathways that lead to changes in the activity of alpha-motoneurons suggests that λ is a complex variable. Whether a particular motoneuron generates an action potential or not depends on its membrane potential reaching or not reaching the threshold value. This threshold property of neurons is very important. In particular, it allows control structures to send a signal to a neuron that will not by itself activate it but will change its reaction to signals that may come from other sources, for example, from peripheral receptors. Changing the average resting membrane potential for a motoneuronal pool leads to a change in the relation between the length of the muscle innervated by the pool and the average activity of all the motoneurons within the pool (tonic stretch reflex). Such a dependence is illustrated in Figure 3.32. For a given state of the motoneuronal pool, when the muscle is short, the activity of its spindles is insufficient to bring about recruitment of motoneurons in the pool. When the muscle is longer, its spindles show higher activity and contribute more to muscle activation via the tonic stretch reflex loop. At some muscle length, signals from the spindles are just large enough for the muscle to shows first signs of activation. So, changing the resting membrane potential leads to changes in the threshold value of muscle length, when the motoneurons start to get recruited. Let us consider several potential contributors to the threshold.

Imagine that the threshold is set at a certain value of muscle length. If the muscle length is held constant, muscle activity is expected when the length is larger than the threshold. If a muscle passes through a certain length value at a nonzero velocity, however, the well-known sensitivity of the primary endings of

$$\lambda = \lambda^* + \mu V + f(t) + \rho$$

Figure 3.32. The dependence between muscle length and the number of recruited alpha-motoneurons (α-MNs) is defined by a parameter λ. This parameter depends on several factors, including a direct central contribution (λ*), a velocity-dependent factor (μ*V*), a history dependent factor, *f*(*t*), and effects from other muscles, ρ. Modified with permission from Feldman AG, Latash ML (2005) Testing hypotheses and the advancement of science: recent attempts to falsify the equilibrium-point hypothesis. *Experimental Brain Research* 161: 91–103 © Springer.

muscle spindles to velocity (see Digression #3) may be expected to contribute to the excitation of the motoneurons. In particular, if the muscle is being stretched, it is more likely to show signs of activation, while passing through a certain length than if the same muscle passes through the same length while shortening. This can be formally expressed as a dependence of the threshold length for muscle activation on its velocity. Feldman (1986) suggested using a linear function to express this dependence with a coefficient μ that can be adjusted by hierarchically higher structures within the central nervous system.

There are other contributors that can change the membrane potential of alpha-motoneurons directly or modulate the efficacy of other signals to the motoneurons. These include, in particular, effects from proprioceptors in other muscles, as well as from cutaneous and subcutaneous receptors. In addition, tonic stretch reflex is known to show effects of history. Equilibrium length reached by a muscle against a load depends not only on the magnitude of the load but also on the time changes in the load leading to its final magnitude (Feldman 1979). To summarize, the threshold for activation of alpha-motoneurons depends on at least four factors: $\lambda = \lambda^* + \mu V + f(t) + \rho$, where the first term on the right side reflects the central contribution to λ, the second term reflects its velocity-sensitivity, and the third and fourth terms reflect its dependence on history and on signals from other muscles, respectively.

This scheme may seem awkward: It creates an impression that the controller that wants to ensure a certain shift in λ has to be able to predict μV, $f(t)$, and ρ to prepare an adequate λ^*. However, this only appears to be a problem. It originates from a rather widespread misconception that one needs to precompute everything to control a movement. Everyday experience suggests that this is not the case. When we drive a car, we do not precompute values of our control variables (angle of the steering wheel and pressure on the gas/brake pedals) that ultimately define where the car moves and at what speed. Our actions on the steering wheel and the pedals have indirect effects on the angle of the wheel and torque applied to the axels. These will depend on the friction between the wheels and the road, wind, incline of the road, and other factors. Do we have in our brain an internal model of how forces applied to the gas pedal translate into torques applied to the axel and to changes in car motion? I would say that this is very unlikely. We are much more likely to have a set of rules, maybe involving an internal look-up table and a set of adjustable gains that tell us what kind of actions lead to what kind of effects, given conditions of driving. If our action leads to an undesirable effect, we introduce corrections. Along similar lines, neural structures do not have to precompute effects of sources other than λ^* on λ. They simply may know, based on experience, what kind of λ^* shifts are adequate to produce required motor effects.

In other words, central shifts in the activation thresholds for muscles are a means the nervous system uses to produce movements (and forces). The threshold values (λ) or their central components (λ^*) should not be confused with internal representations of goals of motor actions. Goals are identified by the central nervous system using adequate physical variables that describe actions of the body in the environment. These goals do not need to be recomputed into internal variables or modeled in any way. Similarly, motor errors are the differences between the actual state of an effector in the environment and its desired state. The difference between the actual and the threshold muscle lengths is not a motor error but a physical cause leading to particular effects of signals from proprioceptors.

3.4.5 The Mass–Spring Analogy and Other Misconceptions

As mentioned earlier, under a particular instruction to subjects, changes in a joint position are graded with the magnitude of the external load. In this respect, the joint seems to behave like a physical mass–spring system. The term *spring-like behavior* refers to this special case and implies a similarity in behaviors of biological and physical systems that otherwise have little in common. In some cases, such an analogy is taken literally, when muscles are modeled as springs with parameters (most often, stiffness and damping) determined by the level of muscle activation.

Spring analogies and models have played both positive and negative roles in motor control research. When the EP-hypothesis was introduced, the spring analogy was applied not to an isolated muscle but to the intact neuromuscular system. This analogy was based on a body of experimental data that could be described with a particular parameter, the threshold of the tonic stretch reflex. This parameter reflects an interaction between descending signals ("command") and sensory proprioceptive feedback on the membrane of the motoneurons. Drawing analogies between this parameter and a change in such spring property as resting length was helpful in introducing notions that had been absent in the motor control lexicon—the concept of an EP, the possibility of spring-like behavior of intact muscles, and the notion of equifinality. However, muscles are not perfect springs, and λ is not the resting length of a perfect spring.

Note that in humans or animals with lost proprioceptive sensitivity, this parameter (the threshold of the tonic stretch reflex) does not exist, and the controller needs to learn a new mode of control. As demonstrated in the classical experiments by Emilio Bizzi and his team at MIT, the central nervous system of a monkey after deafferentation (cutting all the sensory input axons into a number of spinal segments) is able to generate accurate movements (Polit and Bizzi 1978, 1979; Bizzi et al. 1982). The spring-like properties of the deafferented muscles were able to show some of the typical features of EP control, for example, equifinality after short, transient perturbations. However, the mode of control was different from that in intact animals because of the lack of the tonic stretch reflex.

More generally, the spring analogy helped to predict that the intact system, including muscles, reflexes, and the control levels can display equifinality following not-very-large transient perturbations, if the controller does not react to such perturbations. The empirical confirmations of this prediction (Schmidt and McGown 1980; Rothwell et al. 1982; Jaric et al. 1999) should be considered as supporting evidence for the EP-hypothesis. Note that humans are more likely to react to perturbations, which may lead to violations of equifinality. Some of these reactions can be suppressed voluntarily, while others may persist, even if the subject is honestly trying "not to interfere." In particular, the mentioned pre-programmed reactions (see Digression #5) can be modulated by instruction within a certain range, but they are rarely turned completely off. As a result, certain perturbations always cause reactions by the subject that lead to violations of equifinality.

Indeed, violations of equifinality have been reported under particular experimental conditions involving the action of the Coriolis force (Lackner and DiZio 1994; DiZio and Lackner 1995) and other destabilizing force fields (Hinder and Milner 2003). Does this mean that the EP-hypothesis is wrong? No, these findings only mean that the simplified mass–spring analogy has a limited range of applicability to the human neuromotor system.

In recent years, the negative role of the mass–spring analogy has become obvious. In some of the simulations of movements, the spring analogy has been taken literally to represent muscle properties (Gomi and Kawato 1996; Schweighofer et al. 1998; Bhushan and Shadmehr 1999; Popescu et al. 2003), although this leads to misrepresentation of some of the basic properties of the neuromuscular system (reviewed in Feldman et al. 1998; Feldman and Latash 2005). In a number of studies, neuromuscular properties have been oversimplified, resulting in linear second-order models. Most arguments against the EP-hypothesis accepted as an axiom that the hypothesis assumes that the neuromuscular system always behaves like a spring with its characteristic property of equifinality. This resulted in a basic misconception that equifinality is a fundamental property of the neuromuscular system within the EP-hypothesis (e.g. Lackner and DiZio 1994; Popescu and Rymer 2000, 2003; Hinder and Milner 2003).

In some of these papers, violations of equifinality have been claimed to refute the EP-hypothesis. For example, imagine a person sitting in a chair in complete darkness. This person is asked to point at light targets, which turn off immediately after the movement is initiated. So, the subject in this experiment cannot correct the trajectory based on visual information about the ongoing movement. Unexpectedly for the subject, the chair starts to rotate slowly about a vertical axis passing through the subject's trunk. This rotation starts so smoothly and slowly that the subject is unaware of it. When the chair rotates, attempts to move the arm to a target lead to new forces acting on the arm, the Coriolis forces proportional to both the velocity of the arm movement and the angular velocity of the chair. These forces are zero when the arm stops. Hence, they act transiently only during the motion. If the controller specifies an equilibrium position of the hand and does not correct it during the motion, the Coriolis forces should not affect the final position, that is, equifinality should be observed. In experiments, the arm deflected under the action of the Coriolis forces and then tended to move back toward the target, but always showed a residual error, that is, a violation of equifinality.

Does this result mean that the EP-hypothesis is wrong? Not necessarily. Equifinality is not a universally expected behavior within the EP-hypothesis. It is expected if all of the following three conditions are met: (1) The central command signals are unchanged under the perturbation; (2) Inputs into alpha-motoneurons from other sources are unchanged; and (3) The peripheral force-generating capabilities of muscles to activation signals are unchanged. In particular, violations of equifinality have been predicted and observed within the framework of the EP-hypothesis, when the third condition was violated (Walmsley et al. 2001). Those experiments used the so-called catch property of muscles (Burke et al. 1970, 1976). If a muscle is subjected to a strong brief electrical stimulation, over the immediately following short time, it generates higher forces to a standard input. This muscle property allows to predict violations of equifinality when a perturbation is introduced into a quick movement to a target by a brief electrical stimulation of one of the participating muscles. Experiments confirmed this prediction. In each particular case of violations of equifinality, additional analysis is required to understand why it occurred rather than claiming immediately that the EP-hypothesis has been falsified.

The EP-hypothesis was founded on the experimental data that forces produced by intact muscles are position-dependent. Springs were used as an example of a physical system that has an analogous property. Not all systems with position-dependent force generators have the property of equifinality. Experimental observations of violations of equifinality refute the simplified mass–spring model but not the EP-hypothesis.

It is time for me to accept part of the blame and say *mea culpa*! When I came to the United States in 1987, the EP-hypothesis had not been very well understood and was frequently blamed for its inability to handle a variety of topics studied by movement science. In particular, these topics included motor variability, patterns of muscle activation, and control of multi-joint movements. So, for the next few years, I was busy answering these criticisms by demonstrating that the EP-hypothesis was indeed able to handle motor variability (Latash and Gottlieb 1990), electromyographic patterns during fast movements (Latash and Gottlieb 1991a,b), and two-joint movements (Latash et al. 1999).

Another major criticism was that the control variables within the EP-hypothesis were unmeasurable during fast movements and, hence, the hypothesis was applicable only to postural tasks. To answer this criticism, a method was developed that allowed reconstructing patterns of the control variables $\{\lambda_f, \lambda_e\}$ or $\{r, c\}$ as time functions during movements (Latash and Gottlieb 1991c; reviewed in Latash 1993). The method was crude and based on—sorry to admit—a second-order, linear model of the joint. It did allow the reconstruction of time patterns of the control variables (and the *equilibrium trajectory*), but the shape of these patterns during fast movements was nonmonotonic (N-shaped) and was likely affected by the simplified model as was demonstrated later, using a much more sophisticated model (Gribble et al. 1988).

Muscle force–length properties are indeed highly nonlinear. Under certain external forces, nonlinearities (in particular, hysteresis) may be seen in the behavior of sarcomers, muscle fibers, tendons, connective tissues, whole muscle structure, alpha-motoneurons, and sensory feedback signals regulating activity of the motoneurons. In addition, these properties change with shifts of the threshold of muscle activation; such shifts can result from descending signals, as well as from interneurons that are responsible for intermuscle reflex interactions. So, it is a gross simplification to refer to this constellation of factors as defining "stiffness" and "viscosity" in spring models of the muscle-reflex system.

The central concept of the EP-hypothesis is that of threshold control: The relation between muscle force and length changes dramatically when the length becomes greater than the threshold of the tonic stretch reflex (see Figure 3.25). Systems that change their behavior in a threshold-like manner are nonlinear and cannot be considered linear even locally for small changes in variables: A small change in muscle length can lead to a poorly predictable, disproportional change in its force, if the length change happens to cross the threshold value. Since activation thresholds are expected to shift during voluntary movements, the joint angle ranges where muscle behavior becomes essentially nonlinear may appear anywhere in the biomechanical range. Therefore, neither physiologically nor mathematically, even for small changes in variables can one justify the standard decomposition of muscle forces into two additive, position- and velocity-dependent components as acceptable in all mass–spring models.

Note that even seemingly ideal physical springs can show violations of equifinality. Imagine a typical metal spring, for example, a small spring that is commonly used in pens. Under small changes in the external load, it behaves like a proper spring and returns to its resting length, when the external load is removed, that is, it shows the feature of equifinality. However, try to stretch the spring quickly and strongly beyond a certain length, for example, such that its length doubles. It will not return to the previous resting length. It will show a typical spring behavior but about the new resting length. Is it still a spring? Yes. Has this experiment with a violation of equifinality proven that it was not a spring in the first place? No. It has only shown that even the metal spring can change its parameters under certain external conditions.

To summarize this long story of the relations between the EP-hypothesis and the spring analogy, this analogy has been the root of a lot of confusion, especially when taken literally, resulting in simplified, even caricature-like understanding of the EP-hypothesis. After such an "understanding" is accepted, it becomes very easy to disprove the hypothesis. Although the term *spring-like behavior* seems adequate in reference to some motor effects (e.g. changes in the arm posture in response to a smooth unloading), the spring model is highly inadequate with respect to many other aspects of motor behavior, especially those involving a fundamentally nonlinear process—threshold control.

Most of the criticisms of the EP-hypothesis have followed the following scenario. A simplified version of the hypothesis is accepted, for example, viewing the human limbs as perfect damped mass–spring systems with controllable parameters. A prediction is made based on this version and then disproved in an experiment. For example, quick motion of a mass on a spring is possible only if the spring is sufficiently stiff. Data from experiments with perturbations applied during fast movements can be analyzed using such a simple mass–spring model; these data show relatively low stiffness values (Bennett et al. 1992; Popescu et al. 2003). Then, a conclusion is drawn that the EP-hypothesis has been falsified.

There is a deep methodological or even philosophical flaw in such studies. Experimental studies that start with accepting a physical or mathematical model without a good biological rationale are inadequate tools to study biological systems with. Such experiments use biological objects to identify parameters of the *a priori* accepted models, while the goal should be opposite: to study biological objects using models sparingly, only to describe answers to biologically relevant questions in a concise manner and to simplify data analysis. This has been well understood by Gelfand (see section 3.1) who, being a mathematician, was not a great admirer of mathematical and physical models in biology. He has on many occasions emphasized that a model should always be specific to the object of study and closely reflect certain salient features of the object. One needs to have an intuitive understanding of the object before modeling it. Accepting models simply because they have been used widely (and even successfully) in physics, engineering, or control theory is not going to contribute to a better understanding of biological objects.

The neuromuscular system belongs to a class of dynamic systems with position-dependent force generators (muscles and their reflexes). For such a system, it is necessary to distinguish between *state variables* and *parameters*. Variables that express the relations defined by laws of physics are called state variables (e.g. forces and kinematic variables). Any variable that can be computed or otherwise deduced from state variables is also a state variable. For example, in the intact neuromuscular system, variables such as stiffness, damping, and the magnitude of muscle activation are coupled to changes in muscle force and length; hence, all these are state variables. The relations among state variables in physics include parameters, some of which are not conditioned by the laws of physics but define essential characteristics of the system's behavior under the action of the laws. For example, the equilibrium position and the period of oscillation of a simple pendulum (a mass on a rope) in the field of gravity are defined by such parameters as the length of the rope, the coordinates of the suspension point, and the local direction of gravity. These parameters, for example, the coordinates of the suspension point, can be changed causing a transition of the system to a new equilibrium position. This example illustrates a general rule: Although in an equilibrium state all forces are balanced, not forces but system's parameters predetermine where, in spatial coordinates, this state is reached (Glansdorf and Prigogine 1971). Let me recall the mentioned observation of Wachholder that humans can completely relax muscles at different joint positions. In terms of forces, the two positions are equivalent since all muscles are relaxed at both positions. However, as Wachholder concluded about a 100 years ago, they are distinguishable in terms of spring-like muscle properties.

In order to bring a system from one EP to another, the controller must change parameters that are independent of state variables. Our motor skills are thus based on an ability of the neural controller to organize, exercise, memorize, select in task-specific way, and modify parametric control of the neuromuscular system. The EP-hypothesis is currently the only hypothesis of motor control that suggests a set of parameters (muscle activation thresholds) to implement such control.

The EP-hypothesis has proven to be useful not only in the domain of motor control but also in the explanation of sense of effort (Feldman and Latash 1982b; Toffin et al. 2003). As discussed in Digression #3, muscle, tendon, and joint afferent signals carry ambiguous information on the position of body segments. For example, during isometric force production, the activity of muscle spindle afferents increases but the relevant body segments are correctly perceived as motionless (Vallbo 1974). According to the λ-model, a correct position is perceived, based on the measurement of afferent signals relative to their referent values defined by the control signals (muscle-activation thresholds). Different kinesthetic illusions elicited by vibration of muscles or tendons are explained by an interference of the vibration-induced afferent inflow with the central or/and afferent components of positional sense (Feldman and Latash 1982a).

Consider a very simple example of perception of muscle length and force. A motor command to a muscle represents a value of the threshold of the tonic stretch reflex and may be described with a position of the tonic stretch reflex characteristic on the force–length plane (Figure 3.33). So, if a neural structure responsible for perception knows the current value of the motor command (a value of λ), half the problem of perception is solved. A value of λ makes only certain combinations of muscle length and force possible (those corresponding to points on the IC). In more formal terms, defining a command to a muscle cuts a one-dimensional subspace from the two-dimensional state space of the muscle.

Now it is necessary to cut another single-dimensional subspace in the muscle state space, that is, to draw another line on the force–angle plane. This line can be derived from a weighted sum of afferent signals from all the available sources.

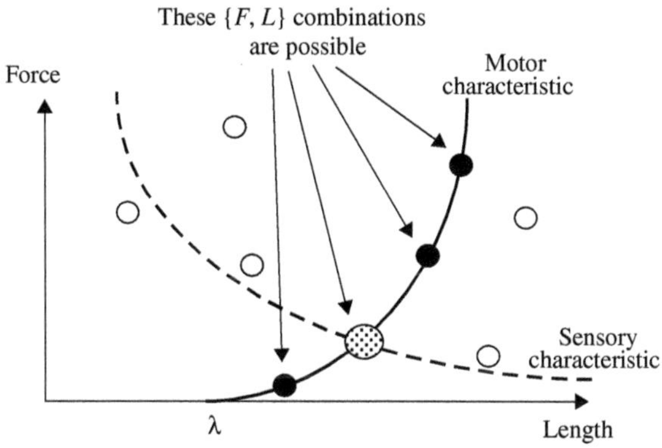

Figure 3.33. Perception of muscle length and force results from an interaction of two processes, motor and sensory. A command to a muscle (λ) defines its invariant characteristic (motor characteristic), which makes only some of the force–length combinations possible (the black dots; the open dots show impossible combinations). Signals from all the proprioceptors form another force–length dependence (sensory characteristic). The intersection of the two characteristics defines an unambiguous combination of muscle force and length (the large, fancy dot).

Note that each point on the tonic stretch reflex characteristic is characterized by different values of muscle force, muscle length, and joint angle. This means that each point has a unique combination of activity levels from the spindle, Golgi, and articular receptors. The information from these sources is redundant, but this redundancy (or shall we call it abundance?) helps to overcome potential problems, if any one of the sources becomes unreliable, for example, as a result of a disease. This seems to be a true synergy at a sensory level. For a more detailed discussion of sensory synergies see section 8.5.

So, specifying a motor command to a muscle results in two characteristics on the force–angle plane. One of them corresponds to a chosen value of the central command (λ), while the other corresponds to a certain level of activity of proprioceptors and signals the state of the peripheral apparatus. The intersection of the two characteristics defines a point, current values of muscle length and force (Figure 3.33).

The λ-model has shown its applicability not only to movements in intact animals and healthy persons but also to certain types of motor disorders. In particular, Mindy Levin and her colleagues (Jobin and Levin 2000; Levin et al. 2000) have hypothesized that a reduction in the range of λ regulation might be a primary cause of weakness, spasticity, and deficits in inter-joint coordination in some patients with neurological movement disorders. This prediction has been confirmed for subjects with hemiparesis and cerebral palsy. There have also been attempts to use the EP-hypothesis to address atypical movements in patients with dystonia (Latash and Gutman 1994) and in persons with Down syndrome (Latash and Corcos 1991).

The EP-hypothesis with respect to control of a single muscle may be viewed as a particular example of a synergy at a level of analysis, when motor units comprising a muscle are viewed as its elements. An EP, which is a stable combination of muscle length and muscle force defined by central command and external load, may be considered a system's performance variable. The combined input of a number of motor units into the muscle activity can be described with their relative contributions equivalent to the sharing pattern. On the other hand, if an EP is viewed as a functional output of the muscle, error compensation will also be observed among the motor units based on proprioceptive feedback (tonic stretch reflex). Imagine that a muscle is at an equilibrium against a constant external load. Now, for some reason, one motor unit stops its activity. This may be expected to lead to a drop in the overall muscle level of activation and force and to its stretch under the action of the external load. The stretch will lead to an increase in the activity of muscle spindles, which will increase the activation level of the whole motoneuronal pool. This will lead to an increase in the activity of the muscle and its contraction toward the earlier EP. Hence, this system satisfies two major requirements to be considered a synergy.

Can it also satisfy the third requirement? In other words, does it have an ability to stabilize different performance variables? If different EPs are viewed as different performance variables, the answer is "yes." The neural controller can change its input into the motoneuronal pool and modify λ^*, that is, change the feature of the synergy in a task-specific way. The multi-motor-unit synergy forming the foundation of EP control suggests that this principle of control and the idea of synergies are closely interrelated.

Neural mechanisms similar in principle to the EP mechanism can hypothetically ensure stable performance of a multi-element system with respect to any performance variable. A synergy is then associated with a neural organization forming the outputs of elements based on the difference between centrally defined reference values of performance variables and feedback on their actual values. The formation of a synergy is in creating adequate feedback loops that ensure a stable EP mechanism. Imagine, for example, a synergy stabilizing the total force produced by four fingers of the human hand. Each finger force output may be associated with output of a neuron (Figure 3.34). All four neurons receive inputs from a controller that defines a required value of the total force and also feedback (inhibitory) on the actual total forces produced by the fingers. If the actual force feedback matches exactly the required force value, no change occurs in the output of the neurons. If the feedback corresponds to a smaller force, all four neurons in Figure 3.34 receive less inhibition (same as additional excitation); if the feedback corresponds to a higher force, all four neurons are inhibited. In both cases, the action of the feedback loops stabilizes the preset value of the total force, while allowing the individual fingers freedom to vary their outputs. This scheme is a simplified version of a model based on central back-coupling loops acting inside the central nervous system (Latash et al. 2005).

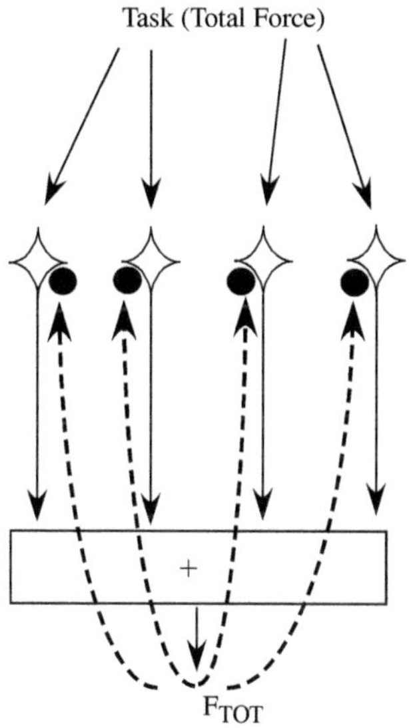

Figure 3.34. A simple scheme that can, in principle, stabilize the total force produced by four elements (fingers). The thick, dashed lines show inhibitory feedback projections from the actual total force to "neurons" that define the forces of individual fingers.

Within the language of dynamic systems, the EP-hypothesis assumes that the central controller sets a trajectory of a point attractor for the dynamic system comprising the neurons, muscles, and feedback loops. One may ask: "And who defines where the point attractor should be? Is there a neuronal apparatus computing the time evolution of point attractors for all the involved muscles?" Such a level of micro-management sounds highly unlikely. Probably, there is another, hierarchically higher controller that generates these trajectories of $\lambda(t)$. But what about the input into that system? Where does it come from? Probably from a system that defines a motor goal at a behavioral level. Ultimately, during the last trip to the supermarket, why did I decide to pick up that particular apple from the bin with 500 seemingly identical fruits? I will answer using Gelfand's words: "This is beyond my comprehension."

Part Four

Motor Variability: A Window into Synergies

If a person tries to repeat the same movement twice, the two actions will never be identical. This was emphasized by Bernstein who addressed the generation of successive movements attempting to solve the same motor task as "repetition without repetition" (Bernstein 1947). Phenomena of motor variability have been viewed by some researchers as "noise," an annoying factor that complicates data analysis. Others, however, have viewed variability as an exciting phenomenon inherent to the production of voluntary movements and potentially very informative about the processes of motor control and coordination. The latter view dates back to the end of the nineteenth century, when Woodworth performed his now classical studies of the relations between accuracy requirements (permissible error magnitude) and speed of movement (reviewed in Newell and Vaillancourt 2001). The appreciation for the wealth of information carried by the apparent "noise" in movements has been growing and resulted in two recent volumes (Newell and Corcos 1993; Davids et al. 2005) and hundreds of papers.

In this book, we are interested in motor variability, primarily as in a phenomenon that can be explored to discover synergies. One of the major features of synergies introduced earlier, namely, flexibility/stability (error compensation), implies that effects of deviations in the contribution of one of the elements of a synergy may be compensated by adjustments in the contributions of other elements. Hence, an analysis of patterns of variability of the elements may show whether a set of elements is united into a synergy and what this synergy is trying to accomplish. A very exciting and promising computational approach to analysis of variability with the purpose of discovering and quantifying synergies has been based on theoretical works by Gregor Schöner (1995). Further this approach has been developed experimentally (Scholz and Schöner 1999) and came to be known as the *uncontrolled manifold* (UCM) *hypothesis.*

4.1 THE UNCONTROLLED MANIFOLD HYPOTHESIS

In this section, a particular approach to quantitative analysis of synergies is described. This approach has been primarily developed for studies of motor synergies, but it can be generalized for other types of synergies as well. The approach is based on accepting the principle of abundance, that is, the combination of axioms 1, 2, and 3b of Gelfand and Tsetlin mentioned in section 3.2. Two central ideas form the core of this approach. First, that biological systems control their important functions using hierarchically organized multi-level structures. Second, when the central nervous system faces a problem of redundancy, it does not select a unique solution based on some optimization or other criteria but facilitates families of solutions that are equally capable of solving the problem within an acceptable margin of error. Please, do not misunderstand me: There is nothing wrong with optimization approaches, and it seems very reasonable that motor actions are organized to optimize certain features of behavior. The important point is that, if and when an optimization criterion is applied, it does not result in a single solution but in a whole family of solutions. To put it slightly less formally, the controller prefers to be sloppy and flexible rather than be precise and prone to complete failure. This approach seems to me to be rather closely related to Gelfand's principle of "reasonable inefficacy" of biological systems mentioned earlier.

To begin with let me introduce two important notions, those of *elemental variables* and *performance variables.* I am going to use the term *elemental variable* to address variables produced by apparent elements of a multi-element system. This is a loose definition; it depends, in particular, on how elements are identified and what variables are selected as relevant to particular tasks. Elemental variables may be complex, for example, when a finger presses on an object, it generates a three-dimensional vector of force and a three-dimensional vector of the moment of force. It is also assumed that the controller can, in principle, change the elemental variables one at a time; the importance of this assumption will become clear later. The term performance variable will be used to describe a particular feature of the overall output of a multi-element biological system, which may be expected to play an important role for a group of tasks. For example, if you are interested in kinematics of a multi-joint limb, independent joint rotations may be viewed as elemental variables, while the trajectory of the endpoint in the external space may be viewed as a performance variable. For the task of holding an object grasped with the five digits of the human hand, forces and moments of force produced by individual digits may be viewed as elemental variables, while the total grip force, total resultant force, and total moment of force produced on the object may be selected as performance variable.

Elemental variables may be "elemental" only at a certain selected level of analysis. For example, force produced by a finger results from changes in the activation levels of many muscles. If one is interested in how muscles are coordinated to produce finger force, muscle activation levels may be viewed as elemental variables, while finger force turns into a performance variable. So, in most cases, we should keep in mind that elements of a synergy are themselves synergies at a different level of analysis.

As will become clear later, choosing a set of elemental variables is not a trivial step. This is partly due to the commonly observed interdependence among such variables. For now, however, let us assume that a set of such variables is somehow selected. To illustrate the basic idea, let me consider a very simple two-element system, with each element producing only one output variable and the whole system facing the task of producing a single performance variable, the sum of the elemental variables produced by each element.

For example, imagine that a person is asked to produce the total peak force of 40 N by quickly pressing on two force sensors with the two index fingers (Figure 4.1). Apparently, this is a redundant system because an infinite number of finger force combinations (F_1, F_2) can satisfy the task requirement: $F_{TOTAL} = F_1 + F_2$. Imagine that, after some practice, the person becomes really good at this simple task and can perform it automatically with relatively small errors. Then, let us collect 100 trials. Each trial can be characterized by two force values—two elemental variables

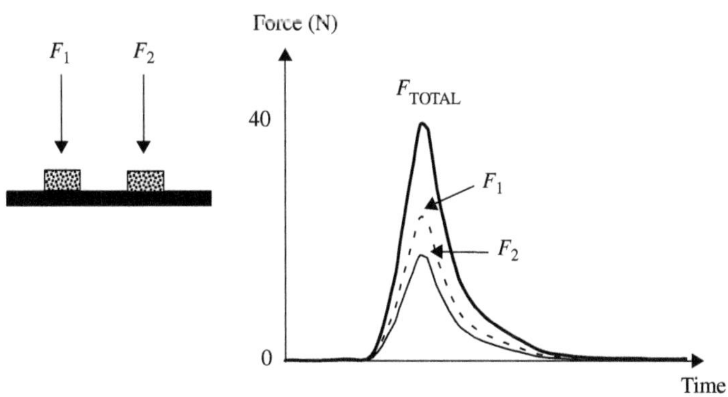

Figure 4.1. A person presses with two fingers on two force sensors (left drawing). The task is to reach the peak total force of 40 N. Forces of the individual fingers (lines on the graph, F_1 and F_2) are not shown to the subject.

produced by the two fingers. The space of elemental variables is two-dimensional; therefore, results of this experiment can be plotted as a cloud of points on a flat piece of paper. This will not be possible for more complex examples of systems consisting of many elements that produce more than two elemental variables that are going to be considered further.

The three panels of Figure 4.2 illustrate three possible outcomes of such an experiment. In the left panel, the data points are distributed evenly about a center corresponding to a certain average sharing pattern of the total force between the two fingers (approximately 50:50 in Figure 4.2). I will address such point distributions as circular. This distribution suggests that if one finger, by chance, produces in a particular trial higher peak force than its average contribution, the other finger will, with equal probability, produce force higher than average or lower than average. In other words, if one finger introduces an error into the total force, the other finger will, with equal probability, reduce this error or amplify it. There is no error compensation between the two fingers. Hence, we can conclude that there is no synergy between the two fingers that would stabilize their total force output. Note that this conclusion is reached, irrespective of the size of the data point cloud: A non-synergy may be sloppy or it may be very precise.

In the middle panel, the average forces produced by the fingers are the same as in the left panel, 20 N each. However, the distribution of data points forms not a circle but an ellipse, which is elongated along a line with a negative slope ($F_1 + F_2 = 40$ N): If a finger produces a higher force than its average contribution, then the other finger is more likely to produce less force. This relation between the two finger forces reduces the error in the total force that could be

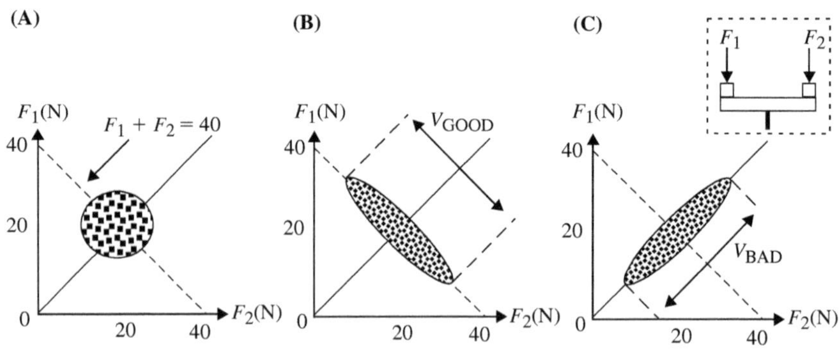

Figure 4.2. Clouds of data points measured in several trials in the experiment are illustrated in Figure 4.1. (A) The data points may form a circular cloud about a certain average sharing of the total force between the two fingers. Alternatively, the data points may form an ellipse elongated along the line $F_1 + F_2 = 40$ (the dashed slanted line, B) or elongated perpendicular to this line (the solid slanted line, C). Variance along the dashed line does not affect total force (good variance, V_{GOOD}), while variance along the solid line does (bad variance, V_{BAD}).

otherwise expected if the two fingers produced forces varying within the same ranges but without such a co-variation. In other words, the co-variation of finger forces shown in the middle panel stabilizes the total force value and may be interpreted as a force-stabilizing synergy. Note that as long as data points stay on the line, the task is performed perfectly (i.e. $F_{TOTAL} = 40$ N); when they deviate from that line, errors in the total force emerge. Apparently, a very narrow elliptical cloud of data points would correspond to a very strong synergy stabilizing the total force, because any accidental change in the force of one finger would be nearly perfectly matched by a change in the force of the other finger in the opposite direction such that the total force is always kept very close to its desired value. A wider ellipse would correspond to a weaker, sloppier synergy.

Another elliptical distribution of data points is illustrated in the right panel of Figure 4.2. This ellipse is elongated along a line with a positive slope. So, if one finger accidentally produces a higher force than its average contribution, the other finger also tends to produce a higher force, which adds to the overall error in the total force. Hence, we can say that this illustration reflects a case when the two finger forces co-vary such that the total force is destabilized. There is no force-stabilizing synergy, but a different synergy may be suspected, for example, stabilizing the total moment of finger forces with respect to a pivot located between the points of force application as illustrated in the insert drawn in the right panel of Figure 4.2.

The straight lines with a negative slope in the three panels illustrate the relation $F_1 + F_2 = 40$ N. These lines are special because they correspond to perfect task execution. As our analysis of Figure 4.2 suggests, when an ellipse of data distribution is elongated parallel to that line, as in the central panel, the two finger forces (two elemental variables) are organized into a synergy stabilizing the total force (the performance variable) at the required level. One can use this example to introduce a quantitative measure of "strength" of a synergy. It is convenient to use a measure of variability across data points called *variance*. This measure is additive in a sense that if several fingers contribute to a task of total force production, and all finger forces vary independently, the variance of the total force across trials is expected to be equal to the sum of variances of individual finger forces. In this case, variance in the finger force space is expected to be of the same magnitude in all directions, which is to correspond to a circular cloud of data points (as in the left panel of Figure 4.2). So, comparing the amount of variance along the line $F_1 + F_2 = 40$ N with the amount of variance orthogonal to that line, for example, the ratio between the two gives a quantitative index that may be used to identify a synergy and to measure its "strength." For example, this index is unity in the left panel (a non-synergy), higher than unity in the central panel (a synergy stabilizing total force), and lower than unity in the right panel (a non-synergy, but a different synergy may be suspected). Certainly, such an index should be used with caution because it reduces the two indices of variability to one and, potentially, can fail to reflect important features of performance such as accuracy of performance, which depends only on the component of variability orthogonal to the line $F_1 + F_2 = 40$ N.

For the task of two-finger force production, it is natural to identify two one-dimensional subspaces in the two-dimensional space of elemental variables (finger forces). One of them corresponds to the constant force value of 40 N, as required by the task, while the other is orthogonal to the first one. Figure 4.2 illustrates these two subspaces (straight lines, solid, and dashed). Note that changes in finger forces that are confined to one of the subspaces, the solid line with the negative slope, do not lead to changes in the total force, while finger force changes along the dashed line (with the positive slope) lead to the largest changes in the total force, given certain values of change in the finger forces. A circular data point distribution, as in the left panel of the figure, will have equal variances within the two subspaces, while an elliptical data distribution may have different amounts of variance within the two subspaces. The distribution of data points in the middle panel of Figure 4.2 will obviously have much more variance within the first subspace, while the distribution of data points in the right panel will have much less variance within the first subspace. Let us call the first and second subspaces UCM with respect to the total force and its *orthogonal complement*, respectively.

The term *uncontrolled manifold* has been criticized quite a few times as misleading. Let me explain where it comes from. When a multi-element system changes its state within a UCM computed for a particular performance variable, for example, total force produced by a set of fingers, this variable is kept at a constant value. So, as long as the system does not leave the UCM, the hierarchically higher controller does not need to interfere and, in that sense, the system of elemental variables does not need to be controlled within that manifold. If the system leaves the UCM and shows an unacceptable error in the performance variable, the controller may have to interfere and introduce a correction. Note that construction of a UCM may reflect quite a bit of control directed at establishing proper relations among the elemental variables.

Several indices have been used to quantify synergies based on the amounts of variance per dimension (per degree-of-freedom) within the UCM and within its orthogonal complement. Computation of variance per dimension is necessary to make these indices comparable across subspaces of different dimensionalities. In our example, the dimensionality of each subspace is unity; so, total amounts of variance within and orthogonal to the UCM may be directly compared. These two variance indices have been referred to using several intuitive terms such as compensated and uncompensated variance, goal-equivalent and non-goal-equivalent variance. (Scholz et al. 2000, 2002; Latash et al. 2001). The most obvious pair is variance within the UCM (V_{UCM}) and variance orthogonal to the UCM (V_{ORT}). Using less precise but maybe more intuitive words, variance can be *good* or *bad* (just like cholesterol). *Good variance* (V_{GOOD}) does not affect the selected performance variable, while *bad variance* (V_{BAD}) does. Synergies ensure that most of the variance is *good*. Stronger synergies make a higher proportion of the total variance *good*. I am going to claim that *good variance* is truly good and not simply "not bad." It gives the system an opportunity to be flexible, a very useful feature if one has to deal with unexpected perturbations or with other tasks that have to be performed concurrently using the same set of elements.

For example, imagine that you carry a cup of coffee in the right hand. Motions of the joints of the arm have to be coordinated to make sure that the vertical axis of the cup does not deviate from the gravity line (assuming that you do not want to spill the coffee). Having a flexible multi-joint synergy (a lot of V_{GOOD}) allows, for example, to open the door handle by pressing on it with the elbow without spilling the contents of the cup. Effects of the unusual elbow action on the cup orientation will be compensated by adjustments in other joints. These adjustments will be taken care of by the multi-joint synergy such that no special intervention from the controller is required.

In each case, I will always assume that the indices V_{GOOD} and V_{BAD} are computed per dimension within each subspace. Depending on the purposes of a particular study, one can use the ratio V_{GOOD}/V_{BAD} or a more complex index such as $(V_{GOOD} - V_{BAD})/V_{TOT}$, where V_{TOT} is total variance per dimension within the space of elemental variables. The former index is always positive, while the latter can be positive (a synergy), zero (not-a-synergy), or negative (not-a-synergy but, possibly, a reflection of another synergy).

The example of force production with two fingers suggests that noncircular (elliptical) distributions of data points may be signs of synergies stabilizing particular performance variables. By itself, however, the presence of such an ellipsoid-shaped distribution does not prove that a synergy exists. For example, imagine that the same experiment resulted in a distribution of data points illustrated in Figure 4.3. The points form an ellipse elongated along one of the axes

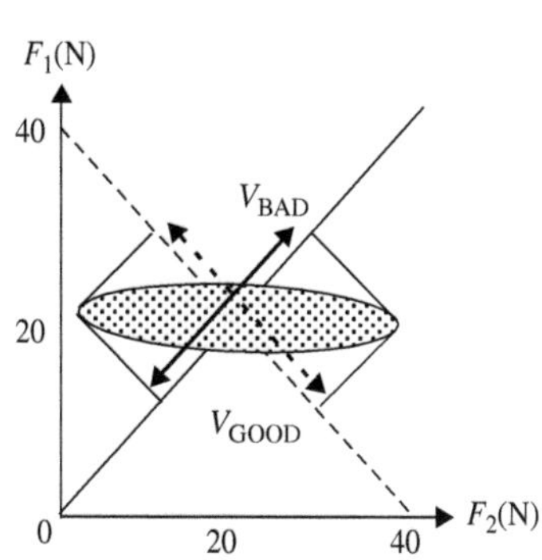

Figure 4.3. An elliptical distribution of data points can correspond to similar values of good and bad variability (V_{GOOD} and V_{BAD}). In this illustration, one finger performs more accurately than the other, while there is no co-variation between their forces.

of finger forces, but the only conclusion one can draw from this ellipse is that one finger showed significantly higher variability in its force than the other finger. Whether a certain data distribution reflects a synergy has to be tested in a rigorous way that reflects relations between elemental variables and potentially important performance variables. In Figure 4.3, variation in the two finger forces across trials does not reduce variability of the total force, $V_{GOOD} = V_{BAD}$; so, this is not a synergy.

The original example of two-finger force production illustrated in Figure 4.2 suggests that one method of testing alleged synergies is looking into correlations between pairs of elemental variables. In this particular example, correlation analysis would allow to distinguish a force-stabilizing synergy from a non-synergy and to quantify such a synergy. When a performance variable depends on more than two elemental variables, pairwise correlations may lose their power. Consider, for example, a task in which the total force produced by three fingers must be kept at a constant level, for example, the same 40 N. If all three fingers have the ability to contribute to the task, then the plane in the three-dimensional space of finger forces on which this task is exactly fulfilled is spanned by three points, each lying on one axis at a distance of 40 N from the origin of coordinates (Figure 4.4). This plane is the UCM for total force stabilization. Perfect error compensation would be achieved when a distribution of data points in the three-dimensional space of finger forces across trials is restricted to that plane and no data points lie off the plane. If data points fill the plane evenly, each

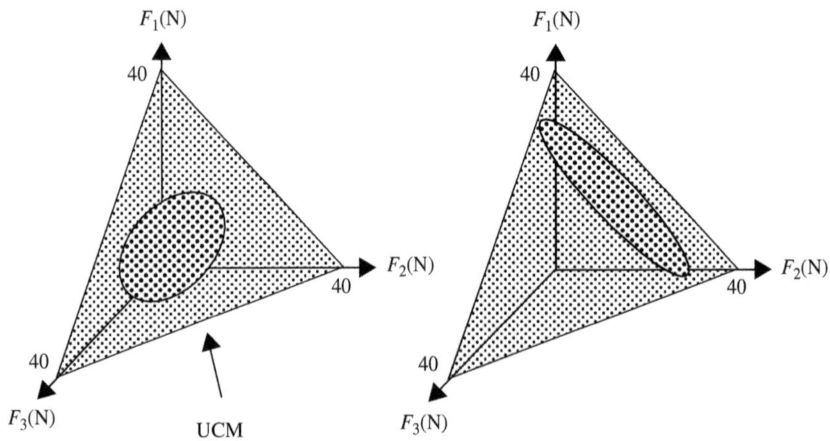

Figure 4.4. If the task of constant total force production is performed by more than two fingers, individual forces produced by any two fingers can show both negative and positive correlation. In both drawings, all data points are within the uncontrolled manifold (UCM) (the gray plane). This means that the total force is always exactly 40 N. In the left figure, all finger pairs show negative correlations. In the right figure, both negative and positive correlations can be observed.

pair of finger forces may be expected to show a negative correlation, but this correlation may be far from perfect. Pairwise correlations may even show opposite results between pairs of fingers; for example, a data distribution may show a positive correlation between the forces of two fingers and a strong negative correlation between the sum of these forces and the force of the third finger, as in the right panel of Figure 4.4.

There is, however, a common feature of the data point distributions shown in the two panels of Figure 4.4: They both belong to the subspace (UCM) defined by equation $F_1 + F_2 + F_3 = 40$ N. If accurate production of the total force is the only important component of this action, multi-finger synergies may be characterized by the amount of variance within the UCM compared to the amount of variance orthogonal to the UCM. If V_{GOOD} is larger than V_{BAD} (per dimension; note that UCM is two-dimensional, while the orthogonal complement is unidimensional), one may claim that the elemental variables co-vary to stabilize the total force, that is, there is a force-stabilizing synergy. The synergy may be strong such that nearly all the variance of the elemental variables is confined to the UCM. It may be weak: More variance is confined to the UCM than what one would predict by chance, but there is still a substantial amount of variance outside the UCM leading to errors in performance.

Typically, in a multi-element system, each elemental variable is expected to produce changes in a selected performance variable. Otherwise, it is not part of the system with respect to that variable and should not be considered as a component of a potential synergy. Including such a nonparticipating elemental variable into analysis may artificially inflate the number of dimensions in the space of elemental variables and lead to wrong estimates of variance within the UCM and orthogonal complement subspaces. Changes produced by an elemental variable in a performance variable may be described using different approaches. In particular, one may assume that small changes in an elemental variable always lead to proportional small changes in the performance variable. Formally, this may be expressed with partial derivatives of a performance variable (PV) with respect to each elemental variable (EV): ∂PV/∂EV. A set of such derivatives may be combined into a matrix, which is addressed as the *Jacobian* of the system (**J**). Computation of the Jacobian for a given multi-element system and a given performance variable may be nontrivial. Sometimes the **J** matrix can be computed based on the mechanical design of the system, and sometimes it reflects neural factors and needs to be discovered experimentally.

Let us consider a slightly more complex example—that of a planar three-joint effector (an "arm"), which performs the task of producing an accurate position (or trajectory) of the tip of the end-effector, for example, a pointer, in two dimensions (Figure 4.5). This is a kinematically redundant system, because multiple combinations of the three joint angles may achieve the same two-dimensional end-effector position. Two such configurations are shown in the figure. The subspace of joint

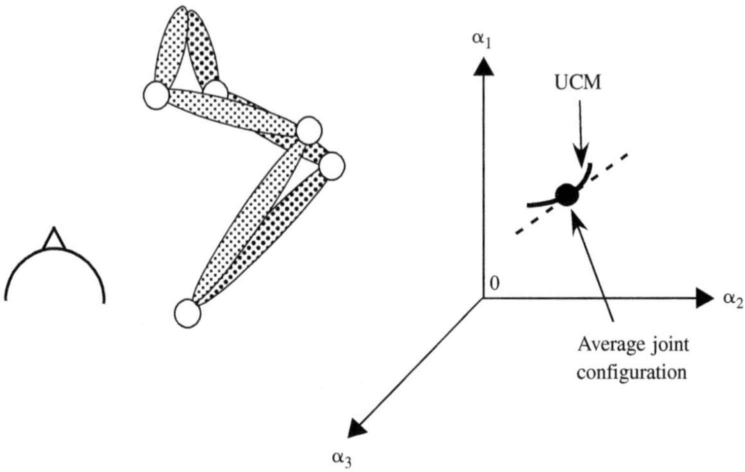

Figure 4.5. During multi-joint arm movement, the joint redundancy allows reaching the same endpoint location with different joint configurations (two are illustrated in the left drawing). The right drawing shows a point corresponding to an average joint configuration and a segment of a curved line corresponding to the same endpoint location—this is the uncontrolled manifold (UCM). The dashed straight line shows a local linear approximation of the UCM.

configurations compatible with the same end-effector position is a manifold, but it is not a plane as in the earlier example of three-finger force production. The curvature of the manifold is due to the geometric relations between changes in the three joint angles and endpoint displacements, relations that contain trigonometric functions. Such a manifold is illustrated on the right panel of Figure 4.5. To compare amounts of variance within such a UCM and orthogonal to it, the manifold may be approximated linearly (dashed line in Figure 4.5), as long as joint configurations are not spread very broadly in the three-dimensional joint space.

In summary, to perform an analysis of the amounts of *good variance* and *bad variance*, we must first commit to a set of elemental variables that form a multi-dimensional space. The controller must be able to change the elemental variables independently of each other, at least theoretically. Otherwise, co-variation of elemental variables may be due to built-in relations that are not modifiable by the controller and, therefore, such co-variation does not reflect a control strategy, a synergy. For instance, as in the very early example of co-variation of forces produced by the four legs of a table, this co-variation by itself does not mean that there is a multi-leg synergy. Rather, it means that the structural design of the table brings about the co-variation. If elemental variables are linked by such built-in relations, then this may by itself lead to distributions of data points of different shapes. Such distributions are unrelated to error compensation but may, by pure chance, show relations between V_{GOOD} and V_{BAD} computed with respect to a certain performance variable suggesting the existence of a synergy. Further, we will consider in more detail several tasks that involve apparent elements whose outputs co-vary irrespective of particular tasks. In such cases, a switch to a more appropriate set of elemental variables may be required.

Once elemental variables have been defined, hypotheses can be formulated on what performance variables might be particularly important for a given task. Let me refer to these hypotheses as *control hypotheses*. This is a very important step in analysis within the UCM hypothesis. It follows an earlier statement that there is no such a thing as an "abstract synergy." Synergies always do something. Therefore, in order to answer the question "Is there a synergy?" one needs to formulate it more exactly: "Is there a synergy within a such-and-such space of elemental variables with respect to such-and-such performance variable?"

To allow quantitative testing, any given control hypothesis has to be formalized by constructing a formal model that determines a value of the corresponding **PV** given values of the **EV**s:

$$\mathbf{PV} = f(\mathbf{EV}) \qquad \text{Equation (4.1)}$$

where bold characters are used for vectors. In the earlier example of two-finger force production, the **PV** is unidimensional (total force) and Equation 4.1 becomes the familiar $F_{TOT} = F_1 + F_2$. To determine the UCM relative to a certain value of the performance variable, this equation must be looked at inversely, that is, solved for a set of elemental variables satisfying this equation. In practice, testing a control hypothesis does not actually require computation of the UCM. Because statistical analysis of variance is done in a linear approximation, the UCM is approximated with a linear subspace or, in other words, the UCM is linearized. This can be done using the Jacobian matrix of partial derivatives, $\boldsymbol{J} = \partial\mathbf{PV}/\partial\mathbf{EV}$, described earlier. This matrix determines the changes, d**PV** induced in the performance variable by small changes, d**EV** in the elemental variables:

$$\mathrm{d}\mathbf{PV} = \boldsymbol{J} * \mathrm{d}\mathbf{EV} \qquad \text{Equation (4.2)}$$

The linearized approximation of the UCM around some point, $\mathbf{EV}_0$ in the space of elemental variables can now be determined as the null-space of the Jacobian, that is, a space within which variations of elemental variables keep the performance variable unchanged. $\mathbf{EV}_0$ is typically chosen as corresponding to the mean values of elemental variables across trials at each point in time.

$$\mathbf{EV}_0(t) = \mathrm{mean}[\mathbf{EV}(t)] \qquad \text{Equation (4.3)}$$

At the next step, at each time sample, the **EV** data can be made mean free ($\mathbf{EV}_0$ is subtracted from their values). This results in a data set (for each moment of time) that contains deviations of elemental variables from their mean, $\Delta\mathbf{EV}$. These deviations are further projected onto the null-space of $\boldsymbol{J}$, which is a linear approximation of the UCM, and onto its orthogonal complement. In future, for brevity, I will sometimes refer to this null-space as the UCM although, in fact, it is the UCM's simplified representation. The variance across all the data points is computed per dimension within the UCM (V_{GOOD}) and perpendicular to it (V_{BAD}). If V_{GOOD} is significantly higher than V_{BAD}, a conclusion can be made that a synergy is present stabilizing the selected performance variable, that is, that the control hypothesis has been confirmed.

In the very first example illustrated in Figure 4.2, the control hypothesis proposed that the total force was stabilized at 40 N by the two elemental variables representing forces produced by the two fingers. In this example, the UCM corresponds to the straight line with the negative slope. It is not necessary to linearize the UCM because it is linear to start with. The orthogonal subspace is represented by the dashed line with the positive slope. The Jacobian corresponding to the sum of the two finger forces is $\boldsymbol{J} = [1, 1]$. Its null-space for the total force of 40 N is defined as $F_1 + F_2 = 40$ (or, after the mean value of finger forces is subtracted, $\Delta F_1 + \Delta F_2 = 0$). In the left panel of Figure 4.2, $V_{GOOD} = V_{BAD}$, which is a non-synergy. In the central panel, $V_{GOOD} > V_{BAD}$—this is a force-stabilizing synergy. In the right panel, $V_{GOOD} < V_{BAD}$—a non-synergy with respect to total force stabilization, but the distribution of the data points may reflect a synergy with respect to another performance variable. Further, we will also consider a possibility that this example represents a control strategy that purposefully destabilizes the total force without caring about other variables too much.

Let me summarize the following features of the UCM hypothesis and the described computational method of assessing motor synergies.

1. The hypothesis assumes a two-level (or multi-level) hierarchical control, a controller that uses an apparently redundant set of elements to ensure stable performance with respect to a task and the elements that are either united or not united into an appropriate synergy.
2. Presence of a synergy and its strength are defined by relative magnitudes of V_{GOOD} and V_{BAD}, not by their absolute magnitudes. Note that V_{BAD} (V_{ORT}) defines the magnitude of variability of the performance variable. This means that the presence of a synergy does not necessarily ensure high accuracy, while a non-synergetic behavior may be very accurate. This is illustrated in Figure 4.6 that shows a huge ellipse (a sloppy synergy) and a very small spherical data distribution (a non-synergy resulting in very accurate, stereotypical performance).
3. Results of a single experiment leading to a single distribution of elemental variables may be analyzed with respect to different performance variables. This requires using different $\boldsymbol{J}$ matrices. In other words, this method allows researchers to ask a multi-element system participating in a behavior the following string of questions: Are you a synergy with respect to performance variable #1? Are you a synergy with respect to performance variable #2? and so on.

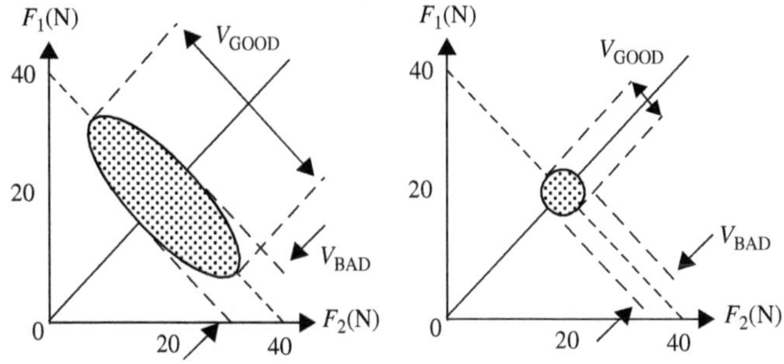

Figure 4.6. Two data point distributions in the task of producing accurate total force with two fingers (as in Figure 4.1). The left panel shows an ellipse corresponding to a synergy ($V_{GOOD} > V_{BAD}$), while the right panel shows a non-synergy ($V_{GOOD} = V_{BAD}$). However, the non-synergy is more accurate than the synergy (its V_{BAD} is smaller).

4. The UCM method of analysis is linear, and it may fail if relations between changes in elemental variables and a performance variable are strongly non-linear. A different computational approach based on creation of uncorrelated, surrogate data sets from the original data has recently been proposed and developed (Kudo et al. 2000; Martin et al. 2002; Müller and Sternad 2003, 2004; Latash et al. 2004); this method is discussed in one of the next sections. This alternative approach has the advantage of being applicable to strongly non-linear systems. This method, however, may have limitations discussed later.

4.2 MODES AS ELEMENTAL VARIABLES

The UCM hypothesis and the associated quantitative analysis of data sets are based on several axiomatic notions. One of them is that of an elemental variable. Choosing elemental variables is an important step that should be based on both the selected level of analysis and range of tasks. For some levels of analysis and some tasks, this choice is relatively straightforward. Even in those cases, however, the seemingly obvious choice may be imperfect. For example, if a researcher is interested in joint coordination during a multi-joint arm movement, rotations of individual joints seem like a reasonable set of elemental variables. In the human body, joints are spanned by bi-articular and sometimes even multi-articular muscles, that is, muscles that cross two or more joints. Such muscles bring about natural mechanical coupling of joint actions: If one joint moves, other joints of the same limb may also move because of the forces produced by multi-articular muscles, not because the controller sent a specific signal for those other joints to move. In addition, there are inter-joint reflexes that originate from force- and length-sensitive receptors in muscles and change activation levels of other muscles crossing other joints of the limb (Nichols 1989, 2002). Can a person naturally move a joint without producing noticeable movement of other joints of the same limb (certainly, without looking at them)? Intuitively, the answer seems to be "yes." There are examples, however, when the answer has been shown to be "no": In particular, during motion of one of the interphalangeal joints of a finger, other joints showed unintended motion (Li et al. 2004).

If joint rotations are coupled, clouds of data points across repetitive movements may be expected to form not spheres but more complex shapes, including ellipses in the joint angle space. This may be expected, even if commands to muscles crossing each joint vary independently of each other. In other words, variance in the space of joint angles may be expected to show preferred directions in the absence of any particular control strategy. This may be wrongly interpreted as a synergy stabilizing a performance variable if the UCM for that variable happens to be elongated along the long axis of the naturally occurring cloud of data points. Nevertheless, it is commonly assumed that joint rotations can potentially be changed independently of each other and that any nonspherical distribution of data points reflects a control strategy by the central nervous system, a synergy (Scholz and Schöner 1999; Scholz et al. 2000).

In other cases, however, this assumption cannot be made because apparent elements are known to be strongly coupled. I am going to consider two such cases, the action of several fingers of the human hand during force production tasks and the action of leg and trunk muscles during tasks performed by a standing person. For consistency, I am going to use the term *mode* for a hypothetical elemental variable that may not be obvious and requires a special experiment to be discovered.

4.2.1 Force Modes

If you try to wiggle quickly the ring finger of a hand, you will notice that other fingers also show wiggling motion although of a smaller amplitude. This lack of finger independence can also be seen in force-production tasks: When a person tries to press down with just one fingertip, other fingers of the hand also show involuntary force production, a phenomenon addressed as *enslaving* (Li et al. 1998; Zatsiorsky et al. 2000) or *lack of individuation* (Lang and Schieber 2003; Schieber and Santello 2004). Enslaving is present in all people, although its magnitude varies, for example, it is smaller in the fingers of the dominant hand (Li et al. 2000) and in persons who practice individual finger control for a long time such as pianists (Slobounov et al. 2002a,b).

Enslaving is due to both peripheral connections among the fingers such as shared muscles and interdigit tendinous connections, and neural factors such as overlapping cortical representations for individual fingers (Leijnse et al. 1993; Schieber and Hibbard 1993; Kilbreath and Gandevia 1994; Schieber 1999; Schieber and Santello 2004). Recent studies have shown that a simplistic view of cortical neuronal maps as distorted drawings of the human body on the cortex of the large brain hemispheres (a homunculus) is wrong. In particular, neuronal projections from the primary motor cortex to the arm show both convergence and divergence (Figure 4.7) (Schieber and Santello 2004). The former term means that signals from a single neuron can induce activation of arm muscles that cross different joints and produce different movements. The latter term means that several cortical neurons that may be rather far from each other can all produce activation of the same muscle and result in the same arm movement. The phenomena of convergence and divergence may be expected to complicate individual finger actions, but they may also be expected to promote co-variation in finger forces that is useful for a whole range of actions; we will discuss this issue a bit later.

Because of enslaving, variation of a command to a finger can lead to variation of forces produced by all fingers. So, if commands to fingers vary independently of each other, the cloud of data points in the space of finger forces is not likely to form a circle (or an ellipse parallel to one of the axes, as the one illustrated in the left panel of Figure 4.8) but an ellipse corresponding to positive

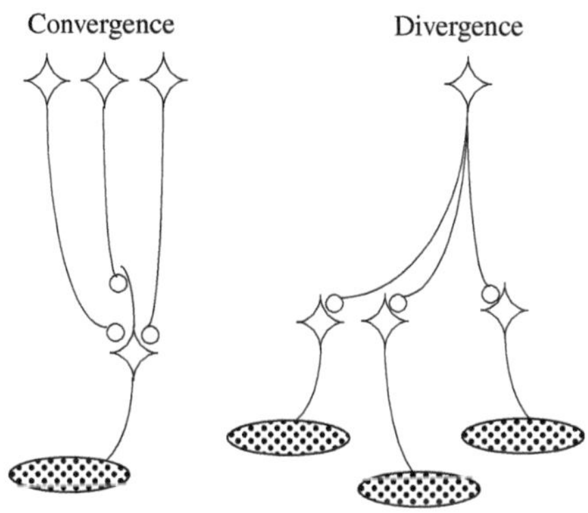

Figure 4.7. An illustration of the phenomena of convergence (left) and divergence (right).

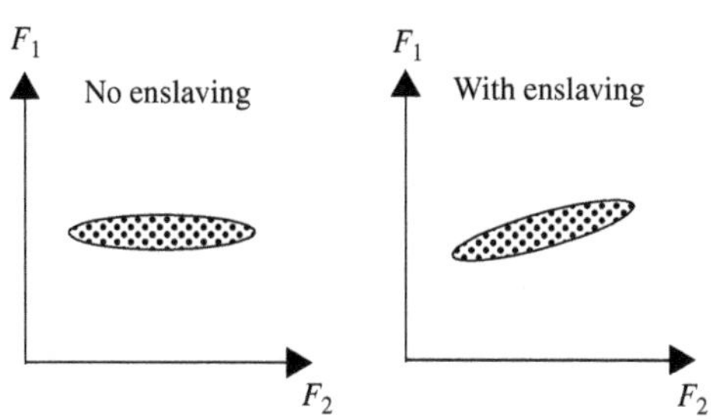

Figure 4.8. Imagine that a person tries to press the same way by two fingers simultaneously. In the absence of enslaving, if commands to two fingers vary independently, the cloud of data points is expected to show no co-variation between the forces (the horizontal ellipse in the left panel). Enslaving is expected to lead to co-variation of finger forces in different trials (the slanted ellipse in the right panel).

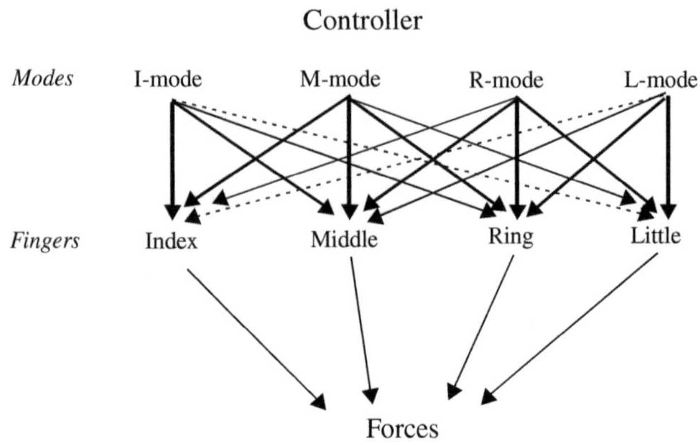

Figure 4.9. The notion of finger modes. The controller specifies the gains at central variables (modes). Each mode leads to force production by all four fingers of the hand (thicker lines correspond to larger forces). Actual forces reflect the superposition of all four modes.

co-variation of finger forces. Such a distribution may be wrongly interpreted as a synergy. For example, a distribution of data points shown in the right panel of Figure 4.8 may be mistaken for a synergy stabilizing the total moment finger forces produced with respect to a pivot between the fingers ($V_{GOOD} > V_{BAD}$).

To analyze the structure of variance that is task specific and reflective of a particular control strategy, a different set of elemental variables has to be introduced, such that these variables may be expected to show spherical distributions of data points when the controller does not attempt to co-vary them. For tasks with multi-finger force production, I will address these as *force modes* and assume that their number is equal to the number of explicitly involved fingers (Latash et al. 2001; Scholz et al. 2002; Danion et al. 2003). This assumption is based on an intuitive consideration that a person is able to try to press with one finger at a time even if this attempt is marginally successful, that is, it leads to finger force production by all fingers. Figure 4.9 illustrates the notion of force modes. Note that each mode induces force production by all the fingers.

Digression #7: Digit Interaction and Its Indices

The muscular apparatus of the hand is rather complex. It involves two types of muscles, intrinsic muscles whose bellies lie within the hand, and extrinsic muscles, whose bellies lie outside the hand, in the forearm (Figure 4.10). The intrinsic flexor muscles are digit-specific in their apparent action, that is, their tendons attach to proximal phalanges of one finger only. The strength of tendinous connections between intrinsic muscles is relatively weak (Kilbreath and Gandevia 1994). These muscles, however, also attach to a connective tissue structure that forms the so-called extensor mechanism, a network of passive

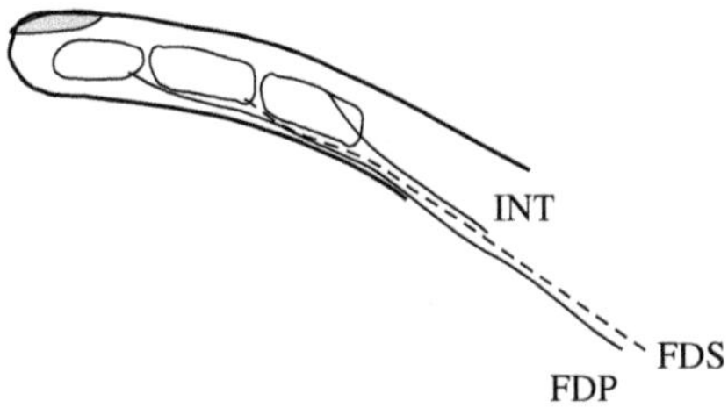

Figure 4.10. An illustration of the muscles that contribute to finger flexion. Digit-specific intrinsic (INT) muscles attach at the proximal phalanx; they also contribute to the extension mechanism at more distal phalanges. Large multi-digit extrinsic muscles, flexor digitorum profundis (FDP) and flexor digitorum superficialis (FDS) attach at the distal and middle phalanges, respectively.

elastic tissues that produces an extensor action at the distal finger joints. As a result, when a person tries to press against a stop with the proximal phalanges, intrinsic muscles produce the required focal action and also contribute to extension of the fingers in the distal phalanges. Extrinsic finger flexors are commonly described as consisting of four muscle compartments dedicated to the four fingers of the hand. The presence of compartments is assumed to facilitate independent movements of the fingers (McIsaac and Fuglevand 2007). Extrinsic finger flexor muscles have four distal tendons, each inserting at the intermediate (flexor digitorum superficialis, FDS) or at the distal (flexor digitorum profundis, FDP) phalanges. The presence of multi-tendon, multi-digit muscles have been viewed as a major factor that makes individual finger actions depend on each other (Leijnse et al. 1993, 1997; Kilbreath and Gandevia 1994; Li et al. 2001).

Three main characteristics of finger interaction have been introduced based on studies of finger force production in pressing tasks (Li et al. 1998; Zatsiorsky et al. 1998, 2000). One of them reflects unintended force production by a finger when the person tries to press down with another finger of the hand—this is *enslaving*. Another index reflects the fact that when a person tries to press with several fingers as strongly as possible, the peak forces reached by the fingers are smaller than those observed when the maximal pressing task is performed by one finger at a time. This phenomenon is called *force deficit* (see also Ohtsuki 1981; Kinoshita et al. 1996). The third characteristic is termed *sharing*. It reflects the fact that, during natural pressing, the four fingers of the hand share the total force in a relatively uniform way over a broad range of forces, that is, each finger tends to produce a fixed percentage of the total force when the total force varies.

Several findings point at central, neural mechanisms as the main source of finger interaction. In particular, let us consider what can be expected when a person tries to press as strongly as possible with the fingertips and with the proximal phalanges (the latter is a rather unusual action but can easily be done). When a person tries to press as strongly as possible with the fingertip of a finger, the FDP muscle has to be activated to produce its maximal force (Landsmeer and Long 1965; Long 1965). Other muscles also have to be involved to balance the mechanical action of the reactive force at all the joints along the finger. In particular the intrinsic muscles have to produce force to balance the rotational action of the reactive force at the metacarpophalangeal joint (the joint where the finger connects to the nonfinger part of the hand). The involvement of those muscles, however, does not have to be maximal; existing estimates suggest that it is under 30% of maximal (Harding et al. 1993; Li et al. 2000).

When a person tries to press against the stop with proximal phalanges, intrinsic muscles have to produce maximal force. This is going to be accompanied by an action of the extensor mechanism at more distal joints. To balance this action, extrinsic muscles have to be involved, but at a submaximal level (An et al. 1979, 1985; Chao et al. 1976). So, the limiting factor during pressing at the fingertips is the FDP (a multi-digit muscle), and during pressing with the proximal phalanges is the intrinsic muscles (which are digit-specific). If the multi-tendon design brings about the phenomena of finger interdependence, such as enslaving and force deficit, these phenomena can be expected to be large during pressing with fingertips and absent or much smaller during pressing with the proximal phalanges. The results, however, show that both enslaving and force deficit are slightly larger when a person presses with the proximal phalanges (Latash et al. 2002a; Shinohara et al. 2003): This seems to be a serious argument against the dominant role of the peripheral muscular design in these phenomena.

There have been several computational models of finger interdependence. The simplest one (Danion et al. 2003) used the notion of force modes and introduced an equation that is able to predict maximal finger forces with high accuracy based on two components, a matrix of enslaving effects and a coefficient related to the number of fingers explicitly involved in a task:

$$\boldsymbol{F} = k^*[\mathbf{E}]^*\boldsymbol{M}^{\mathrm{T}} \qquad \text{Equation (4.4)}$$

where $\boldsymbol{F}$ is the vector of forces, [**E**] is the enslaving matrix, $\boldsymbol{M}$ is the mode vector, T is sign of transpose, and k is a coefficient depending on the number of involved fingers. The coefficient k reflects the phenomenon of force deficit, that is, a drop in finger forces with an increase in the number of explicitly involved fingers. Empirically, this coefficient has been defined as $1/n^{0.7}$, where n is the number of explicitly involved fingers.

The matrix of enslaving effects, [**E**] contains forces of individual fingers in separate trials with an individual finger trying to produce maximal force in each trial. Figure 4.11 shows an example of such a matrix. Note that the numbers on the main diagonal of the matrix are generally larger. These numbers show forces produced by the task fingers (sometimes called "master fingers"). The off-diagonal numbers show forces produced by nontask fingers (addressed as "slave fingers"). The four fingers show different degrees of independence. Typically, the index finger is best controlled independently, while the ring finger shows large indices of enslaving. Control signals to such a system may be described with four numbers united into a vector, the mode vector. Each number varies from zero (the finger is not explicitly involved) to unity (the finger tries to produce its

Enslaving Table

Finger task	*I*	*M*	*R*	*L*
I	40	5	2	1
M	3	38	8	5
R	1	8	30	10
L	1	4	12	25

Figure 4.11. The matrix shows typical finger forces in newtons produced by individual fingers when a person tries to press with only one of the fingers "as strongly as possible." The large numbers on the main diagonal correspond to "direct forces," that is, forces produced by instructed fingers. The off-diagonal numbers correspond to forces by noninstructed fingers, which are also called enslaved forces.

maximal force). In any force-production task (not necessarily maximal force production), the mode vector shows what the subject tries to do.

The mode vector [1,0.5,0,0.5], for example, shows that the person tries to press maximally with the index finger, at 50% of maximal effort by the middle and little fingers, and not pressing with the ring finger. The product of this vector and the enslaving matrix has to be attenuated by k to produce expected forces of all four fingers. For the example under consideration, the number of explicitly involved fingers is three (since the mode for the ring finger is zero). From Equation 4.4, a set of expected finger forces can be computed.

Equation 4.4 suggests a linear relation between muscle forces and modes; this assumption is based on several studies that analyzed enslaving effects and showed that the entries of the [**E**] matrix remained almost unchanged over a wide range of forces and tasks (Li et al. 1998; Zatsiorsky et al. 2000; Latash et al. 2001). So, one does not have to perform maximal contraction tasks to compute entries of [**E**]; an alternative method is to simply ask a person to produce a simple profile of force (e.g. its ramp-like increase within a comfortable range) using one finger at a time in four individual trials. In each trial, all four fingers will show force changes that allow estimating [**E**].

In most human hand actions, the thumb plays an important role. In English, its importance has been emphasized with linguistic clarity: The human hand has four fingers and the thumb. In other languages, however, the human hand is described as having five fingers, and the thumb is one of them. The question as to whether the thumb is a finger has been studied by quantifying indices of interaction between pairs of fingers and between a finger and the thumb (Olafsdottir et al. 2005). The results have both emphasized the special role of the thumb and suggested that it is indeed a finger, a very special finger but still a finger. Indices of finger interaction depended significantly on the thumb involvement; in particular, they were dramatically different when the thumb opposed the fingers as compared to pressing with all five digits in parallel (Figure 4.12).

Nevertheless, in both tasks, fingers interacted with each other, on average, in the same way as they interacted with the thumb. Evolutionary biologists are not surprised to hear that the thumb interacts with fingers in the same way fingers interact with each other (Marzke 1992): Although in humans the thumb has a special muscular apparatus, in nonhuman primates its motion is controlled by the fifth compartment of the extrinsic flexor muscle, so it does look like a fifth finger.

Formally, a mode vector $\boldsymbol{M}$ may be represented as:

$$\boldsymbol{M} = \boldsymbol{E}^{-1} * \boldsymbol{F}^{\mathrm{T}} \qquad \text{Equation (4.5)}$$

where $\boldsymbol{F}$ is a vector of finger forces, T is the sign of transpose, and $\boldsymbol{E}^{-1}$ is an inverse of the enslaving matrix.

End of Digression #7

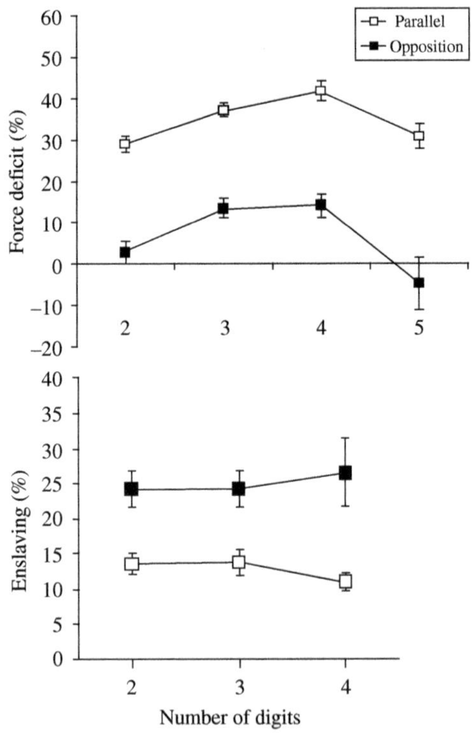

Figure 4.12. Indices of finger interaction, force deficit (top panel) and enslaving (bottom panel) depend on position of the thumb, acting parallel to the four fingers or acting in opposition (open and filled symbols, respectively). However, the thumb interacted with other fingers in the same way the fingers interacted among themselves. Reproduced by permission from Olatsdottir H, Zatsiorsky VM, Latash ML (2005) Is the thumb a fifth finger? A study of digit interaction during force production tasks. *Experimental Brain Research* 160: 203–213, © Springer.

As mentioned in Digression #7, ***E*** matrix reflecting enslaving can be computed for a subject based on data from trials, when the subject tries to press using one finger at a time. For example, if only the index and middle (I and M) fingers of a hand are involved in a task:

$$E = \begin{bmatrix} \Delta f_{\mathrm{I,I}}/\Delta F_{\mathrm{I}} & \Delta f_{\mathrm{I,M}}/\Delta F_{\mathrm{M}} \\ \Delta f_{\mathrm{M,I}}/\Delta F_{\mathrm{I}} & \Delta f_{\mathrm{M,M}}/\Delta F_{\mathrm{M}} \end{bmatrix} \qquad \text{Equation (4.6)}$$

where $\Delta f_{j,k}$ and ΔF_k are the changes of individual finger force j (j = I, M) and the change of total force produced during the task when finger k (k = I, M) was instructed to produce force. This experimentally reconstructed matrix can now be used to compute changes in force modes based on experimentally recorded changes in finger forces:

$$\Delta M = E^{-1} * \begin{bmatrix} \Delta f_{\mathrm{I}} \\ \Delta f_{\mathrm{M}} \end{bmatrix} \qquad \text{Equation (4.7)}$$

The notion of force modes has been relatively well developed only for flexion force production by the digits while pressing perpendicular to the opposing surface. During natural manipulation tasks, each digit produces a three-component force vector and a three-component moment of force vector on the hand-held object. In general, complex enslaving patterns may be expected among all these variables. These have not been studied in sufficient detail. A study of enslaving effects during the tasks with purposeful generation of shear forces (those generated when one tries to spread the fingers or to bring them together) has shown more complex patterns of enslaving among finger forces as compared to those described for fingertip-pressing tasks (Pataky et al. 2007). Relations between variations in voluntarily produced normal and shear forces are unknown, as well as relations between variations in force and moment variables. Therefore, studies of multi-digit interaction during prehensile tasks have commonly used mechanical variables, forces, and moments, as elemental variables. This introduces a possibility of mistakes in identification of synergies such that actual synergies are not noticed (if they are weak) while conclusions on weak synergies may be made in their absence.

4.2.2 Muscle Modes

A similar approach has been applied to analysis of multi-muscle synergies. At the end of the nineteenth century, a great British neurologist Hughlings Jackson wrote: "The brain does not know muscles, it knows only movements" (1889). Similar views were expressed at about the same time by a great French neurologist Babinski (1899, cited after Smith 1993). Since those times, researchers have agreed that the brain does not control large muscle groups by specifying signals to pools of motoneurons innervating each individual muscle but rather by grouping muscles and using smaller sets of control variables. This insight has been confirmed in several studies that showed close to proportional scaling in the activity of muscle groups over a variety of tasks (Gielen and Van Zuylen 1986; Latash et al. 1995; Maier and Hepp-Reymond 1995; d'Avella et al. 2003; Krishnamoorthy et al. 2003a,b; Ivanenko et al. 2004, 2006; Weiss and Flanders 2004; Ting and Macpherson 2005; Tresch et al. 2006; Zehr et al. 2007). Commonly, correlation analysis, principal component analysis (PCA), or similar matrix factorization techniques have been used to discover such phenomena. These studies used electromyography (EMG), a method of quantifying muscle activation that deserves another Digression.

Digression #8: Electromyography

To recall, activation of a limb or trunk muscle starts with the generation of action potentials by alpha-motoneurons located in the front (ventral) horns of the spinal cord. These signals travel along the rather long and fast-conducting axons and make synapses on muscle cells. Each axon typically innervates several muscle fibers that act as a unit: They all respond to an action potential arriving along the axon. The alpha-motoneuron and muscle cells it innervates form a unit of muscle activation termed a *motor unit.* Changes in muscle activation can be produced by two main means. First, the number of recruited motor units may vary. Second, an alpha-motoneuron can generate action potentials at different frequencies, resulting in different contributions of that motor unit to the overall muscle activity.

EMG is a method of recording electrical events during muscle activation. There are two major EMG techniques, intramuscular and surface. Intramuscular EMG involves placing a thin wire with a bare tip into a muscle (Figure 4.13) and recording the difference of potentials between the tip of the wire and another electrode (this other electrode may represent the surface of the needle with the wire or any other conducting object placed on the body). The small dimensions of the wire tip make this method sensitive to local changes in electric potential, while electrical events that may happen at some distance from the site of recording will have little or no effect on the signal. This method allows, in particular, identifying and quantifying action potentials of individual motor units. Motor unit compound action potentials represent the sum of the potentials produced on the membranes of all individual muscle cells comprising the motor unit. Since all muscle cells innervated by a motoneuron receive excitatory inputs virtually simultaneously, their action potentials overlap in time and bring about the compound action potential. The shape and size of a compound action potential depend on many factors, including the number of muscle cells, how close the cells are to the recording site, and the orientation of the muscle fibers with respect to the recording site. So, if one is interested in such issues as recruitment patterns of individual motor units, their changes with aging, development, fatigue, disease and treatment, and similar issues, intramuscular EMG is the method of choice. The drawbacks of this method are that it provides information about activation in a relatively small portion of a muscle besides the fact that most people do not enjoy having wires stuck in their muscles.

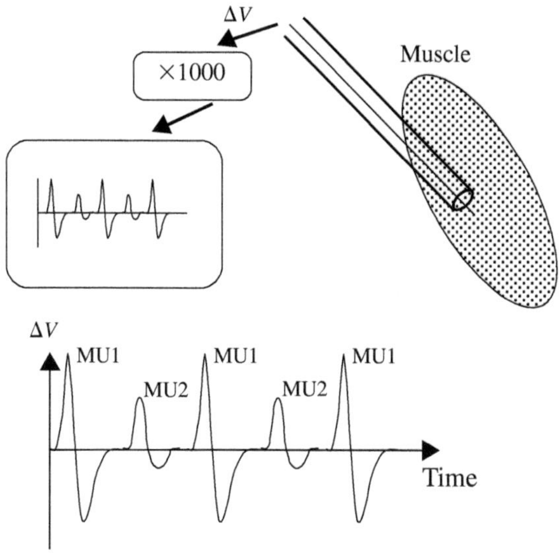

Figure 4.13. Intramuscular electromyography uses a thin wire inserted into a muscle. The difference of potentials (ΔV) between the tip of the wire and another electrode is amplified and recorded. This method allows identifying and analyzing individual motor units (e.g. MU1 and MU2 shown in the lower graph). Modified by permission from Latash M L (1999) *Neurophysiological Basis of Movement.* Human Kinetics: Champaign, IL. © Mark L. Latash.

The other method, surface EMG is much more humane. Electrodes are placed on the skin over the belly of a muscle of interest (Figure 4.14). The skin is cleaned, and the electrodes are usually covered with conducting jelly to ensure good contact. This method typically does not allow identifying individual motor units, but it provides information on the overall level of muscle activity. Electrical signals are amplified, digitized, and stored in a computer for further analysis.

The rest is up to the researcher. EMG processing commonly involves procedures such as rectification, filtering, alignment, averaging, and integration. Rectification involves removing the negative values in the signal (half-wave rectification) or substituting them with positive values of the same magnitude (full-wave rectification). Since the signal is commonly nearly symmetrical about zero line, dealing with both positive and negative values may lead to problems, if one wishes to average such a signal over several trials or to estimate its overall magnitude. Without prior rectification, an integral of an EMG signal over time is predicted to be very close to zero. One should be careful, however, because rectification may lead to errors in identification of the time when a signal shows a quick change in its magnitude (Farina et al. 2004).

EMG filtering is commonly done after rectification. Its purpose is to both eliminate noise and optimize identification of time changes in the signal at a particular time scale. One may be interested in millisecond-to-millisecond changes in the EMG or in changes over hundreds of milliseconds or even seconds.

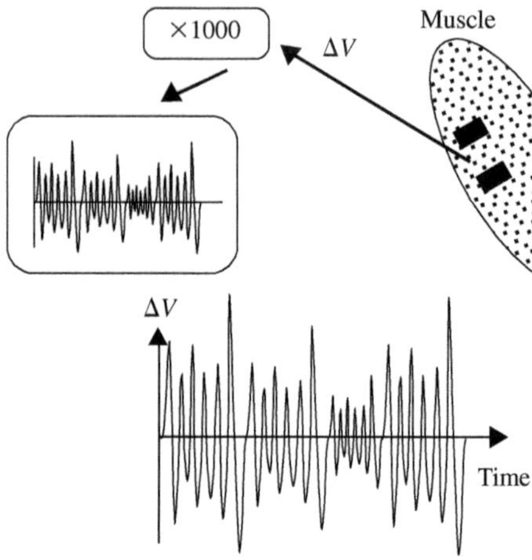

Figure 4.14. Surface electromyography measures the difference of potentials (ΔV) between two electrodes placed on the skin over the muscle belly. The recorded signal reflects the overall level of muscle activation. Modified by permission from Latash M L (1999). *Neurophysiological Basis of Movement.* Human Kinetics: Champaign, IL. © Mark L. Latash.

In the latter case, low-pass filtering allows the visualization of relatively slow changes in the magnitude of the signal (Figure 4.15).

EMG signals are notoriously noisy. Therefore, typically, researchers ask subjects to perform a task several times and then look for similarities in EMG signals across trials. To do this, the trials should be aligned in time by an event that makes the signals comparable. This procedure, known as alignment, is done differently by different researchers. Signals can be aligned by the initiation of an action as defined by an electrical signal (the first change in the EMG) or by a mechanical signal (the initiation of motion or force generation). Most biological signals do not show an abrupt step-like jump but rather a slow build-up. This does not allow giving an unambiguous answer to a question: "When does the action start?"

After the trials have been aligned, they can be averaged to scale down idiosyncratic EMG changes in individual trials and to emphasize similarities across trials. Integration of EMG signals within particular time windows is used to produce quantitative indices of overall muscle activity that can be compared across trials and conditions. This procedure can be applied to both individual trials and averaged signals. Selecting time windows for integration is another individual decision made by researchers based on their particular research questions.

End of Digression #8

The discovery of proportional scaling of activity across muscle groups has resulted in their being called *synergies* (Saltiel et al. 2001; Holdefer and Miller 2002; Sabatini 2002; Ivanenko et al. 2004, 2006; d'Avella and Bizzi 2005; Ting and Macpherson 2005). This may be a linguistic preference; however, earlier we had

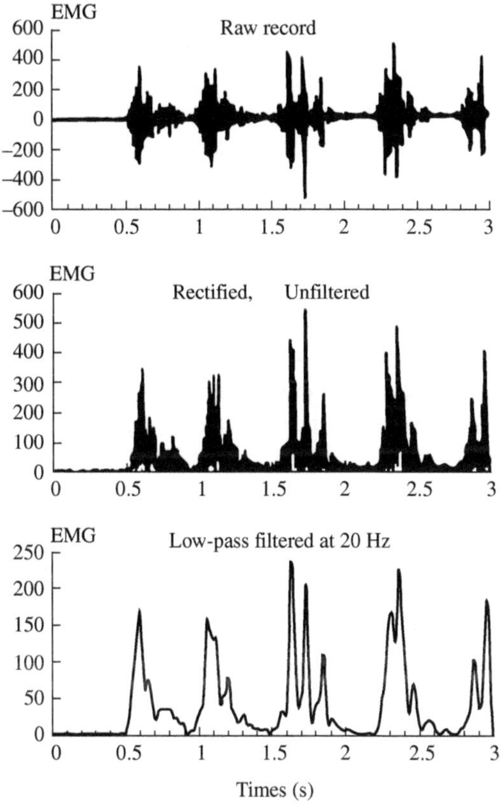

Figure 4.15. A typical example of a surface EMG signal before any additional processing (top), after rectification (middle), and after low-pass filtering at 20 Hz (bottom). Modified by permission from Latash M L (1999) *Neurophysiological Basis of Movement.* Human Kinetics: Champaign, IL. © Mark L. Latash.

agreed that proportional changes in elemental variables by themselves do not allow one to claim that a synergy exists. Otherwise, the legs of a table and the prongs of a fork qualify as a synergy, and we have already agreed that they are not.

To claim that one deals with a synergy, that is, with a particular control strategy leading to purposeful, task-specific co-variation of elemental variables, it is necessary to demonstrate that the observed scaling is related to ensuring stability of a particular performance variable and that it can be modified, if the controller decides to do so. If one and the same proportional scaling is observed across a variety of tasks, one gets suspicious that it reflects a built-in relation among muscle activations that may not be specific to particular performance variables. On the other hand, however, finding a proportional scaling of activity over a group of muscles over a group of tasks may serve as an important indicator of a particular strategy of reducing the number of variables the controller has to deal with.

Take a look at Figure 4.16. It illustrates a scheme of control with the controller uniting muscles into larger groups and then manipulating magnitudes of activation signals (gains) sent to each of the groups. Accepting this scheme implies that multi-muscle synergies should be searched for not in the space of muscle activations but in a lower dimensional space of magnitudes of recruitment of muscle groups. Such muscle groups may be viewed as analogous to finger modes, that is, elemental variables based on a set of nonindependent outputs of apparent elements of the system. Hence, they are called *muscle modes*. (Another term, *stable muscle grouping* was suggested but later dropped because the abbreviation SMUG did not befit reviewers of scientific journals.)

Given the breadth of motor tasks humans encounter during everyday life, it would be unrealistic to expect muscle modes to be universal and applicable across all tasks involving those particular body parts. For example, leg muscles participate not only in standing but also in walking, jumping, kicking a ball, and dancing. The abundance of muscles allows forming numerous sets of muscle modes, each set relevant for a particular group of tasks. Examples of such groups of tasks may include (1) standing, swaying, and preparing for or responding to a perturbation applied to a standing person; (2) walking uphill, downhill, on a level surface, and maybe running; (3) kicking different objects such as a football, a soccer ball, a volleyball, or even a pebble.

Only a handful of studies have used this approach consistently across a variety of postural tasks (Krishnamoorthy et al. 2003a,b, 2004, 2007; Wang et al. 2005, 2006a,b; Danna-Dos-Santos et al. 2007). The tasks involved preparing for a self-inflicted perturbation (releasing a load from extended arms), swaying voluntarily, preparing to making a step, and using an arm to stabilize the body of a person sitting in an unstable chair. Across all the tasks performed by standing persons,

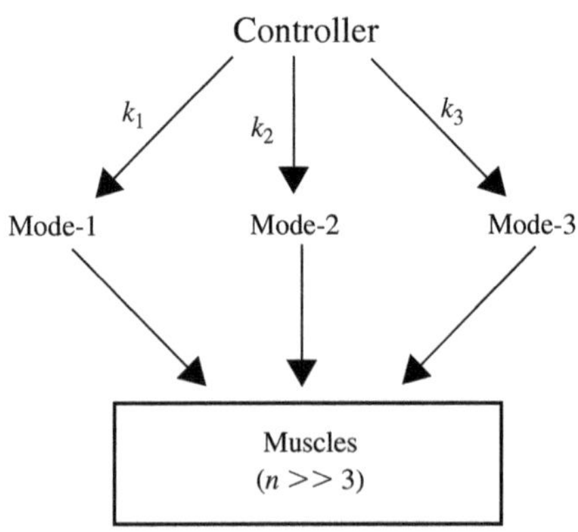

Figure 4.16. An illustration of the notion of muscle modes. The controller unites muscles into groups (modes) in such a way that all muscle within a group show parallel scaling of their activation levels. Then, the controller manipulates gains for each mode. The modes project on muscles whose number is potentially much larger than the number of modes.

factor analysis applied to muscle activation data (integrals of muscle activity were analyzed over reasonable time intervals) resulted in the identification of three to four factors grouping muscles of the leg and trunk in a way that made mechanical sense. Figure 4.17 illustrates results of one of such studies. The table in this figure suggests the existence of two muscle groups (muscle modes) that unite muscles on the dorsal and ventral surfaces of the body. These muscle modes have been called *push-back* and *push-forward,* reflecting their expected effects on location of the center of mass of the body. Another muscle group commonly united muscle with a pronounced lateral action (such as tensor fascia lata and gluteus maximus)—a *push-side* mode.

Formally, muscle modes can be viewed as a set of orthogonal vectors of unitary length (eigenvectors) in the space of muscle activations. To specify muscle activation, the controller uses a set of gains, hypothetical variables that define the magnitude of each of the modes. Figure 4.18 illustrates typical time profiles of activation of a selected subset of postural muscles and typical time profiles of muscle modes when the subject swayed the body voluntarily while being paced by the metronome (Danna-Dos-Santos et al. 2007). Muscle modes showed rhythmic patterns similar to the patterns of muscle activation, and the gains at the three modes changed smoothly as well.

The repertoire of muscle modes was limited to the three mentioned modes when a person stood normally on a flat surface. When the conditions changed, for example, when a person was standing on a board with a narrow support area and/or using a hand for support, the mode composition could change. The *push-back* and *push-forward* modes that unite muscles acting at different joints could disappear, while joint-specific co-contraction modes could emerge (Krishnamoorthy et al. 2004). These modes corresponded to parallel changes in the activation levels of antagonist muscle groups, for example, ankle plantar- and dorsiflexors, knee

Muscle	M1-mode (push back)	M2-mode (push forward)	M3 mode (mixed)
TA	–0.27	0.05	**–0.73**
GL	**0.66**	–0.22	0.36
GM	**0.81**	0.05	0.04
SOL	**0.75**	–0.08	0.11
VL	0.07	**0.77**	0.08
VM	0.29	**0.69**	0.02
RF	–0.25	**0.65**	–0.01
PF	**0.81**	0.20	0.06
ST	**0.79**	0.14	–0.31
RA	–0.31	0.22	**0.69**
ES	**0.74**	–0.05	–0.43

Figure 4.17. An illustration of muscle groups (modes) identified with the help of principal component analysis (PCA). The table shows the loading factors for all muscles; large (>0.5) loading factors are shown in bold. Note that the first two modes unite muscles of the frontal and of the dorsal surface of the body, respectively. Reproduced by permission from Krishnamoorthy V, Latash ML, Scholz JP, Zatsiorsky VM (2003b) Muscle synergies during shifts of the center of pressure by standing persons. *Experimental Brain Research* 152: 281–292, © Springer.

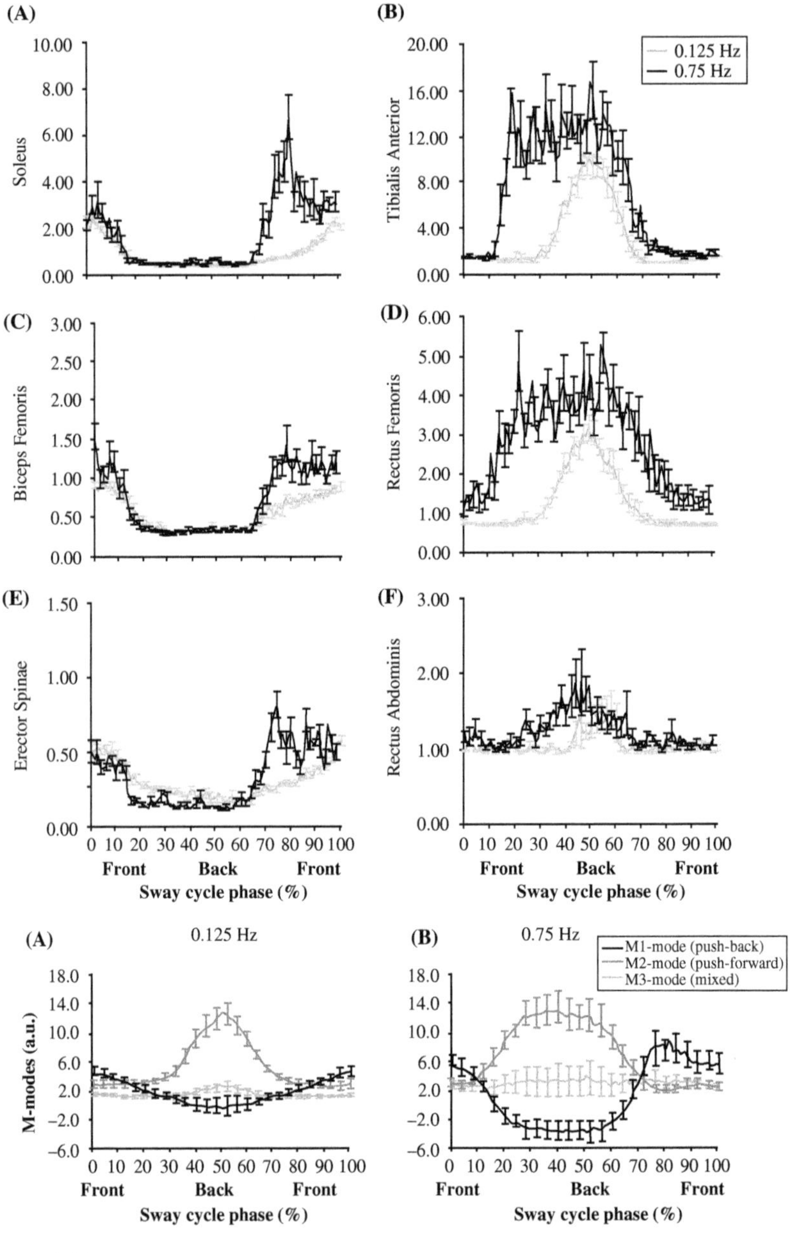

Figure 4.18. Typical time profiles of muscle activation patterns (top) and of the gains at the three modes (bottom). The subjects of this study were asked to sway with a large amplitude at different frequencies while paced by the metronome. Reproduced by permission from Danna-Dos-Santos A, Slomka K, Zatsiorsky VM, Latash ML (2007) Muscle modes and synergies during voluntary body sway. *Experimental Brain Research*, 179: 533–550, © Springer.

flexors and extensors. These are important observations suggesting that modes are not hard-wired universal relations among muscles (like the legs of a table) but flexible groupings, possibly synergies at a different level of analysis.

Taken together, all these observations support the scheme in Figure 4.18 and also show that, if external conditions change, the central nervous system can regroup muscles to form a different set of modes, adequate for the new conditions. How large are such mode libraries? We do not know. How many modes are typically used? Three or four sound right, manipulating a dozen or so of variables may be too much for the controller. How did the modes come about? Probably, by trial and error, by a kind of Darwinian within-a-person selection: Modes are biological creatures that survive if they provide the controller with handy, reliable tools for the control of groups of everyday motor tasks.

4.2.3 Experimental Identification of the Jacobian

By definition, synergies provide stability of performance variables using a pattern of co-variation among elemental variables. Therefore, to analyze synergies, one has to know how changes in elemental variables map on changes in potentially important performance variables. In some cases, such mappings are obvious. To remind, in a linear approximation, such mappings can be represented with a Jacobian, a matrix $\boldsymbol{J}$ that maps small changes in elemental variables on changes in a performance variable:

$$\Delta\mathbf{PV} = \boldsymbol{J}^{*}\Delta\mathbf{EV} \qquad \text{Equation (4.8)}$$

where **PV** stands for a performance variable and **EV** stands for the vector of elemental variables.

For example, during multi-joint movements, location of the endpoint of a multi-joint limb (a performance variable) is related to individual joint rotations through a Jacobian that reflects the geometry of the limb; entries of the Jacobian represent functions of the length of limb segments and trigonometric functions of joint angles. Other performance variables may be built on the same geometrical considerations. For example, in a study of quick-draw Western-style shooting with an infrared gun at an infrared-sensitive target (Scholz et al. 2000), a candidate performance variable was the angle between the pistol barrel and the direction from the back-sight to the target. Relations between small changes in this variable and joint angle rotation could also be obtained from the limb geometry.

Note that performance variables may be complex and have different dimensionalities. For example, the endpoint location represents a three-dimensional performance variable corresponding to the three-dimensional space where we happen to live, the angular deviation of the pistol barrel from the direction to the target is two-dimensional, while the total force or the total moment of force produced by a set of fingers are unidimensional. The dimensionality of a performance variable is expected to be smaller than the number of elemental variables. Otherwise, the system is not redundant (sorry—abundant!), and the notion of synergies is hardly applicable. Hence, in our case, the Jacobian is a matrix with the number of rows smaller than the number of columns.

During multi-finger force and moment of force production tasks, defining the Jacobian seems to be nearly trivial: A small change in the total force is simply the sum of small changes in the finger forces, while a small change in the moment of force can be computed from finger force changes multiplied by corresponding lever arms. The situation is complicated somewhat by the fact that finger forces do not qualify as elemental variables (see section 4.1). Changes in finger modes, on the other hand, do not sum up to produce a change in the total force. This is, however, a relatively minor inconvenience that can be resolved with the help of a corresponding enslaving matrix (see, for example, Equation 4.4 in Digression #7).

In the case of muscle modes, the situation becomes more complex. Muscle activations are measured in microvolts (μV); their integrals are measured in μV*s. In contrast, performance variables are measured in mechanical units such as meter, newton, newton-meter. Most of the postural tasks mentioned in the previous section are associated with changes in the location of the point of application of the resultant force acting from the support surface on the body (the center of pressure, COP). The COP coordinate looks like an excellent candidate performance variable across a variety of postural tasks (reviewed in Winter et al 1996). But how do small changes in μV*s of particular muscle modes map on COP shifts measured in meters? Such relations are not obvious and have to be discovered experimentally.

Fortunately, relations between small changes in the gains at muscle modes and COP shifts have proven to be close to linear within typical ranges of these variables. This allowed using multiple linear regression techniques to identify coefficients that map small changes in the mode gains to small shifts of the COP, which is the Jacobian of this system. In this case, the Jacobian is relatively simple, it is a set of coefficients mapping small changes in the gains at the modes at the unidimensional variable, location of the COP in the anterior–posterior direction.

4.3 STABILITY, VARIABILITY, AND WITHIN-A-TRIAL ANALYSIS OF SYNERGIES

Two related notions have been used for analysis of motor behavior, those of stability and variability. Sometimes, they are used as antonyms: If a motor system shows low variability in a particular behavior, this behavior is declared more stable as compared to a behavior that shows higher variability. However, such conclusions may be wrong since these two notions describe different features of motor behavior that may or may not correlate.

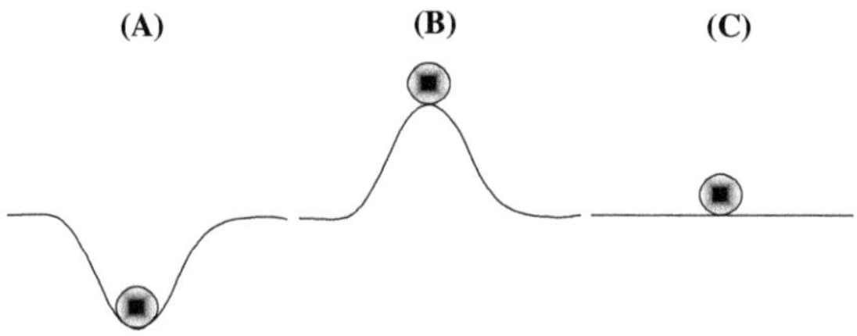

Figure 4.19. A ball is in (A) a stable equilibrium, (B) in an unstable equilibrium, and (C) in an indifferent equilibrium.

A mechanical system is considered to be in a stable equilibrium state if, after a small brief mechanical perturbation, it returns back to the original state. Figure 4.19 illustrates three states of a simple mechanical system (a ball on a surface). The ball is in a state of stable equilibrium in the left panel: After a small push, the ball will tend to move toward the original location and, in the presence of friction, will ultimately stop there. The ball is in a state of unstable equilibrium in the middle panel: A similar push will make it fall down the slope, and it will not come to rest at a new equilibrium (at least, not within the illustration). The right panel illustrates the state of indifferent equilibrium: A small, brief perturbation will induce ball motion from the original equilibrium state but, in the presence of friction, it will find a new equilibrium at a new location.

For moving systems, the notion of stability is somewhat more complex. Imagine that a mechanical object (a ball) rolls downhill along a semicircular tube. If the ball is released several times at the top of the tube, it may be expected to show somewhat different trajectories (depending on the exact point of its release and initial velocity). All these trajectories will lead to the bottom opening of the tube. A small perturbation applied to the ball during its motion will perturb the trajectory of the ball, but the ball will tend to return toward the original trajectory as soon as the perturbation is over (Figure 4.20). This example illustrates a dynamically stable behavior when effects of a small perturbation (or a small difference in initial conditions) on the trajectory of the system are small.

Imagine now a ball on the top of a semicircular hill similar to the one illustrated in the middle panel on Figure 4.19. A small push will initiate ball motion downhill. A small change in the direction of the push will make the ball move along a different trajectory, and the deviation of this trajectory from the originally planned one will increase with time. If a ball is pushed slightly to a side during its motion, its trajectory will change and will start to diverge from the original unperturbed trajectory with the deviation increasing with time. This system is dynamically unstable in a sense that a small change in the initial conditions or a small perturbation induces a deviation from a planned trajectory, and this deviation increases in time.

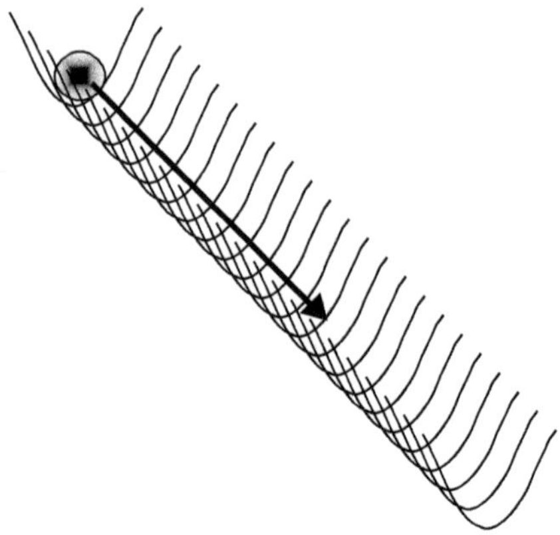

Figure 4.20. A ball rolling down a tube is dynamically stable: It will return close to its trajectory if perturbed (not too strongly) on its way down.

In general, in the presence of perturbing factors (sometimes addressed as "noise"), a dynamically stable system is expected to show lower variability than a dynamically unstable system. Indeed, at any time, the variability of the trajectory of the ball rolling down the tube (Figure 4.20) is limited by the diameter of the tube, while a ball rolling downhill (Figure 4.19) can roll along any route and may be expected to show higher indices of variability of its trajectory.

Until now, I have been using the notion of stability with respect to synergies in a somewhat different meaning. For example, under "force stabilization" I implied that comparison of performance of a multi-finger system across trials shows lower variability of the total force as compared to what one could expect if all the elements were varying their forces independently. This meaning is closer to lay-term expressions such as "This athlete showed stable performance over the last year." The important difference is that this usage of the notion of stability does not refer to time evolution of a system but rather to its ability to repeat its important performance variables with low variability across repetitive attempts at the same task.

By the way, what is "the same task?" It is theoretically impossible to reproduce perfectly initial conditions across trials. It is also impossible to reproduce the initial state of the actor (subject of the experiment). This was already well known to ancient Greek philosophers: "One cannot step twice into the same stream." This becomes too philosophical; so, to simplify life let us assume that, for all practical purposes, performance at "the same task" can be recorded several times. But also let us keep in mind that this assumption is wrong, not too wrong but still…

Analyses illustrated in earlier figures (e.g. Figures 4.2, 4.4, and 4.5) compared snapshots of the states of a system at a particular phase of its action across repetitive trials. Is it possible to address a different question that seems to be more directly related to the classical definition of stability: "If one element of a multi-element system gets off its expected time evolution, will other elements change their time profiles to minimize the effects of the first element's error on an important performance variable?" This question has two subcomponents. First, I will address the question of co-varying changes in the magnitudes of elemental variables. Then, I will also discuss the issue of co-varying changes in the timing of individual elemental variables; I am going to address this issue as that of *timing synergies*. In one of the sections of Part 5, co-variation of directions of vectors will be addressed.

With respect to the issue of co-varying magnitudes of elemental variables, it is not easy to generalize the introduced method for analysis of consecutive states of a system over a single performance at a task. Let me mention a couple of main complicating factors. First, during motion of a multi-element system, its Jacobian matrix, ***J*** may be expected to change. For example, during movement of a multi-joint extremity, a deviation of one of the angles by one degree will have a certain effect on the endpoint location at time t_1. This effect will change with time as the geometrical configuration of the extremity changes. So, if another joint is to correct this deviation at a later time, a change in its action should take into account both the original deviation of the first joint and a change in its effect on the endpoint location with time.

Second, in order for some elements to correct errors introduced by other elements, these errors probably need to be detected, and corrective actions need to become effective. Both these processes may be expected to take time because of the features of the human body described in Digressions #1, 2, and 3. Moreover, these time delays may differ across the elements. For example, a signal generated by the alpha-motoneurons in the spinal cord takes less time to reach more proximal muscles (e.g. upper-arm muscles) as compared to more distal muscles (e.g. intrinsic hand muscles). So, different joints may initiate their corrective actions at different delays after a necessity for such corrections has been established.

The question of error compensation within a trial may be addressed for particular systems that do not show changes in their ***J*** matrices and whose elements may be expected to take comparable times to introduce corrections. In particular, this can be done for the task of force production by a set of fingers pressing in parallel, as long as the finger configuration does not change. During isometric multi-finger force-production tasks, quantitative indices of enslaving remain relatively unchanged within the force range up to the maximal voluntary contraction force (Li et al. 1998; Zatsiorsky et al. 1998, 2000). Hence, the ***E*** matrix introduced earlier (see Equation 4.4 in Digression #7) may be assumed to remain unchanged over the duration of the ramp force-production trials thus making a single-trial UCM analysis possible.

Single-trial UCM analysis follows the same logical structure as the across-trials analysis described earlier. Across-trials analysis is based on comparison

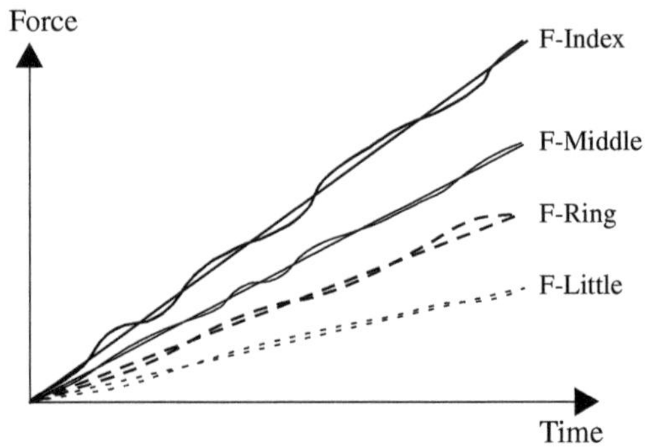

Figure 4.21. When a person is asked to produce a ramp increase in the total force, individual finger forces show ramp-like time profiles with variations about perfect ramp lines.

of the variance in the finger force space with respect to average values of finger forces across many trials. Such comparison is performed separately at each percentage (phase) of the force production task. Consider, for simplicity, the task of producing a ramp time profile of the total force while pressing down with the four fingers of the hand. During such a trial, individual finger forces typically show profiles of their individual forces that are also close to a linear ramp but vary in time about this hypothetically perfect ramp profile (Figure 4.21). Let us assume that the controller selects a particular pattern of sharing the total force among the four fingers reflected in the slopes of such hypothetical individual finger force time profiles. Then, at any time along the force ramp, expected force magnitudes may be computed for each finger. Deviations from these magnitudes may be viewed as "errors." Then, the following question may be asked: "If one finger force at some moment of time along the ramp shows a higher value than what is expected, based on its fair share, will other finger forces, on average, show deviations into lower values also, compared to what is expected from their shares?" A positive answer to this question may be interpreted as indicating a multi-finger force-stabilizing synergy.

Figure 4.22 illustrates the idea with a simple example of force production with only two fingers. The left panel shows finger forces recorded during a trial. The central panel shows the linear functions ("perfect pattern") that the finger forces could be expected to follow, if the system kept the sharing pattern unchanged over the trial. The right panel shows the results of subtracting the "perfect pattern" from the actual data. This panel shows oscillations of finger forces about zero line. Now, the main question formulated in the previous paragraph can be explored quantitatively. If the two residual lines in the right panel show predominantly negative correlation along time, a conclusion on a two-finger force-stabilizing synergy may be drawn. If the correlation is positive or zero, total force is not stabilized by co-varied adjustments of finger forces during the trial.

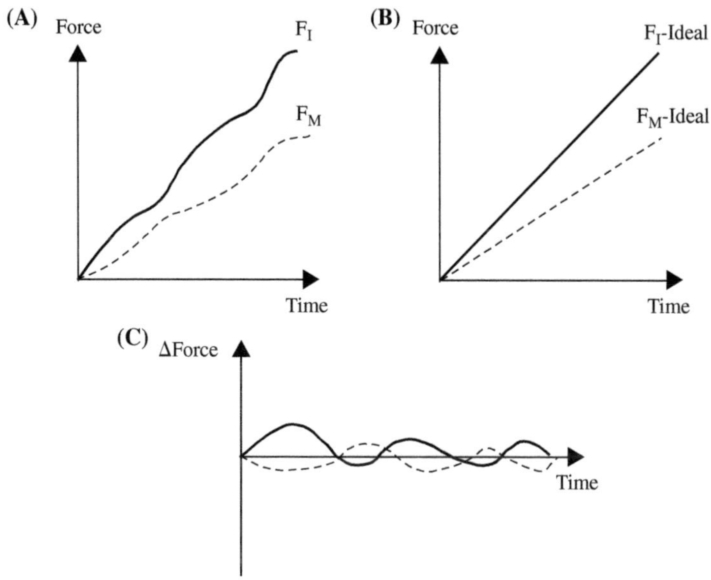

Figure 4.22. During a two-finger ramp force production task, fingers show ramp-like profiles in their forces with variations (A) about an ideal linear pattern (B). If the "ideal" patterns are subtracted from actual ones, the time series of the difference (C) may be analyzed using a method similar to the uncontrolled manifold analysis. This illustration shows predominantly negative co-variations of the two finger forces about the "ideal pattern," which can be interpreted as a force-stabilizing synergy.

To address the same question in a more general way, single-trial UCM analysis may be performed, based on computation of two components of the variance in the finger force space analogous to V_{UCM} and V_{ORT}, as introduced earlier. However, computation of these indices is performed not across repetitive trials at the task but across samples in time over a particular segment of a single ramp trial. The first step is to transform the finger force data into time series of finger modes and then to de-trend the mode ramps to make sure that they vary, on average, about zero across the period of the ramp over which the analysis is performed. Figure 4.23 shows typical force profiles in such an experiment for the four-finger force production before (panel A) and after de-trending (panel B). After the de-trending, finger force variance may be separated into two components, V_{UCM} and V_{ORT}, exactly as described earlier in section 4.1.

A priori, it is not obvious whether such single-trial analysis will lead to results compatible with the earlier introduced across-trials analysis. Figure 4.24 shows the results of both kinds of analysis performed for young, healthy subjects performing a four-finger ramp production test. For both types of analysis, an index ΔV was computed as the normalized difference between V_{UCM} and V_{ORT} (both certainly quantified per dimension in the corresponding subspaces);

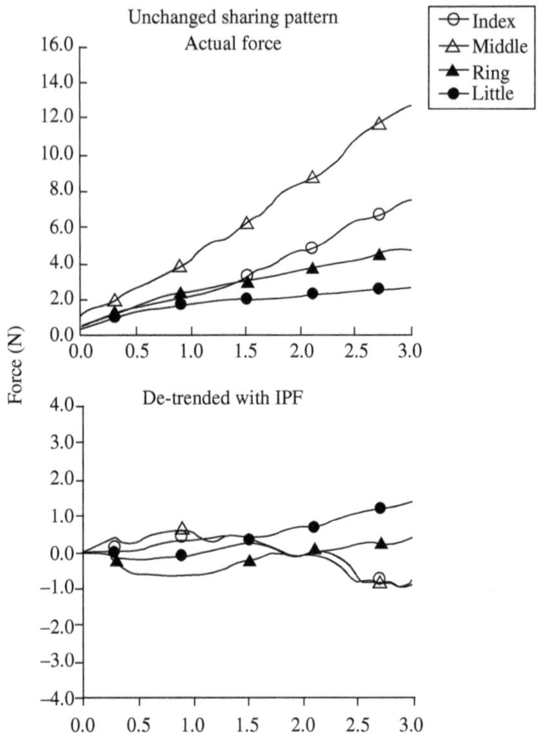

Figure 4.23. Results of a study with four-finger ramp force production. Fingers show close to linear time profiles (top). After de-trending, they show predominantly negative co-variation (bottom). Reproduced by permission from Scholz JP, Kang N, Patterson D, Latash ML (2003) Uncontrolled manifold analysis of single trials during multi-finger force production by persons with and without Down syndrome. *Experimental Brain Research* 153: 45–58, © Springer.

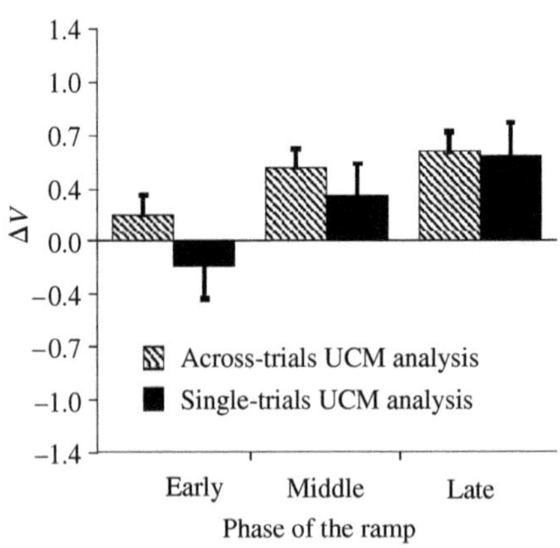

Figure 4.24. Single-trial and across-trials uncontrolled manifold analysis produce somewhat different results, which converge at larger forces (the late phase of the ramp). Reproduced by permission from Scholz JP, Kang N, Patterson D, Latash ML (2003) Uncontrolled manifold analysis of single trials during multi-finger force production by persons with and without Down syndrome. *Experimental Brain Research* 153: 45–58, © Springer.

positive values of ΔV correspond to stabilization of the total force by co-variation of signals to individual fingers. This index was computed (and averaged) separately for the three one-second long segments of the ramp task. The figure shows that both types of analysis lead to comparable results, supporting the existence of a force-stabilizing synergy and its strengthening (an increase in the relative magnitude of *good variance*) at the later segments of the ramp time profile.

Single-trial UCM analysis may potentially become a very helpful extension of this method of quantitative analysis of motor synergies. It can be applied to movements performed by persons who cannot be reasonably expected to produce many trials at a task. For example, the original study that introduced the method (Scholz et al. 2003) used it to analyze motor performance in young adults with Down syndrome. These persons could not perform dozens of trials to make it possible to use the across-trials UCM analysis. The single-trial analysis was able to quantify differences in multi-finger synergies in these persons that happened over a few days of practicing multi-finger force-production tasks (Latash et al. 2002). I will get back to applications of this method to atypical synergies and changes in synergies with practice a bit later.

Unfortunately, as of now, this method has not been developed to deal with multi-element systems, whose Jacobians change during the execution of motor tasks. This problem does not seem to be insurmountable. As long as time evolution of the Jacobian is known, the effects of co-variation among elemental variables on a performance variable can be estimated over a reasonably selected time interval to allow drawing conclusions on the presence or absence of a corresponding synergy and estimating its strength. However, such experiments have not as yet been performed.

Before moving to the promised issue of timing synergies, I would like to introduce other computational tools that have been used to study synergies. They all have advantages and shortcomings (including the UCM analysis). One of these methods has been particularly helpful in addressing the issue of timing synergies.

4.4 OTHER COMPUTATIONAL TOOLS TO STUDY SYNERGIES

4.4.1 Principal Component Analysis and Uncontrolled Manifold

One of the computational methods frequently used to test whether several elements of a system "work together" is the PCA. This method looks at correlations or co-variations between pairs of elements across a series of attempts at a task or across time samples during a single task realization. Further, it analyzes the correlation or co-variation matrix and defines its eigenvectors, which correspond to a set of orthogonal vectors in the space of original elemental variables. The first vector (first principal component or first PC) indicates a direction corresponding to the maximal amount of variance in the original data set. The second PC is orthogonal to the first one and indicates a direction that accounts for the maximal amount of remaining variance (second PC), and so on. In general, each PC is a vector defined in the space of elemental variables. Overall, the number of PCs is equal to the number of elemental variables. However, commonly, a few PCs account for a disproportionately large amount of variance in the data. Then, only a small number of PCs are considered sufficient to represent the data set, resulting in a more economical description of the data compared to using the space of elemental variables.

Let us suppose that a distribution of data points in the space of elemental variables forms a near-perfect ellipsoid. For such a distribution, the first PC will show the orientation of the longest (main) axis of the ellipsoid. The second PC will show the direction of the second longest axis of the ellipsoid, and so on. This method looks rather similar to the UCM: It compares the amounts of variance in different directions in the space of elemental variables. There is, however, a significant difference. PCA is "objective": It describes the directions of the axes and the amounts of variance in those directions without interpreting the directions as related or unrelated to stabilization of particular performance variables. The UCM method is "subjective": It always performs analysis with respect to a control hypothesis, that is, a hypothesis on a performance variable being stabilized or not being stabilized by the co-varying action of the elements.

For example, consider once again the very boring task of constant force production with four fingers pressing on their own force sensors. Consider, for simplicity, that all finger forces may be specified independently (in other words, ignore for now the phenomenon of enslaving). A fixed value of the total force, F_{TOT} corresponds to a single equation $F_1 + F_2 + F_3 + F_4 = F_{TOT}$. This equation defines a three-dimensional subspace (a UCM) in the finger force space. Application of the UCM analysis would result in computation and comparison of two components of the total variance in the finger force space, V_{UCM} and V_{ORT}. Imagine that the total force is indeed stabilized at F_{TOT} by co-variation of individual finger forces. Then, most of the variance is expected to be within the UCM, and a crucial statistical test would compare V_{UCM} with V_{ORT} per dimensions in the two spaces, the three-dimensional UCM and its unidimensional orthogonal complement.

Let us now suppose that PCA is applied to the same data set. The first three PCs are expected to span the UCM, since they account for most variance, while the fourth PC is expected to be orthogonal to the UCM. *Note, however, that for any arbitrary data distribution, the fourth PC by definition accounts for less variance than any of the first three PCs.* Hence, direct application of this method in principle does not answer the question: "Does co-variation of finger forces across trials stabilize the total force magnitude?" As we will see further in this section, a modification of PCA may shed some light on this issue.

PCA has been used in many studies to discover structure in clouds of data points. In particular, it has been used to discover co-variation in joint angular displacements over multi-joint actions by an extremity or by the whole body (Alexandrov et al. 1998; Freitas et al. 2006). A single PC that accounts for nearly all the variance in the joint space (space of elemental variables) is a strong argument for the existence of a control strategy that is responsible for this co-variation. It is up to researchers to interpret this fact and test whether it reflects stabilization of one or more performance variables.

For example, when a standing person performs a whole-body motion (sway) from an initial posture to a final posture, changes in the three main postural joints, the ankle, the knee, and the hip in a sagittal plane (see Figure 4.25) demonstrate strong co-variation across consecutive sway cycles. PCA applied to such data shows a single PC accounting for over 95% of the joint angle variance (Freitas et al. 2006). This means that the data cloud forms a very thin ellipsoid oriented along a line in the three-dimensional joint angle space. Is this a reflection of a synergy? Let us assume that some hypothetical performance variable is a function of the three joint angles and that it is stabilized by co-varied motion of the three joints. Stabilization of a value of this variable corresponds to a two-dimensional UCM in the joint space. PCA, however, shows that variance in the joint space is limited to a one-dimensional, not two-dimensional space. This suggests that there may be another constraint, possibly another important performance variable stabilized by a multi-joint synergy. So, likely, the results of PCA suggest the existence of two synergies, not one.

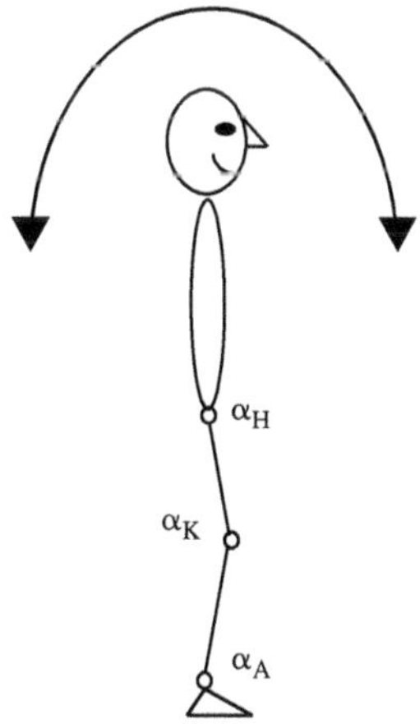

Figure 4.25. When a standing person is asked to sway, the three joint angles in the sagittal plane, the ankle, knee, and hip (α_A, α_K, and α_H, respectively) show strong co-variation. Application of principal component analysis (PCA) to such data results in a single PC that accounts for virtually all the data variance.

Indeed, further analysis using the UCM framework applied to such data showed that the described joint coordination had resulted from simultaneous stabilization of two performance variables during the swaying action, the instantaneous location of the center of mass of the body in the anterior–posterior direction and the orientation of the trunk with respect to the vertical. Both variables may be seen as important for postural stability. The first reflects the fact that in order for the person not to fall down, the center of mass should project onto the relatively small area of support. The second may be interpreted as ensuring relatively constant head orientation, for example, to avoid disturbances of visual perception and/or vestibular signals.

This example illustrates that UCM method can be used for *post hoc* analysis after PCA shows a strongly nonuniform data distribution within the space of elemental variables. But the situation can also be reversed. Imagine that UCM analysis has demonstrated a strong effect with respect to stabilization of a one-dimensional performance variable. This means that most variance is restricted to the UCM, a $(n - 1)$-dimensional subspace in the n-dimensional space of elemental variables. This analysis, however, does not indicate whether the variance within the UCM is also structured. PCA applied to the data point cloud within the UCM can answer this question and provide further insight into the control strategies associated with the task.

There is another potentially very important role for PCA in analysis of multi-element synergies. Recall that UCM analysis always starts with identification of elemental variables. Sometimes, this step is nearly trivial. In situations when apparent elements are numerous, and are unlikely to be specified by the controller independently, identification of elemental variables may benefit from PCA. A typical example would be virtually any multi-muscle action. It is relatively easy to record muscle activity and produce its quantitative index. Although the methods of recording and processing of muscle activation signals are many and varied, they are more of art than science (see Digression #8).

It would be naïve to assume that the brain manipulates as many variables as the number of channels in an EMG system a researcher happens to own. Since the times of the great British neurologist Hughlings Jackson (nineteenth century), researchers have agreed that the brain does not control muscles independently but unites them in groups and modulates the involvement of such groups into motor tasks. Hence, it seems more reasonable to associate elemental variables with such groups than with individual muscle activation levels. The number of such elemental variables is likely to be smaller than the number of muscles involved in typical motor tasks.

How can one define M-modes? Since, by definition, they correspond to groups of muscles, whose activities are scaled together over a family of similar tasks, PCA is the obvious tool of choice. As described in section 4.2.2, application of PCA to identify elemental variables (M-modes) has allowed quantifying multi-muscle synergies participating in a variety of multi-muscle motor tasks. One should, however, recognize limitations of PCA. In particular, this is an essentially linear method of analysis. EMG signals, on the other hand, are essentially nonlinear. In particular, there is no such thing as negative muscle activity: The worst thing a muscle can do is to be silent. This puts limitations on the utility of PCA.

One way to circumvent the problem is to look at deviations in indices of muscle activity from a particular baseline level. This allows the signals to be both positive and negative and, if their variations are assumed to be relatively small, PCA may be an adequate method. Another possibility is to use other methods of finding the best set of variables that describe an original data set most completely and economically: in other words, a minimal number of variables that account for a maximal amount of variance in the data set. Such methods are collectively called "matrix factorization techniques"; PCA is one member of the group, factor analysis is another. To deal with signals that can only be positive, another method has been developed called nonnegative matrix factorization (compared to other methods in Tresch et al. 2006).

4.4.2 Analysis of Surrogate Data Sets

Consider the task of throwing a basketball accurately into the basket from the free-throw line. Trajectory of the basketball after release is defined by its initial state and the laws of physics. If the ball does not spin, relevant initial conditions include its coordinates and the velocity vector at release. If we consider a simplified planar task illustrated in Figure 4.26, four variables affect accuracy of a free throw, the two coordinates of the ball $\{x_0, y_0\}$, the angle β of its velocity vector ***V***, and the magnitude of the velocity vector, *V*. Changes in any of these four variables have predictable effects on the coordinates of the ball $\{x_1, y_1\}$, when it falls down and passes the level of the ring. These coordinates should be

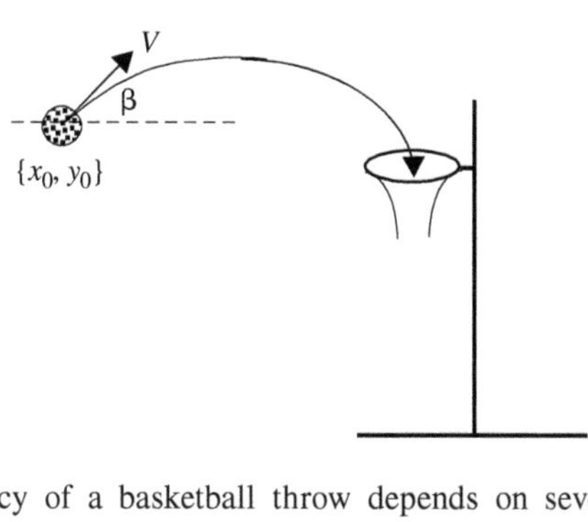

Figure 4.26. Accuracy of a basketball throw depends on several variables such as velocity at release (*V*), its direction (β), and coordinates of the ball release, $\{x_0, y_0\}$. To produce accurate throws in a series of attempts, these variables have to co-vary.

within a permissible error margin from the coordinates of the center of the ring. Relations between small variations in the four variables describing the initial conditions at the release and variations in $\{x_1, y_1\}$ may be strongly nonlinear. This may not allow building a formal linear model using a Jacobian of this system to estimate its UCM. To make things worse, these variables are measured in different units, meter/second, radians, and meters, which make explicit analysis of effects of their co-variation on the outcome variable (e.g. minimal deviation of the ball from the center of the target) problematic.

In such cases, when several elemental variables are expressed in incommensurable units and show nonlinear interactions in producing an important performance variable of the system, another method has been suggested to estimate effects of co-variation among elemental variables across repetitive trials at the task. This method involves creating a surrogate data set, which is supposed to contain no co-variation among the elemental variables. This can be achieved by selecting different elemental variables from different trials and computing predicted outcomes in such fictitious trials. Within such a surrogate data set, all elemental variables would have the same mean values and the same statistical distributions across trials as in the original set. Hence, any differences in the performance of the system may be attributed to co-variation among elemental variables present in the original data set and absent in the surrogate set.

For the described example of the basketball free throw, one can take the coordinates of the ball at release from trial #5, its angle of release from trial #8, and the magnitude of the velocity at release from trial #2. Then, a relatively simple computation allows predicting the coordinates of the ball when it is expected to pass through the ring level. As many surrogate trials can be created as necessary by mixing and matching different elemental variables from different trials. Then, performance across a set of actual trials may be compared to predicted performance across a set of surrogate trials. If the actual trials show lower variability of $\{x_1, y_1\}$, or a smaller percentage of errors, compared to the surrogate trials, one may assume that the actual trials contained co-variation among the elemental variables that was beneficial with respect to the ultimate performance. In surrogate trials, this co-variation was eliminated resulting in worse performance.

Several experimental studies used this method to discover co-variation among elemental variables in tasks such as ball throwing (Kudo et al. 2000; Martin et al. 2001, 2002) and playing skittles (Müller and Sternad 2003, 2004). These studies have been successful in showing better performance in the actual trials, which has been interpreted as a sign of a performance improving co-variation (a synergy) among elemental variables, with respect to the outcome of the performance defined by respective tasks.

The idea underlying this method has been further developed by Müller and Sternad (2004), who have suggested analyzing motor performance in such tasks, using maps between a set of elemental variables and a task variable. Such maps may be essentially nonlinear, which makes this method different from the UCM analysis. If such a map is known, one can analyze it with respect to a variety of questions. The most obvious being "What combinations of elemental variables produce the best performance?" This question is close to the issue of sharing in the view on synergies that is described in this book. A slightly less obvious question is "What pattern of co-variation among elemental variables is most beneficial for performance?" This question is close to the idea of error compensation, and it is also in the center of analysis within the UCM hypothesis. In addition, the method permits one more question that we have not as yet addressed: "How sensitive is performance to small variations in elemental variables?" The latter question has been also addressed using a different computational method close in spirit to the UCM method, the so-called method of *goal-equivalent manifold* (GEM, Cusumano and Cesari 2006). To be fair, it should be mentioned that the issue of tolerance to deviations in elemental variables had also been addressed in one of the first publications, using the framework of the UCM hypothesis for analysis of the sit-to-stand action (Scholz and Schöner 1999).

Using the earlier example of free basketball shooting, the three questions can be reformulated as follows: (1) What are the combinations of coordinates of release, angle of release, and velocity of release that produce accurate shots? In the terminology accepted in this book, this question refers to *sharing patterns* among elemental variables. (2) What co-variation among these variables will ensure that shots are still accurate, even if each of the elemental variables varies from trial to trial? This question refers to error compensation. (3) If there are still unavoidable variations in elemental variables that are not corrected by co-variation, in what area of the space of these variables will the errors have minimal impact on performance? This issue has been addressed as *tolerance*. Modifications of tolerance to errors in elemental variables may be important for producing optimal performance. In particular, a study of the effects of practice on performance in a computer-simulated task of playing skittles has suggested that the improvement of performance in that particular study was mostly due to better sharing (the first component), then to better tolerance (the third component), and only then to better co-variation (Müller and Sternad 2004).

So far, partitioning motor variability into three such slices remains controversial. For example, a change in data distribution with practice can be analyzed sequentially as a consequence of (1) a change in sharing, (2) a change in tolerance, and (3) a change in co-variation. Alternatively, the order of identifying the effects of changes in the three components may be switched, for example, (1) co-variartion, (2) tolerance, and (3) sharing. Results will very likely depend on ordering the three components. This does not allow claiming that changes in one of the three are more important than changes in the other two.

Application of the method of creating surrogate data sets to the task of multi-finger production of quick force pulses at predefined moments of time resulted in a somewhat unexpected outcome (Latash et al. 2004). The surrogate data set was less accurate in reaching the required level of peak force, but it was more accurate in doing this at an appropriate time. This leads us to the next section.

4.5 TIMING SYNERGIES: DO THEY EXIST?

Before addressing this question, let me reiterate what I mean by a timing synergy. A definition I am going to suggest is similar to the previous one used for synergies: A timing synergy is a neural organization that adjusts the timing of individual elemental variables to keep errors in the timing of an important performance variable low. For example, imagine that a pianist is playing a quick, difficult passage. The whole passage is to take 0.5 s. Imagine that the first three key strikes follow each other too quickly. Will the next key press slow down (without an intervention from the supreme controller) to keep the duration of the whole series closer to the required 0.5 s? Certainly, if the whole process is slow, the controller may have enough time to realize that the process needs to be speeded up and adjust the timing of later events, based on timing errors accumulated over the earlier portion of the trial. However, let us remember that we are interested in synergies that ensure stability of performance variables without an intervention from the hierarchically higher controller.

According to a scheme suggested by one of the founders of the UCM hypothesis, Gregor Schöner (1995, 2002), the generation of a purposeful motor action may be represented as a result of the functioning of a hierarchical control scheme that has four major levels (Figure 4.27). At the upper level (the *Task level*), the task and the effectors that will try to execute the task are identified; this is the "What to do?" level. At the next level, the timing of action is defined (the *Timing level*); this is a level that defines when to do the task and how quickly to do it. The next level is the *Synergy level*. At this level, elemental variables are organized to stabilize important performance variables. Finally, at the fourth level, called the *Force level* by Schöner, control of the biological motor—the muscle—is organized: This is a level, at which the tonic stretch reflex is supposed to participate in the generation of muscle forces and displacements according to the equilibrium-point hypothesis (see section 3.4).

Although the scheme in Figure 4.27 looks rather simplistic, it offers a particular general framework for analysis of the motor control system. In particular, it suggests that the timing signal is supplied to the *Synergy level* from a hierarchically higher level, and this signal is common for all elemental variables. Hence, according to this scheme, the level of synergies is in principle unable to organize timing synergies: Speeding up the process for one elemental variable

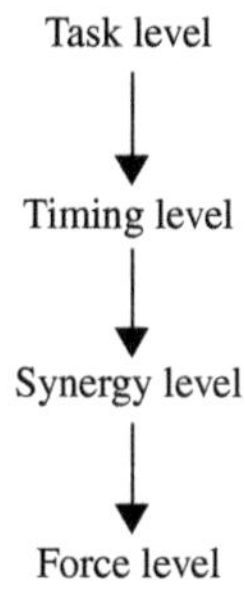

Figure 4.27. A four-level scheme of the organization of a voluntary movement offered by Gregor Schöner. At the upper level (the Task level), the task and the effectors are identified. The next level (the Timing level) defines when to do the task and how quickly to do it. At the next level (the Synergy level) elemental variables are organized to stabilize important performance variables. At the Force level, the control of the muscles is organized.

should be accompanied by speeding up for all other variables and, hence, the overall performance will be sped up as well.

This nontrivial prediction was studied using the method of creating surrogate data sets. The subjects in that study were instructed to produce a very quick pulse of force by pressing with the four fingers on four force sensors; the force pulse was supposed to be accurate in both magnitude and timing. This was achieved by showing the subjects their total force in real time as a signal running at a constant speed across the monitor screen. The target was a cross—its vertical line showed the allowed error in force magnitude, while the horizontal line showed a permissible timing error. Each finger in this test produced its own force pulse (see Figure 4.28). The four pulses summed up to produce the total force pulse with the peak that was supposed to coincide with the center of the cross target. Surrogate data sets were created by selecting at random pulses produced by individual fingers from different trials and then summing them up to generate a total force pulse that would have happened if those finger pulses happened together. If there is a co-variation among timing and/or amplitude characteristics of the actual pulses that contributes to more accurate performance, the actual pulses are expected to be more accurate as compared to the surrogate ones, since this co-variation is expected to be absent in the surrogate data set.

The results of that study looked unexpected; however, they fit well with Schöner's scheme. The surrogate data set was indeed less accurate in producing the required magnitude of the force pulse, but it was more accurate in reaching the peak force at the correct time. The first result suggests that there was a negative co-variation among the magnitudes of the individual force pulses in the actual data set that helped reduce error of the total force magnitude. The second result suggests that there was a positive co-variation among the timings of the individual force peaks. This positive co-variation contributed to the timing

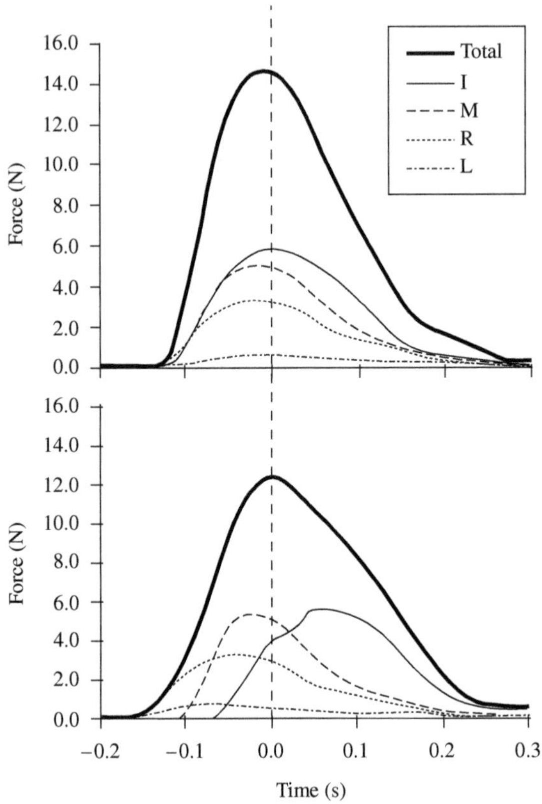

Figure 4.28. If a person produces quick force pulses with four fingers acting in parallel, the individual finger pulses show negative co-variation of magnitudes (stabilizing the total force magnitude) but positive co-variation of timing of their peaks (destabilizing the timing of the peak force) across trials. If individual finger pulses are taken from different trials at random, the result is a total force pulse of a wrong magnitude but accurate timing (the bottom panel). Reproduced by permission from Latash ML, Shim JK, Zatsiorsky VM (2004) Is there a timing synergy during multi-finger production of quick force pulses? *Experimental Brain Research* 159: 65–71, © Springer.

error of the total force. So, when it was eliminated in the surrogate data set, the performance became better (more accurate). In other words, in the actual data set, if one element sped up in a particular trial, all other elements were also likely to speed up, thereby adding to the overall timing error.

This result could be blamed on the formulation of the task, which implied synchronous action by the fingers. So, in a follow-up study, a more complex task was used. In that task, fingers were supposed to produce force pulses in a quick sequence as if playing a quick musical phrase. The task was the same: To reach a required total force pulse magnitude at a required time. This task was indeed hard to learn for the subjects (nonpianists), and many of their trials were not included in analysis because they made mistakes in the finger sequence. When all the acceptable trials were analyzed, the result was the same as in the earlier study: There was a magnitude synergy among individual finger force pulses but there was no timing synergy. In a sense, there was an "anti-synergy," a co-variation that destabilized the timing of the peak of the total force.

This does not mean, certainly, that humans are unable to produce accurately timed actions with several effectors. But such tasks may involve processes at a higher level of the control hierarchy as suggested by Schöner's scheme. For example, there may be a neural organization at the *Timing level* that stabilizes the timing of motor actions using its own set of elemental timing variables different from those that form synergies at the hierarchically lower level. This is all very speculative since no studies have investigated the nature of these hypothetical variables and interactions among them.

In the next part of the book, I am going to illustrate how the computational method of the UCM hypothesis can be applied to discover and quantify synergies in a variety of tasks. One of the reasons this method seems so attractive is that many of the early experimental studies using the method led to unexpected results. These results led to new questions that required new experimental and theoretical studies. Figuratively speaking, the method opened a few new cans of worms.

Part Five

Zoo of Motor Synergies

5.1 KINEMATIC SYNERGIES

During natural multi-joint movements, combinations of rotations in individual joints typically produce smooth trajectories of the endpoint in the external space. Commonly, such trajectories during point-to-point motions follow a straight path from an initial to a final position (Morasso 1981, 1983; Soechting and Lacquaniti 1981; Abend et al. 1982; Atkeson and Hollerbach 1985). This outcome is far from trivial and apparently requires coordinated motion of individual joints. Indeed, rotation in any one of the joints is expected to produce a curved motion of the endpoint of the limb. To get a straight-line trajectory, coupling of joint rotations is necessary.

Natural trajectories of voluntary movements show certain features that are consistent across tasks and effectors. For example, there is a relation between the tangential velocity and curvature of the trajectory known as the *two-third power law*: Movements slow down when they change direction (Morasso and Mussa Ivaldi 1982; Viviani and McCollum 1983; Viviani 1985). Another general kinematic feature of natural trajectories is their smoothness, which has been expressed as the minimum-jerk criterion (Hogan 1984; Flash and Hogan 1985). According to this criterion, an integral measure of the time derivative of acceleration over movement duration is minimized during natural movements.

We are now more interested in issues related to kinematic multi-joint synergies: How are rotations in individual joints coordinated to generate reproducible, task-specific trajectories of the endpoint? Are there preferred patterns of joint rotation during natural movements? Do these patterns show error compensation in a sense that if one joint moves the endpoint too much in a particular direction, other joints would move it less in that direction (compared to their average effects across trials)? In this section, we will only consider joint kinematics and leave issues of force and torque coordination for the next section on kinetic synergies (for recent kinematic models see Dounskaia 2007; Flash and Handzel 2007).

Principal component analysis (PCA) and correlation analysis have been used in a number of studies of joint kinematics during various natural movements, including whole-body sway, pointing, reaching, grasping, and even such complex everyday movements as bringing a morsel of food to the mouth and taking a sip from the mug (Castiello 1997; Alexandrov et al. 1998; van der Kamp and Steenbergen 1999; Latash and Jaric 2002). Typically, this analysis reveals a very strong pattern of co-variation, commonly a single principal component that accounts for virtually all the variance in the joint space. In other words, individual joint rotations frequently scale together in a close-to-linear fashion. Such analyses were performed both across repetitions of a task and along a trajectory during a single realization of a task. The observations of linear co-variation among joint trajectories cannot be a universal rule. It is easy to come up with an example of a movement when joint rotations cannot scale over the whole movement duration. For instance, moving the right index fingertip from an initial position to the right of the trunk, as shown in Figure 5.1, to a final position located to the left of the trunk typically leads to a monotonic change in the shoulder joint and a nonmonotonic sequence of flexion-extension in the elbow joint, a coordination illustrated in the bottom panel of Figure 5.1. Besides, as we have already discussed, PCA has inherent limitations in revealing an important feature of synergies, that is, their ability to stabilize performance variables.

One of the most influential studies of movement kinematics was the earlier mentioned experiment of Nikolai Bernstein on professional blacksmiths. Recall that Bernstein filmed electric bulbs attached at the end of the hammer and at the major joints of the dominant arm of a blacksmith. The blacksmith performed a sequence of standard actions hitting the chisel held by the other hand. The main finding was that the variability of the trajectory of the tip of the hammer was lower than the variability of the trajectories of individual joints. Since the brain was apparently unable to send signals directly to the hammer, Bernstein concluded that joint rotations did not vary independently but were instead united into a multi-joint synergy.

Since that seminal study, a number of experiments have shown relatively accurate performance of a task when contributing elements showed relatively high

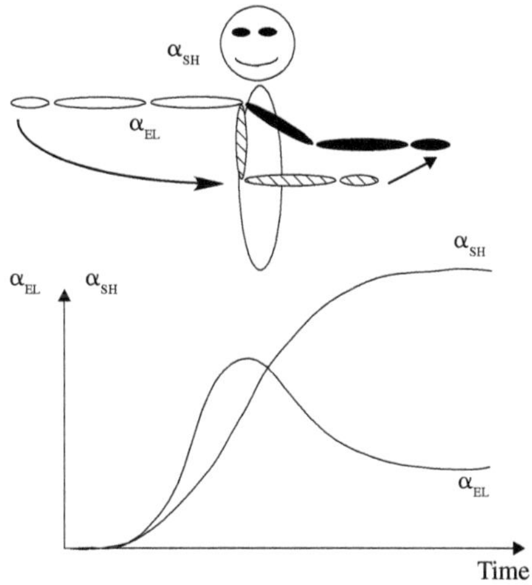

Figure 5.1. An arm movement crossing the midline of the body leads to monotonic changes in the shoulder joint and nonmonotonic changes in the elbow joint (as illustrated in the bottom graph). This example shows that joint kinematics do not co-vary in a linear fashion.

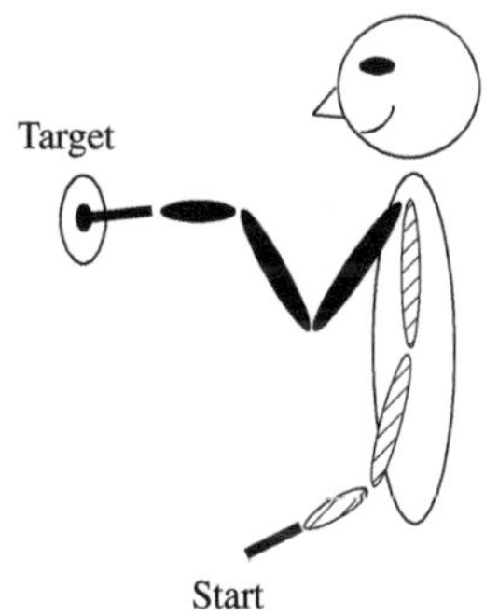

Figure 5.2. A person performs a pointing movement with a hand-held pointer from a starting position to a target. Errors in rotation of proximal joints (shoulder and elbow) are expected to lead to larger location errors of the pointer tip compared to the wrist joint.

variability (reviewed in Latash et al. 2003a, 2007). Suppose that a person is asked to point accurately at the center of a target with a pointer. Suppose also that all movements start from a relatively clumsy posture that requires substantial motion of all major arm joints (Figure 5.2). To makes things easier, consider a simplified case of two-dimensional pointing with three major joints, shoulder, elbow, and wrist contributing to the performance. All joint rotations are expected to show some variability across trials. Variability in the shoulder and elbow joint rotations will lead to variability in the locations of both the wrist joint and the pointer tip. The pointer tip is located farther away from each of the two proximal joints.

Therefore, a small rotational error in one of these two joints is expected to lead to a larger spatial error in the location of the pointer tip than in the location of the wrist. However, experiments have shown that indices of spatial variability for the pointer tip can be smaller than for the wrist (Jaric and Latash 1999). This could happen only if wrist rotation compensated (partly) for variability of the pointer tip location expected from imprecise rotation of the shoulder and elbow joints. As such, these studies implied that a synergy existed among the three joints stabilizing the final location (or maybe even the whole trajectory) of the pointer tip.

The uncontrolled manifold (UCM) approach has allowed addressing such questions more formally and investigating quantitatively trajectories of variables that could potentially be stabilized by multi-joint kinematic synergies during a variety of natural movements. The studies reviewed further in this section used the UCM approach to study the flexibility/stability feature of motor synergies for very different coordination problems. In all studies, it was assumed that individual axes of joint rotation are independent elemental variables. As mentioned earlier, this assumption is not trivial and may even be wrong because of the existence of bi-articular muscles and inter-joint reflexes (reviewed in Nichols 1994, 2002; Zatsiorsky 2002).

5.1.1 Postural Synergies in Standing

The task of maintaining balance while standing is not trivial. The complexity of postural control is due to three major factors. First, the inherently unstable mechanical structure of the human body, which is commonly modeled as an inverted pendulum with several joints along its axis, needs to be balanced over a relatively small support area (Figure 5.3A). Second, there are frequent changes in external forces acting on the body (Figure 5.3B). Third, voluntary actions performed by a standing person are themselves sources of postural perturbations. This is due to changes in the geometry of the body leading to displacements of the center of mass, interaction with external objects (e.g. during such actions as picking up, catching, pushing against, or dropping an object), and reactive forces due to the mechanical coupling of the body segments (Figure 5.3C). Correspondingly, three phenomena related to the three factors mentioned are commonly studied in experiments. These are postural sway, pre-programmed reactions, and anticipatory postural adjustments (APAs).

Postural sway is typically studied as spontaneous changes in the location of the center of mass or in the coordinates of the center of pressure (COP, the point of application of the resultant force acting from the supporting surface on the body), when a person tries to stand quietly in the absence of any overt actions or changes in external forces. There are two views on the nature of postural sway: It has been considered a consequence of "noise" within the system of postural control (Kiemel et al. 2002), as well as a purposeful search process exploring the limits of postural stability (Riccio 1993; Riley et al. 1997). In particular, when

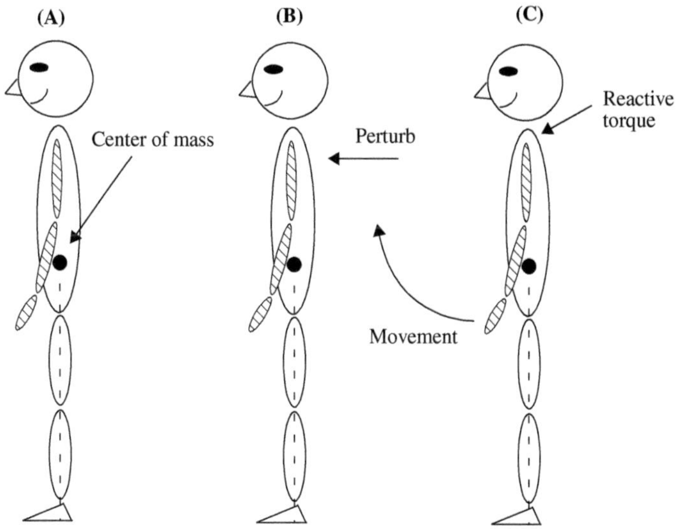

Figure 5.3. Vertical posture is inherently unstable because of the high center of mass location, small area of support where the center of mass should project (the dashed vertical line), and a number of joints along the axis connecting the center of mass to the feet (A). (B) The problem is exacerbated, because of perturbations by external forces that any person experiences frequently. (C) A fast movement by a standing person is by itself a perturbation for balance, because of the mechanical coupling of the body segments.

subjects are required to balance while standing on a narrow support surface, postural sway increases rather dramatically (Latash et al. 2003b; Mochizuki et al. 2006). This sway increase seems counter-intuitive because it apparently leads to higher chances of losing balance. It represents one of the strong arguments that sway is not a reflection of noise but rather a purposeful action by the nervous system. Along similar lines, sway has been shown to depend on the amount of attention the subjects pay to the task of standing (Donker et al. 2007).

A recent study tested the hypotheses that all major joints along the longitudinal axis of the body (six joints were analyzed from the ankle to the neck) are actively involved in balance during quiet standing (Hsu et al. 2007). The study used the toolbox of the UCM hypothesis to analyze effects of co-variation of joint angle displacements on the spatial positions of the head and of the center of mass of the body during natural postural sway. The UCM analysis revealed that the six joints were indeed united into synergies stabilizing the two variables ($V_{GOOD} >> V_{BAD}$). Removing vision increased the joint sway but most of this increase was channeled into V_{GOOD}, that is, it had minimal effect on sway of the center of mass or head.

If a standing person is asked to sway voluntarily, for example, to move the body in a cyclic fashion forward and backward without moving the feet, joint displacements increase even more. We mentioned application of PCA to such tasks earlier (section 4.4.1): PCA shows a single Principal Component (PC) accounting for over 95% of the variance in the three-dimensional space of the main leg joint angles, the hip, the knee, and the ankle. This means that joint variations are effectively limited to a one-dimensional subspace, that is, a line in the joint space. Stabilization of any performance variable that is a function of the three joint angles is expected to lead to the data limited to a two-dimensional subspace (the UCM). The results of the PCA suggest, therefore, that maybe two performance variables are being stabilized at the same time during voluntary sway tasks. One candidate variable is obvious: The horizontal coordinate of the center of mass. And indeed, just like in the mentioned study of spontaneous sway, UCM analysis shows that significantly more variance is limited to the UCM computed for this variable than in a subspace orthogonal to this UCM ($V_{UCM} > V_{ORT}$ or, equivalently, $V_{GOOD} > V_{BAD}$).

But what could the other performance variable be? Observations of subjects performing voluntary body sway tasks suggested that, while they swayed, the leg joint angles changed significantly, but the trunk remained rather accurately aligned with the vertical. Since trunk orientation is apparently a simple function of the three major joint angles, analysis of variance in the joint space was performed with respect to this candidate performance variable, and it indeed confirmed that joint angles co-varied in such a way that they kept the trunk orientation relatively unchanged. There may be different functional reasons to avoid excessive changes in the trunk orientation during such tasks including, for example, minimization of head displacement to preserve constancy of visual perception and vestibular signals. An important lesson, however, is that a combination of PCA and UCM methods allows interpreting such results as a single principal component accounting for a lot of variance as a superposition of two (or potentially more) synergies with separate functional purposes.

In a recent study, the UCM method was applied to analysis of joint angle co-variation during postural sway in subjects who tried to stand quietly (Krishnamoorthy et al. 2005). The analysis has shown that most of the variance in joint angle space was compatible with a fixed horizontal position of the center of mass; in other words, it was within the UCM computed with respect to this particular performance variable. When the subjects closed their eyes, an increase in the postural sway was accompanied by the strengthening of the multi-joint synergies stabilizing the center of mass coordinate. This means that the *good variance* (V_{UCM}) increased more than the *bad variance* (V_{ORT}) under such conditions.

Studies of postural reactions to unexpected perturbations revealed certain preferred patterns of joint involvement. Such reactions termed in different studies as long-latency reflexes, pre-programmed reactions, M_{2-3} responses, or triggered reactions (see Digression #5) have been commonly induced in experiments by unexpected quick translations or rotations of a platform on which the subject is standing. If a young, healthy person stands on such a platform, its small motion typically induces the quickest and largest reaction in the ankle (Nashner and Cordo 1981; Horak and Nashner 1986). Other joints certainly move as well, but the amplitude of their motion is smaller. This corrective pattern has been termed the *ankle strategy*. If a young person stands on a narrow support, a different pattern is seen involving primarily the hip joint—the *hip strategy*. The hip strategy has also been described in the elderly, in response to similar perturbations applied while they stood on the platform without additional instability (Horak and Nashner 1986; Woollacott and Shumway-Cook 1990). More recently, a particular quantitative approach has been suggested that offers a physical foundation for the experimentally found ankle and hip strategies (Alexandrov et al. 2001a,b, 2005). It identifies three eigen-movements in the space of three major leg joints, the ankle eigen-movement, the knee eigen-movement, and the hip eigen-movement. The eigen-movements represent movements along eigenvectors of the equations of motion written for the three-joint system. The advantage of the eigen-movement approach is in representing the inherently coupled dynamic equations of motion of such a system in the form of three independent equations. This approach goes beyond purely kinematic analysis, because it is based on consideration of both joint torques and displacements.

Can ankle and hip strategies be viewed as synergies in our current understanding? There is no clear answer to this question, because joint patterns to perturbations have not been analyzed with respect to specific performance variables that they could potentially stabilize. It is quite possible that the two strategies are indeed two synergies stabilizing the horizontal position of the center of mass and/or some other performance variables. Then, the two patterns of joint deviation corresponding to the two strategies are expected to be mostly within the UCM computed with respect to this performance variable.

APAs are seen prior to expected postural perturbations (reviewed in Massion 1992). They apparently reflect a person's expectation with respect to the upcoming perturbation, rather than actual features of the perturbation. As a result, APAs are typically suboptimal and lead to only approximate corrective actions. APAs have commonly been studied at the level of electromyographic (EMG) variables and as displacements of the COP. There are only a few studies that report measurable joint deviations during the APAs. This is not surprising, since EMG changes during the APAs usually appear about 100 ms prior to the perturbation. These changes need time to produce visible joint motion due to both the unavoidable electromechanical delay (the delay between the electrical and mechanical phenomena within the human body, Corcos et al. 1992) and the high inertia of the body. Several patterns of joint involvement have been described during the APAs, but these patterns are variable and have not been analyzed with respect to stabilization of performance variables. Further, I am going to describe how the UCM method can be applied to analysis of EMG signals and consider APAs as an exemplary phenomenon, to which such analysis can be applied. However, analysis of kinematic synergies during APAs is next to impossible because of the very small joint displacements. Therefore, we move to another well-studied task where joint coordination has been quantified.

5.1.2 Sit-to-Stand Task

Rising up from a sitting posture is a very common motor task. However, mechanically, it is quite challenging: The original sitting posture with the relatively large base of support is expected to become considerably less stable because of two factors, a decrease in the area of support and an elevation of the center of mass of the body above the support level (Figure 5.4). In addition, the task is associated with large transient joint torques required to accelerate and decelerate the body. Any errors in these torques may be expected to lead to errors in associated joint rotations, which may lead to losing balance and failure at the task. Nevertheless, all healthy persons can perform this task easily, suggesting that there are inter-joint synergies that help stabilize important variables related to balance during the sit-to-stand action.

It is natural to assume that the coordination of the major joints involved in the sit-to-stand action stabilizes the trajectory of the center of mass of the body. For now, ignore possible deviations of the center of mass coordinate in the medio-lateral direction; these motions may be considered small as long as the person performs a symmetrical action leading to an even distribution of the body weight

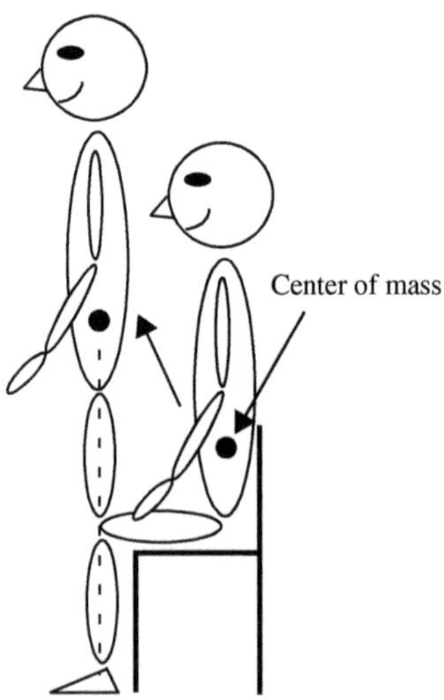

Figure 5.4. A sit-to-stand motion is associated with a displacement of the center of mass (the black dot) in both horizontal and vertical directions. Accuracy of the horizontal displacement is crucial not to move the projection of the center of mass outside the small support area.

between the two legs. The center of mass, however, has to move in both vertical direction and in the anterior–posterior horizontal direction to accomplish the task (Figure 5.4). Only the horizontal coordinate of the center of mass is essential for postural stability, while the vertical coordinate does not appear to be crucial. So, two control hypotheses can be formulated. First, that deviations of the joints from a preferred (average) trajectory across trials are organized to stabilize the horizontal (anterior–posterior) trajectory of the center of mass. Second, that this coordination also stabilizes the vertical trajectory of the center of mass.

Experimental studies of the sit-to-stand task under different constraints, such as changes in the area of support under the feet, variations of the mass of the body, and availability of visual information (Scholz and Schöner 1999; Scholz et al. 2001), have all supported the first hypothesis but not the second one. The instantaneous horizontal position of the center of mass was clearly stabilized by a synergy reflected in an inequality $V_{UCM} > V_{ORT}$, between the two variance components quantified within the respective UCM and orthogonal to it in the space of joint angle displacements. This result was strongest in the middle of the movement, as illustrated in Figure 5.5, where the center of mass was perhaps in its most precarious position. The effect also was stronger when standing up under challenging task constraints such as on a narrow support surface (Scholz et al. 2001).

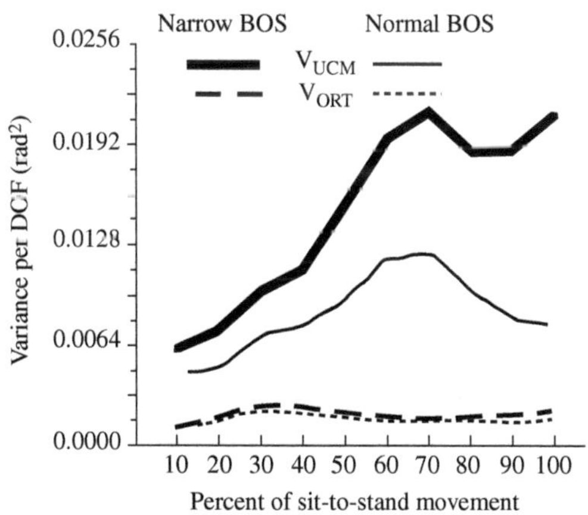

Figure 5.5. Analysis of variability of the joint kinematics across repetitive trials shows that most variability is "good" with respect to the trajectory of the center of mass in the horizontal direction ($V_{GOOD} > V_{BAD}$). This result was strongest in the middle of the movement, where the center of mass was perhaps in its most precarious position. Reproduced by permission from Scholz JP, Reisman D, Schoner G (2001) Effects of varying task constraints on solutions to joint coordination in a sit-to-stand task. *Experimental Brain Research* 141: 485–500. © Springer.

In contrast, the joint motion did not show a synergy stabilizing the vertical path of the center of mass (Scholz et al. 2001), that is, variations of the joint configuration quantified at each phase of the action across repetitions of the sit-to-stand movement were just as likely to lead to a change in vertical coordinate of the center of mass as to keep it stable ($V_{UCM} \approx V_{ORT}$). The effect was quite different, however, for sitting down. In that case, the synergy of joint motion led to stabilization of both the horizontal and vertical position of the center of mass, perhaps to help ensure soft landing on the chair (Reisman et al. 2002). The conclusions are that standing up is associated with a kinematic multi-joint synergy that stabilizes the horizontal trajectory of the center of mass but not its vertical trajectory. During sitting down, however, there is an additional synergy that stabilizes the vertical trajectory as well.

5.1.3 Reaching

When a person produces a reaching point-to-point arm movement within a comfortable space, the endpoint (hand or fingertip) trajectory typically shows a close-to-straight path with a smooth bell-shaped velocity profile (Morasso 1981, 1983). As mentioned earlier, this behavior is nontrivial and requires coordinated rotation in major arm joints. Two types of inverse problems have been studied with respect to reaching movement, those of inverse kinematics and of inverse dynamics. We discussed these problems earlier, in section 3.3. Since we are now discussing kinematic synergies, only the problem of inverse kinematics is relevant.

Let us identify two components in the problem of inverse kinematics. First, how can one select joint rotations that would, on average, ensure motion of the endpoint of the arm to the target? Second, how should joint rotations co-vary across trials to ensure stable performance along the selected trajectory. The former problem is expected to produce a particular *sharing pattern* among the joints, while the latter problem addresses the issue of error compensation or flexibility/stability. One of the computational approaches to the former component of the problem of inverse kinematics is that of *optimization*. Let us first consider possible optimization principles underlying the selection of a trajectory for "an important point" (e.g. the endpoint of a multi-joint limb). Further, we will query: Given a trajectory, how should joint rotations be organized?

Digression #9: Optimization

> A central notion common across all optimization approaches is that of cost function. Cost function is a particular function of the system's performance (sometimes including hypothetical control variables) that the controller tries to keep at an optimal value, commonly at a minimal or at a maximal possible value. Movement studies have used a variety of cost functions related to mechanical performance by the system, hypothetical control processes within the system, and psychological factors such as effort and comfort (reviewed in Nelson 1983; Latash 1993).

Imagine that a movement in a single joint from a certain initial to a certain final position is to be performed as quickly as possible, that is, within the minimal time. In this case, movement time can be used as a cost function that has to be minimized. The optimal solution would involve the so-called *bang-bang control* (Figure 5.6; Nelson 1983), that is, using the maximal levels of acceleration and deceleration. This type of control is also known as the teenager driving principle: Pressing the gas pedal as strongly as possible when one sees the green light, and pressing the brake pedal as strongly as possible, when the light turns red. The velocity profile for this mode of control does not look like a typical bell-shaped pattern observed in experiments. For a purely inertial system, the optimal switching time between the accelerating and decelerating ""bangs"" is exactly in the middle of the movement. For a system with energy dissipation (e.g. due to velocity-dependent damping), the optimal point of switching the "bangs" is shifted. A somewhat similar approach involves minimization of the overall impulse (time integral of force). However, it also fails to generate trajectories that look like those during human movements.

Probably the most influential kinematic cost function in movement studies has been an integral measure of squared jerk. Jerk is the time derivative of acceleration (or the third time derivative of displacement). Application of this cost function has been termed the *minimum jerk criterion* (Hogan 1984; Flash and Hogan 1985):

$$J = \frac{1}{2}\int_{0}^{\mathrm{MT}} (\mathrm{d}a/\mathrm{d}t)^2\,\mathrm{d}t \qquad \text{Equation (5.1)}$$

where a stands for acceleration and MT stands for movement time. The jerk cost is sometimes normalized by movement time squared to compare movements at different speeds. The minimum jerk criterion leads to smooth trajectories with a bell-shaped velocity profile, and a symmetrical double-peaked acceleration (Figure 5.6, bottom). It provides a close fit to endpoint trajectories during fast human arm movements.

A different optimization approach has been suggested based on an assumed relation between muscle force generation and fatigue (Crowninshield and Brand 1981; Prilutsky 2000). In particular, this minimal fatigue criterion has been successfully applied to the problem of defining patterns of individual muscle force production during tasks that involve multiple muscles.

Another group of optimization criteria includes less strictly defined movement characteristics such as comfort and effort (Hasan 1986; Cruse and Bruwer 1987; Bruwer and Cruse 1990; Emken et al. 2007; Guigon et al. 2007). They have sometimes been combined with more mechanistic criteria to form complex cost functions. In general, adding a criterion to a cost function can only lead to an improvement of the fit between predictions based on the cost function and data.

A rather sophisticated optimization approach has recently been developed, based on an idea that movement planning is performed in posture space based on a set of memorized postures (Rosenbaum et al. 1993, 2001). This posture-based

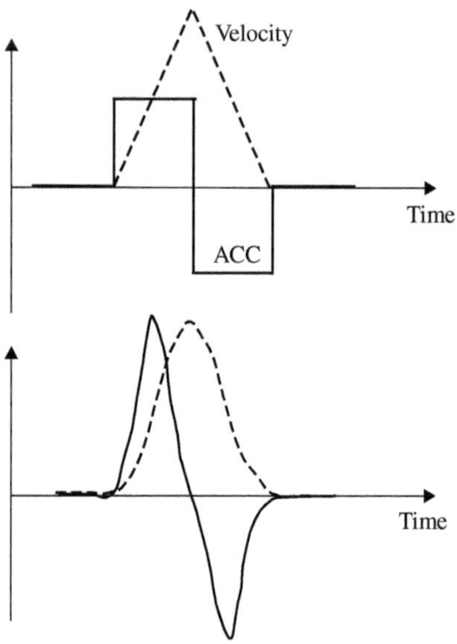

Figure 5.6. Top: The "bang-bang" control leads to acceleration and velocity profiles that do not look like the smooth profiles typical of natural fast movements. Bottom: The minimal-jerk criterion produces movement kinematics that is smooth and similar to those observed during natural movements.

planning approach includes selecting a target posture and a trajectory moving the effector from the initial posture to the target posture based on a number of criteria associated with costs of moving the endpoint, moving individual joints, clearing obstacles, etc.

A typical common feature of all optimization approaches is that they require an estimate of the cost function over movement time before the movement is initiated. This implies, in particular, that movement time should be known to the controller in advance. This condition also assumes that each movement has a clear beginning and a clear end. Both assumptions look rather artificial; since everyday movements are typically smooth, they do not show abrupt starts and finishes, and movement time does not seem to be preplanned but rather emerges as a result of control processes and external force fields. Humans can perform movements faster or slower, but they are not very good at performing them within a given movement time. At least, movement time seems to be the least reproducible movement variable, compared, for example, with final position, distance, or force level.

If a trajectory of the endpoint of a multi-joint limb has been selected, the next step is to select a combination of joint rotations that would bring this trajectory about. If the system has no kinematic redundancy, this step is trivial, since any position of the endpoint has a unique joint combination that brings it about. As a result, a trajectory of the endpoint defines unambiguously trajectories of the joints. In a redundant system—and we are interested in such systems—there is no unique solution for this problem. Optimization techniques can be applied at this step as well, with the cost functions built using joint-specific variables rather than endpoint-specific ones. Some of the mentioned optimization approaches, in particular those by Cruse and Bruwer (1987) and by Rosenbaum et al. (1993), solve this problem together with selecting an optimal endpoint trajectory. Alternatively, joint trajectories corresponding to a particular endpoint trajectory may be selected, using computational techniques that are not based on an explicit cost function.

End of Digression #9

A generalization of the equilibrium-point (EP) hypothesis has been developed to address the generation of multi-joint reaching movements. It has been addressed as the reference configuration hypothesis (Feldman and Levin 1995). This hypothesis assumes that the neural controller has an ability to modify reference joint configurations (joint configurations, at which the system would be at an equilibrium) using control variables to individual muscles that define the thresholds of the tonic stretch reflex for those muscles (see section 3.4). As such, the reference configuration hypothesis assumes movement planning and control in joint space, similar to the mentioned *posture-based planning hypothesis* of Rosenbaum and colleagues (Rosenbaum et al. 1993, 2001). It is more in the spirit of this book to consider two levels of control of a multi-joint movement (actually, more than two, but two would be enough for now).

In section 3.4, I introduced a notion of generalized displacement reflex (GDR) as a hypothetical mechanism that brings about the relationship between external force and endpoint location in space. GDR is not necessarily related to stretch of a certain muscle, or to activity of muscle spindles, but to a more general notion of displacement. It may get contribution from all the receptors, whose firing level changes as a result of displacement including muscle spindles, Golgi tendon organs, articular, cutaneous, and subcutaneous receptors. There are a number of differences between the tonic stretch reflex (TSR) and GDR. In particular, since muscles can only pull but not push, TSR action is unidirectional. GDR always involves pairs of muscles that oppose each other (so-called *antagonist muscle pairs*); therefore, its action is bi-directional, that is, it can both pull and push.

Effects of GDR on the endpoint location may be described with two characteristics, location to which the endpoint tends to move given a constant external force field (its EP, a three-dimensional vector) and how strongly it resists when a change in the external force tries to move it away from the EP. This latter characteristic may be formally represented in a first approximation with an ellipsoid of apparent stiffness (cf. ellipses of stiffness in Flash 1987), which can be described with the directions and lengths of its main axes. Let us assume a two-level hierarchical control scheme where, at the higher level of the hierarchy, the central nervous system (CNS) manipulates a hypothetical control variable, $\Lambda(t)$, that, given an external force field, defines the EP and the orientation and size of the apparent stiffness ellipsoid for the endpoint of the limb. Further, this variable is expected to lead to control variables to individual muscles, namely, their thresholds of the TSR, $\lambda(t)$.

At the next stage, multi-λ synergies may be expected that define patterns of control variables to muscles in such a way that a required $\Lambda(t)$ is stabilized by co-varying changes in $\lambda(t)$. Unfortunately, until now, studies of such control synergies have been elusive due to technical problems. Besides, the story of $\Lambda(t)$ and $\lambda(t)$ is not universally accepted because there is no agreement even on the nature of control signals to individual joints and muscles. At this point, let us skip this obstacle and ask a question that is more directly relevant to the notion of multi-joint *kinematic* synergies: Are there multi-joint synergies that stabilize preferred endpoint trajectories? Under multi-joint synergies we imply co-varied changes in individual joint rotations that stabilize a desired endpoint trajectory, not co-varied changes in control signals. Naturally, the UCM approach can be used to find an answer to this question.

5.1.4 Reaching in a Changing Force Field

John Scholz and his colleagues used the UCM method to analyze multi-joint reaching to a target under varying conditions (Yang et al. 2007). In those experiments, movements were constrained to a two-joint plane, while three major joint angles participated. Therefore, the system was kinematically redundant (Figure 5.7). When subjects performed a series of reaching actions to the target, they moved the endpoint along a nearly straight trajectory. Analysis of joint variability was performed in the same way as described earlier. It involved the following steps.

First, average trajectories of the endpoint and of the joint angles were computed over a series of trials performed by a subject. The whole trajectory was then viewed as a sequence of small steps. For each step, deviations of joint angles from their average values were computed in each trial. These deviations were projected onto the UCM corresponding to the average endpoint location and orthogonal to it. The manifold was computed based on the average joint configuration and the length of arm segments, that is, based on the average Jacobian of the system at that particular phase of the trajectory. Further, variance in the joint space across trials was computed within the UCM and orthogonal to it. These two values (V_{UCM} and V_{ORT}) were compared per degree-of-freedom in each subspace.

The main result was that the amount of joint variance within the UCM was significantly larger than orthogonal to the UCM, meaning that there was a multi-joint kinematic synergy stabilizing that particular endpoint location. This difference was particularly large during the first half of the movement, but it remained high over the whole movement duration.

In those experiments, the authors also used a robotic arm to produce perturbing forces directed orthogonally to the endpoint trajectory (Figure 5.7, bottom). When a trajectory was perturbed by a robotic arm, the endpoint trajectory became curved and showed an increase in its variability. When the same perturbation was repeated several times in a reproducible and predictable fashion, it took the subjects a few trials to start generating straight endpoint trajectories (Figure 5.8A). Later, when the force field was turned off, the trajectories became curved in the opposite direction (Figure 5.8B; also see Shadmehr and Mussa-Ivaldi 1994; Malfait and Ostry 2004). New synergies were elaborated that stabilized the trajectory in the new conditions as reflected by a change in both V_{UCM} and V_{ORT}. The adaptation to reaching in the force field was accompanied initially by an increase in both components of variance, followed by a smaller decrease of V_{UCM} than V_{ORT}. The larger V_{UCM} following adaptation seems a bit puzzling. Indeed, why would the controller increase variance that has no effect on performance? These observations suggest that the CNS makes use of kinematic redundancy to ensure flexibility of motor patterns that may help facilitate adaptation to unusual force conditions. It also speaks against an idea of elaborating an optimal *internal model* during the adaptation (see section 3.3).

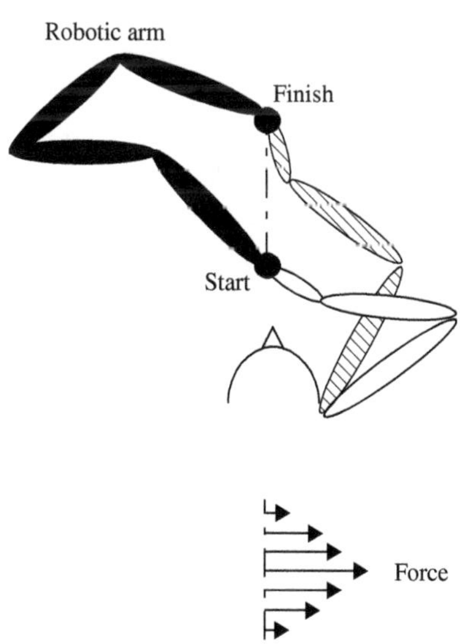

Figure 5.7. In this experiment, the subject was asked to move a handle horizontally along a straight line from a starting position to a target. A robotic arm (black) could apply forces to the handle proportional to its tangential velocity and directed perpendicular to the velocity vector.

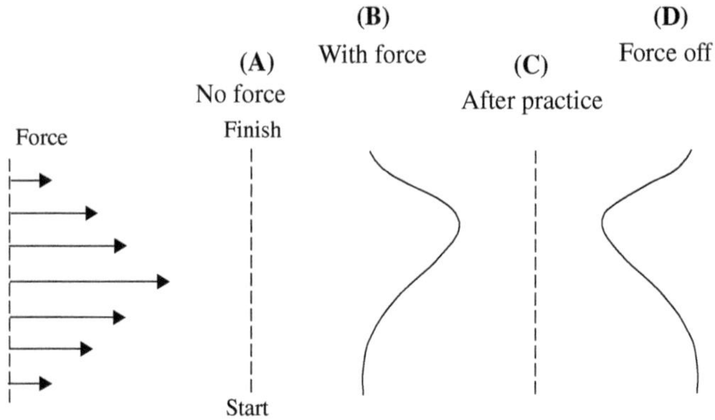

Figure 5.8. (A) In the absence of external forces, the subject shows a straight trajectory to the target. (B) When the force is applied, the trajectory becomes curved. (C) After a few minutes of practice, the subject is able to produce straight trajectories in the presence of the velocity-dependent force field. (D) When the force field is turned off, a few trials are curved in the opposite direction compared to the curvature of the first trials immediately following the application of the force field.

5.1.5 Multi-Joint Pointing

Pointing to objects is a well-practiced task that humans perform daily from the first year of life onward. It is, therefore, reasonable to assume that important characteristics of pointing are stabilized by multi-joint synergies. But what is important for the success of a pointing action? It depends on the action. If you try to point at a distant object with a finger, apparently, the direction from the observer (yourself) to the object should coincide with the direction from self to the finger. Other characteristics, such as, the actual location of the finger in space, are not important. In contrast, when you want to point at an object and touch it (or nearly touch it) with a fingertip or with the tip of a pointer, you need to make sure that the distance between the tip of the pointer (or finger) and the object is very small or even zero. The angle, at which the pointer (finger) approaches the object, may or may not be important. Let us start with the second task of placing the tip of a pointer into a target.

One of such studies has been already described (Jaric and Latash 1999). Recall that when subjects tried to place the tip of the pointer into a small target, the variability of the location of the wrist computed across several trials was larger than the variability of the location of the tip of the pointer. A conclusion has been made that the wrist rotation compensated partly for errors that would otherwise be introduced by imprecise rotations of the elbow and shoulder joints.

In another study (Domkin et al. 2002), the subjects were asked to perform bi-manual pointing "as quickly and accurately as possible." They held the pointer in the dominant hand, and the semicircular target in the nondominant hand. They

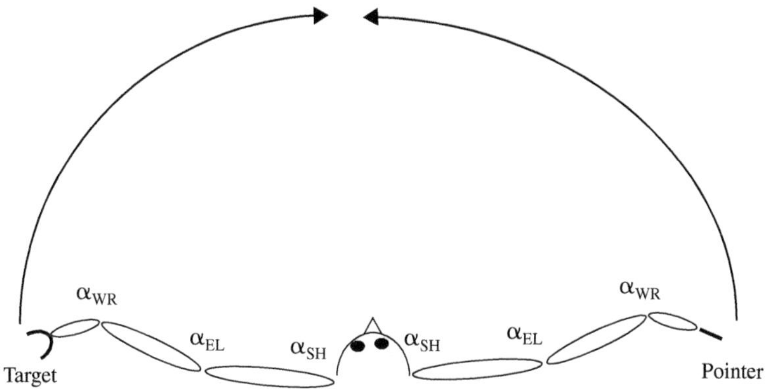

Figure 5.9. In this study, the subject sat initially with the arms spread far apart and held a pointer in one hand and a semicircular target in the other. The task was to make a very fast and accurate horizontal movement of both arms such that the pointer stopped in the middle of the target. The three major joints in each arm make it kinematically redundant with respect to the two-dimensional task.

started each movement from a rather clumsy posture with the two hands spread as far from each other as possible in the horizontal direction (Figure 5.9), and their movements were very close to planar. Several synergies may be suspected for such a task. First, the three major joints of the dominant arm (the wrist, the elbow, and the shoulder) may be united into a synergy stabilizing a trajectory of the tip of the pointer. Second, the joints of the nondominant arm may be united into a synergy stabilizing a trajectory of the target. Third, all six involved joints may be united into a synergy stabilizing a time profile of the distance between the target and the tip of the pointer. Actually, the last possibility can be split into two: Such a two-arm synergy may stabilize the vector distance between the two endpoints or the scalar distance. The difference between the two is illustrated in Figure 5.10. Each suspected synergy may be formalized as a hypothesis leading to an analysis of a set of trials within the UCM hypothesis.

Analysis of the experimental data supported all three hypotheses (from the two two-hand hypotheses, the two-hand vector distance hypothesis was selected). However, when the ratios ($R_V = V_{UCM}/V_{ORT}$) between the amounts of variance within the UCM and orthogonal to it were compared, this index for the two-arm hypothesis was significantly larger than for either of the one-arm hypotheses. This result confirms an intuitive guess that a two-arm pointing is not simply a superposition of two one-arm movements of the target and of the pointer, but a true two-arm, six-joint synergy. These results were further confirmed in experiments with both uni-manual and bi-manual pointing in three dimensions (Tseng et al. 2002, 2003; Domkin et al. 2005).

These results may seem nearly trivial; however, they are not—at least, not in all their aspects. For example, it is nearly trivial that, for accurate pointing, the

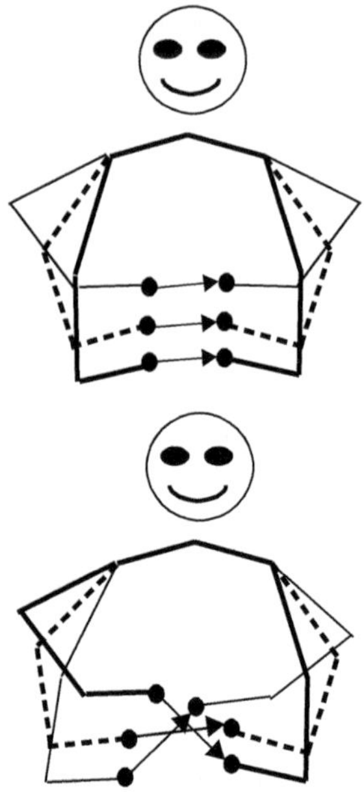

Figure 5.10. The two drawings illustrate three joint configurations (arm segments are drawn with lines of different styles), when the vectorial distance between the endpoints of the two arms was preserved (top) or the scalar distance was preserved (bottom). Reproduced by permission from Domkin D, Laczko J, Djupsjöbacka M, Jaric S, Latash ML (2005) Joint angle variability in 3D bi-manual pointing: uncontrolled manifold analysis *Experimental Brain Research* 163: 44–57. © Springer.

tip of the pointer should get close to the center of the target. This implies that the variability of the distance between the two should be small. But this is true only for the final phase of the movement. Why should the neural controller bother about stabilizing an instantaneous position of the target (or the pointer, or the distance between the two) soon after the movement initiation, when they are still miles apart? The controller, however, does so. Maybe, it is easier to assemble a synergy and use it throughout a movement than to introduce it only in a crucial movement phase. Maybe, it is impossible to ensure high accuracy of an outcome of a fast movement without ensuring high accuracy of a trajectory leading to this outcome, for example, because of the inertial properties of the arm segments. These are only guesses.

5.1.6 Quick-Draw Pistol Shooting

Shooting may be viewed as a particular example of pointing with the pistol at the target. Apparently, success in shooting crucially depends on two angles that define the orientation of the pistol barrel in space, but it does not depend on

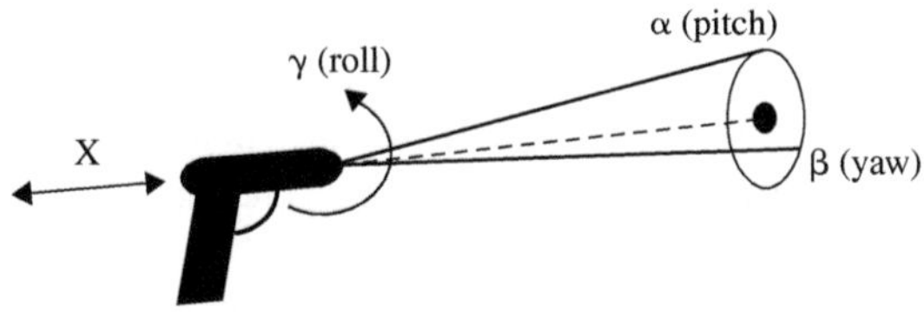

Figure 5.11. During shooting at a target, accuracy depends crucially on two angles that define the orientation of the pistol barrel with respect to the target, pitch, and yaw. It does not depend on the exact location of the pistol along the line of shooting and on the roll angle.

the position of the pistol along the line of shooting or on its rotation about that line (Figure 5.11). Are there multi-joint synergies that stabilize the two apparently important angular variables? Are there synergies that stabilize variables that do not seem to be important? These are main questions that were addressed in an experimental study of quick-draw shooting with an infrared pistol at the infrared-sensitive target (Scholz et al. 2000). In that study, subjects were asked to imagine that they participate in a Clint Eastwood-style shootout. They were not allowed to extend the arm with the pistol and take an accurate aim but rather to shoot as quickly as possible after the initiation of the movement.

After a few trials, all subjects became very accurate shooters (partly due to the large effective size of the target). Then, they were asked to perform a series of trials. Angular trajectories in the seven major arm joints were recorded and analyzed. The seven joint angles were considered elemental variables, while three hypotheses were tested on stabilization of three performance variables by co-varied changes in the joint angles across trials. The three variables were the angle between the pistol barrel and the direction from the back-sight to the target (this variable is two-dimensional), the location of the pistol in space (this variable is three-dimensional), and the location of the center of mass of the whole system "the pistol plus the arm" in space (this variable is also three-dimensional).

The first hypothesis is simply a reflection of the earlier statement that accurate shooting requires accurate production of these two angular variables. The second hypothesis reflects another intuitive consideration: Maybe, the subjects selected a stereotypical pistol trajectory in space and stabilized it by a multi-joint synergy. The third hypothesis is slightly less trivial. Any fast movement by the arm induces reactive forces and moments in the shoulder joint acting on the trunk. The trunk muscles show APAs (reviewed in Massion 1992) to counteract these forces/moments and keep the balance. If these forces are well predictable, the CNS may be able to generate an optimal set of APAs and use it in all trials, thereby minimizing reactive, poorly reproducible trunk motion. The reactive forces depend on the kinematics of the arm, in particular, on the kinematics of its center of mass. Therefore, dealing with a predictable trajectory of the center of mass of the arm may be viewed as beneficial for accurate task performance.

To test the hypotheses, UCMs were computed for each of the three variables based on the geometry of the arm of each individual subject for each phase of the movement. Further, for each phase of the movement, the variance in the seven-dimensional joint space across trials was projected onto the UCM and onto its orthogonal complement. And then, these two variance indices were compared (certainly, per dimension in each subspace). This rather complicated analysis led to the following main findings.

The first hypothesis was supported starting from the initiation of the movement and over its whole duration. This means that significantly more variance in the joint angle space was within the UCM than within the rest of the space ($V_{GOOD} > V_{BAD}$; Figure 5.12A). This result may seem trivial because the subjects were accurate, and accurate shooting required alignment of the pistol barrel with the direction of the target. Although this is true, the hypothesis was also supported in the early phases of the movement when the pistol pointed away from the target. Why did the controller stabilize an angle of say 60° between the

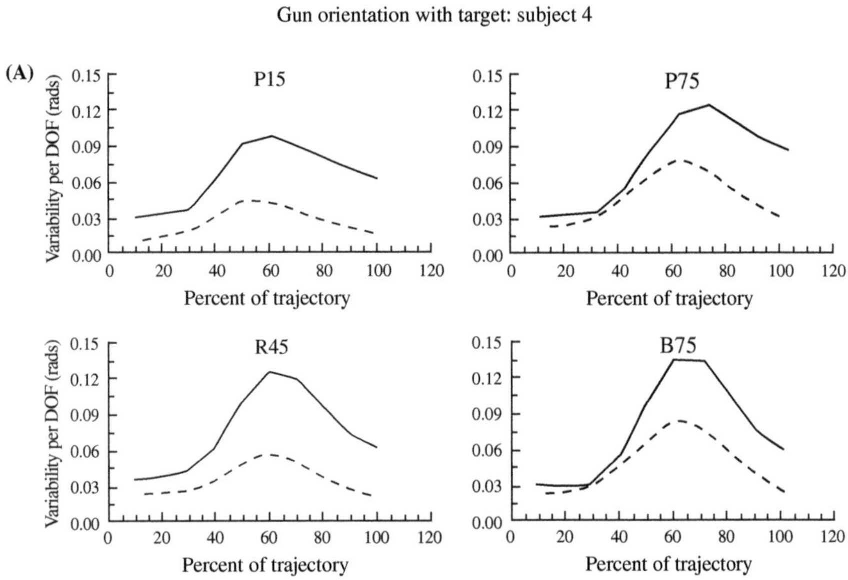

Figure 5.12. An illustration of the patterns of variability within the uncontrolled manifold (UCM) (V_{GOOD}, solid lines) and orthogonal to the UCM (V_{BAD}, dashed lines) computed across successive attempts at shooting with an infrared pistol at an infrared-sensitive target. The components of variability were computed with respect to three performance variables. (A) The angle between the pistol barrel and the direction from the back-sight to the target. (B) The location of the pistol in space, and (C) the location of the center of mass of the whole system "the pistol plus the arm" in space. Note that $V_{GOOD} > V_{BAD}$ over the whole duration of the movement only for the first variable (A); but $V_{GOOD} = V_{BAD}$ during the second half of the movement for the other two variables. Reproduced by permission from Scholz JP, Schöner G, Latash ML (2000) Identifying the control structure of multi-joint coordination during pistol shooting. *Experimental Brain Research* 135: 382–404.

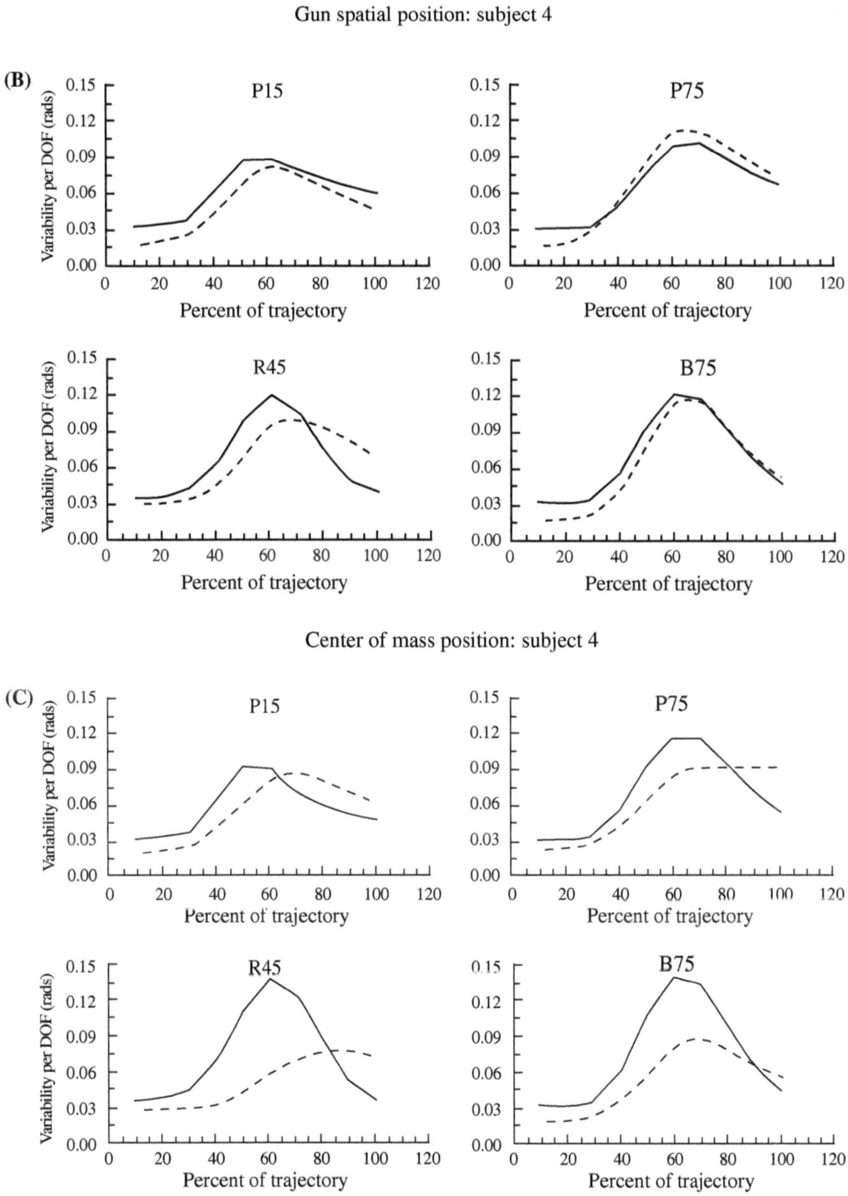

Figure 5.12. (Continued)

pistol barrel and the direction of the target? This does not seem to make sense. In a way, this result is similar to what was observed in the mentioned studies of two-arm pointing. In that study, also, the distance between the tip of the pointer and the center of the target was stabilized when the pointer and the target were still very far from each other. As in the description of that study, I can only speculate that this strategy of control may help simplify the construction of an appropriate kinematic synergy.

The other two hypotheses were only partially supported. UCM analysis showed (Figure 5.12B,C) that both positions of the pistol in space and position of the center of mass of the arm were stabilized by joint angle co-variation early in the movement but not during the second half of the trajectory, closer to the moment of pressing the trigger, which was obviously crucial for success of the task. Thus, joint variance for the arm was structured differently with respect to performance variables with different relevance to task success. The neural controller of the highly abundant (seven degrees-of-freedom) kinematic chain was able to stabilize three different performance variables early in the movement, but it focused on stabilization of only the most relevant variable at the crucial phase close to pressing the trigger.

That study used one more manipulation: Without any additional practice, the subjects were asked to perform shooting at the same target from the same initial position but with a rather stiff elastic band restricting movement in the elbow joint. In this new condition, all joint trajectories changed. However, most subjects produced an accurate shot at the first attempt. Note that the shooting movement was very fast; it took subjects about a quarter of a second from the movement initiation to pressing the trigger. This result resembles somewhat the mentioned observations of the wiping reflex in spinal frogs (section 3.3). To recall, the frogs were able to wipe the stimulus off the back accurately when one of the joints of the moving hind limb was fixed or when an additional load was attached to the hind limb.

One interpretation of the observed accurate shooting in conditions of elbow restraint is that the multi-joint synergy stabilizing the pistol orientation with respect to the target was able to deal with the errors introduced by the elbow joint by changing motion of other joints. This is probably one of the most important features of such synergies, that is, using the abundance of the system, its flexibility to overcome unforeseen complications, including both minor deviations of elemental variables (joints) from their expected path due to natural variability and more significant deviations introduced by external perturbations or constraints.

5.2 KINETIC SYNERGIES

As mentioned briefly in earlier sections, coordination of a multi-joint action, even the simplest point-to-point reaching, is a nontrivial task when analyzed at the level of actions at individual joints. Consider the following simple example. Imagine that you place the upper arm of the right arm on a table, the elbow is flexed at a right angle such that the forearm is vertical. The fingers are extended and the hand is supinated such that the palm faces the body (Figure 5.13). Now make a few quick elbow flexion and extension movements. These are very easy to perform. Note that during these movements, the wrist position does not change

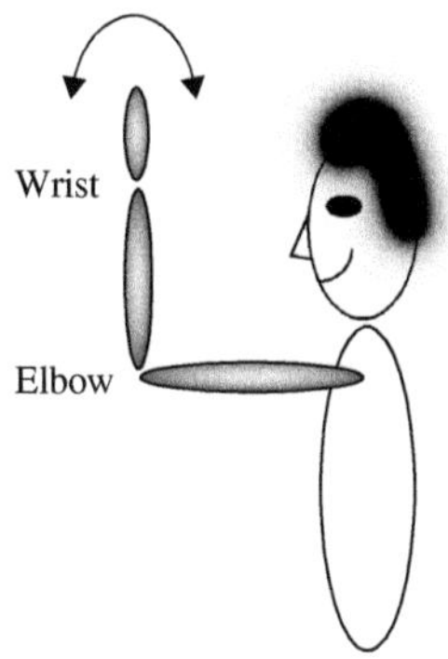

Figure 5.13. When a person moves one of the joints, the elbow or the wrist, the other joint experiences motion-dependent torque perturbations. Moving the elbow or the wrist (in the illustrated posture) is not associated with visible flapping of the other joint. It is prevented by simultaneous commands to muscles crossing the apparently postural joints that counteract the motion-dependent interaction torques.

noticeably, and the hand does not flap. This is not a trivial observation. Quick motion in the elbow joint is accompanied by quick changes in the elbow joint torque that accelerate the upper arm and then decelerate it. These torque changes and associated elbow acceleration should have produced inertial torques acting at the wrist joint, and these torques are expected to move the wrist joint and cause hand flapping. Now repeat the same action, but try to keep the wrist joint as relaxed as possible. The hand will indeed flap. The natural coordination that prevents joint flapping during fast movements of multi-joint limbs may be viewed as synergies of control signals to individual joints stabilizing joint position.

Substantial experimental evidence suggests that multi-joint synergies of the type of the wrist–elbow synergy described in the previous paragraph are of a feed-forward nature (Gielen et al. 1985; Koshland et al. 1991; Latash et al. 1995, 1999). What I mean is that adjustments in control signals to muscles acting at an apparently postural joint occur not in response to sensory signals reflecting deviations of this joint but in anticipation of those deviations. This does not mean that sensory signals play no role in multi-joint synergies. The importance of signals from proprioceptors for multi-joint coordination has been demonstrated in studies of patients with a rare disorder, large-fiber peripheral neuropathy (Cooke et al. 1985; Sainburg et al. 1995; Messier et al. 2003; Spencer et al. 2005). These persons lose the ability to conduct action potentials along the large peripheral neural fibers, commonly as a result of a viral infection. At the same time, conduction of action potentials along motor fibers does not suffer. As a result, these persons cannot feel their limbs but can produce voluntary muscle contractions. Sometimes, they are addressed as "deafferented patients."

A group of researchers asked a person with large-fiber peripheral neuropathy to perform a repetitive horizontal arm movement of slicing a loaf of bread

(Sainburg et al. 1995). When a healthy person performs such a movement, the motion of the hand is indeed close to horizontal. This was true for motion of the hand of the patient, but only when the patient could see the arm. Without seeing, the motion of the hand became grossly curved. Analysis of joint torques suggested that torques produced by the muscles acting at individual joints failed to take into consideration torque components that had been caused by the motion of the other joint, the so-called *interaction torques* (Zatsiorsky 2002). In this context, under "muscle torque" I mean a particular term in the equation:

$$I(d\alpha/dt)^2 = T_M + T_G + T_I$$

where the left side is the total torque in a joint that produced its angular acceleration (I is the rotational moment of inertia), and the right side is the sum of three components, T_M (muscle torque), T_G (torque produced by gravity), and T_I (interaction torque, torque produced by rotation in other joints). T_M is produced not only by muscles but also by other tissues that produce rotational effects about the joint.

A particular simple rule to describe muscle torques generated at individual joints during multi-joint action was suggested by Gerald Gottlieb and his colleagues (Gottlieb et al. 1996). According to this rule, muscle torques at different joints scale linearly. Such close-to-linear scaling was indeed demonstrated in a number of experiments, including studies of motor development (Zaal et al. 1999). However, later exceptions to this rule were discovered (Shemmell et al. 2007). In a way, it is clear that this rule cannot be universal. Otherwise, slow movements where one joint moves only into flexion, while another joint changes the direction of its motion would be impossible. I considered such an example earlier, that is, when you extend an arm to its side and then move it slowly until it crosses the midline and points in the opposite direction (section 5.1). Even when such a rule describes joint torque profiles with considerable accuracy, could we call it a synergy? According to the accepted definition, this can be done if the torque scaling is related to ensuring stability of a particular performance variable. However, no study explored this possibility or even suggested that this might be true.

Until now, we have discussed kinetic synergies in terms of joint torques. This looks natural since these are adequate mechanical variables to describe kinetic effects. However, as mentioned earlier (section 3.3), muscle forces and joint torques are unlikely candidates for control variables. It is not easy to take the next step and analyze multi-joint synergies using control variables, partly because control variables are not directly observable and have to be computed within a particular motor control theory. Let me conclude this section with an example of an experimental study that attempted exactly that.

The two-joint wrist–elbow synergy that opened this section was studied within the framework of the EP hypothesis (Latash et al. 1999). This study used a simplified mechanical model of the joints and natural joint variability to reconstruct equilibrium trajectories in the two joints when only one of the joints was required to produce a quick action, flexion or extension. The natural variability provided trial-to-trial smooth joint perturbations, different in different trials, because of associated changes in interaction torques. Assuming unchanged control process, equilibrium trajectories could be reconstructed for each time slice along the trajectory. The main result was that the equilibrium trajectory of the apparently postural joint showed large amplitude changes timed simultaneously with the equilibrium trajectory of the joint that was instructed to move (Figure 5.14). The effect was particularly strong when the elbow was instructed to move, but it could also be observed when the elbow stayed motionless, and the wrist moved.

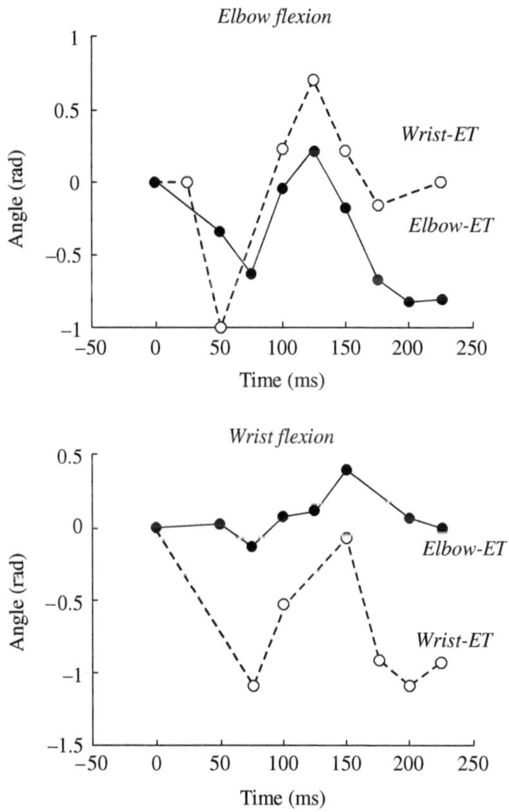

Figure 5.14. Equilibrium trajectories for the wrist and elbow joints were reconstructed, based on a set of trials with quick elbow motion. Note the large-amplitude equilibrium trajectory of the wrist, which was not associated with visible wrist motion. The purpose of this signal was to prevent joint motion that would otherwise be produced by interaction torques. Reproduced by permission from Latash ML, Aruin AS, Zatsiorsky VM (1999) The basis of a simple synergy: Reconstruction of joint equilibrium trajectories during unrestrained arm movements. *Human Movement Science* 18: 3–30. © Elsevier.

Actually, the equilibrium trajectories at the two joints were very much similar to each other showing a nonmonotonic pattern that ended at a new equilibrium position for the focal joint and at the same equilibrium position for the postural joint. This result allows suggesting that voluntary movement of one joint of a multi-joint limb is associated with active control of both joints. The control patterns to the focal joint lead to a new equilibrium position, while for other joints, the control patterns may be viewed as leading to a movement of those joints of zero amplitude. The purpose of such control signals is not to produce motion of those joints but to prevent them from moving under the action of torques generated by motion of the instructed joint(s).

5.3 MULTI-DIGIT SYNERGIES

The human hand is one of the most amazing creations of nature. Despite the impressive progress in technology and engineering, all current robotic grippers look clumsy in comparison to the human hand. Humans can carry a fragile cup of tea and simultaneously open the door handle by pressing on it with the wrist. They can grip a glass of wine and then quickly tap on it with the fingers, one at a time. In other words, they can establish multi-digit synergies that stabilize important characteristics of the hand action (such as the total force and the total moment of force) and then use the abundance of the digits to do "other things" while the synergies take care of the original task. The multi-digit action of the hand may be viewed as a proverbial synergy. Hence, I am going to consider several synergies that unite the digits in tasks that range from very simple, such as producing a pattern of force while pressing down with several fingers, to the rather complex, such as grasping an object with all five digits and then moving it in different directions.

5.3.1 Force and Moment Stabilization during Multi-Finger Pressing

As with any analysis of synergies, a study of multi-finger synergies should start with identification of elemental variables. During pressing tasks, the most apparent elemental variables are forces produced by individual fingers orthogonally to the surface. However, as already mentioned in section 4.2.1, when a person tries to produce force with a finger of a hand, other fingers of the hand also show involuntary force production, a phenomenon called *enslaving* (for more detail see Digression #8). Because of the enslaving, independent variations of commands to individual fingers are expected to lead not to a circular distribution of data points (as the one illustrated by the dashed line in Figure 5.15) but to an elliptical distribution, which may wrongly be interpreted as a task-specific synergy, for example, a synergy stabilizing the total moment produced by the finger forces with respect to a pivot between the fingers. Figure 5.15 illustrates

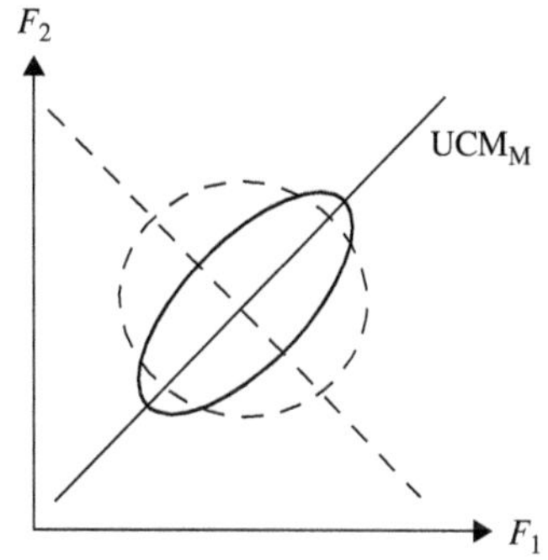

Figure 5.15. Two fingers of a hand produce a certain total force. If the forces of the fingers did not co-vary, a spherical distribution of data points across trials could be expected (the dashed circle). However, even if commands to fingers do not co-vary, their forces are expected to show positive co-variation because of the enslaving (the ellipse). This distribution may be wrongly interpreted as reflecting a control strategy, for example, trying to stabilize the total moment of force with respect to the pivot in-between the two fingers. The solid slanted line shows the corresponding uncontrolled manifold (UCM).

a simple example of pressing with two fingers. The oblique solid line is the UCM for the total moment produced by the fingers with respect to a pivot in-between them. The cloud of data points would show larger variance when projected onto the UCM (V_{UCM}) compared to its projection to the orthogonal complement (V_{ORT}). A formal comparison would lead to a conclusion that there is a two-finger synergy stabilizing the moment of force. To avoid such spurious outcomes, one has to switch from forces to a different set of variables that can at least hypothetically be varied by the controller independently. For tasks of multi-finger force production, such elemental variables have been introduced earlier as *force modes,* and their number has been assumed to be equal to the number of explicitly involved fingers (Latash et al. 2001; Scholz et al. 2002; Danion et al. 2003).

Identification of force modes is nontrivial because the enslaving relations among fingers are different in different persons (Gao et al. 2003). The modes cannot be directly observed and have to be discovered experimentally. This can be done by asking subjects to press with one finger at a time either as strongly as possible or by following a simple force template, for example, a ramp of force that covers a range of forces that one expects to see in the main tasks—the ones used to explore multi-finger synergies. The force mode approach has been used in several experiments that studied whether the magnitudes of force modes to individual fingers during multi-finger tasks co-varied to stabilize such performance variables as total force and total moment of force produced by the fingers with respect to the mid-point between the two most lateral fingers involved in the task.

In some experiments, the subjects were required to produce certain time profiles of the total force (Latash et al. 2001, 2002c; Scholz et al. 2002). It was expected, therefore, that the total force would be stabilized, that is, that the force-stabilization hypothesis would be supported by the UCM analysis. However, the results were rather unexpected. During cyclic force production at a comfortable rate (about 1.5 Hz) with two, three, or four fingers, the finger-mode synergy stabilized total force only within a narrow phase range of the force cycle, close to the peak total force, while the moment-stabilization hypothesis was confirmed over most of the cycle, despite the fact that the subjects received instruction and visual feedback on the total force but not on the total moment. In these studies, analysis was performed at different phases across several cycles aligned in time and considered as separate trials.

In two-finger tasks, the force- and moment-stabilizing synergies are mutually exclusive because their UCMs are orthogonal (see Figure 5.16). However, when three fingers are involved, the abundance of the system allows the stabilization of both force and moment of force at the same time. For example, the forces produced by the two "lateral fingers" could co-vary positively to stabilize the moment of force, while the sum of these two forces could co-vary negatively with the force produced by the "central finger" to stabilize the total force. If four fingers are involved, there are many ways their forces (modes) could co-vary to stabilize both performance variables at the same time.

The seemingly unexpected findings of preferential stabilization of the moment of force were interpreted as reflecting patterns of multi-finger coordination elaborated by the CNS during the lifetime, based on everyday tasks, such as eating with a spoon, drinking from a glass, writing with a pen. Such tasks impose stronger constraints on permissible errors in total moment of force than in total force. For example, while taking a sip from the glass, grip force should

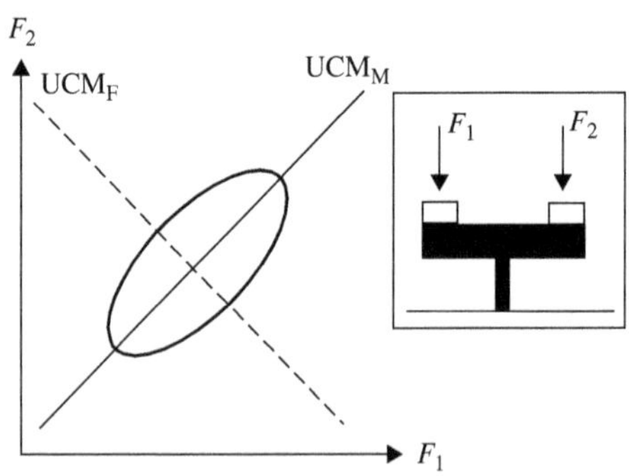

Figure 5.16. When two fingers participate in a force-production task, the uncontrolled manifold (UCM) for the total force (UCM_F) and the UCM for the total moment of force (UCM_M) with respect to a pivot in-between the fingers (see the insert) are orthogonal. This means that good variability with respect to total force production is bad with respect to total moment of force production, and vice versa. Such a system cannot stabilize both the total force and the total moment of force.

only be above the slipping threshold and below the crushing threshold. These are relatively weak constraints. The moment of force, however, needs to be controlled much more precisely if one wants to avoid spilling the contents of the glass.

When the rate of force change was manipulated, a study of both discrete and rhythmic force production tasks (Latash et al. 2002c) showed major effects of the rate of force production on the structure of finger mode variance. In particular, the component of variance that affected the total force, $V_{ORT,F}$, showed a strong relation to the rate of force production and only weak dependence on the magnitude of force. In contrast, the *good variability*, $V_{UCM,F}$, showed minimal effects of the rate of force production and strong effects of the force magnitude. These observations suggest a different interpretation of the findings of apparent synergies stabilizing the moment of force. The interpretation deals with analysis of effects of timing errors on motor performance (see the next section).

Another series of studies explored multi-finger synergies in pressing tasks, when the task was to produce a particular time profile of the total moment of force (Zhang et al. 2006, 2007). Both discrete (trapezoidal) and cyclic tasks were studies. The results have confirmed that humans are much better at stabilizing the time profile of the moment of force; actually, they failed to stabilize the time profile of total force altogether.

5.3.2 The Role of Timing Errors

Consider the following example. A person is trying to produce a simple time profile of a particular variable, for example, finger force, several times. Suppose now that, from trial to trial, the person sets the magnitude parameter of the force profile with small errors and also he or she performs the task each time a bit faster

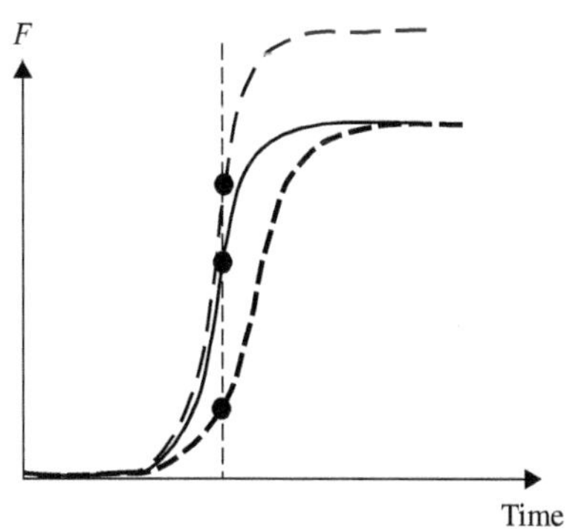

Figure 5.17. Force variability may be related to errors in setting parameters related to force magnitude (compare the solid and thinner dashed lines) and to the rate of force production (compare the solid and thick dashed lines). Both types of errors lead to force variability measured at a certain time after the initiation of force production, for example, at the time shown by the vertical dashed line.

or a bit slower. Figure 5.17 shows that variations in both magnitude and timing can affect force variability measured across trials at different phases of this action. In each trial, the performance, $F(t)$, can be described as a function f of time with two parameters related to the peak magnitude of force and to the time it takes to reach this magnitude; let us address these parameters as A and τ:

$$F(t)=Af\left(\frac{t}{\tau}+s\right)+p \qquad \text{Equation (5.2)}$$

where s and p are constants. One can show that the variance of force, $V(F)$, represents a function of time consisting of two terms. The first term is proportional to the force magnitude squared and depends on the magnitude and variance of A. The second term is proportional to the first force derivative squared and depends on the magnitude and variance of τ:

$$V\left[F(t)\right]=F^2(t)\frac{V_A}{A}+t^2\left[\frac{\mathrm{d}F(t)}{\mathrm{d}t}\right]^2\frac{V_\tau}{\tau^2} \qquad \text{Equation (5.3)}$$

This analytical description was developed and analyzed for human movement by Simon Goodman (Gutman) as a model of motor variability (Gutman and Gottlieb 1992; Gutman et al. 1993). Further, this model was developed for multi-element systems such as a set of fingers pressing in parallel (Goodman et al. 2005). In the latter model, variance of the total force was quantified within two subspaces corresponding to the UCM and its orthogonal complement. This decomposition of force variance showed a peak close to the time of peak rate of force production in the direction of force change (V_{ORT}), while such a peak was absent in directions spanning the UCM. These results correspond well with Equation 5.3. They suggest two important conclusions.

First, even in the absence of any particular control strategy (co-variation of force modes), variance may be expected to show different time profiles in the direction of force change (V_{ORT}, strongly dependent on the rate of force production) and orthogonal to it (V_{UCM}, showing little or no dependence on the rate of force production). Hence, during fast actions, the analysis of synergies within the UCM hypothesis framework may lead to wrong conclusions on the presence or absence of synergies simply because of the natural modulation of the rate of force change.

Second, let me recall the experimental fact that V_{UCM} has been shown to correlate with the magnitude of force while V_{ORT} correlated with the rate of force change (Latash et al. 2002c). Equation 5.3 suggests that a component of variance that shows strong effects of force magnitude but not rate of force change is affected by setting parameter A. In contrast, a component of variance that shows strong effects of force rate but not magnitude is affected by setting parameter τ. So, most of the *good variability* (V_{UCM}) reflects variations in A, while most of the *bad variability* (V_{ORT}) reflects variations in τ. In other words, the controller seems to organize co-variation of A to different fingers rather well such that virtually all this variance is *good*. In contrast, the controller seems to be unable to handle in a similar way variations of τ to different fingers such that most of that variance is *bad*. We are back to a general question discussed in an earlier section: Are there timing synergies? Or, in other words: Is the CNS limited in its ability to organize interactions among fingers such that errors in the timing of individual finger force profiles cancel each other's effects on the total force? As discussed in more detail in section 4.5, as of now, the answer is in the affirmative. Experiments that tried to discover timing synergies have so far failed.

5.3.3 Emergence and Disappearance of Synergies

There is another aspect of synergies that deals with the issue of timing. How long does it take the controller to organize a synergy stabilizing a particular performance variable? This question would not have been asked if not for an experimental observation made in several studies of multi-finger force production.

Imagine that a person sits relaxed with the fingers resting on force sensors. A line runs over the monitor screen showing the total force produced by the fingers. There is a template shown on the screen that requires the person to press with the fingers starting at some point and produce a smooth increase in the total force (Figure 5.18A). Imagine now that after a few practice trials (this is a very easy task to learn), the person performs the task several times. Then, individual finger forces are transformed into force modes (see section 4.2.1) and analyzed over trials at each 1 ms, using the framework of the UCM hypothesis, as described earlier (section 4.1). This analysis will result in a time profile of an index, ΔV, computed in such a way that its positive values reflect predominance of *good variability*, that is, a multi-finger synergy stabilizing the total force.

Figure 5.18B shows a typical time profile of ΔV in such an experiment. When the force produced by fingers is zero, ΔV is obviously undefined. As soon as the finger forces start to change, ΔV shows a drop into negative values (not a synergy), and it takes it some time to start showing consistently positive values. In other words, this task is associated with a multi-finger force-stabilizing synergy, but this synergy does not start from the very beginning of the task. What could be the reason? The very first hypothesis was that, at very low forces, sensory receptors in the finger pads sensitive to pressure are unable to detect deviations of individual finger forces from expected values. Therefore, a feedback-based error correction mechanism does not work. This interpretation is very much in line with a particular model of control (Todorov and Jordan 2002), which will be considered later, in section 8.1.4.

This hypothesis was tested in another experiment when the rate of required force change was modified in different tasks (Shim et al. 2003a). The critical time (the time it took ΔV to turn positive) in that study did not correspond to a fixed force level. Indeed, it was nearly constant for every participant over a range

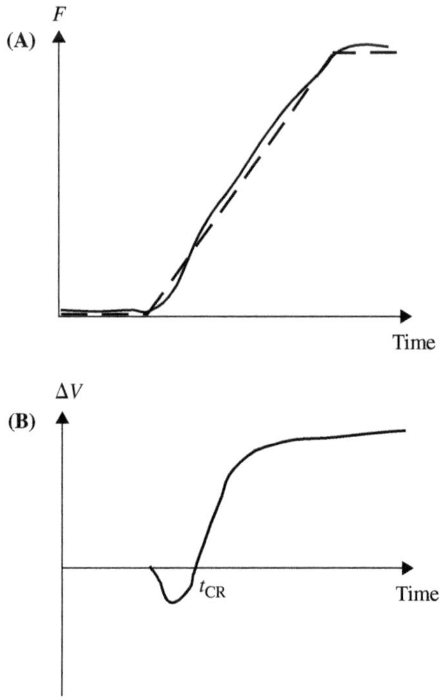

Figure 5.18. (A) A template (ramp increase in the total force, dashed) and a single trial (solid) when the subject tried to follow the template with the total force produced by several fingers of a hand. (B) A time profile of an index of synergy, ΔV, computed in such a way that its positive values correspond to predominance of good variability. Note that it takes time (critical time, t_{CR}) for the controller to organize a force-stabilizing synergy (positive ΔV).

of rates of force production. This means that for faster rates of force production, ΔV turned positive at higher forces, compared to tasks with slower rates of force production. So, the hypothesis on reaching a necessary force level to trigger error correction feedback processes has been refuted. Moreover, in a follow-up study with very fast force production, the critical time was shown to be less than 50 ms (Latash et al. 2004). This is too short to allow any error correction in force based on feedback from peripheral sensory receptors. These results led to the emergence of another model that will also be discussed in Part 8.

If it takes time for the CNS to organize a synergy, does it also take time to destroy one? In other words, are the time profiles of a synergy index (ΔV) symmetrical for a symmetrical task that starts from a relaxed state and ends with a relaxed state? Figure 5.19 shows in its left panel a hypothetical symmetrical profile of ΔV for a symmetrical task of producing a trapezoidal time profile of force (Shim et al. 2005b). The right panel of the Figure 5.19 shows the actually observed time profile of ΔV. The difference between the two panels is obvious in the phase of force decrease. There is no smooth decline in ΔV over the ramp-down phase of

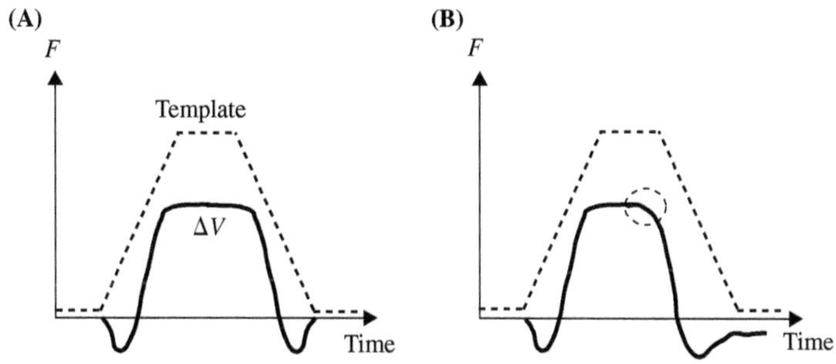

Figure 5.19. The left panel (A) shows a trapezoidal template and a hypothetical symmetrical time profile of ΔV, based on data illustrated in Figure 5.18. The right panel (B) shows a typical time profile of ΔV characterized by an early drop (circled) in anticipation of the planned force decline and a following very steep drop to negative values.

the task but rather an abrupt drop from high positive values (a strong synergy) to zero or even large negative values. There is another phenomenon that may not be that obvious. The time profile of ΔV shows a small, smooth decline in preparation to a change in the total force: See the circled area in the right panel of Figure 5.19. Both results are potentially important and deserve a special subsection.

5.3.4 Anticipatory Synergy Adjustments and Purposeful Destabilization of Performance

What is the major difference between a soaring bird and a flying airplane? Obviously, they are made of different materials and use different sources of energy. But what is different if one simply looks at the external design? Both have elongated bodies and two symmetrical wings. Both have tails. However, only the airplane has a vertical tail fin, while birds do not. Why?

The purpose of the vertical tail fin is to help stabilize the airplane in a horizontal plane. But why did not nature come up with a similar design? Probably, because it carries more disadvantages than advantages. Stability of the airplane means that the pilot does not have to worry about this aspect of control too much and can concentrate on other aspects. However, a design that ensures stability of a particular performance variable makes it harder for the plane to change this variable quickly. The plane is more stable in a horizontal plane but it is less maneuverable. If there were birds with such designs that ensure stability at the expense of flexibility and maneuverability, they would have probably been eaten by less stable and more maneuverable predators. So, stability by design helps some aspects of control but comes at a price. By the way, fighter jets that should be both stable and maneuverable have considerably smaller vertical tail fins. (I am indebted to Ziaul Hasan for this example and for drawing the attention of the motor control community to the excessive emphasis on stability in motor behavior; Hasan 2005.)

Imagine now that a neural controller arranges a synergy that stabilizes a certain steady-state or slowly changing value of a performance variable. If the controller wants to change this variable quickly, the synergy would perceive this change as an "error" and try to counteract it. What would a smart controller do? An obvious answer is "Switch the synergy off during the quick change in the variable." However, biological systems never do anything instantaneously. Then, if the controller knows in advance when it wants to change a variable quickly, a sensible strategy would be to start decreasing the strength of a preexisting synergy in advance, in anticipation of the change in the variable. Can the CNS use this strategy in everyday actions? The illustration in Figure 5.19 suggests that it can.

To test this hypothesis another study was run (Olafsdottir et al. 2005). In that study, the participants were asked to keep the total force produced by the fingers of a hand constant for some time and then to produce a very quick force pulse to a target shown on the screen (Figure 5.20). After several trials, a typical analysis was performed with the computation of $\Delta V(t)$ for total force stabilization. During the steady-state phase, ΔV was always positive reflecting a strong multi-finger synergy stabilizing the total force. When the subjects were free to produce the force pulse at any time, they did indeed show a drop in ΔV 100–150 ms prior to the first detectable change in the total force. In another series of trials, the same subjects were asked to do the same task with a small modification: they were to initiate the force pulse as quickly as possible after hearing a brief beep in the

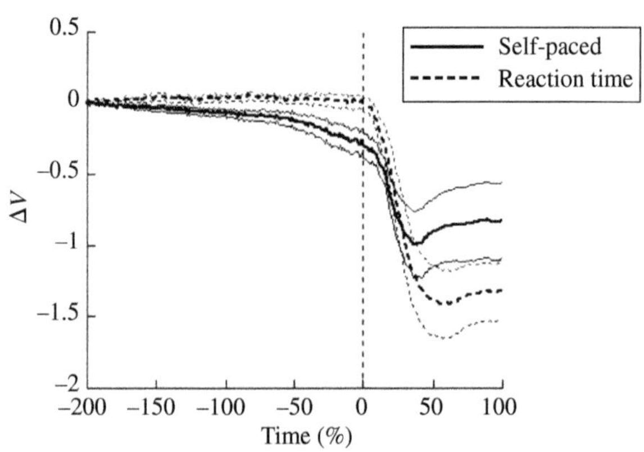

Figure 5.20. A subject was required to hold a constant force level by pressing with the four fingers of the dominant hand and then produce a quick pulse of force to a target. The index of force stabilization ΔV showed a drop (solid, thick line) starting 100–150 ms prior to the initiation of the force pulse (dashed vertical line). Such anticipatory ΔV changes were absent when the same task was performed "as quickly as possible" to an auditory signal (dashed line). Standard error margins are shown with thinner lines. Reproduced by permission from Olafsdottir H, Yoshida N, Zatsiorsky VM, Latash ML (2005) Anticipatory co-variation of finger forces during self-paced and reaction time force production. *Neuroscience Letters* 381: 92–96. © Elsevier.

earphones. This instruction is commonly referred to as *simple reaction time*. Overall, the performance of the subjects did not change much. However, the early, anticipatory change in ΔV disappeared.

Two conclusions can be drawn. First, the CNS can indeed modify multi-finger synergies without changing the overall output of the set of fingers in anticipation to a change in this output. Second, this strategy is a "luxury." It is not necessary to perform the task. When the instruction does not give the controller time to adjust the synergy prior to the force pulse (as in the simple reaction time trials), the controller does not show such adjustments but is still able to perform the task adequately. The phenomenon of such nonobligatory adjustments in an index of a motor synergy in preparation to a quick change in a performance variable has been termed *anticipatory co-variation* or *anticipatory synergy adjustment* (ACV or ASA).

ASA resembles another well-known phenomenon in movement studies. When a person in the standing position performs a very quick action that can potentially perturb the vertical posture (e.g. drops or catches a load or makes a very fast arm movement), the first signs of changes in the muscle activity are seen not in the apparent prime movers, that is, muscles producing the required action, but in postural muscles of the legs and trunk. The phenomenon of early activation (or suppression of activity) of postural muscles is called anticipatory postural adjustments (APAs, reviewed in Massion 1992). APAs have traditionally been interpreted as a reflection of neural signals with the purpose of generating forces and torques that counteract the expected perturbing forces/torques.

The similarities between the APAs and ASAs are rather obvious. Both emerge about 100 ms prior to the generation of an explicitly instructed action. Both become not anticipatory but simultaneous with the action under the simple reaction time instruction (Lee et al. 1987; De Wolf et al. 1998; Olafsdottir et al. 2005). In addition, both have been shown to be smaller and delayed in elderly persons (Woollacott et al. 1988; Olafsdottir et al. 2007a). These three similarities seem too many to be considered simple coincidences. Maybe, APAs (or at least some of the phenomena typically described as APAs) represent a particular example of ASAs. Their purpose is not to generate forces/torques acting against the predicted perturbation but to weaken the preexistent synergy stabilizing a steady-state value of a relevant performance variable in order to facilitate its quick change after the perturbation.

But do ASAs exist as a preparation not to a voluntary action but to a reaction to a perturbation? Feed-forward grip force adjustments in preparation to a predictable (commonly, self-triggered) perturbation have been documented in many studies (Johansson and Westling 1984; Flanagan and Wing 1993, 1995; Danion 2007). All these studies, however, focused on adjustments in the magnitude of the gripping force, not in co-variation of digit forces that may or may not stabilize the gripping force value. The question of whether there are feed-forward adjustments in such multi-finger synergics (ASAs) was explored in a study that required participants to produce a steady-state magnitude of the total force, while pressing with the four fingers of a hand (Kim et al. 2006). The mechanical design was such that the fingers pressed against constant forces directed upwards. One of these forces could be eliminated by pressing a button. The button could be pressed by an experimenter (unexpectedly for the subject) or by the subject himself or herself. As a result, the downward force by the unloaded finger increased, and the total force also showed an increase. The subjects reacted to this increase by changing very quickly the forces of all four fingers and getting back to the required total force level. This is not that surprising, although the adjustments of all four finger forces in response to a perturbation of only one finger are nontrivial.

When the ΔV index for total force stabilization was computed over a series of such trials, it showed an expected high positive value during the steady-state force production. In trials when the unloading was triggered by the subject, ΔV showed a smooth drop starting about 100–150 ms prior to the unloading (Figure 5.21). This drop (ASA) was absent when the unloading was unexpectedly triggered by the experimenter. It is of interest that in self-triggered trials with ASAs, it took ΔV less time to recover to strongly positive values after the perturbation compared to

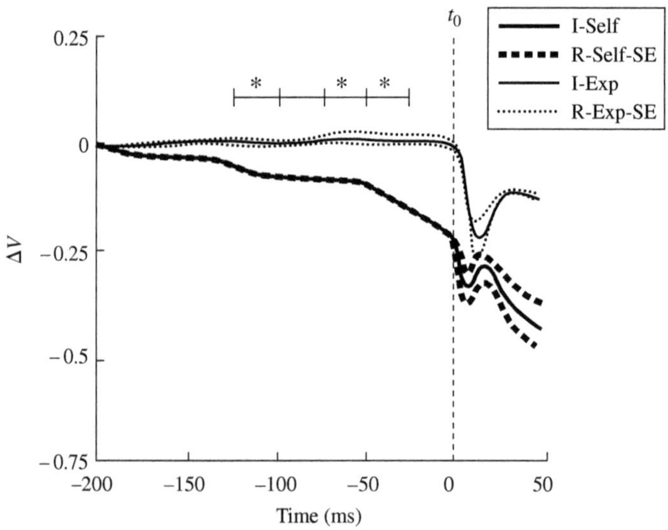

Figure 5.21. Anticipatory synergy adjustments can be seen in preparation to a self-triggered perturbation. The subject produced a constant level of the total force by pressing with the four fingers of the right hand. A perturbation to one of the fingers (the index finger) was triggered by the subject (thick line) or by the experimenter (thin line). There was an anticipatory decline in the ΔV index computed for total force stabilization but only when the perturbation was self-triggered. Standard error margins are shown with thinner lines. Reproduced by permission from Kim SW, Shim JK, Zatsiorsky VM, Latash ML (2006) Anticipatory adjustments of multi-finger synergies in preparation for self-triggered perturbations. *Experimental Brain Research* 174: 604–612. © Springer.

trials with experimenter-triggered perturbations (without ASAs). So, starting to turn a synergy off in anticipation of a quick reaction to a predictable perturbation helps the person restore the synergy quickly after the perturbation.

A similar study was performed later using a more natural gripping task involving all five digits of the hand (Shim et al. 2006). It showed basically the same result. When perturbations were triggered by the subject, there were anticipatory changes in synergies stabilizing the action of the hand on the handle. Such ASAs were absent when similar perturbations were delivered unexpectedly by an experimenter.

Another important aspect of all those studies is that the ΔV index dropped not only to zero but to strongly negative values. This observation suggests that the controller may not simply turn a synergy off but replace a pattern of co-variation stabilizing a performance variable with another pattern that destabilizes it. Figure 5.22 illustrates this idea with the simplest example of pressing with two fingers to produce a steady-state value of the total force and then to change this value quickly. During the steady-state portion of the task (the left panel), data points across repetitive trials form an ellipse corresponding to a negative co-variation of finger forces stabilizing the total force magnitude. Of course, there is nothing surprising there. In preparation to the quick force change (the middle

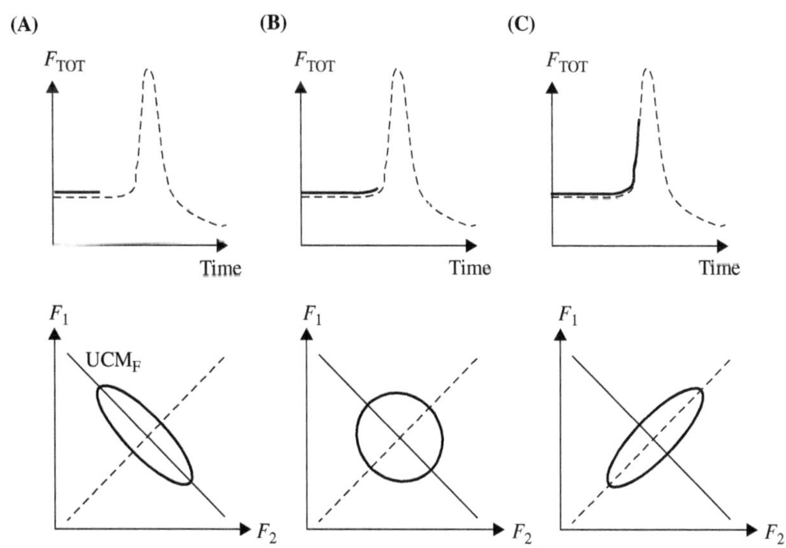

Figure 5.22. A person is required to maintain constant force by pressing with two fingers and then produce a pulse of force (dashed lines in the upper panels). (A) During the steady-state (the solid line in the upper panel), there is a strong two-finger synergy stabilizing the total force [most variance is along the uncontrolled manifold (UCM), bottom panel]. (B) The synergy starts to weaken prior to the force pulse initiation. (C) The synergy disappears, and positive co-variation of finger forces is observed during the pulse.

panel), the ellipse starts to undergo changes in its shape that make it closer to a circle—a weaker synergy. This panel illustrates what happens during ASAs.

Now take a look at the right panel. The ellipse of data points is now turned by 90°. It corresponds to a positive co-variation of finger forces that destabilizes the total force: If in a certain trial, one finger shows higher force than its expected contribution, the other finger is also likely to show a higher force thus adding to the force deviation from its average across trials value. Is this a reflection of a change in the control strategy that facilitates a quick force change by turning a force-stabilizing synergy into an "anti-synergy"? A positive answer sounds most reasonable. However, we should not jump to conclusions. There are two reasons for suspicion.

First, ellipses like the one shown in the right panel of Figure 5.22 were interpreted earlier (see section 5.3.1) as signs of a different synergy stabilizing the total moment of force produced with respect to the midpoint between the two fingers. Why should such a synergy suddenly emerge during fast force changes? I do not know, but it may do so. For example, the controller may try to avoid large changes in the pronation/supination moment of force acting on the hand. This, of course, is just a guess.

Second, let us not forget that fast changes in a performance variable may be associated with an increase in the role of timing errors (see Equation 5.3 in section 5.3.2) that are known to contribute to *bad variance* and thus lead to destabilization of performance. So, maybe, these results are simply reflections of the high rate of force change. This is also possible, although there have been results of studies of multi-muscle synergies (to be discussed later) that do not support this interpretation.

Pressing tasks are an excellent model to study multi-finger synergies. However, it leaves one with a doubt: Are these results applicable to synergies in everyday movements? We rarely perform pressing tasks in everyday life, and the fingers are much more commonly used together with the thumb. Can this framework be applied to synergies involved in hand manipulation of objects? This is the topic of the next subsection.

5.4 PREHENSILE SYNERGIES

There are numerous ways to grasp and manipulate objects. When an object is small and light, only two digits are typically used, most commonly the index finger and the thumb, which contact the object with the finger pads. A typical example would be that of holding a small shot glass filled with a favorite drink with the two digits of the dominant hand. When an object is larger and heavier, more digits are involved. For example, a glass of wine may be handled with the fingertips of three, four, or five digits. When high forces and/or moments

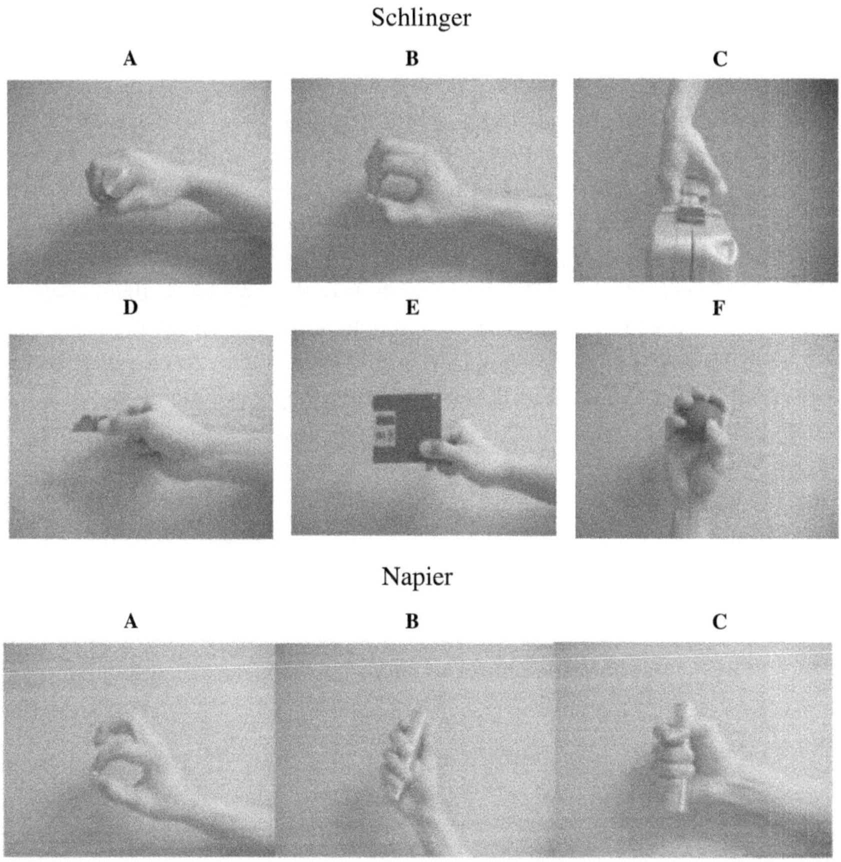

Figure 5.23. Classification of grips according to Schlinger (the upper panels) and Napier (the lower panels). Upper panels: (A) cylindrical grip; (B) tip grip; (C) hook grip; (D) palmar grip; (E) lateral grip; (F) spherical grip. Lower panels: (A) precision grip; (B) power grip; (C) coal-hammer grip. Reproduced by permission from Shim JK, Park J, Zatsiorsky VM, Latash ML (2006) Adjustments of prehension synergies in response to self-triggered and experimenter-triggered load and torque perturbations. *Experimental Brain Research* 175: 641–653. © Springer.

of force have to be applied to a hand-held object, for example when working with a wrench, a power grip is used (Figure 5.23), which increases the contact area between the hand and the object. In this section, I am going to consider manipulation of relatively light objects, such that only the fingertips make contact with them.

When a digit makes contact with an object, it generally can exert on it a vector of force and a vector of moment of force. Both vectors have three components that, in general, can be varied independently. Typically, contact between a finger pad and an external object is viewed as soft and nonsticking. This means, in particular, that the finger pad is allowed to roll over the object such that the point of application of the resultant force can change. On the other hand, the finger can only press on the object and cannot pull on it.

Let us consider, for simplicity, a two-dimensional task when all the points of digit contact with the hand-held object (the handle) are in the same plane—the grasp plane. If a person holds an object statically in the air using all five digits with the thumb opposing the four fingers (in the so-called *prismatic grasp*, Figure 5.24), the digit forces and moments of force have to be balanced to satisfy equations of statics. In particular, the magnitude of the force of the thumb normal to the surface should be equal to the magnitude of the sum of the normal forces of the four fingers (otherwise, the object would start moving to the left or to the right). The sum of the tangential forces of all the digits should be equal to the weight of the object (otherwise, the object would move up or down). And finally, the total torque applied to the hand should be balanced by the moments of force produced by the normal forces and by the tangential forces (otherwise, the object would turn). These three constraints may be expressed as three equations:

$$F_{\mathrm{th}}^{\mathrm{n}} + F_{\mathrm{i}}^{\mathrm{n}} + F_{\mathrm{m}}^{\mathrm{n}} + F_{\mathrm{r}}^{\mathrm{n}} + F_{\mathrm{l}}^{\mathrm{n}} = 0 \qquad \text{Equation (5.4)}$$

$$F_{\mathrm{th}}^{\mathrm{t}} + F_{\mathrm{i}}^{\mathrm{t}} + F_{\mathrm{m}}^{\mathrm{t}} + F_{\mathrm{r}}^{\mathrm{t}} + F_{\mathrm{l}}^{\mathrm{t}} = -L \qquad \text{Equation (5.5)}$$

$$\underbrace{F_{\mathrm{th}}^{\mathrm{n}} d_{\mathrm{th}} + F_{\mathrm{i}}^{\mathrm{n}} d_{\mathrm{i}} + F_{\mathrm{m}}^{\mathrm{n}} d_{\mathrm{m}} + F_{\mathrm{r}}^{\mathrm{n}} d_{\mathrm{r}} + F_{\mathrm{l}}^{\mathrm{n}} d_{\mathrm{l}}}_{\text{Moment of the normal forces} \equiv M^{\mathrm{n}}} + \underbrace{F_{\mathrm{th}}^{\mathrm{t}} r_{\mathrm{th}} + F_{\mathrm{i}}^{\mathrm{t}} r_{\mathrm{i}} + F_{\mathrm{m}}^{\mathrm{t}} r_{\mathrm{m}} + F_{\mathrm{r}}^{\mathrm{t}} r_{\mathrm{r}} + F_{\mathrm{l}}^{\mathrm{t}} r_{\mathrm{l}}}_{\text{Moment of the tangential forces} \equiv M^{\mathrm{t}}} = -T \qquad \text{Equation (5.6)}$$

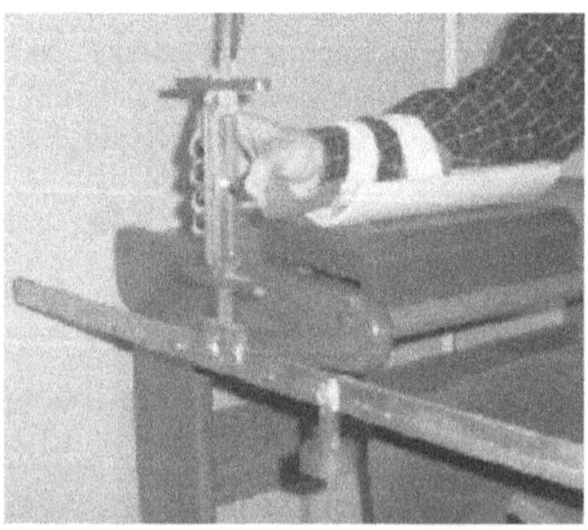

Figure 5.24. An illustration of the cylindrical grip with the four fingers opposing the thumb. Reproduced by permission from Zatsiorsky VM, Latash ML, Gao F, Shim JK (2004) The principle of superposition in human prehension. *Robotica* 22: 231–234. © Cambridge University Press.

where the subscripts th, i, m, r, and l refer to the thumb, index, middle, ring, and little finger, respectively; the superscripts n and t stand for the normal and tangential force components, respectively; L is load (weight of the object), T is external torque, and coefficients d and r stand for the moment arms of the normal and tangential force with respect to a preselected center, respectively.

The first equation reflects balance of normal forces produced by the five digits of the hand. The second and third equations are related to adjustments of digit forces to external task requirements (the load L and the torque T). It is natural to assume that performance variables that are important for a given task are those produced by the hand to meet the task requirements, that is, they represent the total tangential force and the total moment of force computed according to the left sides of Equations 5.5 and 5.6.

However, there is a catch here. All the variables in the three equations are mechanical variables that may be expected to be nonindependent, in a sense that there may be complex patterns of enslaving among them. Enslaving has been demonstrated among the normal forces produced by the five digits (Li et al. 1998; Zatsiorsky et al. 2000; Olafsdottir et al. 2005). In addition, the generation of large tangential forces by a finger has been shown to lead to complex patterns of tangential forces produced by other fingers (Pataky et al. 2007). These are, however, only small pieces of the whole mosaic. Imagine that a person produces a set of forces and moments on an object that does not have to be held in the air, for example, on the same handle that is attached to an external stand. Now let us ask the person to change only the tangential force of the ring finger. What will happen with all the other digit forces and moments of force? There is no answer available. Likely, such relations will be much more complex than the close-to-linear enslaving relations observed during multi-finger pressing tasks. As a result, it is currently impossible to identify digit modes for prehension tasks that would be equivalent to modes introduced earlier for pressing finger forces. This situation forces researchers to look for multi-digit synergies not in the space of modes but in the space of elemental mechanical variables (reviewed in Zatsiorsky and Latash 2004).

5.4.1 Hierarchical Control of Prehension

Grasping an object is commonly analyzed as a task controlled in a hierarchical manner with at least two levels of control (Figure 5.25). The higher level of the hierarchy uses task requirements as the input and produces output signals related to forces and moments generated by the thumb and the virtual finger (VF). VF is a fictitious digit, which produces a mechanical action on the hand-held object that is equivalent to the summed action of all four fingers (Arbib et al. 1985; Mackenzie and Iberall 1994; Baud-Bovy and Soechting 2001). At the lower level, the action of the VF is transformed into variables leading to the generation of individual finger forces and moments.

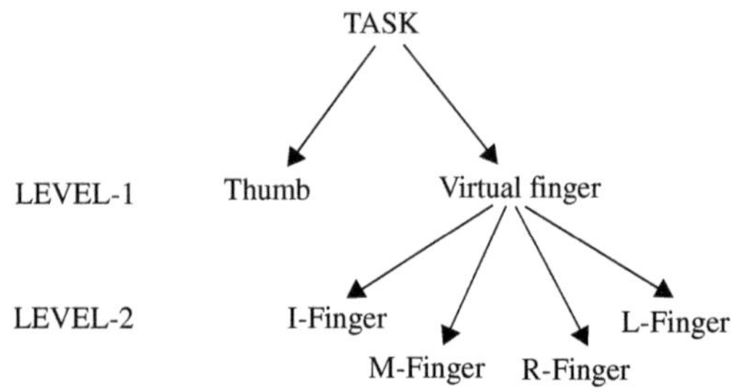

Figure 5.25. Two-level hierarchical scheme of control of prehensile tasks. At the upper level (LEVEL-1), the task is shared between the thumb and the virtual finger. At the lower level (LEVEL-2), the action of the virtual finger is shared among the actual fingers (I—index, M—middle, R—ring, L—little).

There are several reasons to consider the scheme illustrated in Figure 5.25. The most convincing are that the outputs of individual fingers have been shown to co-vary across trials and over time in a long-lasting trial in such a way that the action of VF is relatively undisturbed (Santello and Soechting 2000). If we, for now, forget that this analysis was done using elemental variables that may show unknown patterns of enslaving, these results suggest that fingers are united into a synergy stabilizing the mechanical action of the VF. In particular, it has been shown that the total normal force produced by the fingers (the VF normal force) remains relatively unchanged by co-varied changes in the normal forces produced by the individual fingers (Santello and Soechting 2000; Shim et al. 2004, 2005c). The same is true for the total tangential force produced by the fingers and for the total moment of force (Shim et al. 2004).

Moreover, one of the studies investigated the ability of humans to produce a vector of total force by a set of fingers in a certain direction (Gao et al. 2005a). In that study, there were no equilibrium constraints, but the subjects pressed against thimbles that allowed finger force direction to vary. In different trials, the direction of force produced by any individual finger varied in a rather broad range. However, the direction of the total force showed relatively minor deviations from the target direction (Figure 5.26). This suggests the existence of a multi-finger synergy stabilizing the direction of force produced by the VF.

If we accept the hierarchical scheme of control, issues of synergies may be studies at both levels of the hierarchy. Do the elemental variables produced by the thumb and VF co-vary to stabilize important performance variables such as the total force and the total moment of force? Do the elemental variables produced by individual fingers co-vary to stabilize the output of the VF? In the previous paragraphs, I have already presented a few examples that suggest an affirmative answer to the second question. Now, we are going to consider possible thumb–VF

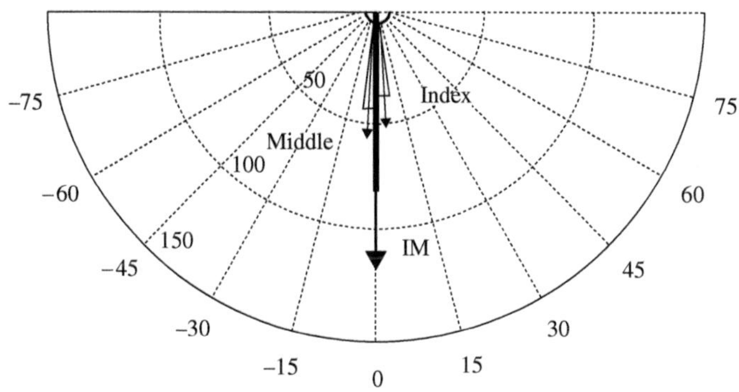

Figure 5.26. An illustration of the direction of the force vectors produced by individual fingers (index and middle fingers in this example) and the total force vector. The task was to point in direction of 0°. Note that the total force vector points more accurately in the required direction compared to the individual finger force vectors and shows smaller variability in the direction. Reproduced by permission from Gao F, Latash ML, Zatsiorsky VM (2005) Control of finger force direction in the flexion-extension plane. *Experimental Brain Research* 161: 307–315. © Springer.

synergies. The three static constraints introduced earlier may be rewritten for the thumb–VF level as

$$F^{\mathrm{n}}_{\mathrm{th}} + F^{\mathrm{n}}_{\mathrm{VF}} = 0 \qquad \text{Equation (5.7)}$$

$$F^{\mathrm{t}}_{\mathrm{th}} + F^{\mathrm{t}}_{\mathrm{VF}} = -L \qquad \text{Equation (5.8)}$$

$$\underbrace{F^{\mathrm{n}}_{\mathrm{th}} d_{\mathrm{th}} + F^{\mathrm{n}}_{\mathrm{VF}} d_{\mathrm{VF}}}_{\text{Moment of the normal forces} \equiv M^{\mathrm{n}}} + \underbrace{F^{\mathrm{t}}_{\mathrm{th}} r_{\mathrm{th}} + F^{\mathrm{t}}_{\mathrm{VF}} r_{\mathrm{VF}}}_{\text{Moment of the tangential forces} \equiv M^{\mathrm{t}}} = -T \qquad \text{Equation (5.9)}$$

where the abbreviations are the same as in the previous set of equations. The number of elemental variables in these equations is larger than the number of constraints. This means that the system is redundant (sorry, abundant!).

5.4.2 Principle of Superposition

In biology, a principle of superposition implies that output of a system composed of a number of elements with several inputs equals the sum of the outputs produced by each of the inputs applied separately. Violations of this principle in neurophysiology are many and varied. They do not come as a surprise because of the well-known highly nonlinear properties of neurons, in particular, their threshold properties and the all-or-none generation of units of information, action potentials (see Digression #1). For example, commonly, individual excitatory synapses cannot bring their target neuron to its threshold for action potential generation, while several stimuli coming simultaneously or at a high frequency can (these phenomena are referred to as spatial and temporal summation). The principle of superposition has been questioned in a number of studies of such diverse processes as motor unit firing patterns and maintenance of vertical posture (Demieville and Partridge 1980; Latt et al. 2003).

Despite the highly nonlinear features of neurons, muscles, and reflex loops, several research teams tried to find regularities of behaviors of large populations of such elements that would behave in a nearly linear fashion and obey the principle of superposition. For example, input–output properties of motoneuronal pools have been reported to be nearly linear in tasks that involved a superposition of steady-state and ballistic contractions (Ruegg and Bongioanni 1989). Studies of neuronal populations in different areas of the brain have also supported applicability of the principle of superposition (Kowalski et al. 1996; Glazer and Gauzelman 1997). In a recent study, Fingelkurts and colleagues (2006) have come up with a conclusion that integrative brain functions can be manifested in the superposition of distributed multiple oscillations according to the principle of superposition.

In the area of motor control, the idea of the principle of superposition has been developed by proponents of the EP-hypothesis (Feldman 1966, 1986; see section 3.3). According to the EP-hypothesis, state variables such as levels of muscle activation, forces, torques, positions are defined by control signals, properties of the tonic stretch reflex loop, and external force field (load). Control of a joint can be described with two variables, a coactivation command (*c*) and a reciprocal command (*r*) (Feldman 1986). Changes in these two control variables affect activation of individual muscles according to the principle of superposition; experimental support for this principle has been obtained in studies of fast arm movements (Adamovich and Feldman 1984; Pigeon et al. 2000) and postural control (Slijper and Latash 2000).

Seminal papers by Arimoto and colleagues (Arimoto et al. 2000, 2001) have shown that the principle of superposition may be applied at the level of mechanical variables for the control of robotic hand action. The main idea of this approach is to separate complex motor tasks into subtasks that are controlled by independent controllers. The output signals of the controllers converge onto the same set of actuators where they are summed up. This method of control has been shown to lead to a decrease in the computation time compared to control of the action as a whole. However, does this principle also apply to the control of digits during object manipulation?

This question is far from trivial. For example, if a hand tries to manipulate a grasped object, it has to produce adequate grasping and rotational actions. However, a straightforward change in the grasping force, in general, leads to a change in the total moment produced by the digits on the object, because of a change in the moment of normal forces (see Equation 5.6). The grasping and rotational components of the action are not independent—this is clear from the sets of equations presented earlier—so, if a controller responsible for the grasping changes its output, a controller responsible for the rotational action also has to change its output. This is an unsolvable problem unless one has a redundant set of elements, which allows decoupling the two action components and using independent controllers for each of them. The possibility of such decoupling is another major advantage afforded by the apparently redundant design of the hand (certainly, if the principle of abundance is followed!).

Several experimental studies have provided support for the existence of two, independent multi-digit synergies corresponding to two commands: "Grasp the object stronger/weaker to prevent slipping" and "Maintain the rotational equilibrium of the object." In one of these studies (Zatsiorsky et al. 2003), changes in the mechanical elemental variables at the VF–thumb level were analyzed as functions of the external load, *L,* and torque, *T.* All the elemental variables showed significant changes with *T* and *L*; however, there were no significant interactions between these two factors.

In another study (Shim et al. 2003a), subjects performed repetitive trials for each of five combinations of *L* and *T.* For a given (*L*, *T*) combination, elemental variables at the VF–thumb level formed two subsets such that the variables within each subset were highly correlated with each other over repetitions of the task, while the variables from different subsets did not show significant correlations. Both mentioned studies used two-dimensional tasks when the external torque acted in the grasp plane. A follow-up study generalized these results to three-dimensional tasks, when the external torque acted perpendicular to the grasp plane (Shim et al. 2005). This happens, for example, when you grasp a heavy book close to its edge and hold it in the air. The principle of superposition has been supported once again.

Still another study explored adjustments in digit forces and moments of force in response to a predictable and unpredictable perturbation applied to a hand-held object (Shim et al. 2006). The perturbation could change both the load and the external torque. Indices of multi-digit synergies were computed for the total grip force and total moment of force as performance variables, ΔV_F, and ΔV_M. Both indices showed high positive values corresponding to strong multi-digit synergies stabilizing both variables, when the subjects held the object steadily in the air. Prior to a self-triggered perturbation, there were anticipatory changes in both ΔV_F and ΔV_M (ASAs, see section 5.3.4). These were seen 100–150 ms prior to the perturbation; ASAs were not observed when the perturbation was triggered by the experimenter unexpectedly. Following a perturbation, there was a drop in both synergy indices, and after about 1–2 s, the indices recovered. It is of interest that ΔV_F decreased in a new steady-state, while ΔV_M increased. This was true for both self- and experimenter-triggered perturbations. These contrasting changes in the synergy indices provide additional support for the relative independence of the two synergies involved in the grasping and rotational components of prehensile actions and show that the principle of superposition holds for multi-digit reactions to force/torque perturbations.

5.4.3 Adjustments of Synergies: Chain Effects

Since all mechanical constraints associated with prehension involve more than one elemental variable and these variables are shared among the constraints, a change in a single elemental variable may be expected to lead to a sequence of cause-effect adjustments ("sequence" does not mean a chronological order) necessitated by those constraints. Such sequential adjustments are called *chain effects* (Shim et al. 2003, 2005a,b; Zatsiorsky et al. 2003).

Imagine, for example, that you are holding a handle using a prismatic grasp (as in Figure 5.24) against an external load and torque. For simplicity, let me consider the task at the thumb–VF level. According to the principle of superposition, elemental variables that contribute to the grasping force (the normal forces of the thumb and VF) and elemental variables that contribute to the total moment of force (such as the point of VF application and the tangential forces of the thumb and VF) are decoupled, that is, elemental variables that belong to different groups are not expected to co-vary across trials at the same task. This was indeed confirmed in experiments that showed no correlation between variations in the VF normal force and the moment that this force produced with respect to the point of thumb contact (Shim et al. 2003, 2005). This is a very nontrivial observation, because the moment of force is the product of the force and its lever arm, and hence a correlation between the two can be expected, unless it is purposefully destroyed by variations in the lever arm (defined by variations in the point of VF force application).

Imagine now that in one of the trials one elemental variable, for example, the tangential force produced by the thumb, shows a small change. To keep the handle from moving in the vertical direction, there should be an adjustment in the tangential force produced by the VF, since the sum of the two should equal the external load. A change in that force would automatically change the moment of force it produces with respect to the thumb contact, because the lever arm cannot change (it is defined by the width of the handle). To satisfy the requirement of rotational equilibrium, the moment produced by the normal forces should adjust, since the two moments should sum up to match the external torque. As a result of this (relatively short) chain, a correlation can be expected between variations in the tangential force of the thumb and the moment produced by the normal force of the VF. This correlation is actually observed in experiments.

Consider another example. Imagine that friction under one of the digits suddenly happens to be low (you grasp an object with an oily spot). Typically, a digit that makes contact with a low-friction surface shows an increase in its normal force and a decrease in its tangential force (to avoid slippage) (Edin et al. 1992; Burstedt et al. 1999; Quaney and Cole 2004). An increase in the normal force is expected to violate two of the three Equations 5.1 and 5.3, while a drop in the tangential force violates Equations 5.2 and 5.3. The abundance of the system allows numerous ways to adjust forces of other digits to keep all three equilibrium equations satisfied. For example, an increase in the tangential force of the finger can be compensated by a drop in the tangential forces of other fingers such that the VF tangential force is unchanged. However, a straightforward drop in the normal forces of other fingers (to offset the increased normal force of the "low friction" finger) will lead to a change in the point of VF force application in the vertical direction. This will lead to a change in the moment of force produced by the normal VF force. This change will require an adjustment in the moment of force produced by the tangential forces of the thumb and VF. Since the lever arm cannot be changed, the tangential forces will have to be redistributed between the thumb and VF. So, the straightforward solution to keep the normal and tangential forces of the VF unchanged does not work.

This chain effect analysis shows that nontrivial adjustments in the forces produced by all the digits are required to keep balance in response to a change in friction under one of the fingers (see Aoki et al. 2006, 2007). Hence, adjustments to changes in conditions for one of the elements (one finger) may be viewed as leading to two groups of effects, local adjustments (in elemental variables produced by this finger) and synergic adjustments in elemental variables produced by the seemingly unaffected elements (digits). In a number of studies, the synergic adjustments have been shown to be large (reviewed in Latash and Zatsiorsky 2007). Their purpose is to keep important performance variables at values dictated by the task.

5.4.4 Hierarchies of Synergies

Control hierarchies have been described for at least half-a-century: A rather complex hierarchical scheme of motor control was suggested by Nikolai Bernstein (1947, 1967, 1996). The second of the four (in some descriptions, five) levels of movement construction was termed "The Level of Muscular-Articular Links," or "Level of Synergies," or "Level B." Bernstein's usage of the term "synergy" was rather loose, as a combined action of large groups of muscles. The development of Bernstein's ideas by Gelfand and Tsetlin (section 3.1) led to a view on synergies (structural units) as neural organizations that produce a stable output variable by a coordinated action of many elements. In this approach, a scheme may be suggested involving a hierarchical set of synergies (Figure 5.27). Any synergy receives an input and produces an output. The input comes from a hierarchically higher synergy, while the output serves as an input into a hierarchically lower synergy. The hierarchy in Figure 5.27 may branch to create synergies involving different sets of elemental variables. Let me assume, for now, that the input into the highest level synergy comes from the "task" and that the lowest level synergy acts on the environment.

Many studies have demonstrated adjustments of the outputs of individual elements to stabilize their combined action. Examples include the coordination

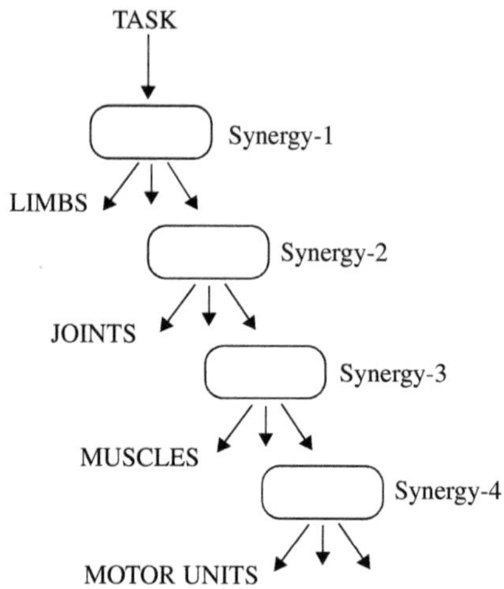

Figure 5.27. A hierarchy of synergies. Input at each level is provided by an output of a hierarchically higher synergy. Outputs of each level serve as inputs into a lower level synergy.

of articulators during speech (Abbs and Gracco 1984), of arms during loading and unloading (Dufosse et al. 1985; Paulignan et al. 1989), of joints during pointing, reaching, standing-up, and swaying (Wang and Stelmach 1998; Scholz and Schöner 1999; Domkin et al. 2002; Freitas et al. 2006), of muscles during standing and stepping (Krishnamoorthy et al. 2003b; Wang et al. 2005), and of digits during pressing, grasping, and holding an object (Scholz and Latash 1998; Santello and Soechting 2000; Gao et al. 2005a; Shim et al. 2005). Very few studies, however, have addressed synergies at different levels of an involved control hierarchy. Those include the studies of digit coordination during prehension (Baud-Bovy and Soechting 2001; Zatsiorsky et al. 2003; Gao et al. 2005a,b; Shim et al. 2005) and of two-arm interaction during pointing (Domkin et al. 2002).

In particular, the coordinated action of fingers during prehensile tasks has been shown to stabilize the action of the VF, while the coordinated action of the VF and the thumb stabilized the gripping and rotational action components on the hand-held object (Zatsiorsky et al. 2003; Shim et al. 2005a). During two-arm pointing, the joints within each arm have been shown to stabilize the trajectory of the endpoint, while across two arms the vectorial distance between the endpoints was stabilized (Domkin et al. 2002, 2005). So, synergies at two (or more) levels of a control hierarchy may coexist.

A recent study (Gorniak et al. 2007a) addressed a seemingly naïve question: Can the CNS organize both two-hand and within-a-hand force-stabilizing synergies in a simple two-hand force production task that involves two fingers per hand? Intuitively, one could expect a positive answer, that is, forces produced

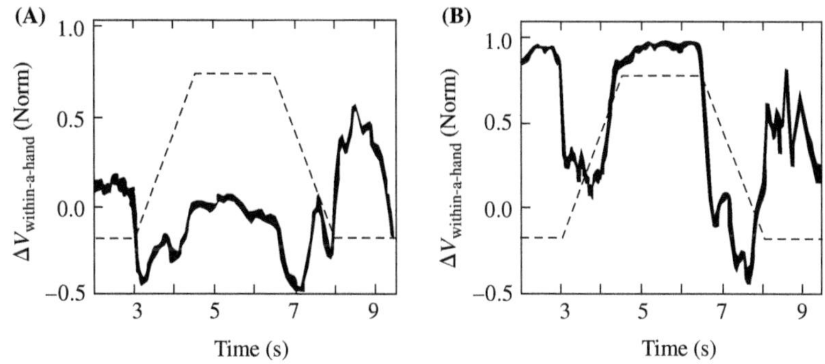

Figure 5.28. An illustration of the index (ΔV) of force stabilization computed for a pair of finger within-a-hand that performed the task of accurate trapezoidal force production alone (right panel) or in combination with two fingers of the other hand (left panel). In the right panel, during the steady-state phase, $\Delta V > 0$, while in the left panel, ΔV is significantly smaller and commonly is not different from zero. Reproduced by permission from Gorniak S, Zatsiorsky VM, Latash ML (2007a) Hierarchies of synergies: an example of the two-hand, multi-finger tasks. *Experimental Brain Research* 179: 167–180. © Springer.

by each hand are expected to co-vary negatively across trials to bring down the total force variability, while forces produced by each finger within a hand are expected to co-vary negatively to reduce the variability of that hand's contribution to the total force. The mentioned studies of two-hand pointing and multi-digit prehension also suggest that this should be the case.

However, the results were quite unexpected. In that study, the subjects were instructed to follow a trapezoidal time profile with the signal corresponding to the force produced by a set of instructed fingers (Figure 5.28). There were two types of tasks. One-hand tasks involved sets of two or four fingers that belonged to the same hand. Two-hand tasks involved sets of four fingers, two per hand, with either symmetrical or asymmetrical finger pairs in the two hands (e.g. IM + IM and RL + RL are symmetrical pairs while IM + RL and RL + IM are asymmetrical; I, M, R, and L stand for the index, middle, ring, and little fingers respectively).

In general, tasks of slowly changing (or steady-state) force production are associated with strong multi-finger synergies stabilizing the total force; these synergies show transient episodes of weakening when a steady-state phase turns into a ramp phase (Shim et al. 2005b). In one-hand tasks, finger forces co-varied across trial to bring down the variability of the total force, while in two-hand tasks, forces produced by the finger groups in the right and left hands also co-varied to stabilize the total force. However, in two-hand tasks, forces produced by individual fingers did not co-vary to stabilize the contribution of that hand to the total force. Moreover, in a follow-up study (Gorniak et al. 2007b), when a single-hand force production task was turned into a two-hand task (the other hand started helping to

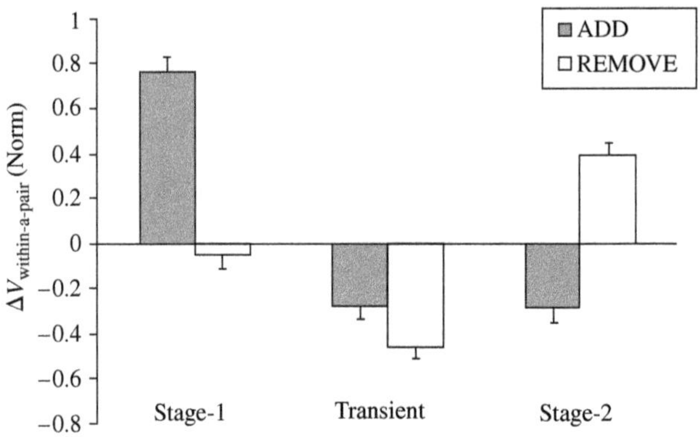

Figure 5.29. ADD: The subject started a constant force-production task with two fingers of a hand (Stage-1). Then, at some point he added two fingers of the other hand, trying to keep the total force unchanged. Note the dramatic drop in the index of force-stabilizing synergy (ΔV in Stage-2) computed for the finger pair that was active throughout the task. REMOVE: The task was started with two finger-pairs (Stage-1). Then, one of them was removed. Note that the ΔV index for the pair that remained active throughout the task was low in the beginning and, after a transient drop, it increased significantly after the other finger pair was removed (Stage-2). Reproduced by permission from Gorniak S, Zatsiorsky VM, Latash ML (2007b) Emerging and disappearing synergies in a hierarchically controlled system. *Experimental Brain Research* 183: 259–270. © Springer.

produce the same total force), the preexistent two-finger force-stabilizing synergy disappeared (Figure 5.29). Why do two-finger synergies disappear when the hand acts not alone but with the other hand?

Currently, there is no answer to this question. It seems that two kinds of synergies may exist in humans. First, there are well-learned synergies that have been based on the lifetime experience and frequently participate in everyday actions. These may be expected to include multi-joint synergies stabilizing the trajectory of the hand (Domkin et al. 2002; Yang et al. 2007), multi-finger synergies stabilizing the action of the VF (Latash et al. 2001; Gao et al. 2005; Shim et al. 2005b), multi-articular synergies stabilizing speech (Abbs and Gracco 1984; Ostry et al. 1996), and numerous multi-muscle synergies (Krishnamoorthy et al. 2003b; Wang et al. 2005; Danna-Dos-Santos et al. 2007). Synergies of this kind can be used as a vocabulary to create other, hierarchically higher synergies. Second, there are synergies that have to be created for each given task. They may be simple but not common enough for everyday tasks to join the first group. For example, pressing with the index and middle finger of the dominant hand is an easy task, but we rarely do this. The results of the mentioned study suggest a surprising limitation of the CNS: It is able to create new synergies only at one hierarchical level at a time. Other supporting synergies may be observed but only if they are well practiced and included in the synergy vocabulary.

Let us consider a two-hand, four-finger task in greater detail. Imagine that a two-hand synergy is organized to stabilize the total force. The presence of such a synergy means that V_{GOOD} (variance parallel to the UCM for the total force

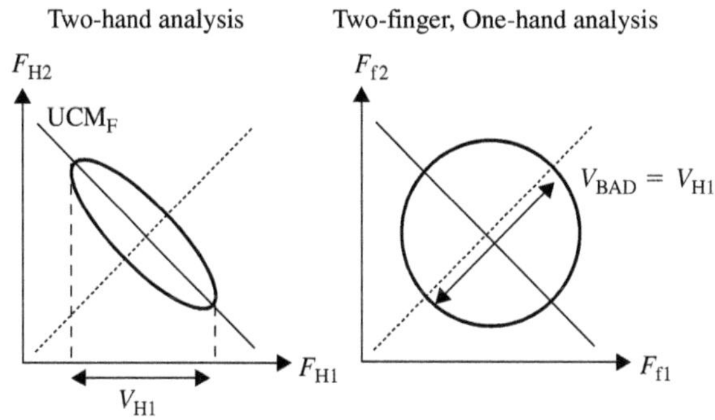

Figure 5.30. A synergy at the hierarchically higher level (two-hands, left panel) leads to $V_{GOOD} > V_{BAD}$. However, high V_{GOOD} leads to high variance of force of each of the hands (V_{H1}, illustrated for Hand-1). At the lower level (two-fingers, right panel), V_{H1} is V_{BAD}, and its increase leads to disappearance of a two-finger synergy. Modified with permission from Gorniak S, Zatsiorsky VM, Latash ML (2007b) Emerging and disappearing synergies in a hierarchically controlled system. *Experimental Brain Research* 183: 259–270. © Springer.

shown as UCM_F in the left panel of Figure 5.30) is substantially larger than V_{BAD} (variance parallel to the dashed line). Note that the magnitude of V_{GOOD} is higher when the range of changes in the forces produced by individual hands in larger (V_{H1} in the left panel in Figure 5.30). However, an increase in the range of force produced by a hand by definition leads to an increase in its *bad variability* (the right panel in Figure 5.30). Hence, the most straightforward method of creating a two-hand force-stabilizing synergy, that is, an increase in its V_{GOOD} is expected to be reflected in an increase in V_{BAD} for each of the participating finger pairs. This may be expected to result in a drop of the index of force-stabilizing synergies for each finger pair (each hand) as observed in the cited experiments.

This simple analysis suggests that there is an inherent trade-off between synergies that stabilize the output of each of the control levels in a hierarchy. This trade-off can be overcome if the controller manages to keep V_{GOOD} relatively low at the higher level of the hierarchy (the two-hand level), while still preserving the inequality $V_{GOOD} > V_{BAD}$. This may require extensive practice as shown by a study in which force-stabilizing synergies emerged after 3 days of practice (Kang et al. 2004).

5.5 MULTI-MUSCLE SYNERGIES

The term *synergy* in studies of movements and movement disorders was originally applied in the combination *muscle synergy* (Babinski 1899; cited after Smith 1993). Bernstein also applied this term to large muscle groups coordinated to perform a task. Typical and atypical muscle synergies are commonly invoked in clinical practice in our times. Can this usage of the word *synergy* be operationalized within the introduced framework? Can one perform a study of muscle activation patterns and come up with an index that would allow to conclude whether there is a muscle synergy or not, and if so, to quantify its strength?

Application of the UCM method to analysis of multi-muscle synergies faces a number of challenges. The first nontrivial step is to determine a set of elemental variables. Since Hughlings Jackson (1889), researchers have agreed that the brain does not control movements by defining signals that have to be sent to each individual muscle. The CNS is likely to unite muscles into groups and modulate the activity of the muscles within such a group with just one parameter (Saltiel et al. 2001; Ivanenko et al. 2004, 2005; Ting and Macpherson 2005; Tresch et al. 2006; van der Linden et al. 2007). This general view suggests applying the notion of modes, similar to how this has been done in studies of multi-finger coordination (see sections 4.2.1 and 4.2.2).

The notion of muscle modes is illustrated in Figure 5.31. The controller (the brain) is assumed to manipulate a relatively small set of variables that serve to define magnitudes of signals to groups of muscles. Not all the muscles in a group are expected to be equally responsive to a given signal but, in a first approximation, their responses may be viewed as scaled linearly within a group. A muscle may be a member of several groups. Formally, each group (we can call it "a mode") represents a linear combination of muscle activations corresponding to a vector in muscle activation space of unitary length (an eigenvector):

$$M_1 = \sum_{i=1}^{n} a_{1,i} E_i$$
$$M_2 = \sum_{i=1}^{n} a_{2,i} E_i \qquad \text{Equation (5.10)}$$
$$M_3 = \sum_{i=1}^{n} a_{3,i} E_i$$

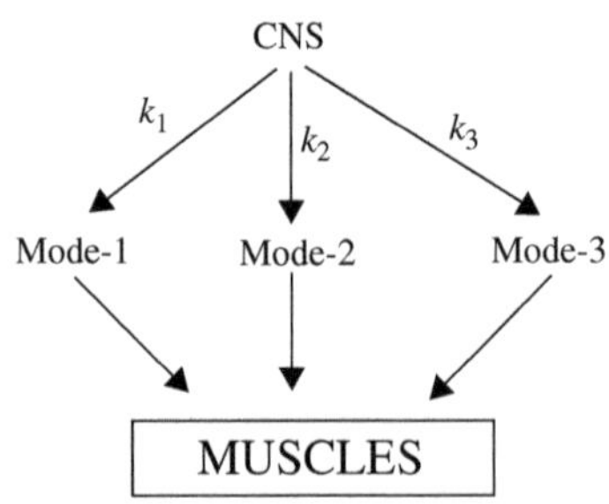

Figure 5.31. An illustration of the notion of muscle modes. The central nervous system (CNS) manipulates coefficients (k_1, k_2, and k_3) of involvement of a few (three in the scheme) variables—M-modes. Each mode leads to changes in activation level of many muscles.

where M stands for a mode, E stands for a measure of activation of a given muscle i (e.g. an integral measure of rectified muscle activity over a small time interval), n is the total number of muscles, and a are coefficients. Note that, by definition, each mode has the length of unity. A control signal defines gains at the modes (k_j) resulting in a vector of muscle activations, **E**:

$$\mathbf{E} = \sum_{j=1}^{m} k_j M_j \qquad \text{Equation (5.11)}$$

Accepting this general scheme of control allows identifying modes in experiments using standard tasks and looking for co-varied changes in muscle activations that could be signs of those muscles belonging to a mode. M-modes were defined, in particular, using PCA applied to indices of integrated EMG activity of a set of postural muscles over small time intervals. In one of the first studies, postural multi-muscle synergies were analyzed during APAs (Krishnamoorthy et al. 2003a,b). Correspondingly, muscle modes were defined using integrals of EMG of a set of postural muscles over a typical time interval for APAs (100 ms) prior to a standard action that could trigger postural perturbations of different magnitude. PCA of EMG integral indices allowed identification of three M-modes that were similar across both tasks and subjects and could participate in APAs in preparation to perturbations in different directions in the sagittal plane (generating a moment of force rotating the body forward or backward about the ankle joints). In a more recent study, EMGs were integrated over smaller time intervals (25 ms), while the subjects swayed voluntarily (rocked the body forward and backward) at different speeds (Danna-Dos-Santos et al. 2007). The modes were similar in composition (similar coefficients at individual muscle activation indices) to those described in the earlier studies; they were also shown to be similar across both tasks (frequencies of sway) and subjects.

There are a couple of important notes. The PCA defined orthogonal axes in the original space (muscle activation space) in an order corresponding to the highest-to-lowest amount of variance along different directions. Imagine, for simplicity, that the data represent a cloud of points that form an ellipsoid. The first PC points in a direction of the largest spread of data points, that is, along the main axis of the ellipsoid. The second PC points in a direction orthogonal to the first one corresponding to the second largest axis of the ellipsoid, etc. In general, the number of PCs is equal to the number of axes in the original space. However, typically, a smaller number of PCs allows accounting for most of the variance, and then the researchers assume that no useful information is contained in the remaining PCs. An important point is that PCs are by definition orthogonal to each other, and one may assume that gains at the eigenvectors corresponding to PCs can be manipulated by the controller independently (at least hypothetically). In a number of other studies of muscle activation patterns, results of similar PCA were interpreted as pointing at multi-muscle synergies (Kutch and Buchanan 2001; Holdefer and Miller 2002; Sabatini 2002; Ivanenko et al. 2004, 2005). We, however, view results of PCA as only the first step in the identification of elemental variables.

The next important step is to define a Jacobian matrix, ***J***, that would link small variations in mode magnitudes (defined by variations of individual gains, k_j, Equation 4.2) to variations in a potentially important performance variable generated by the system. This is also a nontrivial step. Indices of integrated muscle activation are expressed in microvolts or similar units. Important performance variables are typically expressed in mechanical units such as newtons and meters. For example, what does a change in the first mode by 10 μV mean for center of mass displacement in a standing person? This question cannot be answered as simply as in earlier studies, where the total force produced by the hand was equal to the sum of individual finger forces, and displacements of the endpoint of a multi-joint limb were linked to joint displacements by a straightforward geometrical model. Hence, a ***J*** matrix linking changes in magnitudes of M-modes to physical performance variables has to be discovered experimentally.

To do this, one has to select a candidate performance variable that may be important for a given group of tasks. For postural tasks in the cited studies, this variable was associated with coordinate of the COP (the point of application of the resultant force from the support surface acting on the body). This choice was justified by a large number of studies that suggest the importance of controlled COP shifts for postural tasks (Winter et al. 1996; Zatsiorsky and King 1998; Zatsiorsky and Duarte 2000). Relations between small changes in M-mode magnitudes and COP shifts can be discovered using multiple regression analysis across a series of measurements with different COP shifts over the same time interval. This can be achieved by an analysis of APAs to expected perturbations of different magnitude or by an analysis of cyclic postural sway when COP velocity changes naturally within each cycle and also shows variability across cycles.

And finally, after elemental variables (M-modes) have been identified, and a ***J*** matrix linking M-modes to a performance variable has been computed, one can move to the UCM analysis and ask a question: Do M-mode magnitudes across repetitive trials co-vary to preserve a particular trajectory of the COP? In other words: "Is there a M-mode COP-stabilizing synergy?"

5.5.1 Anticipatory Postural Adjustments

The first series of studies explored whether muscle activation patterns during APAs were organized to stabilize a trajectory of the COP. This looks like a reasonable hypothesis. However, it is far from being 100% obvious. Many studies of APAs reported reproducible patterns of muscle activation leading to tiny, not very well reproducible COP shifts (reviewed in Massion 1992). In addition, COP can shift both forward and backward during APAs. It was not *a priori* obvious whether the ***J*** matrix was equally applicable for the two directions of COP shift.

The design of those studies was relatively straightforward. Standing subjects were asked to perform a large series of trials with self-triggered perturbations: releasing a load held in extended arms or making a very fast bilateral shoulder movement. These tasks were accompanied by clear APAs seen about 100 ms prior to the action initiation. The large series of trials with releasing different loads were used to compute M-modes. Further, multiple regression analysis was used to map changes in M-mode magnitudes on COP shifts. And finally, a set of trials was used to perform a typical UCM analysis, that is, variability of EMG signals in the M-mode space (computed across trials) was projected onto the UCM (the null-space of the ***J*** matrix) and onto a subspace orthogonal to the UCM. Since in most studies only three M-modes were accepted, the space of elemental variables was three-dimensional, with the UCM two-dimensional, and its orthogonal complement unidimensional.

The first step, which was the identification of M-modes, showed three M-modes that were reproducibly seen in the subjects—the first two M-modes united muscles of the dorsal part of the body and of the frontal part respectively. If the muscles within a mode were activated together, they would push the center of mass of the body backward or forward, respectively. Hence, the two modes were termed "push-back" and "push-forward." Note that each of these modes united muscles crossing all three major joints, the ankle, the knee, and the hip.

Multiple regression analysis at the second step showed significantly different ***J*** matrices associated with COP shifts forward and backward, $\boldsymbol{J}_{\mathrm{F}}$ and $\boldsymbol{J}_{\mathrm{B}}$. When, at the next stage of analysis, variability in the M-mode space was partitioned into *good* (V_{UCM}) and *bad* (V_{ORT}), the relative amount of *good variability* depended on which of the two ***J*** matrices was used. V_{UCM} was significantly larger than V_{ORT} only during analysis of tasks with COP shifts in the same direction that was used at the stage of ***J*** computation, that is, using $\boldsymbol{J}_{\mathrm{F}}$ to analyze trials associated with forward COP shift and $\boldsymbol{J}_{\mathrm{B}}$—for trials with backward COP shift. Otherwise, there was no difference between V_{UCM} and V_{ORT}. (Note that different tasks were used at different stages of analysis to avoid the circular logic trap.) So, these studies not only confirmed the existence of multi-M-mode synergies stabilizing COP shifts but also showed that there were two different synergies, based on different ***J*** matrices, for COP shifts in the two directions. It is important to emphasize that the two synergies were based on the same set of three M-modes.

In a later study, similar postural tasks were used in combination with postural instability (standing on a board with a decreased area of contact with the floor) and additional hand support (Krishnamoorthy et al. 2004). These experiments have shown that the menu of M-modes during such tasks may be bigger, and include, in addition to the mentioned multi-joint "push-back" and "push-forward" modes, three joint-specific M-modes corresponding to parallel changes in the activation levels of antagonist muscles acting at each major postural joint (cf. ankle and hip strategies, Nashner 1976, 1979; Nashner and Cordo 1981; Horak and Nashner 1986). Within each particular task only three modes were seen, and this set of three out of five possible M-modes could differ across tasks and subjects. Moreover, when subjects were allowed to use their right hand for postural support, muscles acting at the shoulder joint and those acting at the hip joint were commonly united into a single mode.

Important conclusions can be drawn from these observations. First, the idea of M-modes seems to be viable. It allows detecting and quantifying multi-muscle synergies (it is more accurate to address those as multi-M-mode synergies). Second, for a common everyday task like standing, M-modes are similar in composition across subjects and across secondary tasks that may be performed simultaneously. Third, the same muscles can be used by the controller to form different M-modes if the task becomes sufficiently different (e.g. standing on an unstable board). So, M-modes are not hard-wired relations among muscle activations but flexible "beings" that can be brought about in a task-specific fashion. As more recent studies have shown (yet unpublished), M-modes may be viewed as synergies created in the space of muscle activations used as elemental variables.

5.5.2 Making a Step

Postural adjustments can be seen not only in preparation to a perturbation that can violate balance but also in preparation to a voluntary movement involving the lower limbs. Examples include raising one of the legs (Mouchnino et al. 1992, 1998) and taking a step (Elble et al. 1994; Lepers and Brenière 1995; Couillandre et al. 2002; Ito et al. 2003). In particular, prior to taking a step, the person has to prepare the body for a change in the conditions for balance and simultaneously to allow reactive forces acting from the floor to move the body in the required direction. Correspondingly, postural adjustments to stepping involve two simultaneous but functionally different events (Figure 5.32).

First, in order to lift a leg, one obviously has to unload it. During quiet standing, the center of mass projects in-between the two feet, slightly in front of the ankle joints. If a person is standing quietly, the COP coordinate is very close to the projections of the center of mass. Lifting a leg leads to a dramatic decrease in the effective support area. Hence, after lift-off of the stepping leg, the projection of the center of mass should be within the support area of the supporting foot that keeps contact with the floor. This means that the center of mass has to be shifted toward the supporting leg before taking a step. This is achieved by a transient COP shift toward the stepping foot (to create a moment of force moving the center of mass toward the supporting foot), which is followed by a COP shift toward the supporting foot such that ultimately the COP coordinate and the center of mass projection in the medio-lateral direction coincide.

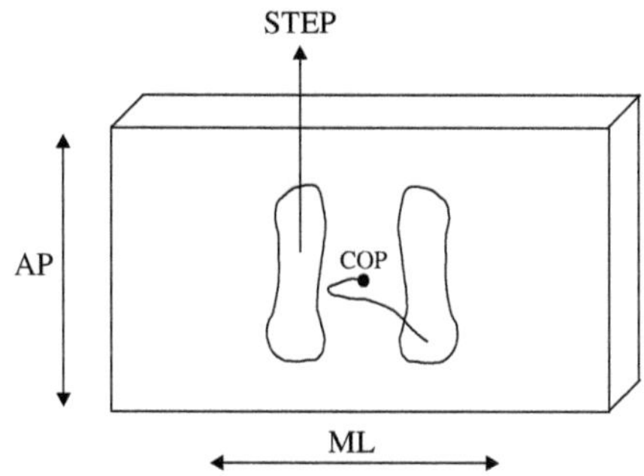

Figure 5.32. Preparation to making a step involves shifts of the center of pressure (COP) in both medio-lateral (ML) and anterior–posterior (AP) directions. The ML shift toward the stepping foot and then toward the supporting foot allows unloading the stepping foot. The AP shift backward allows creating moment of force rotating the body forward—in the direction of stepping.

Second, to destabilize the body and induce its forward rotation about the ankle joint of the supporting foot, the COP shifts backward. Both COP shifts happen simultaneously and take a few hundred milliseconds. This period of postural preparation is also sometimes addressed as an APA, although it does not fit the definition of postural adjustments counteracting expected effects of a postural perturbation, since no explicit perturbation occurs.

Are their multi-muscle synergies stabilizing the COP trajectory during postural preparation to stepping? If so, are they built on a similar set of M-modes as synergies described for APAs? These questions are not trivial. Postural preparation to stepping is a longer process than typical APAs; besides, it involves COP displacement in two dimensions. However, the logic of study remains the same. The first step is to define sets of M-modes that participate in COP shift in the two directions. One can expect these sets of modes to be different; in particular, modes that shift COP to a side (in the medio-lateral direction) are likely to involve muscles with strong lateral action such as tensor fascia lata and gluteus maximus. This was indeed found in experiments (Wang et al. 2005, 2006a,b). "Push-back" and "push-forward" modes were found, but in addition a "push-side" mode emerged that united muscles with lateral action. Second, two Jacobian matrices had to be computed in relation to COP shift in the medio-lateral and anterior–posterior directions. Finally, based on the ***J*** matrices, the UCM and its orthogonal complement were computed, and M-mode data across repetitive stepping were projected onto the two subspaces.

The results confirmed the existence of two multi-M-mode synergies stabilizing COP shifts in both directions. A nontrivial observation was that the index of the two synergies showed a drop 100–200 ms prior to the step initiation, defined as the lift-off of the stepping leg (Figure 5.33). This observation is similar to the

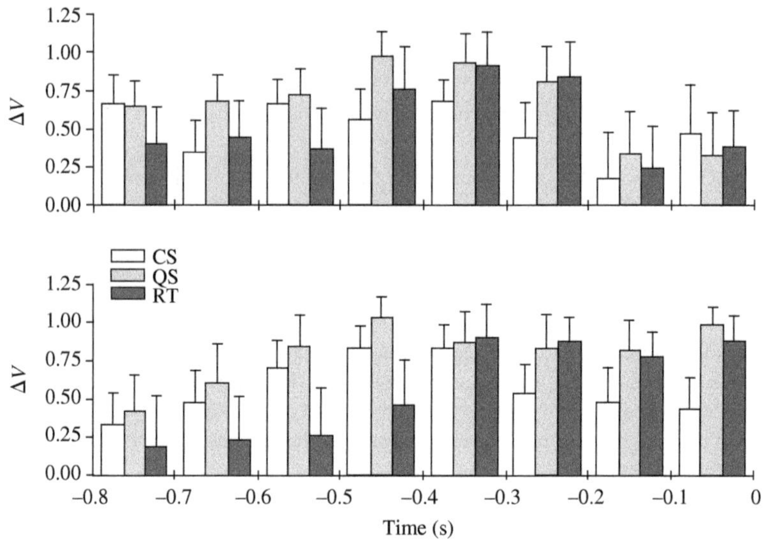

Figure 5.33. An index (ΔV) of a multi-M-mode synergy stabilizing location of the center of pressure (COP) in the anterior–posterior (AP) direction shows a decline prior to the lift-off of the stepping foot. This decline is seen during comfortable stepping (CS), quick stepping (QS), and stepping "as quickly as possible" after hearing a beep signal (RT). The drop is seen only in the stepping leg (the upper graph) but not in the supporting leg (the bottom panel). Modified with permission from Wang Y, Zatsiorsky VM, Latash ML (2005) Muscle synergies involved in shifting center of pressure during making a first step. *Experimental Brain Research* 167: 196–210. © Springer.

phenomena of ASAs described earlier (section 5.3.4). So, they may reflect a similar strategy of turning a synergy off in preparation to a quick COP shift required to land safely on the stepping foot. However, the anticipatory drop in the index of synergies was pronounced in the stepping leg muscles but not in the supporting leg muscles. The controller is probably preparing for the unavoidable loss of contact with the floor by the stepping leg. As soon as this happens, muscles of this leg obviously cannot contribute to stabilization of the COP trajectory. Hence, the anticipatory changes in the index of synergies may reflect a purposeful strategy of gradually phasing out the contribution of the stepping leg to COP stabilization.

5.5.3 Multi-Muscle Synergies in Hand Force Production

A similar approach can be applied to analysis of multi-muscle synergies across a variety of tasks and involved muscle groups. The problem is in recording accurately enough, all, or at least most of the muscles participating in the activity of interest and inventing a way to generate a Jacobian matrix that would link small changes in magnitudes of muscle modes with a potentially important performance

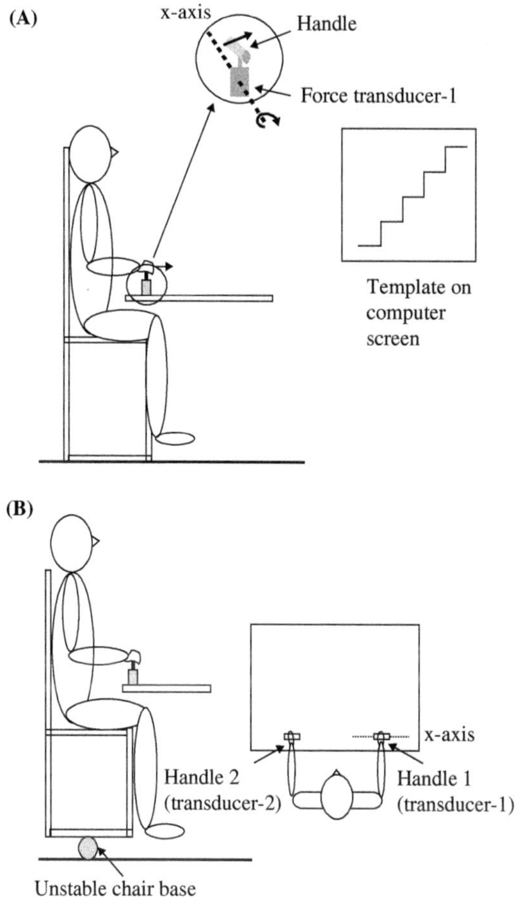

Figure 5.34. This study involved two tasks. (A) The subjects were required to produce forces in different directions by the hand. The data were used to define muscle modes (M-modes) across 12 arm muscles. (B) The subjects used the hand to stabilize themselves while sitting in an unstable (rocking) chair. The M-modes defined in the first task were used by the subjects to build a synergy stabilizing the hand action during the second task. Reproduced by permission from Krishnamoorthy V, Scholz JP, Latash ML (2007) The use of flexible arm muscle synergies to perform an isometric stabilization task. *Clinical Neurophysiology* 118: 525–537. © Elsevier.

variable. Feasibility of this approach has been demonstrated in a study of muscle activation patterns, when a person sits in a rocking chair and uses both hands to balance the chair (Krishnamoorthy et al. 2007). The idea of this somewhat artificial design was to put a person in a situation when quick, properly timed force production by the hand was crucial for not losing balance of the chair.

This study involved two tasks (Figure 5.34). The first one was used to define muscle modes. In that task, the subjects were asked to grasp a vertically oriented handle with the right hand and produce different magnitudes of moment of force about the horizontal axis of the handle. The handle was fixed and the subjects sat in a stable, nonrocking chair. Linear combinations of indices of the integrated EMG of nine arm muscles (M-modes) were defined using PCA; further, relations between small changes in the modes and changes in the moment of force (the Jacobian) were identified.

In the main experiment, the subjects sat in the rocking chair and applied forces to two handles with two arms. Then, they were instructed to release the force applied by the left hand and to try to maintain a sitting position on the chair. The right arm muscles showed anticipatory adjustments in preparation to the left hand release. EMGs during these adjustments were quantified and analyzed using the UCM-framework. The analysis determined the extent to which variance of the M-modes acted to stabilize an average across trials change in the moment of force applied to the handle by the right arm.

This study resulted in identification of three M-modes. Two of them involved opposite changes in the activation of agonist–antagonist muscle pairs crossing individual joints. The third mode involved parallel changes in agonist–antagonist pairs, that is, it was a co-contraction mode. The three M-modes were found to form synergies that produced a consistent change in the moment of force across repetitive trials. Thus, the results of this study provide further evidence for the assumption that functional synergies comprise flexible combinations of activations of ensembles of muscles, organized to stabilize the value of or changes in important performance variables.

This brief overview of a few examples of synergies that have been studied based on the principle of abundance and using methods associated with the UCM hypothesis suggests that these methods are indeed powerful in identifying and quantifying a variety of synergies. The next step is to try to apply this toolbox to more practically relevant issues such as changes in synergies with age, typical and atypical development, neurological disorder, and practice. This is the topic of the next part of the book.

Part Six

Atypical, Suboptimal, and Changing Synergies

6.1 IS THERE A "NORMAL SYNERGY"?

Before even trying to address the title question, let me ask a more general one: Is there such a thing as a *normal movement*? This question has been debated for quite some time, particularly in such areas as physical therapy and motor rehabilitation (see Latash and Anson 1996, 2006). There are several reasons to suspect that *normal movement* is indeed a misnomer, an intuitive but misleading concept. The main reasons are motor redundancy, motor variability, and plasticity within the central nervous system (CNS).

Motor redundancy gives the CNS innumerable options of performing all everyday tasks, and a considerable number of these options are indeed being realized. Bernstein described this property of natural movements with a concise but eloquent expression "repetition without repetition" (Bernstein 1967). What he meant was that repeating a task leads to patterns of performance (kinematic, kinetic, electromyographic (EMG), and even neural) that never repeat themselves. This is partly due to the fact that it is impossible to reproduce in different trials the initial state of the person and all the details of the environment. But the problem is even deeper. It is related to the principle of abundance emphasized throughout this book. This principle suggests that the controller does not look for particular unique solutions when faced with a problem of motor redundancy. Rather, it facilitates groups of solutions that are equally able to solve the task. No wonder that repeated attempts lead to nonrepeating patterns—motor variability—that, however, may belong to the same family. And this would be true even for a purely hypothetical situation, when the same task is performed by the same person in absolutely identical conditions. We will get to the issue of "the same person" in the next section.

So, if an absolutely healthy, *normal* (whatever this may mean) person shows different motor patterns when solving repeatedly an easy everyday task, which one of these patterns is *normal*? Probably, they all are. So, maybe *normal* should refer to a range of motor patterns that are seen in all healthy persons. This is also questionable. Imagine walking patterns of a ballet dancer and a sumo wrestler. Each of them will show variable patterns across successive steps. However, even if we consider their average patterns (across many stepping cycles), they may differ from each other by as much as walking patterns of a healthy elderly person and one with Parkinson's disease. Is this a good reason to consider walking patterns of the dancer and the wrestler as *abnormal*? This makes little sense. Any unbiased person would say that movements of these persons with very specialized skills are *normal for them*, although patterns of these movements may look different from movements of those who do not have those skills. To bring the point home even more obviously, let me ask the following question: Should movements of the great basketball star of Chicago Bulls, Michael Jordan, be viewed as "abnormal" only because no lay person can repeat them? This is ridiculous!

So, I hope that the reader will agree that there is no such a thing as *normal movement*. However, there is an intuitive feeling that most human movements we observe every day have something in common that allows to classify them intuitively (although imprecisely) as *normal*. This feeling cannot be 100% wrong. There is something in natural biological motion that can be recognized by an observer, even if the amount of visual information is drastically reduced. Several studies (Shiffrar et al. 1997; Giese and Poggio 2003; Grossman et al. 2004; Pollick et al. 2005) have shown that human observers can distinguish biological motion from nonbiological motion, even when they do not see the whole moving figure but only a set of small light sources attached at the main body segments. Can it be that biological motion is characterized by particular patterns of co-variation among joint angle trajectories that stabilize important features of the motion? In other words, are all biological motions united by common biological synergies?

Let me get back to two main characteristics of synergies, sharing pattern and flexibility/stability (see section 1.5). The latter characteristic ensures stable performance with respect to a particular set of performance variables. However, the former does not. Figure 6.1 illustrates data sets for two persons, S_1 and S_2, who were asked to perform the task of constant force production with two effectors (e.g. the two index fingers). One of the persons (S_1) shared the forces between the

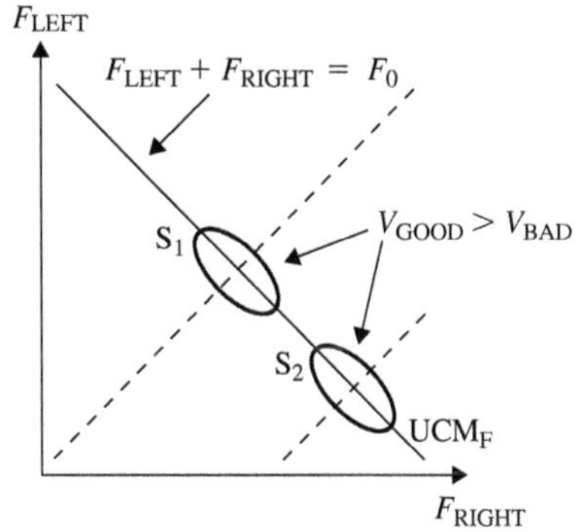

Figure 6.1. An illustration of two data point distributions (shown as ellipses) in an experiment with producing the total force F_0 with two hands. Both subjects, S_1 and S_2, show negative co-variation between the right and left hand forces (F_{RIGHT} and F_{LEFT}). As a result, both ellipses are elongated along the line $F_{RIGHT} + F_{LEFT} = F_0$, which is the uncontrolled manifold (UCM_F). The two subjects show different average sharing of the total force between the two hands. They both show force-stabilizing synergies ($V_{GOOD} > V_{BAD}$).

two fingers nearly equally, while the other one (S_2) produced much more force with the right index finger. Note that both subjects show data distributions elongated along the line $F_{LEFT} + F_{RIGHT} = F_0$. This line is the uncontrolled manifold (UCM_F—see section 4.1) for the total force at the level of F_0. Both subjects show two-finger synergies stabilizing the total force, that is, their *good variability* is larger that *bad variability* ($V_{GOOD} > V_{BAD}$). What could be the reason for the difference in the average sharing of the total force between the two fingers?

Maybe, one of these persons is strongly right-handed. It is also possible that the person with the unequal sharing has a problem in the left hand leading to pain when large forces are produced by the left index finger. Irrespective of the cause, we can conclude that the two persons are similar in their force stabilizing synergies and are different in the sharing pattern. Which one of the two distributions shown in Figure 6.1 is *normal*? Probably, both can be viewed as *normal* but reflecting differences in the states of the persons.

Consider now another couple of data distributions for the same task shown in Figure 6.2. These distributions differ in both average sharing of the total force and pattern of force co-variation between the two fingers. One of them corresponds to a force-stabilizing synergy ($V_{GOOD} > V_{BAD}$), while the other has an opposite relation between the *good* and *bad* fractions of variability ($V_{GOOD} < V_{BAD}$). The ellipse oriented orthogonal to the UCM corresponds to predominantly positive co-variation of finger forces that destabilizes the total force. Such a pattern can be termed a *fork strategy*. It can be expected if one takes a fork by the handle turns its prongs upside down and presses with the prongs on separate force sensors. If in one trial, one of the prongs shows a higher force, other prongs are also

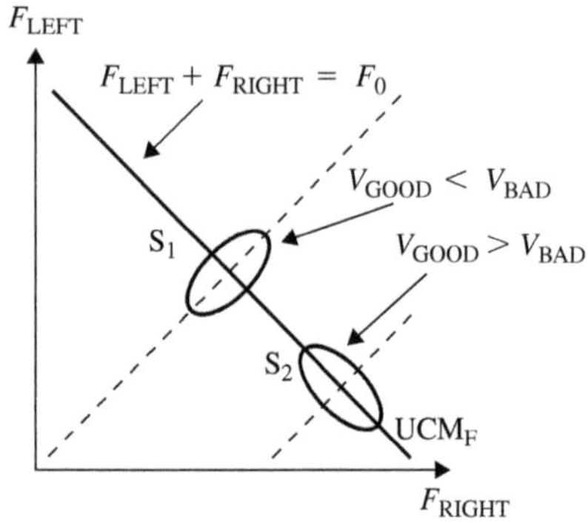

Figure 6.2. An illustration of the same experiment as in Figure 6.1. One of the subjects (S_2) shows the same data distribution as in Figure 6.1. The other subject (S_1) shows the same average sharing of the total force between the hands but a different orientation of the ellipse corresponding to positive force co-variation ($V_{GOOD} < V_{BAD}$)—a "fork strategy."

likely to show higher forces leading to a positive force co-variation across trials. Should such a pattern be considered abnormal? We will return to this question a bit later.

Although the illustrations in Figures 6.1 and 6.2 show data points across repetitive trials, similar arguments can be made with respect to time profiles of elemental variables, for example joint angles, during movements. As mentioned in sections 4.1 and 4.3, analysis of patterns of co-variation can be performed both across repetitive trials and along a trial. If movements cannot be easily classified into *normal* and *abnormal*, maybe patterns of co-variation can. In this case, one can discuss normal and abnormal synergies. At least, in clinical practice, notions of normal and abnormal synergies have been used rather widely. In particular, individuals who have experienced a stroke are described clinically as having abnormal movement synergies, typically characterized as relatively fixed or stereotypic (Bobath 1978; DeWald et al. 1995). Such conclusions have been drawn, based on analyses of the patterns of how the muscles change their activation levels or the joints change their position during the execution of particular tasks. In this context, the word *synergy* means something like "variables that change together." This meaning is rather different from the definition accepted in this book. Parallel scaling of elemental variables may reflect not a control strategy but other factors: In the very first section, we considered an example with objects of different weights placed on the top of a table leading to proportional scaling of forces under all four legs of the table. The table would qualify as a synergy under the definition "variables that change together," but it would not under our current definition. As we will see later, even grossly changed motor patterns may be accompanied by synergies (in the meaning accepted in this book) that are not significantly different from those observed during similar movements performed by healthy persons.

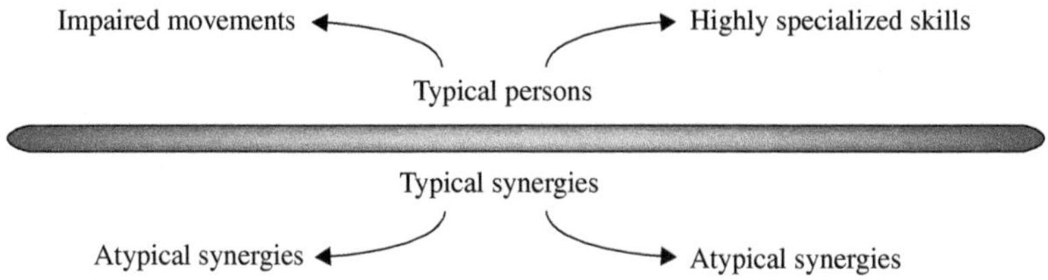

Figure 6.3. A schematic illustration of a spectrum with *typical persons* in the middle, persons with highly specialized motor skills to the right, and persons who would be considered impaired by most clinicians to the left. All persons in the middle of the spectrum are assumed to share the same set of rules that defines what kind of synergies are used in everyday tasks. There is no clear border between *normal* and *abnormal* in this scheme.

So, accepting our current definition of *synergy*, can one classify synergies into *normal* and *abnormal*? This is also unlikely or at least nontrivial. The main reason is demonstrations of changes in synergies with practice in healthy persons (see further in this section). Some studies (Latash et al. 2003c; Scholz et al. 2003; Kang et al. 2004) have shown, in particular, that synergies can show significant quantitative changes and even qualitative changes (e.g. a *non-synergy* may turn into a *synergy*) with relatively short practice. One probably has to step up to a hierarchically higher control level and ask: "Are there sets of rules that govern the processes of formation of synergies and their changes that are common across all healthy persons without major specialized skills?" Accepting an affirmative answer allows the assumption that such *CNS priorities* (a term borrowed from Latash and Anson 1996) define biological motor patterns that we see every day and perceive as *normal*.

What can be expected if the CNS priorities change? Figure 6.3 uses a schematic picture to illustrate a spectrum with *typical persons* in the middle, persons with highly specialized motor skills to the right, and persons that would be considered impaired by most clinicians to the left. All persons in the middle of the spectrum are assumed to share the same set of rules that define what kind of synergies are used in everyday tasks and how these synergies change with changes in external conditions. There is no clear border between *normal* and *abnormal* in this scheme. For example, as we will see further, changes in synergies may reflect not an inability of the person to show more typical synergies but rather a choice by the controller (the CNS) to facilitate synergies that may be considered optimal, given the actual state of the person and the range of tasks that are perceived as important.

To get back to the question formulated in the title of this section, there are no *normal synergies* (or *abnormal synergies*) but there are *typical* and *atypical* ones. The atypical synergies may reflect a genuine inability of the controller to facilitate more typical motor patterns or an adaptive strategy of the CNS to optimize performance in the everyday range of behaviors. How do such atypical synergies come about? What neurophysiological mechanisms can lead to changes in synergies? These questions are going to be discussed in the next two sections, which address an amazing ability of the CNS to rewire itself.

6.2 PRINCIPLE OF INDETERMINICITY IN MOVEMENT STUDIES

Heraclitus is quoted by Plato in *Cratylus*: “No man ever steps in the same river twice, for it’s not the same river and he’s not the same man.” I could not agree more and would like to suggest a corollary directly relevant to experimental studies of movements: “No man ever performs the same movement twice.” This corollary presents a seemingly serious problem for researchers that was already addressed briefly in section 2.7. Research on human movement, similarly to research in any area of physics, assumes that results of an experimental study can be reproduced in another study and in other laboratories if all the details of the experiment are copied with sufficient precision. How can this be expected when even testing the same person twice in perfectly reproduced conditions can lead to different outcomes, because he or she is not the same person anymore?

Actually, the problem is deeper. Most experiments in movement studies involve observation of a system of interest (e.g. a set of muscles or a set of digits), while the subject is trying to perform a particular task (follow a particular instruction) and/or using perturbations applied to the system. In all cases, the system is expected to produce an action and/or a reaction. More and more evidence accumulated over the past 20–30 years has suggested that the CNS reacts to the process of measurement involving external perturbations and to its own activity by changing its state. This means that when one compares data recorded in a system at the initiation of an action and after the completion of the action, the data may refer to two different states of the systems. How can such data be compared in a meaningful way?

The human body can react to typical experimental manipulations that are used to measure its basic properties in major and unexpected ways. For example, application of a quick mechanical perturbation to a moving or stationary joint has been used in many studies (Ma and Zahalak 1985; MacKay et al. 1986; Sinkjaer et al. 1988; Bennett et al. 1992; Blanpied and Smidt 1992) to describe such joint properties as stiffness (assuming that a joint can be assigned this property—a questionable assumption; see Latash and Zatsiorsky 1993) and damping (which is sometimes imprecisely called viscosity, Zatsiorsky 1997). If the perturbation is small and slow, the CNS may not react to it. If the perturbation is quick and large, there will be reactions at different time delays (see Digression #6) that are expected to change the activation level and mechanical properties of muscles crossing the joint. System identification methods are based on comparing system states at different times. But what if over the time period of observation the system changes those very parameters that the method tries to discover?

This situation sounds very similar to the famous principle of indeterminicity in physics described in section 2.7. Although biological systems are much larger than physical systems to which the principle of indeterminicity is usually applied, there is enough evidence to claim that biological systems, including the system for the production of movements, react to the process of measurement. Such reactions can be local and transient or they can be widespread, long-lasting, and even permanent. The local, transient changes may be due to many factors including reflexes and reflex-like reactions (Digression #6), as well as transient changes in the properties of the muscles. For example, there is a well-known so-called *catch property* of human muscles (Burke et al. 1970, 1976). If a muscle is subjected to a strong, short stimulation for a few seconds, after it is over, it produces higher forces to a standard stimulus than prior to the stimulation. This property has been shown to reside in muscle cells and not in more central elements controlling muscle force (Burke et al. 1970).

6.3 PLASTICITY IN THE CENTRAL NERVOUS SYSTEM

In this section, we are interested in more global and long-lasting changes in the state of the CNS that can potentially lead to changes in motor synergies. This so-called *plasticity* of the nervous system has been demonstrated in a variety of animals and humans, resulting from surgical interventions and following practice of motor tasks in healthy persons and in patients with various neurological and peripheral abnormalities, and in different neural structures. Until recently, plastic changes in the CNS were expected to happen only in developing systems (babies and the young ones of animals), not in mature adults. These changes have been viewed as slow and following a particular predefined route. Now it seems that plasticity is everywhere, it happens at every age, and it takes very little time to manifest itself.

At birth, different animals show a broad variety of abilities to function in the environment, ranging from an ability to show independent activity at birth (referred to as *precocial*), to being rather helpless (referred to as *altricial*). Human newborns are definitely altricial; their helplessness is partly due to the insufficiently developed CNS. At birth, the human brain weighs about 300 g. The increase in the brain weight is accompanied by an increase in the number of neurons, their size, myelination of their axons, as well as the growth of supporting glial cells. The cerebral hemispheres are formed by the time of birth but are not fully functional, partly due to the incomplete myelination of neural tracts. Myelination of both sensory and motor axons begins before birth and continues for about 6 months after birth. It proceeds in a cephalo–caudal direction, that is, starting from the brain and moving toward the tail end of the spinal cord. Ultimately, an adult-like set of projections between the brain and peripheral structures emerges, leading to distorted human figures (homunculi) drawn on different brain areas, in particular, on the motor and sensory cortical areas of the brain commonly reproduced in textbooks.

The organization of the mature human brain has been commonly described with such neural maps that represent the body as a collection of human figures drawn on the surface of various brain areas. The recent development of more accurate methods of recording brain activity has, however, led to much less structured results. In particular, the human-like figures are not very much human-like but rather mosaic, resembling drawings by Picasso much more than classical painting of the Renaissance. The neural projections demonstrate plenty of divergence and convergence (Figure 6.4). The term *divergence* means that stimulation of a single brain (e.g. cortical) neuron can lead to reactions (muscle contractions) in several body parts. *Convergence*, on the other hand, means that stimulation of neurons in apparently different brain areas can lead to reactions in the same muscle or in the same part of the body (for review see Schieber and Santello 2004).

Classical works on brain maps were performed in the middle of the twentieth century by a great Canadian neurosurgeon, Sir Wilder Penfield (Penfield and Rasmussen 1950). During surgeries, he used direct electrical stimulation of small brain areas and observed motor responses in different parts of the body. He described, in particular, that the primary motor area produced responses to lower magnitudes of the stimulation compared to other motor areas (such as premotor area and supplementary motor area). Penfield was the first to represent such brain maps using human-like homunculi. A predominant view was that brain

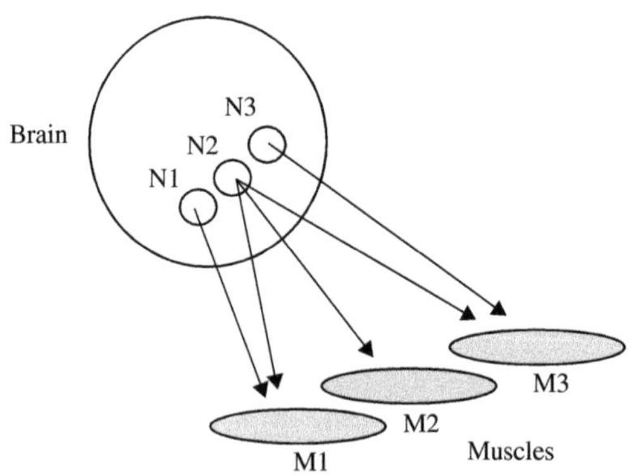

Figure 6.4. Projections from the brain to peripheral structures (e.g. muscles) are characterized by both convergence and divergence. In this illustration, neurons from group N2 diverge to several target muscles. On the other hand, signals from N1 and N2 converge on muscle M1, while signals from N2 and N3 converge on muscle M3.

projections, both motor and sensory, develop during early childhood and then remain unchanged throughout life.

This view was challenged later in now classical experiments of the group of Michael Merzenich (Merzenich et al. 1984) who studied neural maps in monkeys. In one of the first experiments, sensory projections from the digits of a monkey's hand were mapped on the somatosensory brain area. This was done by stimulating small areas of the skin and recording responses in the cortex. Then, one of the digits was amputated. Obviously, cortical neurons that used to receive excitatory signals from that particular digit became orphaned. However, they did not stay orphans for long. After about 2 months, the somatosensory cortex contralateral to the amputation site showed a major reorganization. In particular, the "vacant" neurons started to respond to stimuli applied to adjacent digits of the hand. Further, comparable reorganizations of the somatosensory cortex were shown in less invasive experiments that involved digit fusion (attaching two digits to each other so that they can only move together) and specialized hand training in monkeys (Allard et al. 1991; Recanzone et al. 1992).

Soon after the pioneering studies by Merzenich, similar results were reported in humans. A group of researchers in the National Institutes of Health used the method of transcranial magnetic stimulation (TMS, see Digression #10) of the motor cortex to study changes in neuronal projections after a limb amputation (Cohen et al. 1991a). In studies of patients after a below-the-knee leg amputation, stimuli at optimal positions of the stimulating coil recruited a larger percentage of α-motoneurons that sent their axons to the muscles in the residual portion of the leg (Fuhr et al. 1992). These muscles could also be activated from larger areas of the scalp than the muscles at the intact side. Similar results were also reported in a person with congenital absence of the distal part of the left arm (Cohen et al. 1991b). Taken together, these observations suggest that descending corticospinal projections in humans are likely to be reorganized after an amputation.

Digression #10: Transcranial Magnetic Stimulation

Stimulation of excitable tissues within the human body has been a powerful tool for studies of the function of the CNS and its interactions with muscles. Until the last quarter of the twentieth century, the only available tool was electrical stimulation. If one has direct access to the neural structures, as for example during surgery on an open brain, this method can be used, and was indeed used, to map neuronal projections (as in the mentioned classical studies by Penfield and Rasmussen). However, if the skull is intact, electrical stimulation can only be applied to its surface. One can apply high currents to stimulate neurons in the cortical areas directly below the site of application of the stimulation. Unfortunately, these currents are likely to produce a lot of undesirable effects such as contraction of facial muscles and pain. These side-effects limited the applicability of direct brain stimulation in studies of the human CNS and its possible plastic changes.

A major step in overcoming these problem has been the development of the method of TMS (reviewed in Rossini et al. 1994; Ellaway et al. 1999). The method allows one to stimulate deep structures inside the body (in particular, brain structures) using a quickly changing magnetic field. Here, I present a quick reminder from physics: When electric current runs along a circular wire, it generates a magnetic field, which is particularly strong at points along the lines passing through the wire orthogonal to the plane of the circle (Figure 6.5). If a brief pulse of current travels along such a wire, the magnetic field it creates quickly emerges and then disappears. However, a changing magnetic field produces induction electric current. So, if a bagel-shaped coil is placed close to the skull and a brief, strong pulse of current runs along the wire in the middle of the "bagel," and currents (of a much smaller magnitude) will be produced in tissues at a distance from the coil. Note that the coil is completely isolated from the body; it does not have to touch it. Nevertheless, this method permits the stimulation of neural structures in the cortex and avoids the unpleasant side-effects associated with direct electrical stimulation applied to the surface of the skull. Using coils of particular shapes, for example, the figure-of-eight coils, focuses the stimulation on structures at a particular depth under the skull, avoiding stimulation of structures that are closer to the coil, such as muscles.

This method has been used in both basic research and clinical studies. It is noninvasive, safe (if one follows the guidelines, of course), and provides

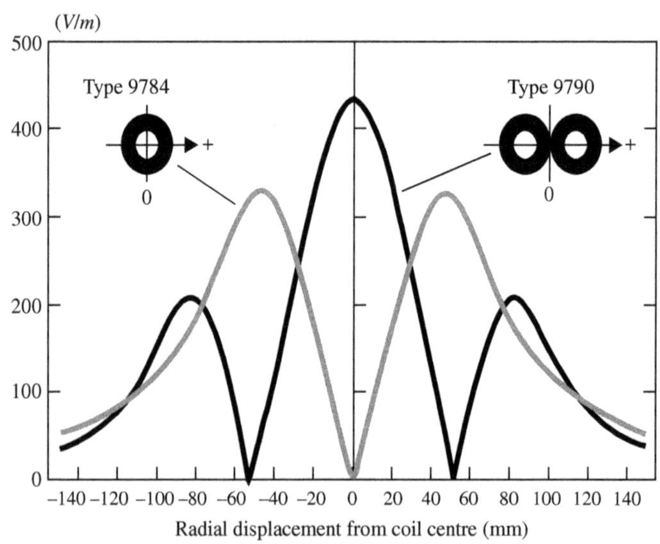

Figure 6.5. An illustration of the method of transcranial magnetic stimulation. Typical circular (type 9784) and figure-of-eight (type 9790) coils and representations of their induced electrical fields. Modified with permission from Jalinous R (1998) Guide to magnetic stimulation. The MagStim Company Ltd. © R. Jalinous.

information on interactions among brain structures and between these structures and the spinal cord. However, as with most indirect methods, interpretation of results is commonly ambiguous. In particular, a typical TMS pulse is likely to produce currents that excite many neurons trans-synaptically (Paus et al. 1997; Wasserman et al. 1997). These neurons, in turn, are likely to excite many other neurons which they project to. A response, commonly a muscle contraction, may reflect the spread of activity along various neural routes and through various neural formations. However, this crude method has proven invaluable in providing information about neuronal maps and their changes in healthy humans.

End of Digression #10

Changes in neuronal projections happen not only after an injury to the peripheral apparatus but also following an injury to the brain itself. The best known example is changes in neural projections from the brain to the spinal cord and between brain areas following a stroke. Stroke commonly leads to muscle weakness and movement deficits most evident in limbs contralateral to the stroke site (Bourbonnais and Vanden Noven 1989). After a stroke, motor function may show recovery reflecting, in particular, brain plasticity (for reviews, see Johansson 2000; Rossini and Pauri 2000; Hallett 2001). Such plastic changes show reorganization of surviving neural elements within the affected cerebral hemisphere (Jenkins and Merzenich 1987; Nudo et al. 1996; Xerri et al. 1998), as well as functional and structural reorganization within the unaffected hemisphere (Colebatch and Gandevia 1989; Jones et al. 1989; Desrosiers et al. 1996; Cramer et al. 1997). In particular, there have been reports of an increased activation in the motor cortex of the unaffected hemisphere during movement by the impaired hand (Cramer et al. 1997; Cramer 1999). This region is not the same as the one used by the unaffected hemisphere to move the unimpaired hand. In addition, stroke has been shown to lead to an increase in interhemispheric inhibitory projections from the unaffected hemisphere to the affected hemisphere (Duque et al. 2005).

Major plastic changes in the CNS can also happen in the absence of any injury whatsoever. I would like to mention here a very ingenious series of experiments by Jonathan Wolpaw and his group who used operant conditioning in monkeys to study learning in the most seemingly simple structure in the body, the monosynaptic reflex arc (see Digression #6). Operant conditioning includes rewarding an animal with a small portion of a favorite food for correct behavioral responses. The response may be under control of the animal's brain, as for example in finding all the right turns during running in a maze, or it may not be under direct brain control, for example, during spinal reflex responses. However, even the simplest, monosynaptic spinal reflexes can learn and store memory traces in operant conditioning experiments (Wolpaw 1987; Wolpaw and Carp 1993). This may require thousands of repetitions, but the animal's CNS eventually learns to modify an apparently uncontrollable phenomenon such as the amplitude of a monosynaptic response. These fantastic studies have shown that plasticity can happen in the spinal cord and that it can allow the animal to control a variable that the animal had not been able to control before.

Recently, a human study has demonstrated that skill training can induce a change in synaptic efficacy at the synapses between Ia afferents and alpha-motoneurones innervating the soleus muscle (Meunier et al. 2007). These results support the view that the spinal cord is able to encode a local motor memory (Latash 1979).

While most earlier studies used major disruptive factors, such as amputation and stroke, in bringing about neural plasticity, more recent work has focused on plastic changes that may happen within the CNS in the absence of any dramatic events but as a result of practice. In most studies, researchers focused on the human hand and its cortical projections. This is understandable, given two factors. First, humans are very good at learning new skills involving the hand. Second, the hand has large cortical representations that are easier to map.

Using TMS allowed researchers to demonstrate plastic changes in neuronal projections from the motor cortex to the spinal cord in healthy persons after the long-lasting specialized training involved in learning such skills as reading Braille and playing musical instruments (Pascual-Leone et al. 1995; Cohen et al. 1997; Sterr et al. 1998; Pascual-Leone 2001). On the other hand, plastic changes can also be seen following a relatively brief practice limited to 1 or 2 hours (Classen et al. 1998; Latash et al. 2003c). For example, thumb movements induced by a standard TMS have been shown to change after the thumb practiced a movement in a particular direction (Classen et al. 1988). The changed response was closer to the practiced movement direction than the response observed prior to the practice.

All the presented information suggests that the human CNS is always in the process of rewiring itself. What can be expected from these processes beyond the demonstrated changes in responses to standard stimuli? Do they have functional importance? Can it be quantified? Through the introduced method of quantifying motor synergies, we could attempt answering these questions, at least in the area of neural control of movements. I am going to discuss changes in synergies that happen during natural, healthy aging, atypical development, following a neurological injury, and resulting from practice of a novel task.

6.4 CHANGES IN SYNERGIES WITH AGE

6.4.1 Effects of Age on Muscles and Neurons

Aging is something most people on the planet hope to experience, while most of them may not look forward to the experience. A number of disorders, including neurological ones, show increased prevalence in older persons. Among the best known are stroke, Parkinson's disease, and Alzheimer's disease. However, even in the absence of diagnosed diseases, the so-called *healthy aging* is still commonly associated with a decline in many of everyday functions. Since here we focus mostly on the motor function of the human body, let me summarize briefly changes in the neuromuscular system that happen with age. In the following text, I am going to use the word *elderly* to describe persons older than 70 years of age.

It is common knowledge that people get weaker with age. This drop in muscle strength during both voluntary contractions and contractions induced by electrical muscle stimulation is accompanied by a decline in muscle mass, called *sarcopenia* (Winegard et al. 1997). This decline begins at about the age of 50–60 years, and it varies greatly among individuals (Narici et al. 1991). Muscle force is known to correlate strongly with the cross-sectional area of muscle fibers. It is not surprising, therefore, that in elderly persons smaller cross-sectional area correlates with a decline in voluntary muscle force. This results from both a drop in the average size of muscle fibers and a drop in their total number (Bemben 1998; Kirkendall and Garrett 1998). There are also changes in the mechanical properties of peripheral tissues. In particular, connective tissue has been shown to replace contractile proteins with aging (Zimmerman et al. 1993).

With age, muscle strength becomes a limiting factor in a variety of everyday activities such as rising from a chair, operating hand-held tools, and even performing some of the everyday personal care actions. Not all the muscles in the human body show changes in their force-generating abilities at the same rate. For example, there are reports on differences in the loss of force in different muscle groups with age. These changes may be expected to lead to important consequences for issues of coordination (as discussed in the next section). For example, some authors reported significant differences between the force loss in the proximal and distal hand muscles (Nakao et al. 1989; Shinohara et al. 2003b). These muscles are expected to generate properly scaled forces during everyday activities with hand-held objects to ensure that torques in all the hand and digit joints scale properly to produce a required force and moment of force by the fingertips.

Recall from section 4.2.1 that the muscular structure of the hand is rather complex (Figure 6.6). For example, to squeeze a hand-held object, the person has to activate both two relatively large muscles with the belly in the forearm (the so-called *extrinsic flexors*) and the relatively small muscles with the belly in the hand (intrinsic hand muscles). The extrinsic muscles are multi-digit, each of them having four distal tendons that attach to the most distal and to the intermediate phalanges of the fingers. In contrast, the intrinsic muscles are digit-specific, their tendons attaching at the proximal phalanges. In addition, the intrinsic muscles contribute to the complex connective tissue network called the *extensor mechanism*. So, to press with the fingertip of a single finger, one has to activate both extrinsic flexors and the appropriate intrinsic muscle to make sure that the

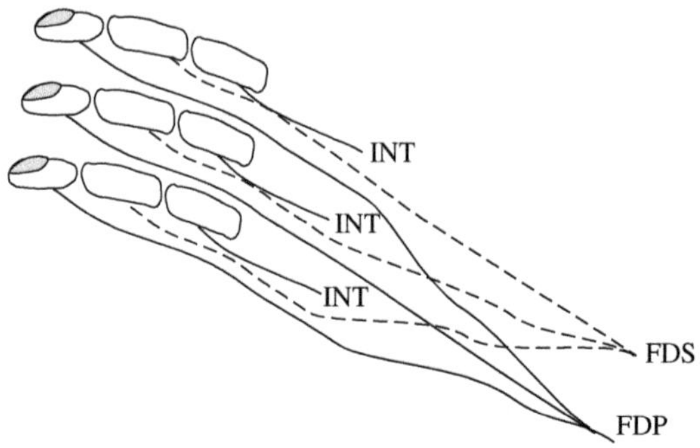

Figure 6.6. The muscular apparatus of finger flexion involves both multi-tendon extrinsic muscles with the bellies in the forearm (FDS—flexor digitorum superficialis and FDP—flexor digitorum profundis) and digit-specific intrinsic (INT) hand muscles. INT also contribute to digit extension at the distal phalanges (not illustrated).

torques in both interphalangeal joints and in the metacarpophalangeal joint are scaled adequately. Likely, over a lifetime, the CNS learns to scale commands to different muscle groups that are involved in typical actions. This scaling may differ across individuals and reflect the differences in the hand anatomy and in the force-generating capabilities of different muscles.

Imagine now that the intrinsic hand muscles start to lose strength with age faster than the extrinsic muscles (as reported in Shinohara et al. 2003b; Cole 2006). The combinations of neural commands well learned over a lifetime may be expected to lead to poorly coordinated torques produced by different muscle groups and result in poorly controlled forces at the fingertips. As a result, objects may be dropped, broken, and mishandled in a variety of other ways. Obviously, adjustments at a neural level are required to be able to use the hand efficiently, given the mentioned changes in the muscular apparatus of the hand.

The CNS is, however, plagued by its own problems. In particular, there is a decline in the number of neurons in many parts of the nervous system. One of the best known examples is the death of neurons in the substantia nigra that produce one of the most important neurotransmitters in the brain, dopamine. This may ultimately result in Parkinson's disease. There is also a decline in the number of cortical neurons, in particular, those that form the corticospinal tract—a major descending tract contributing to the control of voluntary movements. And finally, there is a decline in the number of alpha-motoneurons that send their axons to muscles. As a result of the death of alpha-motoneurons, the number of motor units decreases, and groups of muscle fibers become denervated, that is, they lose excitatory neural inputs.

Not all muscle fibers within any given muscle suffer to the same degree. The fastest and thickest fibers are the first to lose neural excitation. The process of partial muscle denervation is accompanied by sprouting and reinnervation, when the axons of remaining alpha-motoneurons grow additional terminal fibers that make synapses on some of the vacated muscle fibers (Figure 6.7). This saves some of the muscle fibers from complete degeneration and leads to an increase in the innervation ratio—the average number of muscle fibers per motor unit. As a result of the processes of denervation and reinnervation, there are fewer motor units, but they are, on average, larger in size and slower. These changes become particularly pronounced after the age of 60.

Some of the documented decline in the motor behavior of elderly persons may be rather directly linked to the mentioned processes. For example, the relative lack of smaller motor units, due to the reinnervation, may be blamed for the poor control of low forces, high force variability, and poor smoothness of movements, because all these processes depend on the recruitment and derecruitment of motor units, and the larger motor units in the elderly may be expected to lead to larger variations in the total muscle force when a motor unit is turned on and off. However, most everyday human actions depend on many more factors such as, in particular, the state of muscle reflexes and reflex-like actions (see Digression #6) and the sensory function.

Many elderly persons complain of poor vision. However, this is not the only sensory modality that suffers because of aging. There is a loss of vestibular receptors and cutaneous receptors (in particular, Meissner corpuscles; Mathewson and Nava 1985), with age; there is also a loss of the number of sensory neurons innervating peripheral sensory receptors—*peripheral neuropathy*. The decline of the somatosensory feedback is so pronounced that old people show higher reliance on visual information during motor actions compared to younger persons, despite the mentioned decline in visual acuity.

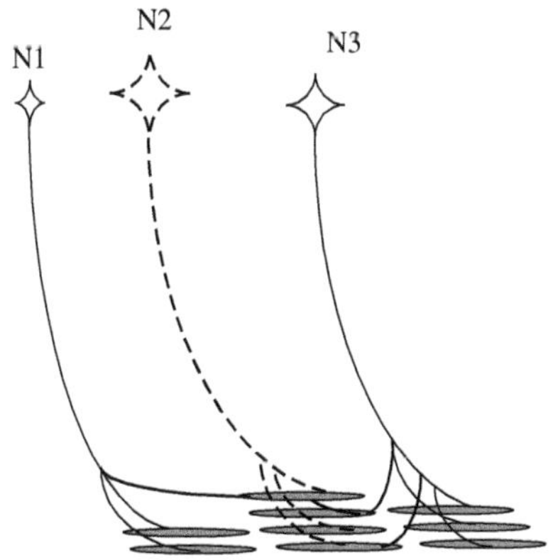

Figure 6.7. With age, larger alpha-motoneurons (N2) degenerate leading to denervation of the muscle fibers they used to innervate. Some of these muscle fibers are reinnervated by other alpha-motoneurons (N1 and N3).

Reflexes and reflex-like actions play a major role in a variety of everyday motor activities such as walking and standing. Some of the changes in muscle reflexes, for example, the documented drop in the amplitude of the monosynaptic H-reflex (Vandervoort and Hayes 1989), may be viewed as direct consequences of age-associated processes, while others may be viewed as adaptive, that is, reflecting a purposeful change in motor control strategies given the changed state of the body (see section 6.1). In particular, older people are more likely to play it safe and use muscle activation patterns that may look suboptimal or mechanically wasteful but ensure stability of important behaviors such as vertical posture. The penalty for falling down is too high to try to use more frugal or beautiful motor patterns. A typical example is co-contraction of antagonist muscle pairs, that is, muscles producing opposing actions at a joint (e.g. a flexor and an extensor acting at the same joint). Old people are more likely to use co-contraction patterns in both adjustments in anticipation of and response to postural perturbations (Inglin and Woollacott 1988; Woollacott et al. 1988; Tang and Woollacott 1998). They are also more likely to involve actions at more proximal joints (the so-called *hip strategy*, Horak et al. 1989) compared to younger persons, who may show adjustments at the ankles to similar postural perturbations.

Given all the mentioned changes that happen with age, what can be expected from motor synergies? This question is to be explored in the next section.

6.4.2 Effects of Age on Motor Coordination

Motor coordination is a very broad term. Within the framework of this book, motor coordination is synonymous with the notion of motor synergy. Using the introduced glossary, it is possible to expect aging to lead to changes in the composition of elemental variables (modes), in their preferred average time profiles (sharing), and in their patterns of co-variation (flexibility/stability) across a variety of tasks and levels of analysis. However, before moving to analysis of motor synergies, let me mention a few typical differences in general patterns of motor coordination between young and old people. Some of these differences lead to apparently suboptimal motor patterns, while others may be viewed as adaptive to the changed neuromuscular apparatus. The differences from the latter group may be summarized as "Older persons prefer to make movements that are safe."

Let me start with a couple of examples related to manipulation of hand-held objects. When a person grasps an object and moves it, the grip force adjusts to the expected load force (that depends on the mass of the object, gravity, acceleration of the object, and maybe other forces) and friction between the fingertips and the surface of the object. The friction defines the largest tangential force magnitude that can be applied for a given normal force (grip force). If the friction coefficient is k, and normal force is F_N, one cannot apply tangential forces (F_T) larger than kF_N. So, if a person wants to keep a vertically oriented object with the weight W stationary in the air (Figure 6.8), he or she has to apply the total normal force of at least $F_{N,MIN} = W/k$. If the object is to be moved, W should reflect not only gravity but also the acceleration of the object. For example, for vertical motion with acceleration a, $W = m(g + a)$, where m is mass and g is gravity constant. Application of $F_{N,MIN}$ would allow the application of a tangential (vertical) force of exactly W to counteract the force of gravity (and inertia). In fact, grip force is typically substantially larger than the minimal necessary one, and the difference between the two values is sometimes called the safety margin (Westling and Johansson 1984):

$$S = 100\%*(F_N - F_{N,MIN})/F_{N,MIN} \qquad \text{Equation (6.1)}$$

The grip force increases with object weight (Johansson and Westling 1984; Winstein et al. 1991; Kinoshita et al. 1995; Monzee et al. 2003), external torque (Kinoshita et al. 1997), and decreased friction (Cole and Johansson 1993; Cadoret and Smith 1996; Burstedt et al. 1999). The relation between the grip force and the load is typially linear, suggesting that the safety margin does not depend on the weight of the grasped object (Kinoshita et al. 1995; Monzee et al. 2003; Zatsiorsky et al. 2005).

Equation 6.1 presents a very simplified description of the notion of safety margin. In particular, when several digits participate in holding an object, and the object is oriented nonvertically, this equation becomes inapplicable (Pataky et al. 2004a,b). Consider, for example, a situation when the grasped object is not oriented vertically (Figure 6.9). In this case, the normal and tangential forces act together to keep the object stationary. When the object is horizontal, the weight is supported only by the normal forces, and the tangential forces do not act against any apparent external force. In such a case, $F_{N,MIN}$ is zero, and Equation 6.1 becomes meaningless.

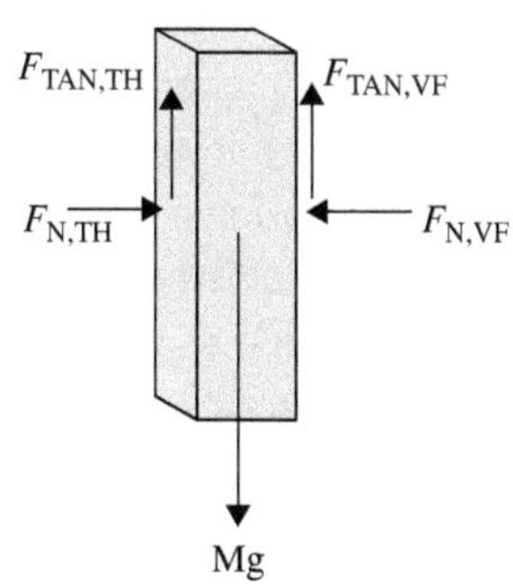

Figure 6.8. Holding a vertically oriented object requires that the sum of the tangential forces applied by the thumb and virtual finger equal the weight of the object ($F_{VF,TAN} + F_{TH,TAN} = Mg$), while the normal forces are equal to each other ($F_{VF,TAN} = -F_{TH,TAN}$). The normal forces should be sufficient to allow the production of required tangential forces, given the friction coefficient k ($F_{TAN} \leq k*F_N$).

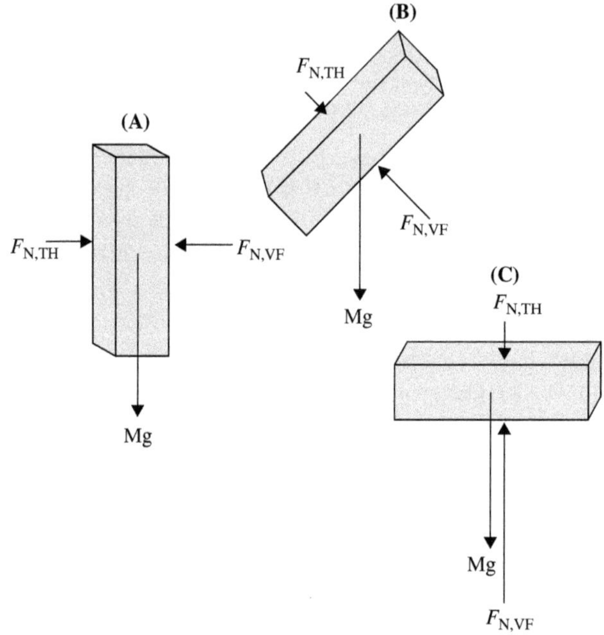

Figure 6.9. When an object is oriented nonvertically, the normal forces play two roles. First, they allow for the generation of required tangential forces (A, as in Figure 6.8). Second, one of them acts against the weight of the object (as in B and C). When the object is horizontal (C), the top digit does not have to produce any force, while the bottom digit holds the whole weight of the object.

However, let us consider a less controversial example of holding a vertically oriented object. In such tasks, young, healthy persons show large individual differences in the safety margin values, which also depend on the properties of the object. The average value of S is about 30% (Cole 1991). Several groups compared grip force production during prehension in young and elderly subjects. They found that elderly subjects showed lower skin friction, higher safety margins, and larger fluctuations in the grip force (Cole 1991; Kinoshita and Francis 1996). The safety margin values in the elderly were about 50%. The large safety margin magnitudes were first interpreted as related to changes in skin friction and/or to production of comparably strong sensory signals in the elderly (Cole 1991). In more recent studies, however, the same group (Cole et al. 1998, 1999) challenged the hypothesis that the decline in the ability of older persons to grip and lift objects is solely due to their impaired tactile sensitivity. A more straightforward interpretation of higher S values in the elderly seems to be related to their less accurate control of force magnitude (maybe due to the mentioned age-related changes in muscles). This can lead to two consequences. On the one hand, if the tangential force differs from the force of gravity, the object will accelerate, leading to a possible increase in $F_{N,MIN}$. On the other hand, if the normal force fluctuates, it may accidentally go below $F_{N,MIN}$, and the object will be dropped. Both these factors make application of additional grip force reasonable as a safety precaution, even though this strategy may look wasteful or fatiguing to an external observer.

Here is another example. If a person holds the same object and wants to keep its orientation constant (e.g. holding a full glass of beer, Figure 6.10), the total moment of force applied by the digits should be equal to the total external torque applied to the object (see Equations 5.4–5.6 in section 5.4). Imagine, however, that forces applied by individual digits are variable and not very well predictable. The total moment of force may be expected to vary leading to disturbances of the rotational equilibrium. For a given magnitude of unbalanced moment of force, kinematic consequences will depend on the rotational moment of inertia (which is not under the control of the CNS) and on opposing forces that the hand generates. The latter factor may be modulated by the controller leading to another seemingly wasteful strategy seen in the elderly (Shim et al. 2004b).

When a person holds an object with an external torque applied in a certain direction (e.g. into supination), the person has to apply a net moment of digit force in the opposite direction (in pronation). In this task, normal forces produced by the index and middle fingers act in the required direction. However, normal forces produced by the ring and little fingers act in the direction of the external torque. One may say, therefore, that these fingers produce antagonist moments of force that add to the task rather than help solve it. Such antagonist moments of force are observed in all persons and in all tasks (Zatsiorsky et al. 2002a,b; Gao et al.

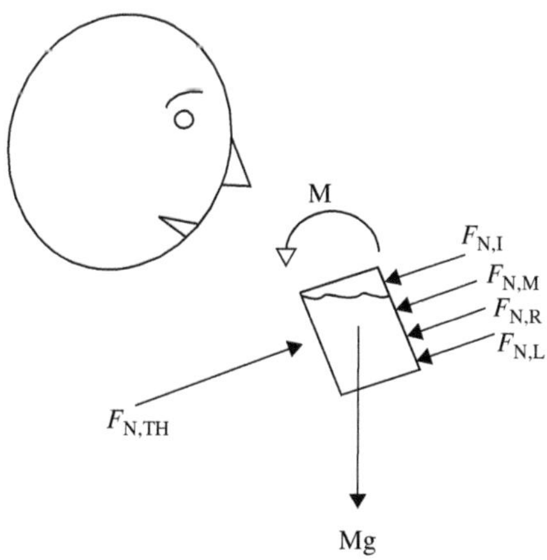

Figure 6.10. When a person takes a sip from a glass of beer, the digits should produce sufficient gripping force (to avoid slippage of the glass) and an adequate moment of force (M). This is achieved, in particular, by a proper distribution of forces across the four digits, I—index, M—middle, R—ring, and L—little.

2005b). They may be partly due to the enslaving effects (see Digression #10) that lead to unintended force production by those fingers when the other fingers of the hand produce force (Zatsiorsky et al. 2000). When elderly people perform such tasks, they produce significantly larger forces by fingers that generate antagonist moments of force. A parallel increase in the normal forces produced, for example, by the index and little fingers does not help generate a net moment of force, but it may be helpful in increasing the apparent rotational stiffness of the hand/wrist system. This may be seen as an adaptive strategy that allows elderly persons to diminish actual deviations of the hand-held object from the vertical, when the less precisely controlled muscles generate rotational perturbations.

Studies of synergies in the elderly following the definition advocated in this book are few. Nevertheless, they show that the operational definition and the associated computational apparatus are able to identify and quantify changes in certain types of synergies that occur with healthy aging. I am going to describe results of a few studies that quantified multi-digit synergies stabilizing such potentially important performance variables of the hand as the total force and the total moment of force. Recall (see section 5.4) that young persons show co-variation of commands to fingers (finger modes) that can stabilize the total force produced by the fingers and also the total moment of force in both pressing and prehension tasks (reviewed in Zatsiorsky and Latash 2004). In other words, if in a particular trial, a mode (command) to a digit deviates from its average (typical) performance, commands to other digits are also likely to show deviations from their average values, such that the expected effects of the original "error" on the total force and/or the total moment of force are decreased.

Experiments with multi-finger pressing tasks in elderly persons (Shinohara et al. 2004; Olafsdottir et al. 2007b) showed a number of differences from typical results in younger subjects. In particular, the indices of both force- and moment-stabilizing synergies were lower in the elderly. In an earlier section, I mentioned that force-stabilizing synergies take time to emerge in young, healthy persons, who participate in multi-finger force production tasks starting from a relaxed state (Shim et al. 2003b; section 5.3.3). This *critical time* was longer in elderly persons. When a person keeps the total force produced by several fingers constant, there is a strong force-stabilizing synergy. If the person produces a quick change in the force, the index of this synergy shows a drop before the total force starts to change. These anticipatory synergy adjustments (ASAs) can be seen in young persons 100–150 ms prior to the action initiation (see section 5.3.4). Elderly persons also can show ASAs, but these adjustments occur later (closer in time to the action initiation) and are smaller in magnitude (Olafsdottir et al. 2007a).

Prehensile tasks—holding a vertically oriented object steadily in the air (as in Figure 6.11)—have also shown weaker synergies in the elderly (Shim et al. 2004b). These synergies were analyzed at two levels. At the level of the thumb and *virtual finger* (see section 5.5), the moment produced by the normal forces

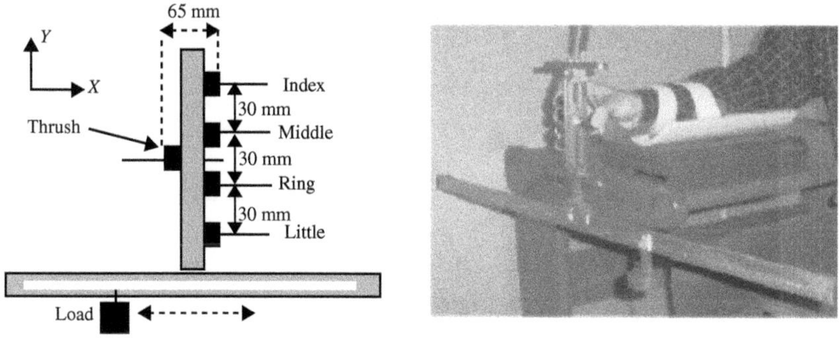

Figure 6.11. An illustration (and a photo—right) of the handle instrumented with six-component force/torque sensor that was used to study prehensile synergies in young and elderly persons. Reproduced by permission from Zatsiorsky VM, Latash ML, Gao F, Shim JK (2004) The principle of superposition in human prehension. *Robotica* 22: 231–234. © Cambridge University Press.

and the moment produced by the tangential forces co-varied to stabilize the total moment of force applied to the hand-held object. In other words, there was a moment-stabilizing synergy. Elderly individuals showed significantly lower indices of this synergy compared to younger persons. At the level of individual fingers, the normal forces produced by the four fingers co-varied to stabilize the normal force of the virtual finger; there was a force-stabilizing synergy. Once again, elderly persons showed significantly lower indices of this synergy compared to young subjects.

To summarize these observations in more intuitive terms, the elderly have problems creating synergies quickly, showing strong synergies, and adjusting the synergies in anticipation of a planned action. All these features may be viewed not as unavoidable negative consequences of aging but rather as adaptive strategies of the CNS (for more examples of adaptive neural strategies in the elderly see Christou et al. 2007; Lee et al. 2007). In a way, patterns seen in young subjects may be viewed not so much as necessary for performance but as a luxury. Recall, for example, that one can perform reasonably accurate actions in the absence of synergies stabilizing important performance variables by using stereotypical, very accurate patterns of elemental variables. One can also change a variable acting against one's own synergy. These strategies may be wasteful and fatiguing, leading to movements that appear clumsy. However, let us look at the other side of the coin. Imagine that a person with an increased variability of muscle performance (which seems to be truly unavoidable in the elderly, Cole et al. 1999; Burnett et al. 2000; Enoka et al. 2003), compromised postural stability (Maki et al. 1990; Melzer et al. 2004; Fujita et al. 2005), and decreased ability to generate quick correction of actions (Horak et al. 1989) participates in various everyday tasks. Using strong synergies and changing such synergies quickly (both in the course of their emergence and their destruction) may be viewed as optimal, if the CNS is confident in the consequences of its own actions and in its ability to generate quick, adequate corrective actions when the task or the environment changes. If this is not true, it makes sense to use less optimal but safer control strategies.

To conclude, age leads to changed motor synergies, but these changes should not be viewed as contributors to the overall decline in motor performance with age. Rather, the changed synergies help maintain an acceptable level of performance over a range of everyday tasks in spite of the suboptimal properties of the elements, the neuronal apparatus and the muscles.

6.5 SYNERGIES IN PERSONS WITH DOWN SYNDROME

Down syndrome is a complex of consequences produced by a change in the genetic material of chromosome-21; this genetic change affects both mental and physical development. Down syndrome was originally described in 1866 by a British physician, J.H.L. Down as "mongolism," a not so well-accepted term in our politically correct times because of its obvious racist connotations. Down syndrome is not very rare, and persons with this condition are born at the rate of 1:600 to 1:1000 of all live births. The majority of children born with Down syndrome have three rather than two copies of chromosome-21, a condition called trisomy-21. Sometimes, the third chromosome is present only in some of the cells; this condition is called mosaic Down syndrome. The extra chromosomal material results in significant changes in the functioning of many of the vital subsystems of the body, including the brain. Down syndrome is also associated with changes in the anatomy of the body, in particular with relatively low height, shorter extremities, and a tendency toward obesity.

In many lay persons' minds, the term Down syndrome is associated with mental retardation. This common opinion is supported by the fact that persons with Down syndrome, on average, are more likely to have problems solving relatively simple problems and to show lower scores on standardized tests such as IQ. However, this condition does not necessarily mean low intelligence (whatever this term may mean). A substantial number of persons with Down syndrome show IQ scores within the typical range; many of them graduate from regular high schools and even from colleges. Over my research career, I met persons with Down syndrome who worked as actors and tourist guides, who were excellent figure skaters and gymnasts, and who could express their thoughts much better than many of the undergraduate students who have taken my classes over the past quarter of a century.

More and more people with Down syndrome live independently, have professional careers, and in general do not show any obvious deficiencies expected from a person with mental retardation. Persons with Down syndrome may have to apply more effort to reach the same level of competence compared to persons without Down syndrome, but ultimately many of them can achieve personal goals typical of the general population.

There are many nonmotor problems associated with Down syndrome. The most important are congenital heart disease, obstruction of the intestinal tract, and increased susceptibility to infection. Recent progress in medical care, in particular, in neonatal heart surgery, programs of rehabilitation and special education has led to a dramatic increase in the life expectancy of persons with Down syndrome, from under 10 years in the 1940s to over 60 years at the beginning of the third millennium. This progress has also led to the emergence of new challenges, in particular, those related to the fact that more and more people with Down syndrome reach an advanced age (Connolly 2001). After 40, people with Down syndrome are likely to show signs of Alzheimer's disease and require extra care. People with Down syndrome tend to be very close to their parents and depend on their care throughout their lifetime. Since babies with Down syndrome are more commonly born to relatively older parents, when a person with Down syndrome reaches 50 or 60 years of age, his or her parents are likely to have passed away. Substituting for the care provided by the parents is a major challenge.

Problems with motor coordination in Down syndrome are not life threatening and may even be viewed as secondary compared to many other problems these persons face during their lifetime. However, dealing with problems of motor coordination is arguably the most frequently encountered challenge in everyday life. This is one reason why specific features of movements performed by those people deserve attention. Another reason is a more selfish one: Persons with Down syndrome are commonly described as "clumsy." There is no accepted definition for this word but it implies something opposite to "coordinated." So, if one understands what clumsiness is by studying movements of persons with Down syndrome, this may lead toward a better understanding of coordination. This is the topic of the next section.

6.5.1 Movements in Persons with Down Syndrome

Differences between movements of persons with and without Down syndrome may be classified into two groups. First, there are obvious differences that are seen across a variety of motor activities and are easy to observe and measure. Second, there are also differences that escape identification through the naked eye. They require special experimental approaches and may be task specific.

In the first group, I would like to mention slowness and hypotonia. Persons with Down syndrome take more time to initiate an action after a command to do so as quickly as possible (Anson 1992). They also take more time to complete an action when asked to do so at the highest possible speed. Another distinctive feature of persons with Down syndrome is that their limbs offer very little resistance when another person tries to move them. This is sometimes called hypotonia (Rarick et al. 1976; Morris et al. 1982; Cioni et al. 1994), a term implying the existence of a "normal tone." Unfortunately, there is no definition for "normal muscle tone" beyond "This is what a neurologist or a physical therapist perceives as normal resistance when he or she tries to move a joint of a limb." Joint resistance to externally imposed motion may be due to a lot of factors, including resistance provided by passive external tissues (including inertia, which is likely to be changed in persons with Down syndrome because of their typically shorter limbs), an inability to relax muscles completely, changes in muscle activation produced by peripheral receptors activated by the motion, an inability to suppress a natural reaction of resisting an externally imposed motion, etc. A decrease in the apparent resistance to external motion (hypotonia) in persons with Down syndrome may be due to changes in any of the mentioned factors.

Among less obvious features, there is one that seems to be present across a variety of motor tasks. This is preference for muscle co-contraction patterns. When a person without Down syndrome makes a fast movement, muscles acting at the involved joints typically show alternating bursts of activity, which are sometimes called reciprocal muscle activation patterns. For example, a fast elbow flexion movement is initiated with a burst of activity in elbow flexors (biceps brachii and brachioradialis), which apparently accelerates the joint into flexion. After a brief delay, this activity subsides (or even disappears), and the extensor muscle group (triceps brachii) shows a burst of activation that helps to slow down the movement. The extensor activation burst is also brief and is followed by a drop in the activity in the extensor muscles, sometimes accompanied by a second burst in the flexor muscles (possibly to avoid oscillations in the final position). This sequence is commonly referred to as the tri phasic EMG pattern (reviewed in Gottlieb et al. 1989; Figure 6.12, the top panel).

Reciprocal patterns of muscle activation are also seen in other tasks. For example, when one joint of a multi-joint limb performs a fast movement, mechanical joint coupling brings about torques that tend to perturb other joints of the limb. If the person does not want the other joints to flap, he or she has to send adequate commands to muscles controlling the other joints. Indeed, when only one joint of a limb performs a fast movement, tri phasic, reciprocal patterns of muscle activity are seen in muscles controlling both this joint and other joints of the limb (Koshland et al. 1991; Latash et al. 1995). Analysis within the equilibrium-point hypothesis (see section 3.4) has shown that the control variables to the wrist and elbow joints of an arm show similar time profiles, even though only one of the two joints performs a fast movement (Latash ML et al. 1999). The goal of changes in the commands to the other joint is not to produce a movement of

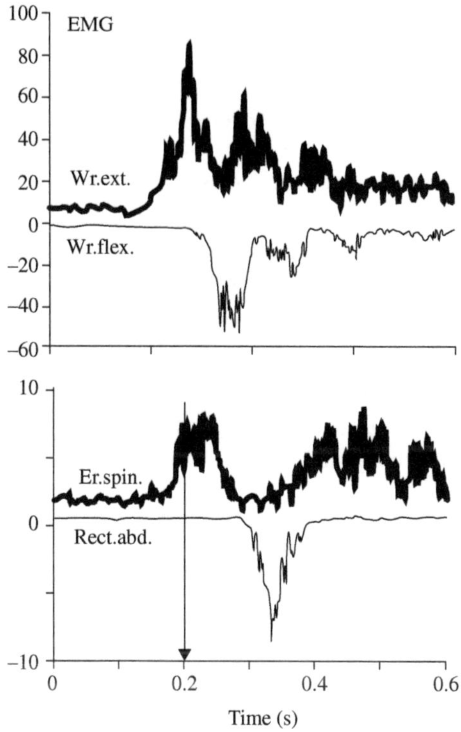

Figure 6.12. Examples of tri phasic electromyographic (EMG) patterns seen during fast single-joint voluntary movements (top, wrist flexor and extensor EMGs are shown) and during anticipatory postural adjustments (bottom, trunk muscles, erector spinae, and rectus abdominis EMGs are shown) to fast shoulder movements performed by a standing person. Reproduced by permission from Aruin AS, Latash ML (1995) Directional specificity of postural muscles in feed-forward postural reactions during fast voluntary arm movements. *Experimental Brain Research* 103: 323–332. © Springer and Latash ML, Aruin AS, Shapiro MB (1995) The relation between posture and movement: A study of a simple synergy in a two-joint task. *Human Movement Science* 14: 79–107. © Elsevier.

that joint but to avoid motion of that joint under the influence of joint coupling torques; one can call this a command for joint movement of zero amplitude.

Reciprocal patterns of muscle activation are also commonly seen in postural tasks. In particular, anticipatory postural adjustments (APAs, reviewed in Massion 1992; see the bottom panel of Figure 6.12) prior to self-triggered perturbations and preprogrammed postural corrections to external perturbations commonly show alternating bursts of activation in muscles that produce opposing actions at postural joints (agonist–antagonist muscle pairs). For example, if a person holds a position in a joint against an external torque, for example, by activating elbow flexors, a sudden increase in the external torque will lead to joint motion into extension. If the person tries to return the joint to the original position as quickly as possible, this will be accompanied by a burst of activity in the elbow flexor and a drop in the activity of the extensor (agonist and antagonist in Figure 6.13A). If the external torque decreases, an opposite pattern will appear: an increase in the activity of the extensor and a drop in the flexor activity (Figure 6.13B). All the mentioned observations seem very reasonable because they favor activation of a muscle, whose action is required by the task and lack of activation of a muscle that opposes this action (antagonist muscle).

Persons with Down syndrome, however, solve all such problems differently. They are much more likely to show simultaneous parallel changes in the activity of muscle groups with opposing actions—the so-called co-contraction patterns (Latash et al. 1993; Almeida et al. 1994; Aruin et al. 1996). Such patterns are seen across all the mentioned tasks: in joints that produce an intended quick motion and in joints of the same limb that stay motionless, as well as in postural

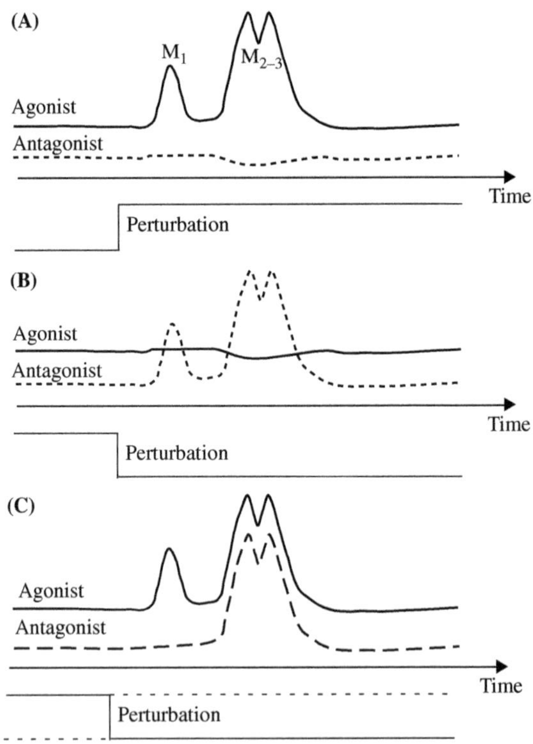

Figure 6.13. In typically developing persons, a perturbation of a joint leads to a reciprocal pattern of medium-latency changes (M_{2-3}) in the activity of two muscles acting at the joint (agonist—solid lines and antagonist—dashed lines) that depend on the direction of the perturbation (A and B). Persons with Down syndrome are more likely to show unidirectional changes in the activity of the two muscles (C) independent of the direction of the perturbation.

joints during preparation to a self-triggered perturbation and in response to an actual perturbation. Using the same example as the one described in the previous paragraph, if a joint of a person with Down syndrome is perturbed, a quick increase in the activity of both the flexor and the extensor muscles is typically seen, irrespective of the direction of the perturbation (Figure 6.13C). A simultaneous increase in the activity of muscles that oppose each other's action seems wasteful. Why would the CNS of persons with Down syndrome send signals that translate into such apparently suboptimal muscle activation patterns?

First, let me ask the following question: What would happen if a person used a typical reciprocal pattern of changes in the muscle activity (as in Figure 6.13A) but misjudged the direction of the perturbation? In other words, what would be the motor consequences if a perturbation into flexion were quickly followed by a burst of flexor activity and a drop in the extensor activity? Obviously, such a response would exacerbate the effects of the perturbation on the limb, produce a very quick movement in the direction of the perturbation and, if the person is unlucky, the limb may hit something, break something, or hurt itself. So, the typical reciprocal pattern could be optimal, but only if the central controller never makes a mistake in judging the direction of a perturbation. What if it sometimes makes such mistakes? What would be the lesson it learns? How would it cope with such a situation?

Muscle co-contraction is indeed wasteful if one considers only its effects on the net torque produced at a joint. It may, however, be purposeful. According to the equilibrium-point hypothesis, two basic commands to a joint represent a reciprocal command (*r*) and a coactivation command (*c*). The former changes the joint angle ranges of activation for the opposing muscles (the agonist–antagonist pair) such that the range for one of the muscles becomes larger while that of the other shrinks. The *c*-command defines a range of joint angle where both muscles are active. As a result, if joint position is fixed, a change in the *r*-command leads to an increase in the activity of one of the two muscles and a drop in the activity of the other (compare the slanted thin solid and dashed lines, Figure 6.14). A change in the *c*-command leads to a parallel increase (or a parallel decrease) in the activity levels of both muscles.

Note that a change in the *c*-command leads to a change in the slope of an overall joint angle–torque relation (compare the slanted solid and thick dashed lines in Figure 6.14). In other words, it leads to a change in the apparent stiffness of the joint (for those who are interested why I use *apparent* before *stiffness*, I recommend reading the following paper, Latash and Zatsiorsky 1993). Changing the apparent joint stiffness modifies how much it can be moved by an external torque of a fixed magnitude. In other words, *stiffening* a joint makes it less responsive to any perturbation, irrespective of its direction.

The co-contraction patterns of changes in muscle activity typical of movements of persons with Down syndrome start to make sense: They are universal and safe. They may never be able to perfectly compensate the effects of a perturbation of

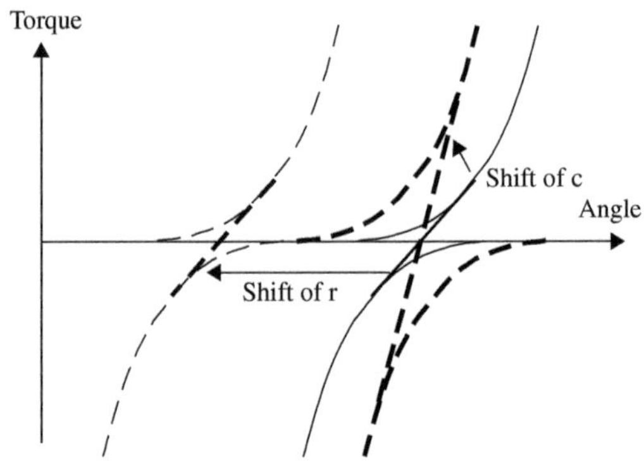

Figure 6.14. A shift of the reciprocal command (*r*, the thin dashed lines) leads to a shift of the overall joint characteristic (the straight, slanted lines) without much change in its slope. A shift in the coactivation command (*c*, the thick dashed lines) leads to a change in the slope of the joint characteristic without a major change in its location. The graph shows the dependences of muscle torque on joint angle (curved lines) with the agonist torque positive and antagonist torque negative.

a joint, but they will also never exacerbate effects of such a perturbation, rather always attenuate them. Now, we are ready to offer a hypothesis on why persons with Down syndrome use such atypical and apparently wasteful motor patterns.

The everyday experience over the first years of life has taught children with Down syndrome that complete failure at motor tasks is much worse than its slow but reliable accomplishment. Therefore, their CNS has learned to use from a variety of possible neural control patterns that can hypothetically solve everyday problems—those that lead to safe motor behavior rather than those that are mechanically, energetically, or aesthetically optimal. Is this strategy a safety catch or a safety haven? Can it be reversed? And, getting back to the main topic of the book, is it reflected in atypical synergies?

6.5.2 Multi-Finger Coordination in Down Syndrome

One of the very first studies of hand action by persons with Down syndrome showed different motor patterns compared to those observed in control subjects (Cole et al. 1988). In that study, the participants were asked to grasp an object, lift it, and hold it in the air (Figure 6.15). This action is associated with parallel changes in the vertical force applied by the digits on the object and the grasping force, that is, the force normal to the surface of the object. Changes in the two force components happen virtually simultaneously. While changes in the vertical force component are necessary to move the object as instructed, changes in the grasping force have been discussed as resulting from a parallel feed-forward control process with the purpose of making sure that the fingers do not slip off

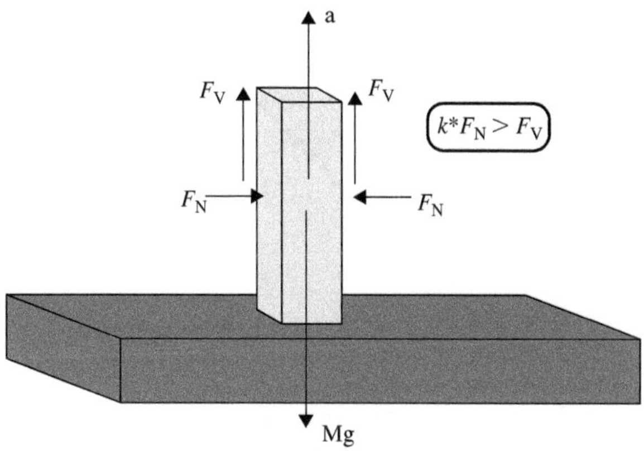

Figure 6.15. When a person lifts a hand-held object upward, the vertical force components (F_V) should sum up to $M^*(g + a)$, where M is mass of the object, g is the gravity constant, and a is acceleration. To produce a vertical force, each digit should produce a normal force that satisfies the inequality $F_V < k^*F_N$, where k is friction coefficient.

the object. As mentioned in the previous section on movements in the elderly, permissible magnitudes of the vertical force (F_V) are defined by the magnitude of the normal force (F_N) and the friction coefficient k:

$$F_V < F_N{}^*k$$

Typically, normal force is higher than the minimal required force ($F_{N,MIN}$), and the difference is termed the safety margin, $S = 100\%^*(F_N - F_{N,MIN})/F_{N,MIN}$.

Persons with Down syndrome have been shown to apply considerably higher grasping forces (F_N) than control subjects leading to substantially larger safety margins. This may be viewed as another example of the "playing it safe" strategy. This example also suggests that patterns of digit involvement in typical tasks may be changed in these persons.

A study of multi-finger force production has supported this assumption (Latash et al. 2002a; Scholz et al. 2003). In that study, young adults with Down syndrome sat in front of a computer screen and placed the four fingers of the dominant hand on four force sensors. The screen showed them a target line and a cursor that moved along the screen at a constant rate (Figure 6.16). The cursor moved up and down when the total force produced by all four fingers increased and decreased, respectively. The participants in that experiment were asked to slowly increase the total force produced by the four fingers in such a way that the cursor matched the target line as accurately as possible. This is a relatively easy task for persons without Down syndrome who very quickly understand the simple mapping between their actions and motion of the cursor. Persons with Down syndrome take more time to master the task.

To help them understand the task and to make the experiment a little bit less boring, two different types of more intuitive visual feedback were used (shown as

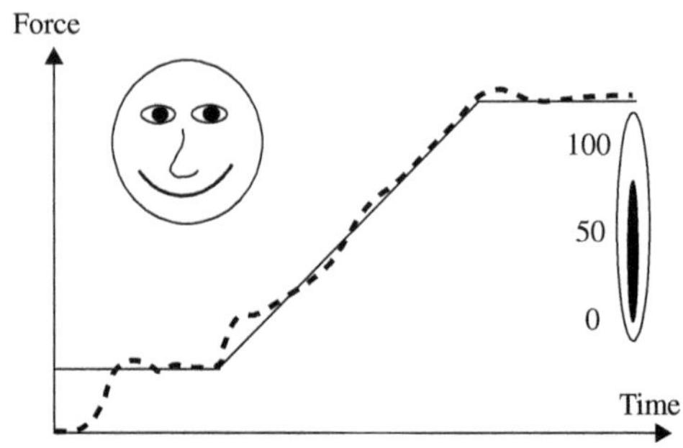

Figure 6.16. In studies of persons with Down syndrome, the task was to produce a smooth total force profile (dashed line) following a ramp-like template (thin solid line). Two types of additional feedback were used illustrated as inserts. These are described in the text.

inserts in Figure 6.16). First, the participants were shown a smiley face on the screen with the mouth curve computed as a parabolic function with the parameters depending on the quality of their performance. If they pressed on the sensors accurately, the face smiled a lot. If they did something wrong, for example lifted a finger off the sensor, the smile inverted, and the face frowned. In addition, the participants were shown a thermometer-looking image on the screen. After each trial, the "thermometer" showed them a score computed as the difference between 100 and an estimate of their deviation from the target line. They were told that only Michael Jordan (the great basketball star of Chicago Bulls) could score 100 points. So, they competed against Michael Jordan, which made them feel very proud when they got over the 90 mark. Actually, the score was computed in such a way that even the most bizarre performance got scored about 50, not to discourage the participants.

After a set of trials were collected, they were analyzed using a modified UCM method. Recall that the UCM analysis computes two components of variance in the space of elemental variables, *good variance* (variance within the UCM, V_{UCM}) that does not change an important performance variable and *bad variance* (variance orthogonal to the UCM, V_{ORT}) that does. In this particular study, the elemental variables were finger modes (see section 4.2.1), while the performance variable was, naturally, the total force. A four-finger synergy would be manifested as proportionally more *good variance* in the total variance in the space of finger modes.

In the traditional UCM analysis, variance components are computed for every time sample over the period of task execution across repetitive trials. To get reliable estimates of the variance components, one has to have a sufficient number of trials (typically about 20) that are assumed to be performed under the same control strategy. This is not a problem with subjects from the general population. Persons with Down syndrome, however, changed their strategy from trial to trial depending on their success in earlier trials, sense of competitiveness, and general arousal. Moreover, performing more than 12 trials in a row was tiring for them,

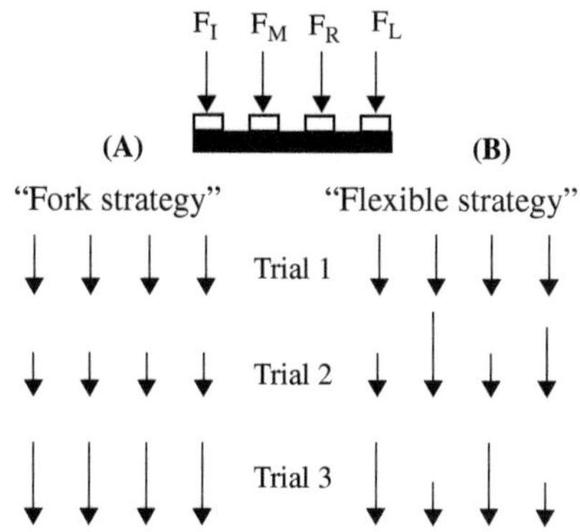

Figure 6.17. An illustration of the (A) "fork strategy" when all four finger forces scale proportionally from trial to trial and of (B) the "flexible strategy" when the forces vary with predominance of negative co-variation. Given similar trial-to-trial variations in the individual finger forces, higher accuracy may be expected in (B) compared to (A).

and their performance dropped. Therefore, a method was developed that permitted the analysis of variance not across different trials at a given time, but across different times for a given trial. This approach has been termed *one-trial UCM analysis*; it has been described in an earlier section (section 4.3).

Application of this analysis revealed two aspects of finger coordination in persons with Down syndrome. First, they were more likely to show positive co-variation among finger forces compared to control subjects. This naturally failed to stabilize the total force since this requires negative co-variation among finger forces. In a sense, these persons used their fingers as the prongs of a fork turned upside down changing the force of all four fingers in parallel (Figure 6.17A). This *fork strategy* (see also one of the very first sections describing the example of load sharing among the four legs of a table) was effective in avoiding excessive moments of force in pronation–supination, but it failed to stabilize the total force. Persons without Down syndrome need only a few trials to show predominantly negative co-variation among finger forces—a force-stabilizing synergy (Figure 6.17B).

The predominance of the *fork strategy* in Down syndrome seems like the reflection of a major difference in the neural control of the hand action. Is this alternative strategy an unavoidable consequence of the genetic difference typical of Down syndrome? Can it be avoided or modified?

6.5.3 Effects of Practice on Movements in Down Syndrome

Practice is known to be an effective way of improving motor performance over various populations and tasks. There is a whole area of study called *motor learning*, which tries to understand the mechanisms of improved performance with practice and also to discover optimal methods of practice, given specific goals.

Obviously, there are limits to what can be achieved with practice: It would be unreasonable to expect practice to help a person grow a new extremity after an amputation. So, the question of how much practice can improve motor performance in persons with Down syndrome remained open for quite some time. On one hand, it has been obvious that practice can and does lead to a significant improvement of motor performance, for example in the athletes who competed in Special Olympics, and in many others who simply mastered novel skills, such as playing musical instruments, after putting lots of effort into practicing them.

However, until relatively recently, studies of movement improvement with practice in persons with Down syndrome were limited to documenting changes in overall motor performance without looking into mechanisms underlying the observed improvements. One of the first studies that looked at changes in muscle coordination investigated a seemingly simple motor task, moving the elbow joint from one comfortable position to a target corresponding to another comfortable position as quickly and accurately as possible (Almeida et al. 1994). The instruction was typical of such studies and purposefully vague. What does it mean to move "as quickly and accurately as possible"?

There are well-known relations between speed and accuracy called speed–accuracy trade-offs. They form two large groups. First, when a person tries to move to a distant and small target, there is an increase in movement time. This relation was formalized by Fitts, based on aspects of the information theory (Fitts 1954; Fitts and Radford 1966) in the form of a logarithmic relation known currently as *Fitts' law*: $MT = a + b*ID$, where MT stands for movement time, a and b are constants, and *ID* stands for *index of difficulty*, the logarithm of the ratio of distance to target size, $ID = \log_2(2D/W)$. This relation has withstood the test of time and been confirmed in a variety of experiments with movements performed by different effectors, in different force fields, and by different populations (Knight and Dagnall 1967; Welford et al. 1969; Flowers 1975; Langolf et al. 1976; Crossman and Goodeve 1983; MacKenzie et al. 1987; Corcos et al. 1988; Smits-Engelsman et al. 2007; Zahariev and MacKenzie 2007). There have been several attempts to offer models for Fitts' law based on engineering, physiological, or control-theoretical considerations (reviewed in Meyer et al. 1988a,b; Plamondon and Alimi 1997). The least controversial model suggests that this relation originates at the level of movement planning (Gutman et al. 1993; Latash and Gutman 1993; Duarte and Latash 2007). Simply put, people are scared of moving quickly to far, small targets. Fitts' law is a reflection of human psychology, not of inherent limitations of the neuromotor system involved in the production of fast movements.

There is another type of speed–accuracy trade-off: When a person is asked to move over different distances at different speeds to an impossibly small (point) target, a measure of the scatter in the final position depends on both the distance and the movement time. Sometimes, this relation is described, using the notion of an effective target width, W_e, commonly estimated as four times the standard deviation of the final position. Unlike Fitts' law, this relation is more commonly represented as a linear relation between movement time and the ratio of the movement amplitude to the effective target width: $MT = a + b*D/W_e$. This kind of speed–accuracy trade-off obviously reflects all the processes that happen between planning a movement and actually performing it.

Because of the speed–accuracy trade-offs, the instruction "to move as quickly and accurately as possible" is ambiguous and depends on what the subject perceives as an acceptable error. Persons with Down syndrome are very much concerned with accuracy in reaching the target and are less responsive to the "as quickly" part of the instruction. Early in practice, they seem to be much happier to make rather slow movements that ultimately land in the very center of the target than to move faster and risk missing the target. This attitude is quite understandable. In everyday tasks, moving too slowly rarely leads to breaking objects, hurting oneself, etc. However, moving fast and missing a target may lead to exactly those consequences: hitting a finger rather than the nail with the hammer, breaking a glass, and such.

Analysis of movement kinematics and muscle activation patterns shows substantial differences between persons with and without Down syndrome. Control subjects produce movements that are very similar to each other, with smooth trajectories, bell-shaped velocity profiles, and triphasic patterns of muscle activation (see Figure 6.12). Persons with Down syndrome show much less well-defined bursts of muscle activation with a lot of co-contraction of muscles with opposing actions at the joint (the agonist–antagonist muscle pairs). As a result, their movements are slower, with multiple velocity peaks. Repeated attempts at the same task frequently lead to rather dissimilar patterns, both kinematic and EMG.

It takes time for a person with Down syndrome to realize that the equipment is safe, nothing is going to be broken, and he or she is not going to be hurt. Then, the performance changes dramatically. The muscles start to show typical alternating bursts of activation, the peak velocity increases, and the trajectories become smooth and reproducible across trials. The improvement observed in persons with Down syndrome in such a simple task is truly stunning: After a few hundred practice trials (spread over 3 days of practice), these persons nearly doubled the peak velocity. For comparison, a comparable amount of practice in control subjects (university students) leads to an increase in the peak velocity only by 10–15%.

Three days of practice are not enough to lead to any major changes in the peripheral structures, for example in the muscles. So, all the changes leading to the improved performance happened at a neural level. These changes apparently led to a modification of the safety–efficacy trade-off allowing for higher efficacy at the expense of performing "less safe" movements. I put "less safe" in quotation marks to emphasize that it relates to perception of safety by the participants in that study, not to actual safety.

Whether an agonist–antagonist muscle pair acting at a joint forms a synergy is still unknown, at least if one accepts the definition of synergy advocated here. Another study explored whether persons with Down syndrome can modify synergies with practice; it used the earlier described task of four-finger accurate force production (Latash et al. 2002a). Prior to practice, persons with Down syndrome commonly showed a *fork strategy*, described earlier. The relative amount of *bad variability* (V_{ORT}) was particularly high at low force magnitudes, and it gradually decreased with an increase in the total force. However, even at the relatively high forces, *bad variability* was larger than *good variability* ($V_{ORT} > V_{UCM}$); there was no force-stabilizing synergy.

After a couple of days of practice, the participants of that study all improved their performance: They became more "like Michael Jordan" in achieving higher accuracy scores. There was also a significant change in the indices of finger force co-variation. These indices shifted toward more positive (less negative) values at all magnitudes of the total force, and at higher forces, *good variability* started to dominate. Figure 6.18 illustrates changes in an index of synergy, $\Delta V = (V_{UCM} - V_{ORT})/V_{TOT}$, where the amounts of total variance (V_{TOT}), V_{UCM}, and V_{ORT} are all normalized by the number of degrees of freedom in corresponding subspaces: The space of finger forces is four-dimensional, the orthogonal space is unidimensional, and the UCM is three-dimenional. At higher forces, ΔV became positive, corresponding to a force-stabilizing synergy that had been absent prior to practice.

The improvement or, more precisely, the emergence of the force-stabilizing synergy with practice was associated with unchanged indices of synergies

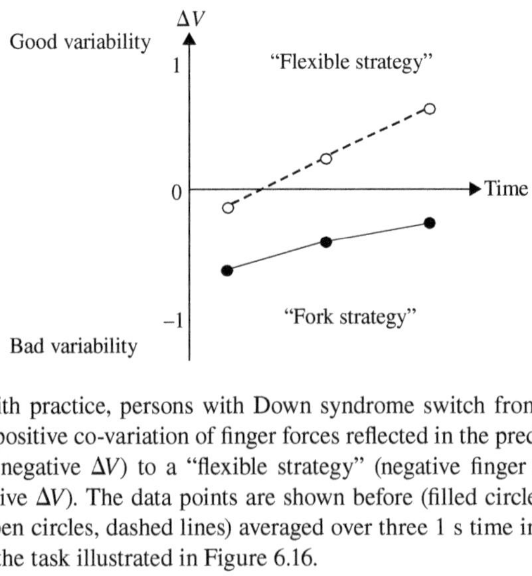

Figure 6.18. With practice, persons with Down syndrome switch from a predominantly "fork strategy" (positive co-variation of finger forces reflected in the predominance of "bad variability" and negative ΔV) to a "flexible strategy" (negative finger force co-variation reflected in positive ΔV). The data points are shown before (filled circles, solid lines) and after practice (open circles, dashed lines) averaged over three 1 s time intervals during the ramp portion of the task illustrated in Figure 6.16.

stabilizing the rotational action of the fingers, that is, the total moment their forces produced with respect to a horizontal axis passing through the midpoint between the middle and ring fingers. This is not an obvious result. Stabilization of the moment of force requires positive co-variation of the combined forces produced by the index and middle fingers and the combined forces by the ring and little fingers. Stabilization of total force requires predominantly negative co-variation among finger forces. The two requirements are not mutually exclusive as long as one performs the task with more than two fingers (see Figure 4.2 in section 4.1) but combining the two is not a trivial undertaking.

There was one more result of that study that seems worth mentioning. The participants formed two groups. One of them practiced the main task only, while the other group practiced both the main task and a whole range of other tasks that involved different single-finger and multi-finger tasks. In a sense, the second group played with the setup. The total amount of practice time was the same for the two groups. The second group showed much greater improvement in the index of force-stabilizing synergy in the main task despite the fact that its participants practiced the task less than those in the first group. This observation was not surprising, because several earlier studies had shown that variable practice is much more effective in improving motor performance in persons with Down syndrome compared to massed practice of a single task (Edwards et al. 1986; Edwards and Elliott 1989).

Taken together, the two studies suggest that persons with Down syndrome, indeed, have a lot of room for improvement of both general indices of performance and patterns of coordination that make performance more reproducible and flexible. This is an optimistic result supporting the idea that most of the apparent "clumsiness" seen in persons with Down syndrome is adaptive and can be reversed with adequately designed practice. However, should it be reversed? This is another question that looks nearly rhetorical but is in fact not trivial. If the "clumsiness" is indeed adaptive, it serves a purpose: to ensure optimally safe motor functioning in the everyday environment. In a friendly and predictable laboratory environment, the safety constraints can be lifted, leading to a more typical performance and coordination patterns (synergies). But then the subjects step out of the laboratory into the real world that may be less friendly and full of unexpected forces and changing goals. Would it not be wiser for the CNS to return to the safe strategies elaborated over a lifetime rather than suddenly switch to less "clumsy" but more risky patterns of control?

6.5.4 Relation of Atypical Synergies to Changes in the Cerebellum

One of the brain structures that has been traditionally considered crucial for motor synergies is the cerebellum. We will discuss the role of the cerebellum in a later section (section 7.3). Now it is important to note that the mass of the cerebellum has been reported to be lower in persons with Down syndrome compared to the general population (Crome and Stern 1967; Molnar 1978; Bellugi et al. 1990). There have also been reports on changes in the cerebellum in other groups of atypical persons (Berntson and Torello 1982; Keele and Ivry 1990), whose movements are sometimes described as "clumsy." These observations suggest a potential link between the atypical synergies in persons with Down syndrome and their lower cerebellum mass. If the mechanism of synergies is malfunctioning, the everyday motor repertoire, which people from the general population take for granted and perform without any visible effort, may start to pose major problems affecting such motor tasks as walking, standing, or reaching for an object.

Down syndrome is not the only condition associated with both changes in the cerebellum and atypical movements. Another example is movements performed by persons with autism. *Autism* was first defined in 1943 by Dr. Leo Kanner. It is estimated that babies with autism are born at a rate of 1 out of 250 live births. Autism is a spectrum disorder. This means that it is characterized by a spectrum of signs that may not necessarily be linked to a single neurophysiological or other mechanism. Persons with autism commonly show resistance to change, distress for unclear reasons, difficulty in mixing with others, lack of responsiveness to words, and difficulty in verbal expression, sometimes reflected in repeating words. Their motor behavior may be characterized by such dissimilar features as physical over-activity or extreme under-activity. They may show stereotypical, repeated movements, and sustained odd play. Their gross and fine motor skills may show very uneven development.

There is increasing evidence that autism is associated with abnormalities in the cerebellum, including reduced cerebellar gray matter and neuronal loss in the cerebellar nuclei (Courchesne 1997; Palmen et al. 2004; McAlonan et al. 2005). These abnormalities may be related to problems in assembling synergies. Here, under synergies, I mean not only motor synergies but a more general notion of elements that work together to stabilize an important feature of a body function. The role of elements can be played by words; then, one observes difficulties with verbal expression such as agrammatisms. The role of elements can also be played by persons, then problems with communication and interacting with others may be expected.

There is one more childhood disorder that is characterized by "clumsiness." This is *developmental coordination disorder* (DCD, reviewed in Gillberg and Kadesjo 2003). It is more common in boys than in girls. DCD is a rather vague grouping, based mainly on observable behaviors that might be the outcome of a range of underlying problems. So, there may be a number of possible subgroups of persons who are diagnosed with DCD. Roughly 1 out of 20 of school-age children have some degree of DCD. Children with this disorder may trip over their own feet, run into other children, have trouble holding objects, and have an unsteady gait. These children have difficulties in mastering gross motor coordination skills such as crawling, walking, jumping, standing on one foot, catching a ball, and fine coordination task such as tying shoelaces. Some children also demonstrate expressive speech problems.

Children with DCD show developmental delays in sitting up, crawling, and walking, deficits in handwriting and reading, and problems in gross and fine motor skills. They show problems with postural control that become obvious in challenging tasks, such as one-leg standing (Geuze 2005). In a variety of motor tasks, they show atypical features that should look familiar to the reader. In particular, during fast movements, these persons show increased levels of co-contraction of agonist–antagonist muscle pairs, while during manipulation of hand-held objects, they show higher grip forces and increased safety margins (Pereira et al. 2001; Raynor 2001; Zoia et al. 2005). The commonalities of these features with those demonstrated by persons with Down syndrome and by the healthy elderly suggest that these features may indeed be adaptive and have different causes. Just like in other cases of atypical development described in this section, motor abnormalities in DCD have been linked to a cerebellar dysfunction (O'Hare and Khalid 2002; Ivry 2003).

6.6 SYNERGIES AFTER STROKE

Cerebro-vascular accidents, commonly called stroke, involve an interruption of the normal blood supply to a brain area as a result of the rupture of a blood vessel or blocked blood flow. The clinical picture depends strongly on the area affected by stroke and the extent of the accident. Strokes affecting large brain areas typically lead to both motor and nonmotor consequences, but here let me consider primarily motor consequences of stroke. General motor consequences include partial loss of voluntary control over muscles on the contralesional side of the body (addressed as hemiparesis), emergence of poorly controlled reflex muscle contractions and spasms (spasticity), and dyscoordination affecting all motor actions, from reaching and grasping to posture and locomotion (Dietz and Berger 1984; Diener et al. 1993; Levin 1996; Kautz et al. 2005).

Most fibers in the descending and ascending pathways that connect the brain and the spinal cord cross the midline such that the left half of the body is represented mostly in the right brain and vice versa. There are uncrossed fibers in the brain–spinal cord pathways; these have been mostly described as carrying information to and from the trunk. Because of the organization of the neural pathways, a stroke localized in one hemisphere leads to a strongly asymmetrical clinical picture. In particular, limbs on the two sides of the body typically show dramatically different degrees of impairment. Sometimes, one side of the body (ipsilateral to the side of the stroke) is even referred to as healthy or normal. This is imprecise for two reasons. First, the relatively unimpaired limb also shows changes in motor function (Colebatch and Gandevia 1989; Jones et al. 1989; Cramer et al. 1997). Besides, stroke is followed by plastic changes of projections both within the brain and between the brain structures and the spinal cord (reviewed in Johansson 2000; Rossini and Pauri 2000; Hallett 2001). In particular, the role of uncrossed fibers in the descending tracts may increase leading to a shift in the balance of limb control from the injured ipsilesional to the spared contralesional hemisphere.

Clinical features of strokes commonly show a proximal-to-distal gradient of symptoms with distal muscles (those that are farther away from the trunk) more affected. In particular, hand function is likely to be more affected than arm movements such as reaching that are controlled by more proximal muscle groups (Shelton and Reding 2001; Mercier and Bourbonnais 2004; Michaelsen et al. 2004). The impaired control of the leg muscles leads to problems with posture and locomotion. In particular, patients frequently show narrow stance (the feet placed too close to each other in the medio-lateral direction) and scissor gait (with the leading foot placed such that it blocks the path of the trailing foot). Both features contribute to the postural instability of patients after stroke and slow down their recovery.

Some individuals who have experienced a stroke are described clinically as having "abnormal movement synergies," typically characterized as relatively fixed or stereotyped (Twitchell 1951; Bobath 1978; DeWald et al. 1995). This conclusion is commonly based on analyses of the patterns of how the muscles change their activation levels, or torques about different joint axes co-vary, or the joints change their positions together during the execution of particular tasks. In this context, the word *synergy* means something like *variables that change together.* Although this meaning is commonly accepted in clinical literature, it is rather different from the definition that is used in this book. Recall that parallel scaling of elemental variables may not necessarily reflect a control strategy (see section 1.3).

Among the variety of motor synergies (in the sense of "variables changing together") seen after stroke, commonly described patterns involve generalized flexion actions across a set of joints within an affected limb and strong co-contraction patterns of agonist–antagonist muscle pairs. In addition, there are changes in typical torque patterns across the major joints of the arm, when a person tries to produce a particular mechanical effect by the hand (Beer et al. 2000; Ellis et al. 2005, 2007; Neckel et al. 2006), changes in the kinematics of joint involvement in arm movements (Cirstea et al. 2003; Micera et al. 2005), increased coupling between arm and leg movements (Kline et al. 2007), and changes in the patterns of muscle activation during natural movements (Ellis et al. 2007; Neckel et al. 2006).

Reaching movements by the contralateral arm after a unilateral stroke are characterized by irregular trajectories that deviate from the nearly straight trajectories typical of reaching in healthy persons (Levin 1996). When an object is placed farther away from the body, patients after a stroke are more likely to involve trunk motion to assist reaching. These observations point at a potentially important reorganization at the level of kinematic synergies (in the meaning accepted in this book) such that the restriction on distal joint motion is at least partly compensated by involvement of less affected proximal effectors (the trunk). In our vocabulary, these observations suggest that the patients find a new sharing pattern in the joint space that allows all the joints to contribute to the task and provide for stability of performance, given their changed state.

A kinematic study of reaching in persons with mild-to-moderate impairments after stroke (Reisman and Scholz 2003) resulted in somewhat unexpected findings. On the one hand, the average kinematic profiles in the patients were different from those in healthy subjects. The patterns of joint coupling also showed significant differences. However, when variance in the joint space was partitioned into two components according to the method developed within the UCM hypothesis, both groups show considerably larger *good variance* (V_{UCM}) compared to *bad variance* (V_{ORT}). The two variance components were computed, assuming that the controller tries to stabilize the endpoint trajectory in external space. There were no obvious differences between the patients and the healthy subjects in the ability to structure the kinematic variance. It seemed that only the sharing pattern feature of multi-joint synergies changed after stroke while the flexibility/stability feature (which I sometimes call *error compensation*) did not.

In a follow-up study (Reisman and Scholz 2006), the object that the subjects reached for could be placed at different points of the workspace. Sometimes, it was placed beyond arm's reach, and the subjects were forced to use trunk motion to reach for the object. In that study, patients with mild consequences of stroke showed much more *bad variance* compared to the healthy, control subjects when reaching toward objects placed in the hemispace of the impaired part of the body but not when reaching toward the contralateral hemispace. This was confirmed for both hand trajectory in space and hand trajectory relative to the trunk. So, there seems to be a deficit in the error compensation feature of multi-joint synergies during reaching movements. It may result from a combination of neural and biomechanical factors secondary to the stroke.

A lot of questions remain unanswered, and the cited studies only whet the appetite of clinical researchers. What features of synergies suffer and what features remain unchanged after stroke? Do changes in synergies show a proximal-to-distal gradient similar to the clinical signs? Can rehabilitation restore lost features of synergies? Can practice lead to emergence of novel, adaptive synergies? What neural reorganization can form the basis for improved and/or adaptive synergies? Unfortunately, there are no answers as yet.

6.7 LEARNING MOVEMENT SYNERGIES

Why do people practice? The obvious answer is "To perform better." Indeed, in many cases, the goal of practice is to improve a particular characteristic of performance, for example force, speed, accuracy, consistency. There are obvious exceptions. For example, it is hard to single out what exactly people try to improve when they practice Yoga. At least for now, we have no physical measure that would be adequate to estimate improvement with practicing Yoga (or, for that matter, many other types of exercise whose purpose is to improve the functioning of the whole body). In this book, however, my goals are much less ambitious. So, let me now focus on effects of practice assuming that there is an unambiguous physical measure of performance that the person tries to improve. Moreover, I will limit analysis for now to tasks where the main requirement is to produce a certain magnitude or a time profile of a physical variable accurately and consistently. This seems reasonable, since the main purpose of the following analysis is to link practice with possible changes in synergies, and synergies by definition are related to such features of motor performance as consistency and stability.

6.7.1 Traditional Views on Motor Learning

One of the most influential traditional views on motor learning was suggested by Nikolai Bernstein over 50 years ago and presented in detail in his magnum opus *On Dexterity and Its Development* (Bernstein 1996). Bernstein took as an axiom that the presence of numerous degrees-of-freedom presented a problem for the controller, a problem that was solved by eliminating redundant degrees-of-freedom and finding a unique solution. Note that this axiom is quite different from the one introduced as the principle of abundance (see section 3.2), the one that suggests that no degrees-of-freedom are eliminated but they are all used in flexible combinations to ensure stability of important characteristics of performance. Let us, however, follow the logic of the Bernstein approach.

When a person tries to perform a truly novel action with a redundant set of effectors (and this is the case for an overwhelming majority of human motor actions), the first attempts typically look clumsy. The motion is fragmented, some of the effectors seem not to contribute to the motion at all, while others join in a staggered fashion. Bernstein called this early stage of skill acquisition *freezing degrees-of-freedom*. His opinion was that the controller tried to simplify the control of the task by reducing the number of variables he or she had to manipulate.

There is a problem with this reasoning. Imagine, for example, that you try to perform a fast novel movement of the arm. Motion of any joint is expected to be associated with torques in other joints of the body (interaction torques, see section 3.3). If you would like to keep a joint motionless, this would require nontrivial control of muscles crossing the joint. In fact, "not controlling" a joint in a sense of not changing control signals to muscles crossing the joint may be expected to lead to its substantial flapping, not to the lack of motion. Earlier, I mentioned a few studies that have shown nontrivial, precisely tuned control of muscles acting at a seemingly "frozen" joint when other joints of the arm move (Koshland et al. 1991; Latash et al. 1995). One of the studies showed large-amplitude patterns, at both the moving joint and the apparently postural joint, not only of muscle activation but also of hypothetical control variables reconstructed within the equilibrium-point hypothesis (Latash ML et al. 1999).

According to Bernstein, the second stage of motor learning involves releasing the previously frozen degrees-of-freedom. This is expected to make motor patterns more flexible and adaptable to possible changes in the conditions of movement execution. The next and third stage was expected to lead to changes at the control level that allow the person to use external forces as contributors to the planned motion rather than to predict and compensate these forces.

This three-stage scheme of motor learning has not been challenged for many years. A number of studies described changes in motor patterns as a sequence of freezing and releasing degrees-of-freedom (Newell 1991; Vereijken et al. 1992; Newell et al. 2003). Virtually all those studies analyzed movements at the level of mechanics and associated degrees-of-freedom with individual joint excursions, torques, or such. A feeling of dissatisfaction with such an attitude to motor learning was reflected in one of the earlier Editorials written for *Motor Control* (Latash 1997): "Just as an example: Imagine that, as a result of an intervention, e.g., a training protocol or a surgery, a joint involved in a multi-joint movement shows a decrease in its movement amplitude below the lowest level detectable by the goniometer, or the trajectories of two joints start to show a high correlation. Does this mean that the number of degrees of freedom in the system dropped? Certainly not! An inadequately formulated question cannot be adequately answered" (p. 207).

Recently, the issue of what adequate degrees-of-freedom are has been actively discussed in the literature. Such terms as *functional degrees-of-freedom*, *eigenmovements*, and *modes* have been introduced based on computational analyses applied to outputs of apparent elements of the system under consideration (Alexandrov et al. 1998, 2001a; Zatsiorsky et al. 1998; Krishnamoorthy et al. 2003a,b; Li 2006). Unfortunately, so far, analysis of degrees-of-freedom at a control level has been only a dream, mostly thanks to the stubborn resistance of the motor control community against accepting the only theory of motor control that specifies control variables in a physiologically sensible and noncontradictory way (if the reader still wonders what theory I am talking about, he or she should re-read section 3.4).

However, even during analysis of elemental variables at the level of output of the elements (mechanical or EMG), one can consider possible effects of motor learning using a scheme that is more compatible with the principle of abundance. This is exactly what I plan to do in the next section.

6.7.2 What Can Happen with a Synergy with Practice?

Imagine that a person demonstrates a synergy stabilizing a certain performance variable by co-varied changes in a set of elemental variables. Within the framework of the UCM hypothesis, this means that the amount of *good variance* per degree-of-freedom within the UCM is larger than the amount of *bad variance* that lies in the orthogonal to the UCM complementary subspace ($V_{UCM} > V_{ORT}$). What can happen with the two components of variance and with the synergy with practice? To illustrate three main scenarios, I am going to consider the same simple example of two-finger force production as earlier.

So, imagine that the task is to press with the two index fingers on two separate force sensors and produce the total force of 20 N. The subject in this mental experiment performs the task many times, and the data for each trial are plotted as a point on a two-dimensional force–force plane. The left panel of Figure 6.19 illustrates the shape of such a distribution with an ellipse. The ellipse is elongated along the line $F_1 + F_2 = 20$, which is the UCM for this task. The shape and orientation of the ellipse mean, in particular, that the projection of the variance of this data set onto this line is larger than the projection on the orthogonal subspace (the dashed line in Figure 6.19). In other words, there is a two-finger force stabilizing synergy, which can be characterized quantitatively, for example, with an index $\Delta V = V_{UCM}/V_{ORT}$.

If a person practices this simple task under the instruction to try to be as accurate as possible in reaching the required level of total force, practice is naturally expected to lead to a drop in the variance of the total force, which corresponds in Figure 6.19 to V_{ORT}. However, what can be expected from V_{UCM}? This component of variance by definition has no effect on total force; so, changing this component is not expected to help or hurt the performance. In general, V_{UCM} can drop proportionally to the decrease in V_{ORT}, drop less (or stay unchanged, or even increase!), or drop more that V_{ORT}. These three scenarios are illustrated on the other three panels of Figure 6.19 (panels A, B, and C). Note that in each of these three panels, V_{ORT} is the same and it is smaller than in the left panel.

The improvement in performance is the same across the three scenarios. However, changes in the two-finger force-stabilizing synergy are different. Different changes in V_{UCM} lead to different changes in the index ΔV. In panel A, both variance components change proportionally leading to no change in ΔV.

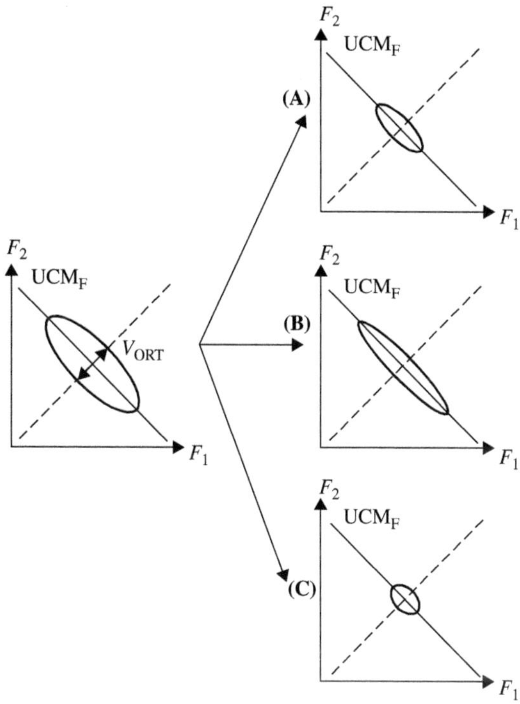

Figure 6.19. Practice is expected to lead to a drop in "bad variability" (V_{ORT}). However, it can lead to different changes in "good variability" (V_{UCM}). (A) V_{UCM} can decrease proportionally to V_{ORT}, (B) it can decrease less (or remain unchanged or even increase), or (C) it can decrease more. These three scenarios may be associated with an improvement in performance accompanied by an unchanged synergy (A), a stronger synergy (B), or a weaker synergy (C).

In panel B, V_{UCM} stays constant leading to an increase in ΔV. In panel C, the seemingly irrelevant component of variance, V_{UCM} changes more than V_{ORT} (the data distribution becomes more spherical) leading to a drop in ΔV. The three scenarios illustrate an improvement of performance with no changes in the synergy, an increase in the index (strength) of the synergy, and weakening of the synergy. As we will see in the next few sections, synergies can show changes with practice that fit all three scenarios illustrated in Figure 6.19.

6.7.3 Practicing Kinematic Tasks

One of the first studies of changes in kinematic synergies with practice led to unexpected results (Domkin et al. 2002). In that study, the task was to make a very fast movement of the pointer held by the right hand and a semicircular target held by the left hand such that the tip of the pointer stops in the middle of the target. This study was described in section 5 as an example of applying the

UCM analysis to multi-joint kinematic tasks. Here, let me describe what happens with practice of that task. The subjects adopted an initial posture illustrated in Figure 6.20 that forced them to use all three major joints of each arm during the movement, and the movement was limited to a horizontal plane. This task makes it possible to test hypotheses on at least four kinematic variables that can potentially be stabilized by co-variation of joint rotations across trials. The first two test whether joint rotations within each arm co-vary to stabilize the endpoint trajectory. Rotations of the three major joints of the right hand may form a synergy stabilizing the trajectory of the tip of the pointer. Similarly, rotations of the left arm joints may form a synergy that stabilizes the trajectory of the target.

However, the task is formulated in terms that make relative motion of the pointer tip and the target center more relevant. Hence, two more hypotheses may be offered. The first is that the scalar distance between the pointer tip and the target is stabilized across trials. The second is that the vector distance is stabilized. The difference between these two hypotheses is illustrated in the top two drawings in Figure 6.20. Simply put, if the controller cares only about keeping the two endpoints at a certain distance (to perform an accurate pointing movement, it should be zero at the end of the movement), stability of the scalar distance may be viewed as a more adequate variable. However, if the controller wants also to control the direction along which the two endpoints approach each other, the vector distance is a more relevant performance variable.

Prior to practice, the subjects showed synergies with respect to both within-an-arm synergies and with respect to the hypothesis on stabilizing the vector distance between the endpoints (the hypothesis on the scalar distance was not

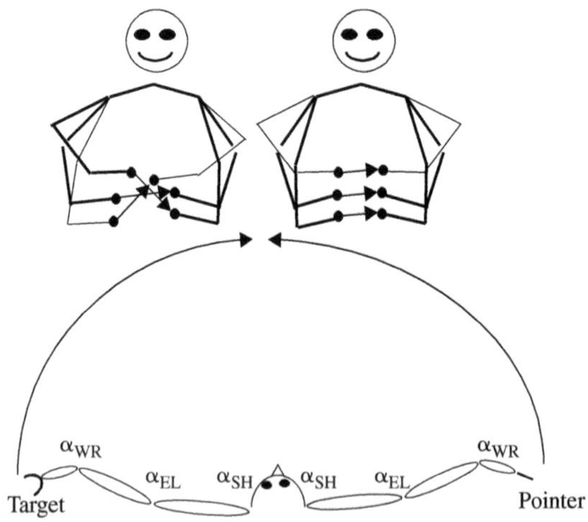

Figure 6.20. An illustration of the two-hand multi-joint pointing task performed from an initial joint configuration shown in the bottom drawing. The upper drawings show three combinations of joint angles, each corresponding to an unchanged scalar distance (left) and an unchanged vector distance between the pointer and the target (right).

formally tested in that study). After 3 days of practice, the participants showed an expected improvement in the accuracy of performance. It was associated with a drop in the variance of the trajectories of both the pointer tip and the target (computed across repetitive attempts at the task). However, *good variance* dropped more than *bad variance* with practice for all three performance variables. The multi-joint synergies became apparently weaker corresponding to scenario C illustrated in Figure 6.19. In other words, the subjects became more accurate but they also became more stereotypical in their joint trajectories.

This was an unexpected outcome. Why would the controller stop using its ability to channel most of the joint variance into the *good variance* component? Indeed, if the controller kept the synergy index unchanged, the same drop in the total joint variability would be associated with better performance! One possibility is that there is a limit in how accurate a movement can be. For example, if a person is unable to detect an error, there is nothing to correct, and no further improvement can be expected. In the two-arm pointing task, this may correspond to a situation when the subject cannot see any discrepancy between the final positions of the pointer tip and the center of the target.

Imagine now that this person continues to practice the task despite the fact that no detectable improvement can be observed. It is natural to assume that, in such a situation, the person will try to improve other movement characteristics that are not related to higher success in meeting the explicit goal. In particular, these characteristics may be related to such aspects as comfort and lack of fatigue. Improvement in these "other aspects" of movement may be expected to lead to selection of a particular family of joint characteristics from the abundant set that had been used before these new practice goals emerged. Consequently, a drop in the component of variance that is unrelated to the explicit task, V_{UCM} may be expected without a change in V_{ORT} leading to a drop in the index of the multi-joint synergy.

A follow-up study (Domkin et al. 2005) used a more complex two-arm pointing task that included making more natural three-dimensional movements. To discourage the subjects from using stereotypical arm trajectories, the left hand held a pole with three targets attached, and the subjects were asked to point at different targets in different trials. The effects of practice on multi-joint synergies were ambiguous: The participants became more accurate without significant changes in the synergy index. In other words, variability in the joint space decreased proportionally in all directions, and the results were similar to scenario A in Figure 6.19.

Scenario B has also received support in studies of kinematic multi-joint synergies. One of the studies investigated kinematic synergies during Frisbee throwing (Yang and Scholz 2005). In that study, the task was to throw a Frisbee with the dominant hand into a rather small circular target; the diameter of the target was just under 30 cm, and the distance to the target was over 6 m. The experimenters tested several hypotheses related to stabilization of such performance variables as hand trajectory, hand orientation with respect to the target, and hand velocity. Even before practice, there were kinematic multi-joint synergies stabilizing all these performance variables. In other words, the amount of variance in the joint space corresponding to a change in a performance variable was smaller than the amount of variance that did not lead to a change in the variable ($V_{UCM} > V_{ORT}$).

As expected, after 5 days of practice, the participants became more accurate in the task. This was accompanied by a drop in the variability of hand motion and a drop in the total variance in the joint space. However, the *bad variance* (V_{ORT}) showed a relatively larger drop that the *good variance* (V_{UCM}) for two performance variables, direction of hand movement and hand orientation with respect to the target. In other words, the synergies stabilizing these two performance variables strengthened with practice (scenario B in Figure 6.19).

One more study used the framework of the UCM hypothesis to investigate how humans adapt to unusual force fields. A large number of studies of such adaptations have formed the foundation for an idea that the formation of new internal models, both direct and inverse, underlies improvement in performance with practice in novel conditions. This line of research has been discussed in section 3.3. Most of those studies used kinematic (reaching) tasks performed with nonredundant joint sets. So, they could not possibly address the issue of changing synergies because synergies, by definition, are based on redundant (sorry, abundant) sets of elements.

In the study by Yang and colleagues (2007), the subjects performed a planar reaching movement against a resistance provided by a programmable set of motors (a robotic arm). The robotic arm was originally programmed not to produce any force, and the subjects simply got used to moving quickly and accurately from an initial position to a target. Then, unexpectedly for the subject, the robot started to produce force acting orthogonal to the vector of endpoint velocity and proportional to the magnitude of the velocity (Figure 6.21). So, the force was zero in the initial state (since velocity was zero), and it was zero at the final position. During the movement, however, the force perturbed the trajectory. The main goal was to determine whether a change in the use of joint motion redundancy is associated with the adaptation process to this novel force field.

As mentioned earlier (Digression #4), natural reaching movements show relatively straight trajectories and bell-shaped velocity profiles. The subjects in the described study were explicitly asked to try to move in a straight line. This was easy in the absence of the external forces. However, the robot made the trajectory during the first few trials rather curved, and the subjects had to correct their control to be able to move along a straight line in the presence of the perturbing force. They adapted successfully over a few dozen of trials. When the robot was turned off unexpectedly, the adaptation was manifested as a markedly curved movement despite the fact that external conditions returned to natural.

When the external field was turned on, the subjects showed an increase in joint variance, and then, over about 300 practice trials, the variance dropped. During this adaptation process, the *good variance* (V_{UCM}) dropped less than the *bad variance* (V_{ORT}); the two variance components were computed for the endpoint trajectory as the performance variable. V_{UCM} was also relatively higher after the motors were turned off, and the subjects returned to the original natural force field. The disproportional changes in the two variance components (illustrated as the V_{UCM}/V_{ORT} ratio in the bottom graph of Figure 6.21) suggest that the CNS makes use of the motor abundance in a purposeful way. This conclusion is not easily compatible with the idea of developing new optimal internal models with practice, based on the mechanical interactions between the limb and the environment (see section 3.3). Why would such an adaptive system care about the apparently irrelevant components of performance reflected in V_{UCM}?

Taken together, the reviewed studies of effects of practice on kinematic synergies suggest that there is a need to reevaluate the traditional Bernstein's hypothesis of freeing and freezing degrees-of-freedom with learning. It seems that no degrees-of-freedom are ever frozen (and, correspondingly, none is freed). The way they are used changes with practice, potentially leading to stronger, unchanged, or weaker synergies.

6.7.4 Practicing Kinetic Tasks

A couple of studies of multi-finger synergies explored whether such synergies can change with practice. Humans are very good at learning new tricks with their fingers. It takes a teenager a few minutes to learn a novel, artificial combination of key presses to be able to control a fighter jet (or another aggressive object) on a computer screen and exterminate other fighter jets or aliens with amazing agility. Since most analyses of synergies require having a set of trials (at least a dozen) performed under an assumption of an unchanged control strategy, one has to invent a task that is not that easy to learn. Otherwise, the learning is complete by the third or fourth trial, and very little can be said about associated changes in synergies. On the other hand, the task cannot be next to impossible to learn, since most subjects do not want to come to the laboratory to be tested over a period of months.

One of the studies used a seemingly simple task that was artificially made a bit more challenging (Latash et al. 2003c). The subjects of that study pressed on three force sensors with the ring, middle, and index fingers of the right hand (they were all right-hand dominant); they were required to track a ramp profile shown on the screen (Figure 6.22), with a signal corresponding to the total force produced by the three fingers (similar to the tasks described in section 5.3.1). This is indeed a very easy task and, in the absence of complicating factors, the subjects would have probably learned it within a few trials.

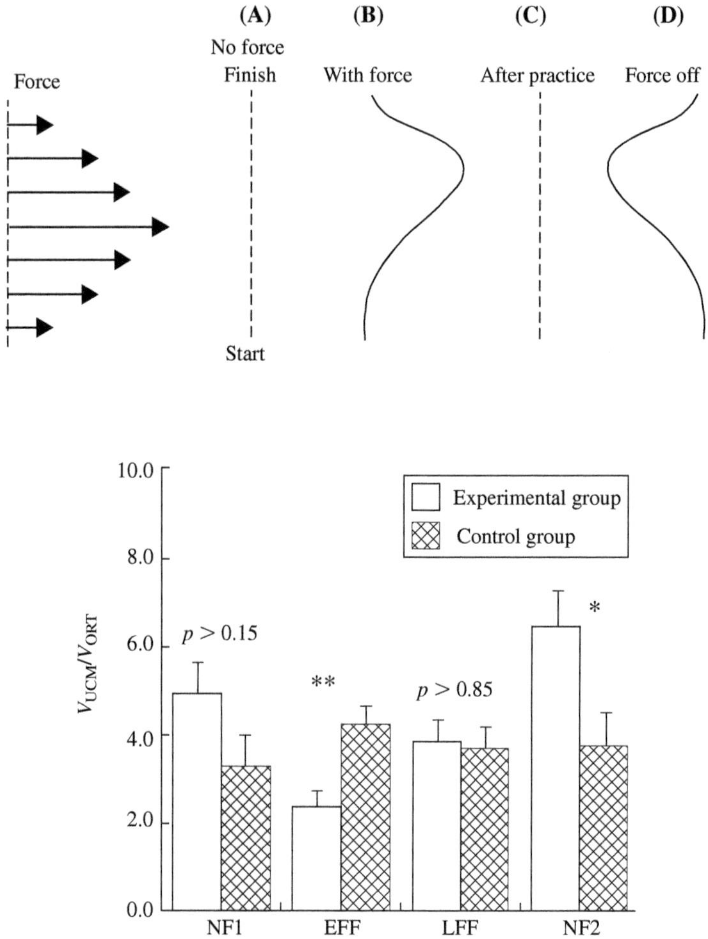

Figure 6.21. (A) In the absence of external forces, the subject shows a straight trajectory to the target. (B) When the force is applied, the trajectory becomes curved. (C) After a few minutes of practice, the subject is able to produce straight trajectories in the presence of the velocity-dependent force field. (D) When the force field is turned off, a few trials are curved in the opposite direction compared to the curvature of the first trials immediately following the application of the force field. The bottom graph shows the ratio of *good variability* to *bad variability* (V_{UCM}/V_{ORT}) for the four illustrated segments of the experiment. For comparison, changes in V_{UCM}/V_{ORT} are also shown for a control group that did not experience any external force field (hatched bars). Reproduced with modifications by permission from Yang J-F, Scholz JP, Latash ML (2007) The role of kinematic redundancy in adaptation of reaching. *Experimental Brain Research* 176: 54–69.

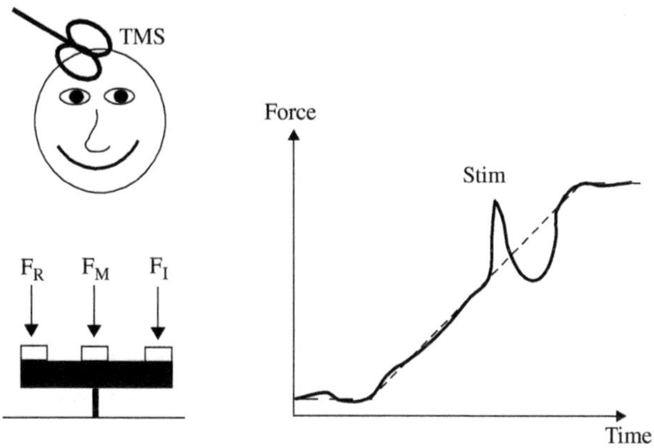

Figure 6.22. In this study, the subject was required to press on three sensors with the index, middle, and ring fingers and produce an accurate ramp force profile. The frame with the sensors rested on a narrow support. At an unpredictable time, a single transcranial magnetic stimulus (TMS) was applied over the contralateral primary motor area of the cortex, leading to a quick force increase followed by a period of force suppression.

There were two complicating factors. First, the frame with the force sensors rested not on the table but on a narrow supporting surface, only 1 mm wide. This means that the subjects were expected to keep the balance of the frame or, in other words, to make sure that the total moment of force computed with respect to the midline of the supporting surface was within a rather narrow range. Second, in most trials, the subjects received a standard stimulus to their left cortical hemisphere delivered with a transcranial magnetic stimulator (see Digression #10). Only one stimulus per trial was delivered at an unexpected time. The stimulus produced a very quick flexion jerk in all the fingers followed by a short silence in the flexor muscles. As a result, both the total force and the total moment were perturbed. Note that the magnitude of a response to TMS depends strongly on the background muscle activation level. As a result, the response was small for relatively low finger forces, and it was stronger for the larger background forces. Since the stimulus was delivered unexpectedly, the magnitude of the perturbation to the total force and the total moment also changed unexpectedly from trial to trial—depending on the actual finger forces during the ramp task when the stimulus was applied. The subjects felt helpless, since they were unable to predict when the stimulus would come, and the force response to a stimulus started at too short a delay (under 30 ms) to allow any kind of sensible reaction.

The two complicating factors made the task rather challenging, and the subjects showed a gradual improvement over a couple of hundred trials that were all performed within a single 1.5-hour-long session. To estimate what happened with multi-finger synergies, three brief series of unperturbed trials were performed (without TMS, and the subjects knew that no TMS would be applied) at the beginning of the session, in the middle, and at the end. The experimenters were lucky: Changes in the synergies between the midpoint test and the initial test were qualitatively different from those between the end-test and the midpoint test. This sequence of synergy changes allowed the authors to suggest a different two-stage hypothesis on synergy changes with practice.

There were two constraints in the task. First, the subjects were supposed to produce a time profile of the total force. Second, they were supposed to keep the total moment of force within a rather narrow window. Since the subjects used only three fingers, the total variance in the finger force space could be easily viewed as consisting of three components: *Bad variance* for force stabilization (V_F), *bad variance* for moment stabilization (V_M), and *good variance* that affected neither mechanical variable (V_{GOOD}). All variance indices were computed across trials for each time sample and then averaged over time intervals corresponding to the early, middle, and late segments of the target ramp force profile.

Figure 6.23 illustrates what happened with the three components over the first half of practice and over its second half. Not surprisingly, the subjects decreased V_F over the first half of practice; in other words, they improved their performance with respect to the explicit task of producing an accurate total force time profile. They also showed an improvement in their ability to avoid large moment of force variations reflected in smaller V_M. Meanwhile, *good variance* did not change much. If one introduces a measure of force- and moment-stabilizing synergies, for example, the ratio between *bad* and *good* variances, the results suggest that both synergies showed an increase in their indices with practice. By the way, note the difference in the scales in the three panels in Figure 6.23. Most variance was *good*—both synergies existed—even when the subjects just started to practice the task.

The second half of practice, however, led to very different results. The *bad variance* components, V_F and V_M did not change much, while V_{GOOD} dropped significantly. This means that both synergies weakened. This was pure luck indeed: Over 1.5 hours of practice, the subjects showed two stages of synergy changes. First, there was strengthening of relevant synergies and later the synergies weakened. In a sense, this study showed both scenario B and scenario C within a single relatively short experiment. Why would the synergies become weaker later in the study? I will attempt answering this question in the next section.

The other study with learning accurate multi-finger force production purposefully used a very unusual and challenging task, although, on the surface, it looked similar to the task in the first study: The participants learned to accurately track a ramp-shaped template by pressing with the fingers of both hands on force sensors (one per finger). However, the trajectory of the cursor on the screen corresponded to a variable that was computed on-line as the sum of forces of four out of eight fingers, two per hand, from which the forces of the remaining four fingers were subtracted. The most challenging finger combinations were selected; the total

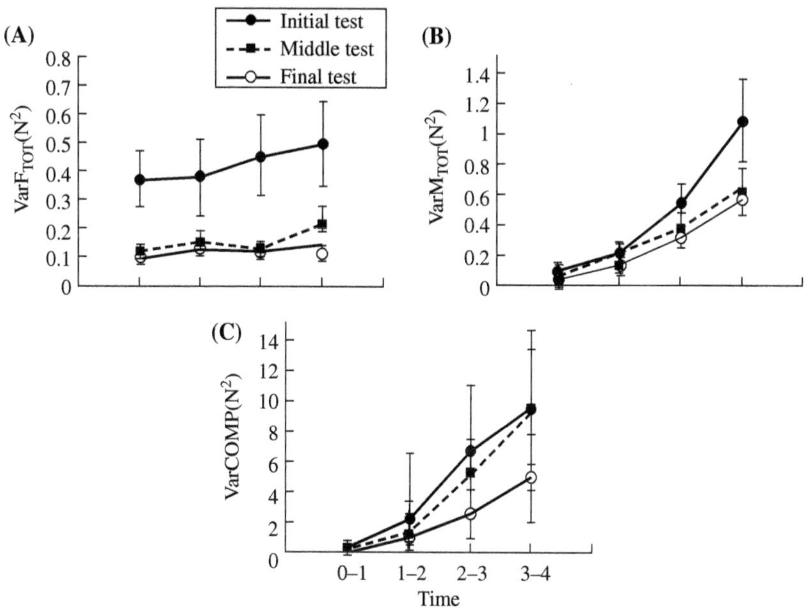

Figure 6.23. An illustration of the changes in three components of finger force variability with practice. (A) Variability related to changes in the total force. (B) Variability related to changes in the total moment of force; (C) Variability that does not change either the total force or the total moment of force. Note a drop in A and in B from the initial test to the middle test with little changes from the middle test to the final test. In C, there was little change from the initial test to the middle test and a significant change from the middle test to the final test. Reproduced by permission from Latash ML, Yarrow K, Rothwell JC (2003c) Changes in finger coordination and responses to single pulse TMS of motor cortex during practice of a multi-finger force production task. *Experimental Brain Research* 151: 60–71. © Springer.

force signal was computed as $F_{TASK} = (F_{IR} + F_{RR} + F_{ML} + F_{LL}) - (F_{IL} + F_{RL} + F_{MR} + F_{LR})$ or as $F_{TASK} = (F_{IL} + F_{RL} + F_{MR} + F_{LR}) - (F_{IR} + F_{RR} + F_{ML} + F_{LL})$. Here the first letter in the subscripts refers to the finger (I—index, M—middle, R—ring, and L—little), and the second letter designates the hand (R—right and L—left). This is indeed a very tough task. I challenge the reader to try to press with only the index and ring fingers of one of the hands and the middle and little fingers of the other hand while not pressing with the other four fingers.

The first few attempts showed that graduate students, who served as subjects were very good at finding cheating strategies. For example, since the total force was never very high (to avoid fatigue), they started to press with only one hand or even with only one finger from the set of explicitly involved fingers. To counteract such cheating, additional constraints were introduced, in particular, the participants got high negative points for producing less than 25% of the total force by one of the hands and for producing less than 25% of a hand's force by one of the fingers.

It took participants of this tough study about 2–3 days to learn the task. Over this time course, major changes were seen in within-a-hand synergies. Note that the task may be viewed as built on a control hierarchy. At a higher level, the total force is supposed to be shared by the two hands, and force-stabilizing between-hands synergies may be expected. At a lower level, forces produced by individual hands are shared among the fingers. Although the task required force production by only two fingers, the other two also contributed in a negative fashion because of the way F_{TASK} was computed. It was impossible for the subjects to avoid force production by those "antagonist" fingers because of the phenomenon of enslaving (see Digression #10). Hence, four-finger force-stabilizing synergies may be expected in each hand. In addition, at both between-hands and within-a-hand levels, commands to individual fingers (finger modes, see section 4.2.1) could be organized to stabilize the moment of force with respect to a midline between the two hands and between the ring and middle fingers, respectively. Such moment-stabilizing synergies were not explicit parts of the task; however, they could be expected based on earlier studies (Latash et al. 2001, 2002c; Scholz et al. 2002; see also section 5.3.1).

Let me skip the computational details and get directly to the main results. Between-hands synergies stabilizing F_{TASK} were seen at the very onset of practice. However, there were no within-a-hand force-stabilizing synergies (Figure 6.24). These results are similar to those mentioned earlier for tasks involving force production by sets of fingers distributed between the two hands (Gorniak et al. 2007a,b; see section 5.4.4). Somewhat unexpectedly (and not very much politically correctly), female participants of the study showed particularly low indices of within-a-hand force-stabilizing synergies. The males were also far from perfect, but at least they showed marginally positive values of the synergy index ΔV (predominance of *good variance*). The participants had no problems stabilizing the moment of force, which was not part of the task, just like in earlier studies (section 5.3.1).

With practice, not surprisingly, all participants improved their performance as reflected in lower indices of force deviations from the ramp template. This was accompanied by the emergence of within-a-hand synergies stabilizing the hand's input into F_{TASK}. The female participants caught up with the males, and their results after practice were indistinguishable. So, synergies can not only strengthen but also emerge over a relatively brief time period.

When summarizing all the available data on synergy changes with practice, one has to admit that the classical Bernsteinian scheme of freezing-freeing degrees-of-freedom does not seem to be supported. Rather, all degrees-of-freedom are used at all times. However, the patterns and strength of their co-variation with respect to particular performance variables may change. As of now, it seems that there are two stages of synergy learning.

First, for a novel, challenging task, synergies related to explicit task variables emerge and strengthen. This stage continues until the performance improves to a

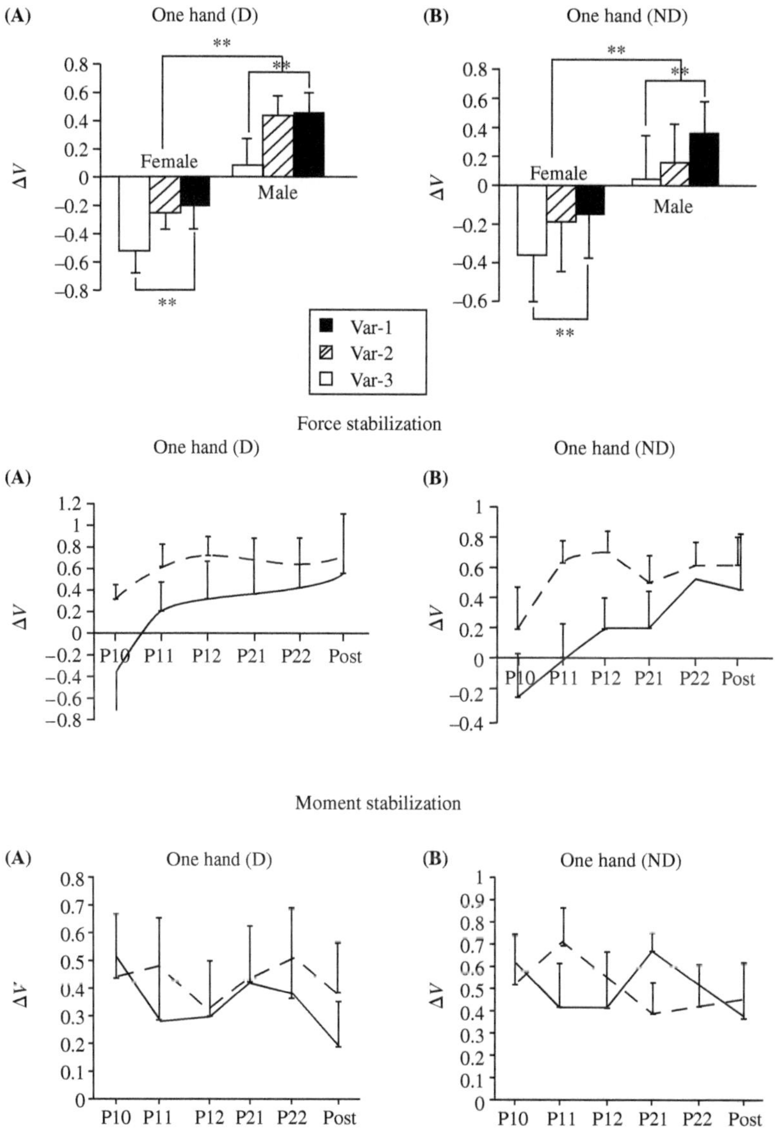

Figure 6.24. Top: Prior to practice, the subjects failed to stabilize the total force produced by each hand (D—dominant, ND—nondominant) by co-variation of commands to the fingers of the hand. The data are shown separately for male and female subjects and for three 1-s time intervals of the ramp task. Bottom: With practice, the index of force stabilization (ΔV) increased, resulting in force-stabilizing synergies. Moment of force was stabilized prior to practice, and the ΔV index for the moment did not change with practice. Reproduced by permission from Kang N, Shinohara M, Zatsiorsky VM, Latash ML (2004) Learning multi-finger synergies: An uncontrolled manifold analysis. *Experimental Brain Research* 157: 336–350. © Springer.

perfect level or at least one good enough from the subject's point of view. Then, the second stage begins when the subject pays more attention to not just improving performance with respect to the task criteria but with respect to "other things." These "other things" may come from various sources and reflect a host of criteria. Such seemingly secondary issues as improving personal comfort during the task execution, avoiding fatigue, or optimizing other aspects of performance, for example, movement aesthetics may start to dominate and dictate further changes in the patterns of co-variation among elemental variables. This second stage is expected to lead to selection of particular subsets of trajectories from a larger family containing trajectories that are all equally capable of solving the explicit task. In other words, the *good variability* is expected to drop while the *bad variability* is expected to stay unchanged. Figure 6.23 illustrates the two stages. It combines the two scenarios from the earlier Figure 6.19, namely, scenario B and scenario C.

Most of the reviewed studies of kinematic and kinetic tasks focused on only one of the two stages leading to an increase or a drop in the synergy indices computed with respect to the explicitly required performance variable. Only one of the mentioned studies (the last one) was lucky enough to capture both stages within a relatively brief practice session (Latash et al. 2003c).

6.7.5 Plastic Neural Changes with Learning a Synergy

Is there anything measurable within the CNS that correlates with the improved (emerged) synergies? Two of the studies reviewed in the previous section both used TMS as a tool of testing excitability of neural pathways involved in the earliest response to stimulation over the primary motor cortex. Despite the fact that TMS is a very crude tool that is likely to produce changes in various within-the-brain and descending pathways, the results of those studies show that learning a new motor skill is indeed associated with measurable changes in neural projections.

As described in section 6.2, TMS-induced motor responses show changes with practice. Most of the earlier studies (Pascual-Leone et al. 1995; Cohen et al. 1997; Pascual-Leone 2001) showed rather general changes in the response pattern such as an increase in the area from which stimulation could induce a motor response in an effector and a decrease in the strength of the stimulation that was able to induce a visible response (TMS threshold). Such changes have been reported in studies of both very long-term practice, such as learning Braille or playing a musical instrument (Sterr et al. 1998; Pascual-Leone 2001) and relatively short-term practice limited to a single session (Classen et al. 1998). In particular, the last of the cited studies showed that practicing a movement of the thumb in a particular direction led to a change in the direction of a jerky motion induced by a single TMS toward the practiced direction. Recently, more direct evidence has suggested that practice sculpts the response properties of neurons in the primary motor cortex (Matsuzaka et al. 2007).

Studies of changes in TMS-induced responses during learning multi-element motor tasks showed more subtle changes. In particular, in an earlier mentioned study of three-finger ramp force production with additional instability (Latash et al. 2003c), practice induced a drop in the magnitude of the TMS-induced force increment produced by all three fingers (Figure 6.25A)—a result opposite to the more commonly reported facilitation of TMS effects with practice. This result made sense: With practice, somehow, the subjects learned to decrease the magnitude of the force perturbation produced by the standard TMS. I was the first subject in that study, and I felt absolutely helpless with respect to the effects of the TMS. I was sure that no changes in the TMS effects happened over the 1.5 hours of the experiment. However, when the data were processed they showed that my response to the standard stimulus dropped, just like in all other participants.

In addition, in that study the drop in the TMS-induced force increment was accompanied by a drop in the difference between the force responses on the index and ring fingers. Recall that in that experiment the frame with the force sensors rested on a 1 mm wide support, which introduced an additional constraint on the total moment of force produced by the three fingers. Since the middle finger was positioned directly above the support area, its contribution to the total moment of force was limited. In a first approximation, TMS induced a change in the total moment, ΔM, proportional to the difference between the changes in the forces of the index and ring fingers (see Figure 6.25B), $\Delta M = d(\Delta F_I - \Delta F_R)$, where d is the lever arm of each of the finger forces. The drop in the difference between

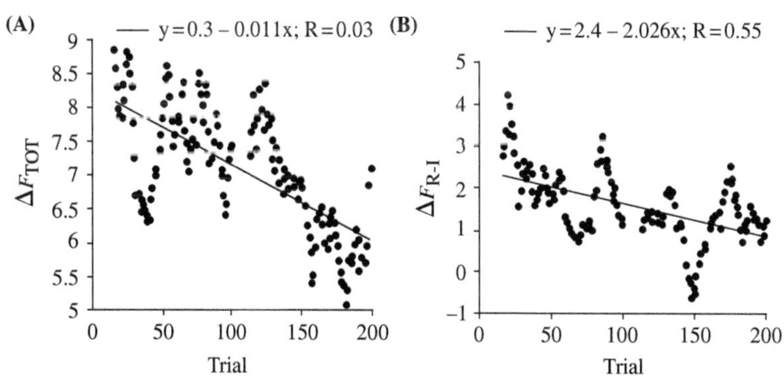

Figure 6.25. (A) In the experiment illustrated in Figure 6.22, the total force responses produced by the transcranial magnetic stimulation (TMS) (ΔF_{TOT}) dropped with practice. (B) This was accompanied by a drop in the difference between responses to the TMS of the ring and index fingers ($\Delta F_{R\text{-}I}$), resulting in a smaller change in the moment of force with respect to the pivot. Reproduced by permission from Latash ML, Yarrow K, Rothwell JC (2003c) Changes in finger coordination and responses to single pulse TMS of motor cortex during practice of a multi-finger force production task. *Experimental Brain Research* 151: 60–71. © Springer.

the responses of the two fingers means that the moment of force perturbation produced by the TMS also decreased with practice.

The explicit task for the subjects was not to decrease the responses to the TMS. This would have been a meaningless instruction because TMS came unexpectedly, and responses to those stimuli took under 30 ms to come about. The task was to learn accurate total force production despite the perturbing effects of the TMS. The improvement in performance with practice and the associated changes in force- and moment-stabilizing multi-finger synergies (see the previous section) were associated with changes in TMS-induced responses that made sense. These changes helped the subjects perform the explicit task better.

This study showed two stages in synergy changes. The first one was associated with strengthening both force- and moment-stabilizing synergies, while during the second stage the synergies became apparently weaker. This was interpreted as a consequence of the controller paying more attention to factors other than the explicit task. In this study, at least one "other factor" was explicit: It represented dealing with the perturbing effects of the TMS. Indeed, during the second half of the practice session, when the synergies apparently weakened, the subjects showed a faster drop in the magnitude of the TMS-induced response than during the first stage.

Another study (Kang et al. 2004) did not use TMS during practice, but only during relatively short testing sessions with the purpose of estimating the effects of practice. To remind, in that study, the subjects learned the very unusual task of producing an accurate time profile of a force signal with two pairs of fingers distributed between the two hands. To test possible effects of practice on interactions between the two hemispheres (recall that the left hemisphere controls the right hand, and the right hemisphere controls the left hand), the subjects were required to produce a constant force by the four fingers of one hand and different levels of constant force by the four fingers of the other hand. TMS was applied over the motor cortex contralateral to the first hand, that is, to the one that always produced the same force level. The logic was that any changes in the TMS-induced responses in that hand could only be due to different effects from the other hemisphere that sent different commands corresponding to different force levels produced by the other hand.

It has been well established that activating motor areas of one of the hemispheres leads to a modification in responses produced by standard TMS applied to the other hemisphere (Liepert et al. 1998; Stedman et al. 1998; Muellbacher et al. 2000; Weiss et al. 2003). Observations of inhibitory effects have been interpreted as resulting from interhemispheric inhibitory pathways. The study of Kang and her colleagues confirmed this general finding. However, it also showed that the strength of the inhibitory effects associated with the production of force by one of the two hands on the responses in the other hand decreased with practice (Shim et al. 2005a). Moreover, this decrease was pronounced in the *task fingers* (the two fingers that were explicitly involved in the main task) but not in the other two fingers. This is a very subtle effect indeed. It fits very well with the results of another recent study of musicians suggesting that the asymmetric activity of the two hands during playing an instrument was associated with a drop in inhibitory interhemispheric effects (Ridding et al. 2000).

The original phenomenon of inhibitory interhemispheric projections may be interpreted as reflecting aspects of everyday hand action in bi-manual tasks. When an object is handled with two hands, an increase in the contribution of one hand may be expected to lead to a decrease in the contribution of the other hand, as in the two-hand synergy described in experiments with holding a "heavy beer glass" (Scholz and Latash 1998). In that experiment, an instrumented glass-like object was grasped by the fingers of one hand, while the other hand could be used to apply supporting force to the bottom of the object. When the subject applied such supporting force himself or herself, the grasping force dropped in anticipation of an increase in the supporting force, that is, without a time delay from the moment of supporting force application. When the supporting force was released, the grasping force increased, also in an anticipatory fashion.

The main task of the study with four-finger, two-hand force production required the subjects to do something different, that is, to increase the forces produced by the two hands in parallel. Inhibitory interhemispheric projections are not useful for such a task; actually they have to be fought to produce the required parallel force increase by the two hands. So, the drop in the strength of these projections, but only for the explicitly involved fingers, may be viewed as a reflection of practice that is specific to and beneficial for performance.

There have been only a few studies that quantified changes in synergies with practice and simultaneously tried to relate these changes to possible plastic changes within the CNS. However, despite the paucity of such data, the first results have been very promising. Apparently, neural changes that happen with learning new synergies or improving existing ones are detectable and quantifiable. This offers a window into the neural organization of synergies. However, before getting overexcited, let me turn to what one could reasonably expect from studies of the neural and computational mechanisms that underlie multi-effector synergies.

Part Seven

Neurophysiological Mechanisms

7.1 NEUROPHYSIOLOGICAL STRUCTURES AND THE MOTOR FUNCTION

This section opens with a digression that is probably going to discredit the rest of what is going to be written here. This digression is likely not going to be accepted benevolently by many of my friends and colleagues who perform excellent, highly sophisticated studies of different neural structures in an attempt to link patterns of neural activity to functions of the human body. This seems to be an undisputable goal of neurophysiology: To find how interactions within and among neural structures lead to all the variety of body functions.

Digression #11. What Is Localized in Neural Structures?

The question how external functions of the body are represented in the brain has been hotly disputed for literally centuries. Until the middle of the nineteenth century, the predominant view was that different brain structures were responsible for different functions and represented their control centers. This view led, in particular, to phrenology, a science of bumps on the human skull that were supposed to reflect features of one's character.

The view on strictly localized human functions was based, in particular, on observations of violations of functions in patients with brain injuries and tumors in different brain areas. The logic was straightforward: If an injury to a brain area leads to a severe deterioration of a brain function, this means that that particular area contained a control center for that particular function. This logic is faulty, of course. Nobody would claim that the control center of a television (TV) set is in its power cord, even though cutting the cord leads to elimination of the function of the TV set. Along similar lines, I would like to remind the following classical story:

> Researchers decided to learn where the cockroach's hearing organs are located. They used the method of conditioned reflexes and taught the cockroach to run to a remote corner of the box when a bell rang. Then, a particularly bright member of the research team tore away the legs of the most unfortunate cockroach. After performing this cruel intervention, the inquisitive scientist rang the bell. Unsurprisingly, the cockroach did not run to the far corner of the box. Conclusion? The cockroach did not hear the bell meaning that the organ of hearing was in the legs.

This story sounds ridiculous, and it is. However, conclusions on the functions of brain structures not uncommonly follow the cockroach-logic. Examples are: If an injury of a brain area leads to deterioration of a function, the area was responsible for that function. Or, if an area shows a lot of neural activation when the person is involved in a particular functional activity, this area is important for (controls) that activity. As already mentioned, according to this logic, most important political decisions are made on stadiums and at rock concerts, where a lot of people shout together.

Enough for this digression within a digression! At the end of the nineteenth century, an alternative view gained prominence. It was based on numerous observations of functional recovery after severe injuries to brain areas that had been assumed to contain those functions' centers. Besides, observations of children with inborn brain abnormalities had also suggested that even very severe pathologies failed to eliminate some of the basic functions apparently controlled by the brain. The alternative view started to dominate that the neural structure of the brain was nondifferentiated and any neuron could participate in any function.

The first book by Nikolai Alexandrovich Bernstein was dedicated to the issue of localization of brain functions (see also sections 2.5 and 6.7.5). Bernstein wrote this book in the early 1930s, and it was ready to be published by 1935. In that book, Bernstein argued with the dominant view espoused by Pavlov's school that all behaviors were based on combinations of inborn and conditioned reflexes. However, in 1935 Pavlov died, and Bernstein decided not to publish a book that argued with an opponent who did not have a chance to retort. Fortunately for us, the proofs of the book were saved, and it was ultimately published nearly 70 years later, in 2003 in Russian (Bernstein 2003, not yet translated into English; see Latash 2006).

In that book, Bernstein reviewed all the pros and cons of the two mentioned polar attitudes to the neural basis of body functions and concluded that both views were flawed. I cannot help but present a few quotations from that book:

1. It would be incomparably harder to imagine how the neural process in all the nerves could be the same (as claimed by academician I.P. Pavlov), disregarding all the morphological differences among the nerves, their peripheral apparata and intrinsic systemic interactions, than to accept that more diverse morphology is expected to produce more diverse dynamic phenomena. This sounds as if someone would try to persuade me that the bassoon, the violin, the double-bass, and the tam-tam produce absolutely identical sounds, and the only difference is in that some of these instruments are to the left of the conductor, while others are to the right, and they play different notes. (p. 315)
2. No area of the cortex can currently be viewed as the origin or the final destination of a neural process...Every area and every layer of the cortex represent only transit points of the neural process. (p. 326)
3. Neither atomism, nor wholism (accepted as postulates) can provide a comprehensive interpretation of the nervous system, primarily because they can exist separately only as abstractions. The nervous system will be explained only when we are able to express the factual inseparability of the two principles and find terms to express processes reflective of the dynamics of their interaction. (p. 325)

Bernstein uses "wholism" to address the opinion that all neural elements have the same properties and only an interaction among a subset of neurons produces functions. The word "atomism" is used in about the same meaning as contemporary "reductionism." That is, a view that a function of a set of elements is a direct result of and hence can be understood based on properties of the elements. Another word used by Bernstein, "dynamics," sounds very modern. Bernstein actually implied under this term what we now call "plasticity," that is, an ability of neural connections to emerge, disappear, and modify their strength.

Let us dwell a bit on the second of the three quotations. This statement is anything but obsolete.

On the one hand, many studies have investigated patterns of neural activity and intensity of metabolic processes in cortical zones including the motor cortex and tried to relate those patterns to external actions by the animal. Some of those studies showed striking results. I would like to mention here the remarkable experiments pioneered by the group of Apostolos Georgopoulos (Georgopoulos et al. 1982, 1986, 1989; Georgopoulos 1986). In those studies, monkeys were trained to move the right arm in different directions of the workspace. Many electrodes were implanted into the projection of the arm in the primary motor area of the cortex; the electrodes were used to record the activity of many cortical neurons. Each neuron showed a baseline level of activity, which could go up or down depending on the direction of the arm movement. This modulation

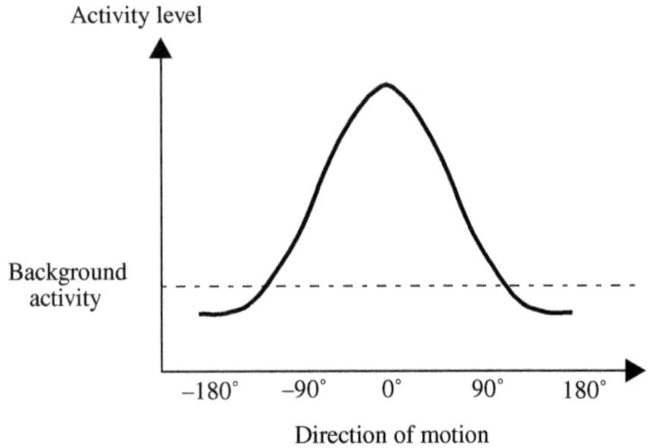

Figure 7.1. A schematic illustration of a cosine-tuning of the activation level of a neuron in a brain structure, for example, in the primary motor cortex. The *y*-axis shows the number of action potentials generated by the neuron within a fixed-time window about the time of limb motion in a particular direction (shown on the *x*-axis). The neuron is assumed to show a background level of activity (the dashed horizontal line). The neuron has a preferred direction (shown as 0°), which corresponds to the highest activity level.

followed a cosine function, that is, it had a peak for a particular direction (*preferred direction* of that neuron), and it decreased with angular deviation from the preferred direction (Figure 7.1). When preferred directions of a large number of neurons were identified, the activity of the whole ensemble of neurons was analyzed for a single movement in the following way. Each neuron was assigned a vector of a unitary length pointing in its preferred direction. This vector was multiplied by a coefficient equal to the total number of action potentials that neuron generated during a brief time interval about the movement initiation. When all these vectors were summed up, they pointed in the movement direction (Figure 7.2). I will get to these exciting studies and their follow-ups a bit later. For now, let me only mention that the described experimental approach was unique in providing a link between modulation of activity within a large *neuronal population* and natural movements produced by the animal.

On the other hand, most evidence shows that motor cortex is not where a decision to move occurs (reviewed in Kandel et al. 1999; Zigmond et al. 1999). It is also not the ultimate generator of commands to muscles. In the neural network parlance, cortical neurons are likely to form a hidden layer of a hypothetical network responsible for the control of a motor action. Figure 7.3 illustrates a very simple neural network with the task formulated at the input layer, the action being produced by the output layer, and the information being completely garbled at the intermediate hidden layer. So, if one records from neurons in the hidden layer, chances to see information that encodes the task are next to nil.

So, what is localized in brain structures? Let us listen to Bernstein one more time: "As such, an increase in the morphological, localizational separation of brain structures leads to a more intensive development of the foundations for the development in the brain of nonlocalizable processes" (p. 317). So,

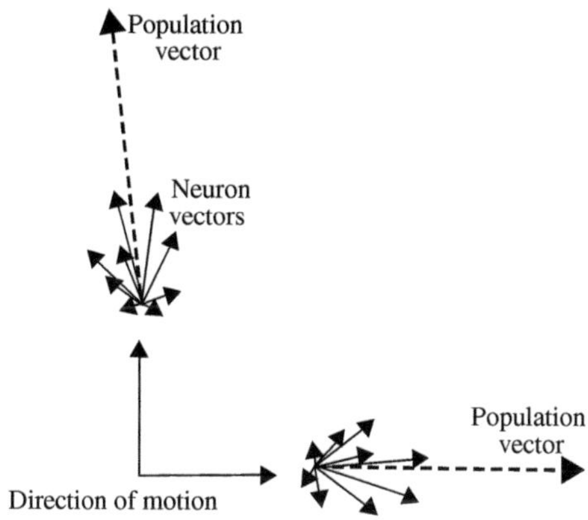

Figure 7.2. The animal moves the limb in one of the two directions. Each neuron in the contralateral area of the primary motor cortex is assigned a unitary vector in its preferred direction. This vector is multiplied by a coefficient equal to the number of action potentials the neuron generates about the time of the movement initiation. When all the vectors are summed up, the resulting population vector (the dashed lines) points close to the movement direction.

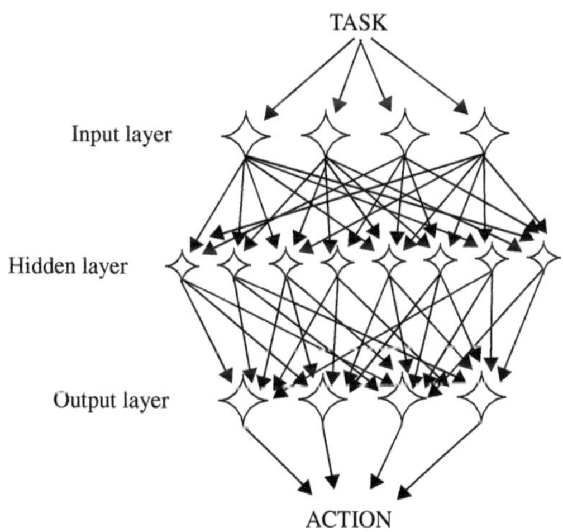

Figure 7.3. An illustration of a simple neural network. The task signals are projected on a set of neurons in the "input layer." These neurons project in a rather chaotic way on neurons in the "hidden layer." Those neurons project on a set of neurons in the output layer that define the action. Gains of all the shown projections can be modified, when the network learns how to project a task input onto a desired action.

in higher animals, certainly including the humans, the brain is supposed to contain nonlocalizable processes. In a later paper, Bernstein with the help of two of his younger colleagues (Bassin et al. 1966; Latash LP et al. 1999, 2000) elaborated on this issue. They suggested, in particular, that no function could be localized in the brain, but operators shared by different functions could.

For example, there may be localizable neural circuits that perform mathematical or logical operations, but these circuits are not prewired parts of a larger circuit that performs a function that may require these operations, for example estimation of movement velocity, or elapsed time, or chance of being hit by a car. The circuits may be shared by many such functions such that no single function can be associated with a particular neural structure.

Currently, I am unable to answer the question formulated in the title of this digression. Maybe, some of my colleagues can. Those beautiful bright red spots that make papers reporting results of functional magnetic resonance imaging (fMRI) studies so attractive beg for a functional interpretation. However, this interpretation cannot be simplistic; the time has come to abandon the cockroach-logic, read old books, and try to get new ideas.

End of Digression #11

After this pessimistic introduction, I nevertheless plan to discuss the potential role of different neural structures in synergies. Synergies are rather abstract neural organizations that are likely to be used in many functions, motor and nonmotor. As such, they or their parts may well qualify for being examples of Bernstein's operators. If this is true, attempts at finding neurophysiological mechanisms of synergies may not be completely futile.

To summarize the rest of this section, let me admit upfront that neurophysiological mechanisms underlying synergies remain basically unknown. On the one hand, many studies focused on such conspicuous brain structures as the cerebellum and the motor cortex as probable sites of synergy formation (Houk and Gibson 1987; Bloedel 1992; Lemon et al. 1998; Schieber 2001; Thach and Bastian 2003). On the other hand, a variety of coordinated multi-joint synergies have been observed in spinal animals (Fukson et al. 1980; Berkinblit et al. 1986a; Bizzi et al. 1995; Field and Stein 1997). Plastic changes with extended practice and with adaptation to an injury have been documented for many structures within the central nervous system (CNS) ranging from the motor cortex to the segmental spinal apparatus (Jenkins and Merzenich 1987; Cohen et al. 1991a; Wolpaw and Carp 1993; Hallett 2001; Nudo et al. 2001; Wolpaw and Tennissen 2001; Meunier et al. 2007). These observations suggest strongly that the neural substrate of synergies can be distributed among many structures within the CNS.

7.2 SYNERGIES IN THE SPINAL CORD

The spinal cord is sometimes viewed as a relatively simple and even somewhat dimwitted servant of the brain, a servant that takes care of simple reflexes and cannot do much on its own. I am not going to discuss a deeply philosophical and probably meaningless issue of whether neural structures of the spinal cord possess mind, will, or intelligence since none of these notions is defined. This book is primarily about movements, and the spinal cord is anything but dimwitted when it comes to movements. There is ample evidence that some of the movements that are vitally important for survival, for example locomotion, are controlled primarily by the spinal cord. Sir Charles Sherrington (see section 2.4) was a big fan of muscle reflexes and the spinal cord. He considered spinal muscle reflexes as building blocks for the whole repertoire of movements (for an update see Nichols 1994), and viewed the role of the CNS as integrative—turning the building blocks into functional structures.

At this point, it is very tempting to present another quotation from the same Bernstein's book (2003): "Integration and unification may be necessary only for something that is not integral by itself, which is obviously wrong with respect to the nervous system and its functions. Maybe, there is an integrative function inherent to the nervous system, but only as the most ancient, elementary function...that may be leading only at the earliest stages of the evolutionary process. Where can one observe this function? Only in a decapitated animal, in its orphan spinal system...Integrative function of the nervous system reflects either its deep pathology or the pre-Cambrian antiquity..." (p. 317). And Bernstein concludes with a piercing coup de grace: "Hence, the functioning of the contemporary nervous system of a highly developed vertebrate is not in integration but in a struggle against the prehistoric integration" (p. 318).

Let us leave it up to the two great minds of the twentieth century to argue whether *integration* is a good or bad term to describe the CSN. Likely, Bernstein and Sherrington used the same word in different meanings. Sherrington used the word *integration* as an antireductionist statement, while Bernstein seemed to imply a more direct mathematical meaning of this word, that is a fixed procedure of producing a common result from a set of input signals. Most likely, Sherrington had never read Bernstein's papers. Otherwise, in addition to the now famous Pavlov–Bernstein controversy (Meijer 2002), we would have witnessed a Sherrington–Bernstein one.

There is plenty of evidence that the spinal cord is not only able to produce neural patterns forming the basis of locomotion (Shik et al. 1966, 1967; Grillner 1975; Orlovsky et al. 1999) but that it can display all the basic features of motor synergies, sharing, error compensation (leading to the flexibility/stability property), and task specificity. Let me first consider the classical wiping reflex that was studied in the nineteenth century in the spinal frogs. I already mentioned some of the striking features of this seemingly very simple action in section 3.3. Recall that a frog with the spinal cord surgically separated from the brain (a small transversal cut is typically made in the upper portion of the spinal cord), sits quietly in the absence of external stimuli to the body. If a small piece of paper soaked in a weak acid solution is placed on the body, the frog makes a quick, nicely coordinated action of the hindlimb and wipes the stimulus off the skin, commonly with a series of repetitive wiping actions (Fukson et al. 1980; Berkinblit et al. 1986a). To be more exact, the action is quick only at high ambient temperature and it becomes rather slow if the frog is cooled down. A frog taken from the refrigerator shows very similar wiping actions but performed as if they were filmed in slow motion.

A typical wiping action has several distinct phases that follow each other in a fixed sequence. Each action starts with a complex hindlimb motion that brings the foot close to the site of the stimulus (placing). It is followed by an adjustment (aiming) that defines the exact location of the toes with respect to the stimulus and the angle at which the toes will approach the stimulus (the attack angle). Then, a wiping action occurs that turns into an extension of all the major hindlimb joints throwing the stimulus far away from the body (see Figure 2.10 in Part 2). This typical pattern is observed when the stimulus can be reached by the toes given the anatomy of the frog's body. If a stimulus is placed close to the caudal (tail) portion of the back, the toes cannot reach it, and the whole action changes leading to wiping the stimulus off with more proximal portions of the hindlimb. The aiming phase is obviously very important for the success of the action. It typically leads to changes in the limb joint configuration that adjust the orientation of the foot with respect to the target with minimal changes in the endpoint location. The first and last phases of the wiping reflex look like rather crude components of a stereotypical action. In contrast, the aiming phase looks like a finely tuned motor adjustment, for which I cannot find a better word than *intelligent*.

Apparently, the frog's spinal cord generates the wiping actions without any help from the brain because the two portions of the CNS cannot exchange signals (actually, the first observations of the wiping reflex in the nineteenth century by Pflüger were made in decapitated frogs whose brains were very far away from the spinal cords). I would like to emphasize a number of features of these actions that allow them to be viewed as controlled by synergies. Each action represents a complex, multi-joint, multi-muscle action. For a single location of a stimulus, actions within a series of wiping movements are quite different. In particular, the toes that wipe the stimulus off the skin perform the quick wiping motion in different directions, with different attack angles. Qualitatively different patterns of the wiping action are observed depending on the exact location of the stimulus on the body or the forelimb. If one of the joints of the hindlimb is blocked (with a lose loop or with a splint), other joints modify their action such that the toes get to the target area and perform accurate wiping movements (reviewed in Latash 1993). Anecdotally (these observations have never been published), if one of the major hindlimb muscles (the knee extensor) is stimulated electrically at the initiation of the wiping action leading to a sudden knee extension, the whole pattern of the movement is adjusted and the frog is still able to reach the target and perform the wiping.

Taken together, the observations fit the famous Bernstein's expression: "repetition without repetition." In different wiping actions and in different conditions, the frog repeats performing the task of wiping the stimulus off its body with nonrepetitive, flexible motor patterns. So, the frog's spinal cord seems to be able to organize complex motor synergies.

There have been several attempts to explore the neurophysiological basis of spinal synergies involved in frog hindlimb movements. In a series of studies, a group from MIT led by Emilio Bizzi explored the effects of electrical stimulation applied directly to structures within the spinal cord on hindlimb endpoint action (Giszter et al. 1993; Mussa-Ivaldi et al. 1994; Bizzi et al. 1995). Stimulation at a mid-thoracic level led in most cases to a hindlimb motion to a new position or, less frequently, to an anatomically extreme posture. At a first glance, these results are readily compatible with the equilibrium-point hypothesis and the idea of multi-joint synergies. However, the results are not that easy to interpret because of a number of factors. First, the exact structures that were subjected to stimulation were unknown. Second, it was not clear whether stimulation at "the same point" could produce in successive trials different joint configuration changes leading to a relatively reproducible endpoint motion. In other words, the stimulation could produce either a multi-joint synergy stabilizing the endpoint trajectory (or endpoint force vector, see the next paragraph) or a stereotypical movement in all the joints due to changes in activation levels of alpha-motoneuronal pools and/or reflex loop gains (by changing the excitability of the corresponding interneurons).

The same group studied the effects of spinal cord stimulation on the force produced at the endpoint of the hindlimb when the limb was prevented from moving. In different trials, the hindlimb could be positioned in different points of the accessible workspace. These experiments resulted in force maps, that is families of force vectors produced by a standard stimulation applied when the limb was held at different spatial locations. All the force fields could be classified as belonging to three groups, converging to a point in space, forming a circular pattern about a point in space, and leading in a certain direction. Further modeling (Mussa-Ivaldi and Giszter 1992) has shown that the set of three groups of force fields is sufficient to generate any arbitrary force field that may be necessary to produce any endpoint action. The force fields produced by electrical stimulation of the frog spinal cord could show effects of linear superposition (Mussa-Ivaldi et al. 1994; Mussa-Ivaldi and Bizzi 2000). Figure 7.4 shows that the force fields and the resulting equilibrium trajectories of the hindlimb endpoint produced by the stimulation at two locations in the spinal cord (A and B) show effects of summation when the two stimulations are applied simultaneously (compare the panels labeled "&" and "+" in Figure 7.4).

These studies culminated in an idea of the spinal cord containing *motor primitives*, that is, neural structures producing relatively simple blocks for

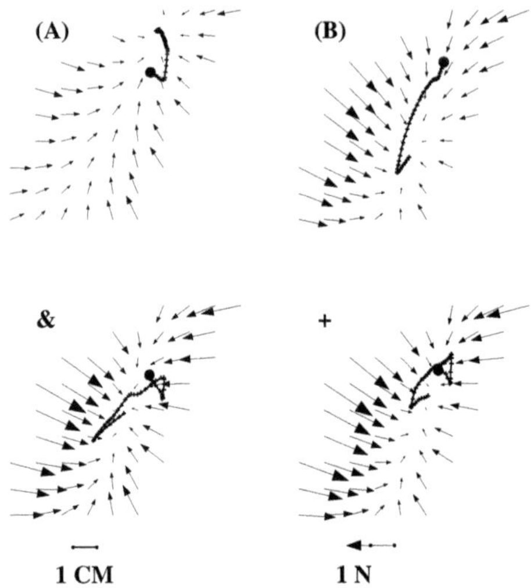

Figure 7.4. Panels (A) and (B) show force vector fields recorded during the stimulation of the spinal cord of the spinal frog at two different locations. Panel "&" shows the results of simultaneous stimulation at both locations, while panel "+" shows the results of summation of the force fields shown in (A) and (B). Reproduced by permission from Mussa-Ivaldi FA, Giszter SF, Bizzi E (1994) Linear combinations of primitives in vertebrate motor control. *Proceedings of the National Academy of Science USA* 91: 7534–7538. © 1994, National Academy of Science, United States.

complex actions. Any action was supposed to be built of a number of motor primitives recruited with corresponding scaling coefficients. Are motor primitives synergies or modes in our current parlance? In general, they can be both when considered at different levels of a control hierarchy. A motor primitive may be a synergy of individual muscle actions, while a complex, multi-joint action may be built as a synergy of primitives. Unfortunately, the authors of those studies were less interested in the problem of motor redundancy and never asked questions that could shed more light on the issue of spinal synergies. For example, what does happen with the stimulation-induced force field if the same endpoint location of the hindlimb is achieved with different joint configurations? Can producing a single motor primitive in different trials be associated with variable muscle involvement? Can the same endpoint motion (or force production) be produced with variable combinations of motor primitives? The lack of answers to these questions makes it rather hard to interpret these intriguing observations within the framework of motor synergies.

All the studies mentioned so far were performed on frogs, animals that may be viewed only as rather remote relatives of humans. What about the spinal cord of higher animals, in particular the human spinal cord? Does it contain synergies? In a sense, this question is rhetorical. Since synergies have been introduced as a major distinctive feature of biological systems, certainly this notion is applicable to the neuronal apparatus of the spinal cord. A more appropriate question would be: What kind of motor synergies can be organized by the spinal cord?

A study with microstimulation of the spinal cord at a cervical level was performed on monkeys (Moritz et al. 2007). Typically, many muscles were coactivated at threshold currents needed to evoke movements including agonist–antagonist pairs acting at forelimb joints. These findings may be related to the idea of muscle modes (see section 4.2.2).

Locomotion is a proverbial synergy whose basic patterns are organized at the spinal level. For thousands of years, people have known that a beheaded chicken can run. In the twentieth century, the spinal origins of locomotion in mammals have been proven with more scientifically acceptable means. Many of these studies were preformed on the so-called *chronic spinal animals*, that is, animals who recovered after the surgery that had separated the spinal cord from the brain. Such an animal does not display locomotion in the absence of external stimuli or special drugs. However, if the paws of a spinal cat are placed on a treadmill, movement of the treadmill at a constant speed can induce stepping of the limbs with a coordinated motion of individual limbs typical of a gait. Changing the speed of the treadmill leads to changes in the stepping frequency, and when the speed reaches a certain threshold, the gait changes from walking to trotting and to galloping. Stepping cycles can also be observed in response to certain chemicals, for example DOPA, which is better known as the main component of several drugs often used in therapy of Parkinson's disease. On the other hand, similar locomotor movements of the hindlimbs can also be observed in a spinalized animal in which all the dorsal roots of the spinal cord innervating the hindlimbs were cut, that is, without any afferent inflow (Goldberger 1977; Atsuta et al. 1991).

These observations prove that the spinal cord is capable of producing locomotor-like activity even when it receives no sensory feedback from the stepping limbs. Hypothetical neural structures in the spinal cord that are responsible for the production of such patterned neural activity are commonly addressed as central pattern generators (CPGs). Further experiments, particularly by a Swedish research group led by Sten Grillner (reviewed in Grillner 1975; Grillner and Wallen 1985), have demonstrated that individual CPGs are likely to exist for each limb. During natural locomotion, all the individual limb CPGs are coordinated so as to produce a coherent interlimb pattern.

Until relatively recently, the existence of spinal CPGs for locomotion in humans (and nonhuman primates) has been a matter of debate. Attempts to induce locomotion after a complete spinal cord transection using the same means as in experiments on cats and dogs were unsuccessful. Only at the end of the twentieth century, a few studies described locomotor-like patterns in apes and humans following a complete spinal cord injury (Vilensky et al. 1992; Calancie et al. 1994; Shapkov et al. 1995). Along similar lines, high-frequency muscle vibration applied to the leg of a healthy person has been shown to lead in some cases to cyclic, locomotor-like leg movement (Gurfinkel et al. 1998).

Recently, the existence of a spinal locomotor generator at a lower thoracic–upper lumbar level in humans has been suggested based on observations in patients with a severe spinal cord injury (reviewed in Shapkova 2004). In certain conditions, these patients could demonstrate involuntary stepping movements of the legs while they were unable to produce such movements voluntarily. A group of patients with clinically complete spinal cord injury (this means that, at a clinical examination, these patients showed no signs of muscle activity during voluntary attempts at moving the legs and reported no sensations when their legs were touched or moved) were implanted with an array of electrodes directly placed over the back portion of the lower thoracic–upper lumbar spinal cord. The stimulation could induce alternating activity in the muscles of the two legs bringing about motor patterns that resembled those seen during locomotion, walking or running.

Figure 7.5 illustrates such a movement pattern. There are large artifacts of the electrical stimulation and also clear alternating bursts of muscle activity accompanied by cyclic changes in the joint angles. An important feature of the induced cyclic motion was that its frequency could be different from the frequency of the stimulation. This means that the stimulation provided a nonspecific input into spinal neural structures (a CPG?), which then generated a rhythmic pattern. Changing the strength and the frequency of the stimulation could change the pattern of

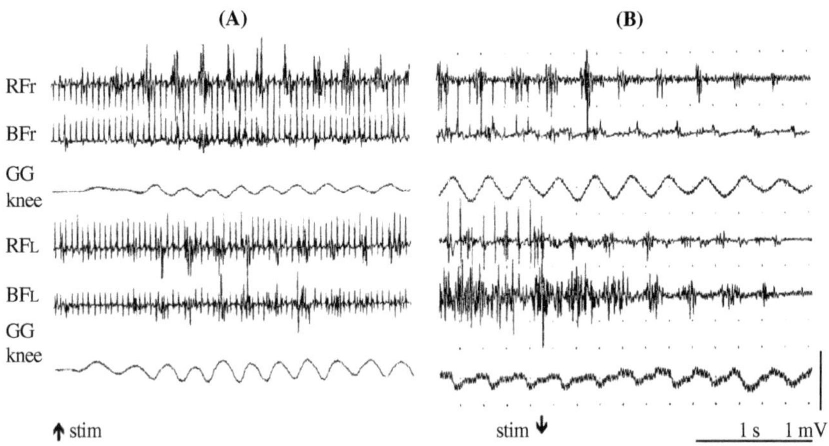

Figure 7.5. An illustration of alternating muscle activity induced by electrical stimulation of the spinal cord in a patient with no voluntary movements in the lower extremities. Note the cyclic motion of the legs at a frequency different from the frequency of the stimulation. Reproduced, by permission, from Shapkova EYu (2004) Spinal locomotor capability revealed by electrical stimulation of the lumbar enlargement in paraplegic patients. In: Latash ML, Levin MF (Eds.) *Progress in Motor Control-3*, pp. 253–290, Human Kinetics: Champaign, IL. © Human Kinetics.

the induced movement similarly to effects described earlier in experiments with stimulation of midbrain structures in cats (Shik et al. 1966, 1967).

Depending on the exact location of the stimulating electrodes, the stimulation could produce different patterns of locomotor-like activity such as rhythmic movement of one leg only or even of one of the major leg joints only. The rhythmicity of such isolated movements suggests that the spinal CPG for locomotion can be based on an hierarchy of CPGs for each extremity and maybe even for each joint. However, what is the relation between the idea of CPG and the notion of synergy? To explore possible synergic interactions between CPGs at a lower level of the hypothetical hierarchy (elements) Elena Shapkova performed experiments with loading or blocking motion in selected joints. If elemental CPGs were indeed united into a synergy stabilizing a particular aspect of performance, one could expect to see effects of such manipulations not only in the joint/limb that is being perturbed but in other joints and limbs.

Here is a summary of the main observations. When the stimulation produced bi-pedal stepping of the two legs, the manual blocking of one leg movement led to a higher amplitude of the stepping motion of the other leg. Sometimes, the two legs could show different frequencies of stepping during the stimulation (typically, one leg stepped at twice the frequency of the other leg). In such a case, blocking movement of one of the legs could completely abolish the stepping of the other leg. When motion of a joint within a leg was blocked, the other joints of the same leg increased the amplitude of their motion. Sometimes, only two homonymous joints (joints of the same name in the two legs) showed rhythmic motion during the stimulation. In such cases, blocking one of the two joints could produce an increase in the amplitude of motion of the contralateral joint and/or rhythmic motion of another joint within the same limb.

Let us take a deep breath and try to think what all these observations may mean. They all seem to point at interdependences among the stimulation-induced motions of elements, joints or limbs. It is very hard to assess what such interdependences (co-variations) could mean with respect to potentially significant physical variables during actual locomotion. In all the mentioned experiments, stepping-like movements were produced in patients who were supine with the legs suspended in the air. However, the mere presence of such relations allows one to suspect that CPGs responsible for single-joint rhythmic motion are organized to stabilize a feature of the whole limb rhythmic motion, while rhythmic motions of the two legs form a synergy stabilizing a feature of the gait (a schematic illustration in presented in Figure 7.6).

I would like to mention one more observation from the same study. When the step cycle was divided into two parts based on the activity of the hip flexor and extensor muscles, an interaction between the episodes of flexor and extensor activity was reflected in indices of variability of the duration of step phases. The coefficient of variation of the duration of the hip flexion and extension phases

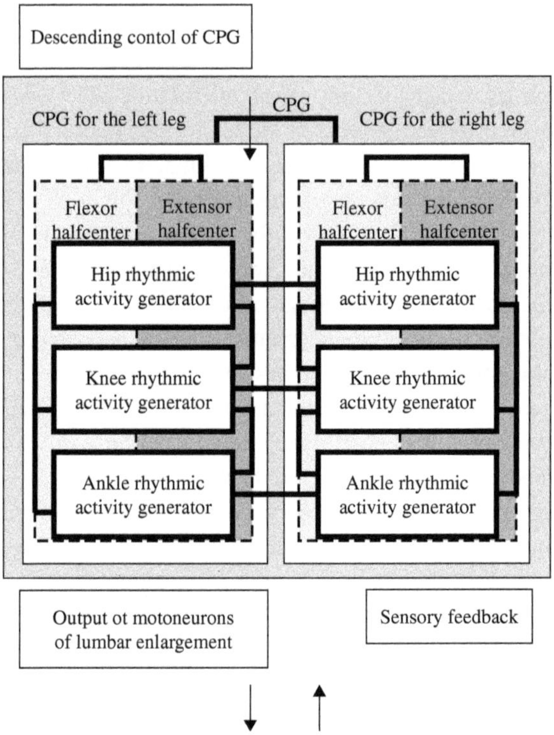

Figure 7.6. A hypothetical scheme of the organization of a central pattern generator (CPG) for locomotion. Reproduced by permission from Shapkova EYu, Latash M L (2005) The organization of central spinal generators in humans. In: Gantchev N. (Ed.) *From Basic Motor Control to Functional Recovery–IV*, pp. 141–149, Marin Drinov Academic Publishing House: Sofia, Bulgaria. © Marin Drinov Academic Publishing House.

computed across successive cycles varied across the patients, but it was always higher than the coefficient of variation of the step duration as a whole. This can be interpreted as a synergy stabilizing the mean step cycle duration by co-varied changes in the duration of the flexion and extension phases.

The activity produced by a spinal CPG is insufficient by itself to ensure meaningful locomotion because of a number of reasons. First, the animal needs to know in what direction to move, that is, it needs signals from visual (or other) receptors that carry information about the environment. Second, locomotion is always intimately tied to the control of posture in the field of gravity. In all the experiments with locomotion of spinal mammals, the animals were suspended with a system of belts so that they did not have to support their own weight or worry about losing balance. Third, normal locomotion is always associated with perturbations (e.g. stepping on an uneven area of the surface) that may require urgent corrections.

Studies of quick reaction to perturbations delivered in different phases of locomotor cycle have provided evidence for flexible combinations of muscle reactions that seem to qualify as synergies according to our definition. In particular, if a mechanical or electrical stimulus is applied to a paw of a walking cat, reactions to this stimulus are dramatically different depending on whether the paw is in contact with the ground or in the air (Duysens and Pearson 1976; Forssberg et al. 1975, 1977). When the stimulus is applied in the stance phase, it produces a quick muscle response in extensor muscles leading to an accelerated step. If the same stimulus is applied in the swing phase, it leads to a similarly quick muscle reaction in flexor muscles leading to a movement that resembles stepping over an obstacle.

Similar responses and associated patterns of changes in muscle activation in human subjects have been described as functionally relevant muscle synergies (van der Linden et al. 2007). The word *synergy* in that study was used in a meaning of synchronized, reproducible patterns of muscle activations—different from the meaning accepted in this book. Recently, a study of persons with spinal cord injury has suggested that neuronal circuits within the spinal cord deprived of normal supraspinal input respond to perturbations applied in the swing phase in a manner that is similar to that of the intact spinal cord (Field-Fote and Dietz 2007). It is not easy to single out a performance variable that could be stabilized by these reactions. It may represent a combination of avoiding pain and maintaining both balance and body progression.

I have already discussed some of the problems related to using the term *reflex* with respect to such corrective actions (Digression #6), even though they come at a relatively short latency. The switching of muscle responses between antagonist groups have been referred to as reflex reversal (reviewed in Latash 1993). Reversals of ipsilateral (Lisin et al. 1973; Duysens and Pearson 1976; Forssberg et al. 1976, 1977), contralateral (Duysens and Loeb 1980; Duysens et al. 1980), and propriospinal (Halbertsma et al. 1976; Miller and van der Meche 1976; Miller et al. 1977) responses to stimulation have been reported in cats and dogs as well as in humans (Lisin et al. 1973; Latash and Gurfinkel 1976). Some of these studies were performed on spinalized animals suggesting that the spinal cord contains the neural apparatus capable of producing the described phenomenon of reflex reversal.

I would like to end this section with two examples of spinal synergies that may represent basic synergies shared by many motor tasks. One of them was mentioned earlier in section 3.4. It refers to the equilibrium-point hypothesis and the tonic stretch reflex as a feedback mechanism ensuring a synergy of motor units that tries to stabilize a certain level of muscle activity. To remind, the tonic stretch reflex is a spinal mechanism that leads to changes in the activity of an alpha-motoneuronal pool induced by changes in the activity of length- and velocity-sensitive receptors located in spindles within the muscle innervated by this pool (Figure 7.7). Imagine that a number of motoneurons are generating action potentials at their preferred frequencies leading to a certain level of muscle activation. One can say that muscle

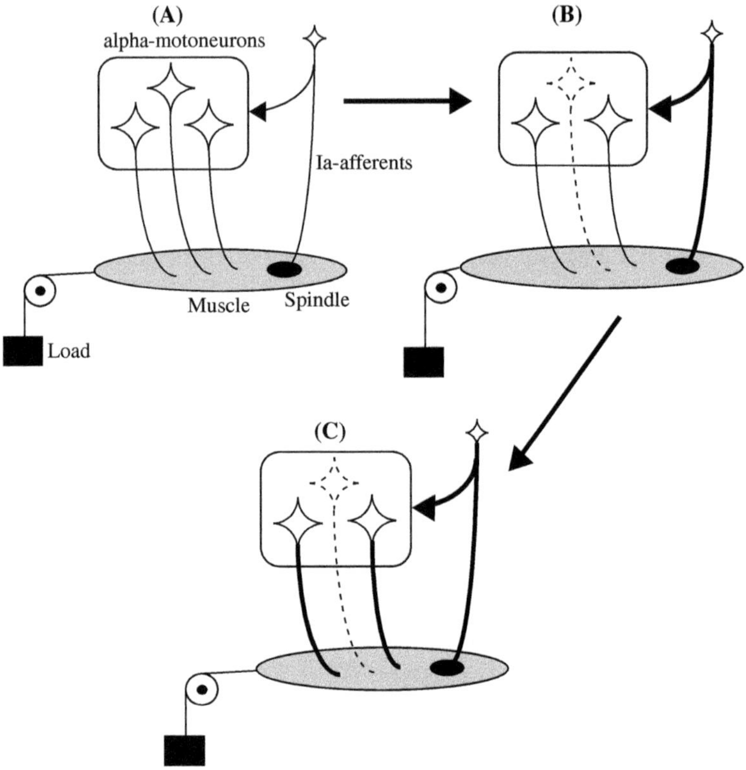

Figure 7.7. (A) A muscle is in an equilibrium with an external load, which corresponds to a certain level of activity of its alpha-motoneurons. (B) If one alpha-motoneuron stops firing (the one shown with dashed lines), the muscle will generate less force and stretch under the action of the load. (C) This will lead to an increased activity in the stretch–reflex loop, resulting in higher activation of other alpha-motoneurons, partly compensating for the mechanical effects of the original cause.

activation is shared in a certain way among the motoneurons. The muscle is in an equilibrium against a certain external load (panel A). The central command to the muscle stays unchanged meaning, in the framework of the equilibrium-point hypothesis, that there is a fixed relation between muscle actively produced force and length, mediated by the tonic stretch reflex.

Imagine now that, for an unknown reason (a perturbation), one of the motoneurons stops producing action potentials (the one shown with dashed lines in panel B). This would lead to a drop in the muscle activity and a drop in the muscle force. Since the external force does not change, a drop in the muscle force would lead to the muscle being stretched. The stretch is expected to lead to an increase in the activity of muscle spindle receptors (thick lines) and an increase in the excitation the alpha-motoneuronal pool receives via the tonic stretch reflex loop.

This additional excitation will be spread over the whole motoneuronal pool. It can lead to an increase in the frequency of already recruited motoneurons and/or to recruitment of additional motor units (thick lines in panel C). Muscle activation level will increase compensating (partly) for the effects of the original perturbation. Hence, the tonic stretch reflex mechanism compensated, at least partly, for the effects of the original error.

For the motoneurponal pool, the muscle it innervates, its sensory receptors, and the tonic stretch reflex arc to qualify as a synergy, one more requirement should be met. The strength of projections onto alpha-motoneurons via the tonic stretch reflex loop should be modifiable such that the synergy could be adjusted in a task-specific way. This is indeed true. There is experimental evidence that descending signals from brain structures are able to modify the sensitivity of many interneurons in the spinal cord (Feldman and Orlovsky 1972) that are likely to contribute to the tonic stretch reflex. There is also direct evidence that the gain of the tonic stretch reflex can be changed (Nichols and Steeves 1986).

The other example is dealing with an even better described system in the spinal cord (the loop of the tonic stretch reflex is unknown), namely the system of Renshaw cells. Figure 7.8 illustrates the role of Renshaw cells in the spinal cord. These small, inhibitory neurons are located in the ventral part of the spinal cord, close to the alpha-motoneurons. They receive excitatory input from alpha-motoneurons and project back to the motoneurons of the same pool (and some other pools). This well-known system provides for recurrent inhibition of alpha-motoneurons; it has recently become incorporated into several hypotheses on the control of movement (van Heijst et al. 1998; Uchiyama et al. 2003).

Imagine that a pool of alpha-motoneurons produces a certain level of muscle activity (panel A in Figure 7.8). Imagine now that one motoneuron for an unknown reason stops generating action potentials (panel B). The input Renshaw cells receive from this motoneurons would drop. Those Renshaw cells will decrease their activity and, since they are inhibitory, this will lead to a net increase in

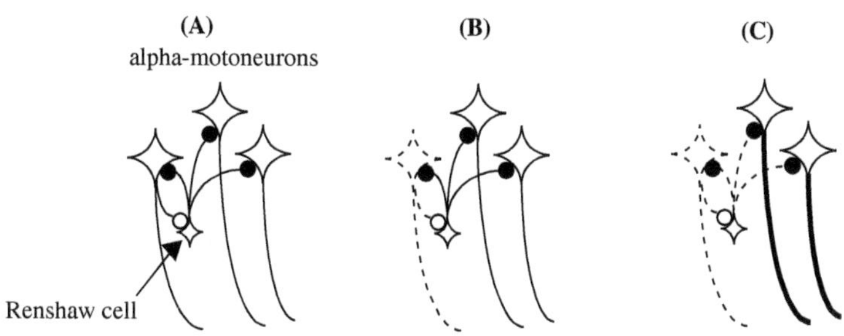

Figure 7.8. (A) Renshaw cells are excited by alpha-motoneurons (open circle) and inhibit all the motoneurons of the pool (closed circles). (B) If one of the motoneurons stops firing (dashed lines), the Renshaw cell becomes less active. (C) As a result, other motoneurons are disinhibited and show an increase in their activity (thick lines).

the amount of excitation all the motoneurons of the pool receive (panel C). This will lead to an increase in the activity of the pool partly compensating for the original "error"—the drop in the activity due to switching off one of the active motoneurons.

Such organization favors a role of the Renshaw cells in stabilizing the output of a motoneuronal pool in a way that could be muscle and task specific (Hultborn et al. 2004). Note that there is substantial variability in the organization and the strength of inhibitory projections mediated by Renshaw cells in different muscles (Katz et al. 1993). These projections can be modulated pharmacologically and by descending projections (Mattei et al. 2003; Hultborn et al. 2004).

To summarize, the spinal cord is likely to contain many neural structures that play the role of synergies. These structures may be relatively simple, like the system of recurrent inhibition, or completely unknown, like the CPGs involved in locomotion. Let me now get to a structure that has been the favorite part of the brain where researchers placed synergies—the cerebellum.

7.3 SYNERGIES AND THE CEREBELLUM

The cerebellum has been a favorite structure for a variety of models addressing issues ranging from memory to motor coordination. As compared to other, even much smaller and less conspicuous brain structures, at the first, superficial look, the cerebellum looks like a network whose function can be understood relatively easily. It contains only five types of neurons. It receives two very different types of input signals. Its output is produced by only one type of neurons, Purkinje cells. These are very large inhibitory neurons that project on cerebellar nuclei that, in turn, project to a variety of both more caudal (e.g. the spinal cord) and more rostral (e.g. the thalamus and the cortex of the large hemispheres) structures within the CNS.

The two inputs into the cerebellum are provided by the so-called *mossy fibers* and *climbing fibers* (Figure 7.9). The mossy fibers carry signals from many different structures within the CNS. They project onto small granule cells (the number of these neurons is estimated to be larger than the number of all other neurons within the CNS). The granule cells send their axons to the surface of the cerebellum and there they form a network of long (by neural standards) parallel fibers. These fibers form relatively weak synapses on the huge dendritic trees of the Purkinje cells. A single Purkinje cell can receive about 200,000 synaptic inputs from parallel fibers (Figure 7.9 fails to show all the 200,000 synapses). This design favors nonlocal, fuzzy effects of the mossy fibers on the cerebellar output.

The other major input is formed by the climbing fibers that are very different from mossy fibers. They originate from two paired (left and right) structures in

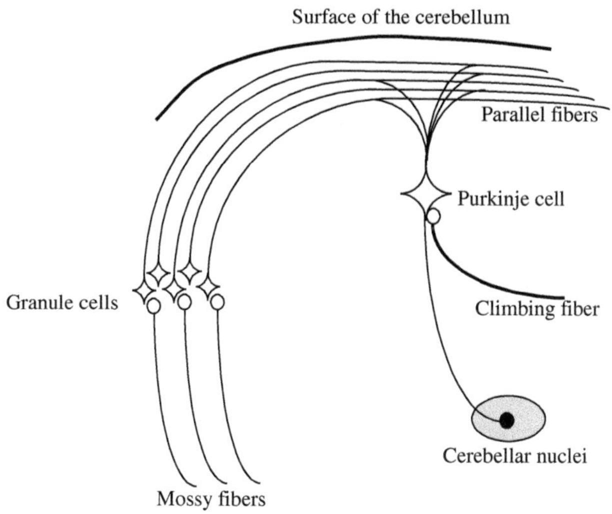

Figure 7.9. Two inputs into the cerebellum are delivered by the system of mossy fibers and climbing fibers. The mossy fibers excite the granule cells, whose axons form the system of parallel fibers. The climbing fibers terminate on Purkinje cells, which are the output cells of the cerebellum. The Purkinje cells project on the cerebellar nuclei.

the brain stem, the inferior olives. These fibers project directly onto the Purkinje cells, and their synapses are very powerful (Figure 7.9). Actually, a single climbing fiber can make a target Purkinje cell produce an action potential in response to a single incoming action potential. The climbing fibers are also responsible for rather unusual, complex spikes (action potentials of a longer duration and more complex shape) whose function remains mysterious.

The cerebellum is part of two major loops involving other brain structures (Figure 7.10). One of these loops involves the thalamus and cortical structures of the large hemispheres: The output fibers of the cerebellar dentate nuclei (the cerebellar nuclei are paired symmetrical structures, hence the plural) project onto the ventro-lateral part of the thalamus, that makes projections onto a variety of cortical structures including the motor cortex (for a detailed, updated report see Dum and Strick 2003). The cortical output is being fed back to the cerebellum via nuclei in the pons through the system of mossy fibers. The other loop is mediated by the interposed cerebellar nuclei; it passes through the red nucleus (the origin of the rubrospinal tract, a very fast-conducting tract with a not very well understood role in motor control) and the inferior olives, and is being projected back to the cerebellum via the system of climbing fibers. A third major output of the cerebellum mediated by the fastigial nuclei is directed at the vestibular nuclei of the brain.

The involvement of major brain structures that are known to contribute significantly to the control of posture and movement, such as the motor cortex, the red

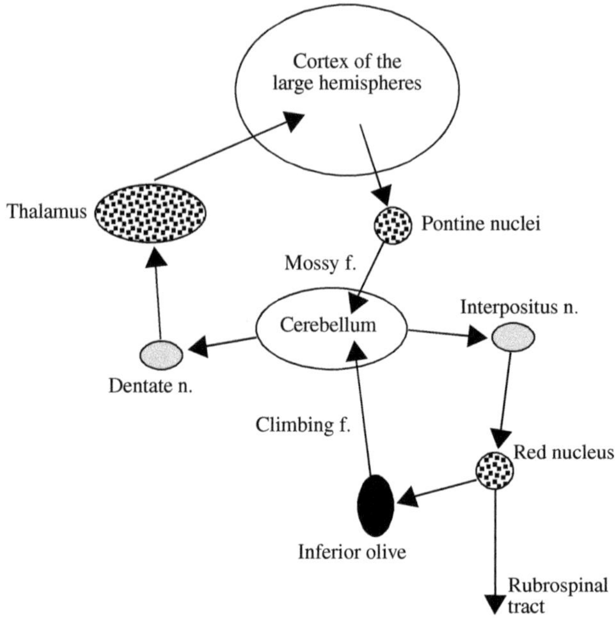

Figure 7.10. The cerebellum is part of two major loops. The first one involves the dentate nucleus (Dentate n.), the thalamus, different areas of the cortex of the large hemispheres, and the pontine nuclei, whose axons contribute to the system of mossy fibers (Mossy f.). The second loop involves the interpositus nucleus (Interpositus n.), the red nucleus, and the inferior olives, whose axons form the system of climbing fibers (Climbing f.).

nucleus, and the vestibular nuclei, suggests that the cerebellum is a "middle-man" (Bloedel 1992) of motor control. What does this middle-man do? (Actually, a more appropriate question would be: Does this middle-man do anything in particular that can be considered meaningful outside its interaction with other brain structures?) There have been several major hypotheses on the role of the human cerebellum in the production of movements. These hypotheses have been mostly based on observations of patients with cerebellar disorders and on changes in metabolic processes within the cerebellum during motor learning and adaptation. Indirect evidence has also been supplied by studies of animals with injuries to the cerebellum.

In particular, observations of timing errors typical of movements of patients with cerebellar disorders (Braitenberg 1967; Beppu et al. 1984; Llinas 1985; Miall 1998; Topka et al. 1998; Ivry 2003; Ivry and Spencer 2004) have led to a suggestion that the cerebellum is a time-keeper, an internal clock. It has also been assumed to generate and contain both inverse and direct internal models based on brain imaging experiments with learning motor tasks in unusual conditions (Bastian et al. 1996; Miall 1998; Kawato 1999; Imamizu et al. 2003). The cerebellum has been assumed to play a major role in motor learning and adaptation based on problems demonstrated by patients with cerebellar injuries (Ojakangas and Ebner 1992; Thach et al. 1992a; Contreras-Vidal et al. 1997; Thach 1998; Bastian et al. 1999). It has also been assumed to participate in more general processes associated with memory acquisition, retention, and retrieval (Marr 1969; Albus 1971; Ito 1989, 2005; De Schutter and Maex 1996; Linden 1996).

However, the most relevant to this book hypothesis on the cerebellar function is probably one of the oldest ones: It was proposed at the end of the nineteenth century by a great French neurologist Joseph Francois Felix Babinski. According to Babinski, the cerebellum plays an important role in controlling movement synergies (Smith 1993). Babinski implied under synergies coordinated actions of large muscle groups and did not worry about an operational definition for this term. So, in our current vocabulary, he could mean either modes or synergies.

Actually, ancient Greeks, many centuries before Babinski, had suggested that the cerebellum was the site of the human soul. I like this hypothesis very much! At least it is honest in using the openly nonscienific term *soul* where later researchers used more acceptably sounding *synergy* or *coordination*—also without a definition. If, for the sake of discussion, we define *soul* as something that is present in biological objects and absent in inanimate nature, it becomes very close to how a definition for *synergy* has been built in the early sections of this book.

A number of recent studies have suggested that the cerebellum does have something to do with the coordination of large, redundant groups of effectors such as muscles, joints, and limbs (Bloedel 1992; Thach et al. 1992a; Houk et al. 1996). A series of studies on monkeys by the group of Thomas Thach have suggested that signals from the dentate nuclei are more closely related to control of muscle synergies rather than prime movers of the explicitly required action (e.g. Thach et al. 1992b). This view has been supported in other studies, also on monkeys, that showed that the electrical stimulation of the dentate nucleus could elicit complex movements of several body segments, although the authors of that study emphasized the stereotypical patterns of the observed activity (Rispal-Padel et al. 1981). These observations suggest that maybe the output of the dentate nucleus defines modes rather than synergies. This tentative conclusion is in line with a view that the inferior olives create conditions that favor contractions of discrete collectives of muscles at specific times during movement (Welsh and Llinas 1997)—a description that fits the earlier introduced definition of muscle modes (see section 4.2) very well.

An idea that ensembles of Purkinje cells can facilitate activation of large muscle groups (whether these are modes or synergies) has been developed by a team from Northwestern University led by James Houk. Earlier versions of this model focused on interactions between the cerebellum and the red nucleus (Houk and Gibson 1987) while later the focus has shifted toward large-scale networks of cerebral cortical areas that are individually regulated by loops through subcortical structures, particularly through the cerebellum (Houk 2005). It has been hypothesized that in the course of learning the cerebellar neural structures learn to modulate activity of Purkinje cells in a predictive fashion, and further, this signal from Purkinje cells is combined with a cerebral cortical signal in the production of natural movements (Miller and Houk 1995; Barto et al. 1999; Miller et al. 2002).

A few recent studies have provided rather direct physiological evidence for a role of the cerebellum in coordinating a variety of multi-joint actions. In one of those studies, principal component analysis of the activity patterns of a large set of neurons within the doral spinocerebellar tract was performed during hindlimb motion simulating walking (Bosco and Poppele 2002). The two principal components that accounted for most of the variance of the neuronal activity were found to be related not to individual joint movements but to the whole limb length and orientation changes during the leg movement cycle, which can be considered important performance variables for locomotion. Thus, the pattern of sensory input from those cells into the cerebellum seems to provide information related to the synergic action of the joint motions rather than to individual joint motions.

Projections of the cerebellar nuclei onto vestibular nuclei make the cerebellum a likely contributor to functional synergies among neck muscles that have been studied by Sugiuchi et al. (2003). These authors describe branching of single long axons in the medial vestibulospinal tract that innervate sets of neck muscles possibly contributing to head movement synergies. This organization may be viewed as a way to decrease the number of variables the controller manipulates while dealing with the complicated system for head stabilization, an organization that defines modes of this system.

Learning a novel movement is commonly associated with creation of novel motor synergies, which may involve elements that have also been involved in other task components. For example learning how to kick a soccer ball accurately requires coordinated activity of leg and trunk muscles, the same muscles that are also involved in postural stabilization. Potentially, creating a novel synergy may be expected to interfere with pre-existing synergies built on the same set of elements. Fortunately, this does not happen too often. For example, in an example of learning a complex four-finger task (Kang et al. 2004; see section 6.5), the emergence of within-a-hand force stabilizing synergies did not destroy the pre-existing synergies stabilizing the total moment of force. The cerebellum may be involved in resolving conflicts among synergies that share elements and are supposed to work in parallel (Bloedel 1992; Thach et al. 1992a; Houk et al. 1996).

Altogether, researchers tend to agree that the cerebellum is a crucial organ for motor coordination across a variety of movement and postural tasks. However, recently, the idea that the cerebellum is a mostly motor organ of the brain has been challenged. In particular, many nonmotor consequences of cerebellar disorders have been described and united under the name cerebellar cognitive affective syndrome (Schmahmann and Sherman 1998; Schmahmann 2004). In particular, patients with cerebellar lesions demonstrate an impairment of such general abilities as planning, abstract reasoning, working memory, and spatial cognition. They have problems copying complex pictures and drawing pictures of familiar objects that are characterized by regular spatial features, for example, the face of a clock. Language deficits in persons with cerebellar disorders include improper use of grammar and unusual voice inflections. Many of these abnormalities can be described as reflecting an inability to put several "things" together properly. The "things" may be parts of a complex picture, words, or muscles. In later sections, I am going to return to a more general understanding of synergies that will go beyond motor control and include such diverse functions as creation of a coherent picture of the world and self based on sensory signals, and language.

7.4 SYNERGIES AND THE BASAL GANGLIA

Another conspicuous structure of the brain that does not send signals directly to muscles or to the spinal cord but is still viewed as crucially important for the control of movements is the basal ganglia. Actually, the basal ganglia is not a single structure but a constellation of paired (right and left) structures that connect to each other and also to other brain structures. When the function of the basal ganglia is discussed, their role in the cortico–basal–thalamic–cortical loop is typically emphasized. Dysfunctions of the transmission in this loop have been implicated in such disorders as Parkinson's disease, Huntington's chorea, ballism, and dystonia.

A very much simplified scheme of connections involving the basal ganglia is shown in Figure 7.11. In this scheme, excitatory connections are shown with solid arrows and open circles, while inhibitory connections are shown with dashed arrows and filled circles. The scheme shows two loops through the basal ganglia, direct and indirect loops, that both lead to inhibitory projections onto thalamic nuclei, which in turn project onto the cortex of the large hemispheres. Gains of both loops are controlled with a neuromediator, dopamine, produced by an area within the substantia nigra—one of the structures of the basal ganglia. The lack of dopamine may be caused by massive death of neurons within the substantia nigra causing dysfunction of the cortico–basal–thalamo–cortical loop that results in excessive inhibition of the thalamus. This leads to poverty of movements, slowness, and difficulty with movement initiation and modification typical of Parkinson's disease (reviewed in Fahn et al. 1998; Paulson and Stern 2004).

Despite the obvious motor problems in patients with a dysfunction of the basal ganglia, their role in motor control is still rather unclear. There seems to be an agreement only on the importance of the basal ganglia for movement initiation (Evarts et al. 1981; Sanes 1985; Stelmach et al. 1986; Cheruel et al. 1994).

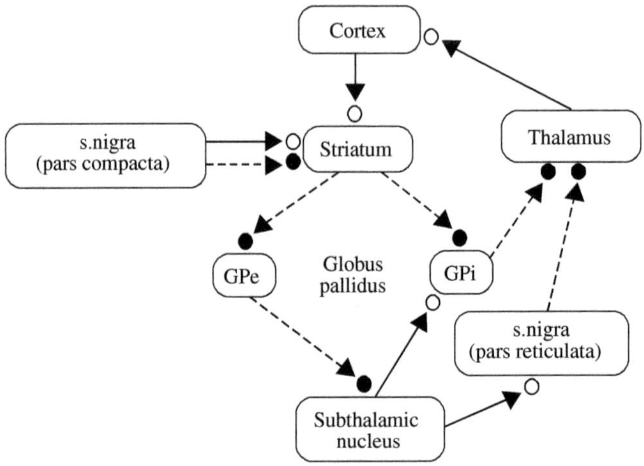

Figure 7.11. The basal ganglia are part of a neural loop involving the cortex of the large hemispheres and the thalamus. This simplified scheme shows two loops, direct and indirect. Solid arrows and open circles show excitatory projections, while the dashed arrows and filled circles show inhibitory projections. S.nigra = substantia nigra; GPe and GPi = external and internal parts of the globus pallidus.

However, attempts to correlate changes in the activity in different parts of the basal ganglia with potentially important characteristics of voluntary movements such as its speed and amplitude have been only marginally successful. So, the basal ganglia seem to be another "middle-man" of the brain structures involved in motor control: They do not define movement parameters but facilitate tasks performed by other structures that may be more directly involved in defining movement characteristics. In particular, the basal ganglia may participate in the creation of motor synergies, either directly or by forming modes, that is groups of potentially independent elemental variables that decrease the number of variables that the other parts of the brain have to manipulate.

Several groups of researcher have explicitly linked activity of the basal ganglia to the formation of synergies or modes. In particular, the basal ganglia have been implicated in uniting the postural and the locomotor synergies into a single functional synergy (Mori 1987) and in the control of multi-joint reaching (Jaeger et al. 1995). Studies of a bi-manual synergy in primates have shown a relation between activity in about one-third of neurons in the striatum and pallidum (anatomical structures within the basal ganglia) to the synergy indices (Wannier et al. 2002). The importance of the basal ganglia for the formation of a grip-lift synergy (scaling of grip force with load force—see section 5.6.3) has also been hypothesized (Forssberg et al. 1999).

There are obvious similarities and even overlaps between hypotheses on the role of the cerebellum and of the basal ganglia in motor control and learning. In particular, similarly to the cerebellum, the basal ganglia have been suggested to be essential for some forms of learning-related neural plasticity. A series of studies by Ann Graybiel and her group have suggested that the basal ganglia may be parts of a brain-wide set of adaptive neural systems contributing to optimal motor and cognitive functions (Graybiel 1995, 1997, 2005). In particular, the striatum has been viewed as a key structure in the learning circuitry of the brain, a major site for adaptive plasticity in cortico-basal circuits, affecting a broad range of behaviors (Blazquez et al. 2002). Plastic changes in the basal ganglia do not necessarily lead to better performance; some of these changes may be maladaptive resulting in disease states involving the basal ganglia (Graybiel 2004).

7.5 SYNERGIES AND THE CORTEX OF THE LARGE HEMISPHERES

Once again, I am going to start with a quotation from Bernstein; this one is from his classical paper published in 1935 in the Archives of Biological Sciences in Russian: "In the higher motor centers of the brain (very probably in the cortex of the large hemispheres) one can find a localized reflection of a projection of the external space in a form in which the subject perceives the external space with motor means" (p. 80 in Bongaardt 2001).

If this statement were known to and paid attention by researchers of the West in the second half of the twentieth century, maybe many arguments about what variables are represented in the cortex would have been avoided. As it happened, however, several groups studied the effects of brief pulses of electrical stimulation applied to different cortical areas, particularly to the primary motor cortex, and ended up with seemingly contradictory reports. Seminal studies by Evarts and Asanuma (Evarts 1968; Asanuma 1973) led to reports that the strength of stimulation encoded the amount of electrical activity seen in a target muscle or the magnitude of force the target limb produced. These reports matched well the dominant view on neurons in the motor cortex as "upper motoneurons," that is neurons that send signals to the spinal cord that define the activation level of the "lower" alpha-motoneurons.

Further studies, however, cast doubt on this simplistic interpretation by showing contributions to muscle activation from sources other than the primary motor cortex (Schieber and Rivlis 2007) and correlations of the strength of stimulation applied to the primary motor cortex with a variety of physical variables such as displacement and/or velocity of the target effector, the rate of force production, and even more complex variables such as joint and limb impedance (reviewed in Schieber 1999; Krakauer and Ghez 2000; Amirikian and Georgopoulos 2003). It is not easy to compare results of different studies because different animals were trained to perform different actions, and external conditions also varied rather broadly.

Please, re-read the opening quotation from Bernstein. It contains an answer to the seeming puzzle of different results reported in different studies. Indeed, if a localized reflection depends on how the subject perceives the external space, one should expect reflections of different variables from the external space if the task and conditions vary.

In human studies, only recently, the introduction of the method of transcranial magnetic stimulation (TMS, see Digression #10) has allowed to explore effects of stimulation applied to the primary motor cortex on mechanical and electrophysiological characteristics of peripheral motor actions. Typical effects of TMS reported across many studies, motor tasks, and effectors have included changes in muscle activation nearly proportional to the background muscle activity (within a reasonable range, that is from about 5% to about 50% of the maximal activation level) and changes in the force produced by an effector in isometric conditions nearly proportional to the background force of the effector (again, within a reasonable range) (Hess et al. 1987; Ravnborg et al. 1991; Kiers et al. 1995; Ugawa et al. 1995; Taylor et al. 1997).

However, when a number of effectors participates in a task, this general rule starts to break down. For example, if only one finger produces force, TMS induces an increase in the force in proportion to the background force level up to about 50% of maximal finger force (Danion et al. 2003a). However, if several fingers produce force simultaneously, there may be little effect of the background force level on the TMS-induced force increments (Latash et al. 2003c).

These observations suggest that neurons in the primary motor cortex are not "upper motoneurons" that define unambiguously what happens with the target alpha-motoneuronal pools. Their signals may lead to nontrivial effects at the level of alpha-motoneurons; in particular, these effects may be task-specific.

7.5.1 TMS and the Equilibrium-Point Hypothesis

Before running any experiments, one should ask himself or herself a question: What kind of variables can be encoded in signals from the motor cortex? An answer will depend on a motor control hypothesis one accepts. I hope that Part 3 of this book has persuaded the reader that currently there is no viable alternative to the equilibrium-point hypothesis. If this is so, let us assume, for the sake of discussion, that stimulation applied to the primary motor cortex results in a shift of the control variable λ (threshold of the tonic stretch reflex, see section 3.4) by a particular value for participating muscles. This change may depend only on the strength of the stimulation or it can be proportional to the background value of λ. The latter possibility sounds more likely because of the following reasons.

For simplicity, let us consider the level of activation of alpha-motoneurons as defined by two factors, descending signals from the brain (reflected in λ) and tonic stretch reflex feedback. Length-sensitive receptors in a shorter muscle show a lower level of activity as compared to a longer muscle. Hence, their contribution to the activation of alpha-motoneurons via the tonic stretch reflex arc is expected to be smaller. To reach the same level of muscle activation, this drop in reflex excitation has to be balanced by a stronger excitatory descending command reflected in a different value of λ. A standard TMS applied over the primary motor cortex may be expected to lead to a response reflecting the overall excitability of cortical neurons under the stimulation coil. Excitability of these neurons is expected to increase with an increase in the level of activity of those neurons. In other words, effects of TMS on alpha-motoneurons (and muscle force) depend on the state of only one of the two major inputs into alpha-motoneurons, the descending input, but not on the reflex input. Hence, the equilibrium-point hypothesis makes a counter-intuitive prediction that a shorter muscle (with a lower stretch reflex contribution) will show a higher TMS-induced response to a standard stimulus as compared to a longer muscle, a prediction that has been confirmed recently (Burtet et al. 2006).

Values of λ should certainly be computed in a reasonable way. Since λ is measured in units of muscle length, let me measure it as the difference between its current value and its "rightmost" value (the longest possible muscle length given the biomechanical constraints). In other words, $\lambda = 0$ means that the muscle is very unlikely to be excited at any length because λ is at the edge of the range of its anatomically possible length changes. Larger values of λ correspond to higher muscle activation at a given length.

As a straw-man alternative, let us also consider what can be expected if one assumes that the TMS-induced change in the activity of cortical neurons encodes a change in the force produced by an effector against an external object or a change in the activation level of a particular muscle.

Figure 7.12 illustrates the whole range of λ values and the smaller mechanically accessible range of muscle length. Imagine now that λ has a certain value (λ_1 in Figure 7.12), and the muscle is at a certain length (L_1) in isometric conditions. A single TMS pulse is expected to shift λ by $\Delta\lambda_1$ for a short time interval, thus leading to a transient increase in both muscle electrical activity and muscle force. If the same muscle has a different background length (L_2) while $\lambda = \lambda_1$, the same stimulus is expected to lead to the same shift in λ, but its effects on force and electromyogram (EMG) are expected to differ. This is due to the nonlinearity of the tonic stretch reflex characteristic. The slope of the characteristic increases with an increase in muscle length. Therefore, a standard shift in λ is expected to lead to larger changes in muscle force and in EMG in a longer muscle ($\Delta F_1 < \Delta F_2$).

Imagine now that the subject in this hypothetical experiment increased the muscle force voluntarily (using λ_2 as compared to λ_1) while the muscle was kept at the same length (isometric conditions). This may be illustrated as a λ shift to the left (Figure 7.13). In this case, if $\Delta\lambda$ is the same, it is still expected to lead to

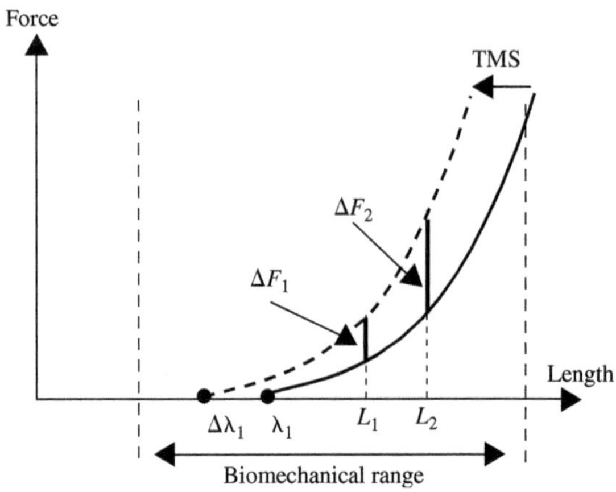

Figure 7.12. If a pulse of transcranial magnetic stimulation (TMS) leads to a shift of the control variable λ, by a certain value (Δλ), changes in the muscle force induced by the TMS may be expected to increase with the background length (and force) because of the nonlinearity of the force–length characteristic.

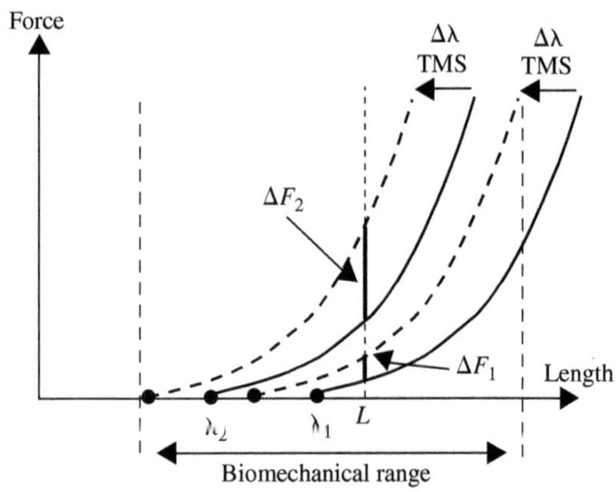

Figure 7.13. In isometric conditions, at a certain length L, commands to the muscle have to be changed (λ_1 and λ_2) to produce different background force levels. A standard shift in λ produced by a transcranial magnetic stimulation (TMS) is expected to lead to a larger force change for the larger background force (compare ΔF_1 and ΔF_2).

a larger increment in force/EMG (compare ΔF_1 and ΔF_2). So far, expected results are the same as in the scenario where the cortical neurons increased muscle activation and force in proportion to the background level of these variables.

Let us now consider the following modification of the experiment. A person is asked to produce a certain background force at a certain muscle length (L_1) in isometric conditions (e.g. to press with a finger on a force sensor, Figure 7.14). A

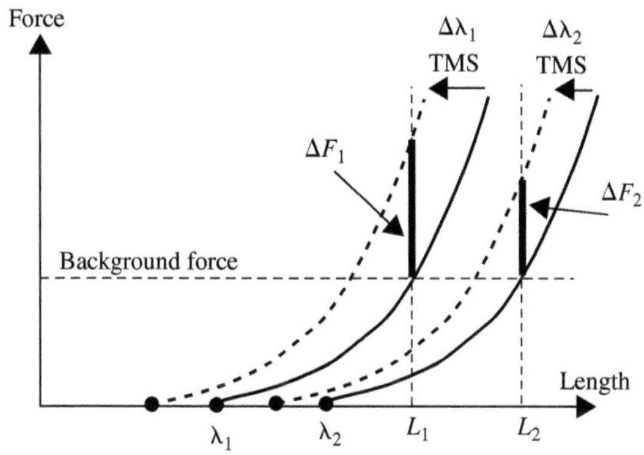

Figure 7.14. If the effects of transcranial magnetic stimulation (TMS) on the control variable λ depend on its background value, a standard stimulus may be expected to produce different shifts in λ ($\Delta\lambda_1$ and $\Delta\lambda_2$), leading to different force increments (ΔF_1 and ΔF_2).

single TMS applied over the primary motor cortex leads to a change in the force of ΔF_1. In several trials, a range of background forces is produced, and a relation between the background force and TMS-induced force increment is defined. Assume, for simplicity, that it is linear and can be adequately characterized with one number, the slope of this relation.

Now let us ask the subject to produce the same range of forces at a different joint position corresponding to shorter flexor muscles. To produce the same force at a shorter muscle length, the subject will have to shift λ to the left (λ_1 in Figure 7.14). If this procedure is repeated several times using different background force levels, another dependence between the background force and TMS-induced force increment will be obtained, $\Delta F(F_{BG})$. If TMS produces a shift in muscle force proportional to the background force, the $\Delta F(F_{BG})$ relationships for the two joint positions should be identical. If, however, as argued before, TMS produces larger λ shifts for larger background values of λ (λ_1 is larger than λ_2 in Figure 7.14 if measured from the point corresponding to the largest muscle length), a stronger response can be expected when the same force is generated voluntarily at a shorter muscle length (as illustrated in Figure 7.14). Note that this prediction goes against expectations based on the known dependence of muscle force on muscle length (see Digression #1)!

There have been a couple of recent attempts to run such experiments using wrist action (Burtet et al. 2006) and finger flexion action (F. Danion, A.G. Feldman, and M.L. Latash, unpublished data). In the latter study, passive changes in joint position were used under the "do-not-intervene-voluntarily" instruction. However, the logic of the study and the predictions remain the same. Both studies resulted in findings of different dependences between the TMS-induced change in force and background force, $\Delta F(F_{BG})$ depending on the joint position. Figure 7.15 illustrates

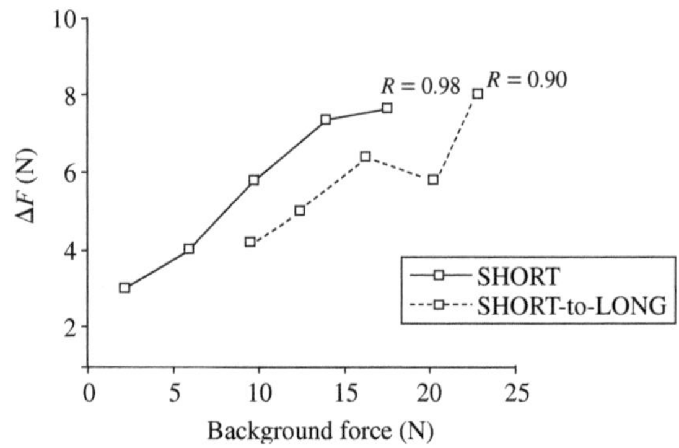

Figure 7.15. In this experiment, a standard transcranial magnetic stimulation (TMS) pulse was applied over the contralateral primary motor area to produce an increase in the finger flexion force. The stimulus was applied at different levels of force. The figure shows the magnitude of the force response to the stimulation for two conditions: SHORT—when the fingers were more flexed and the muscles were shorter and SHORT-to-LONG—when the fingers produced force at a more flexed position and then moved by the experimenter to a more extended position. On average, the background forces increased by this manipulation, but this increase did not lead to an expected increase in the response to the TMS. In the second case, smaller-than-expected responses were observed.

such two dependences from the study of finger flexion: It is clear that the same forces are associated with larger responses to the same standard TMS when the forces are generated at a position corresponding to shorter flexor muscles.

There are two conclusions from these studies. First, the equilibrium-point hypothesis is able to make nontrivial predictions with respect to effects of TMS, and these predictions have been confirmed experimentally. Second, a standard TMS pulse applied over the primary motor cortex is likely to produce a shift of λ that depends on the background level of descending activity (background λ) rather than a standard increment of muscle activity or end-effector force.

7.5.2 Studies of Neuronal Populations

In recent years, studies on the role of different cortical structures have focused more on characteristics of activity of neuronal populations rather than single neurons. Earlier (Digression #9), I mentioned studies of neuronal populations pioneered by the group of Apostolos Georgopoulos (Georgopoulos et al. 1982, 1986; Georgopoulos 1986; reviewed in Amirikian and Georgopoulos 2003). The basic idea of these experiments has been that individual neurons may show substantial variability in their firing patterns when an animal performs the same task repetitively, but that there is enough co-variation in the individual firing levels such that, as a population, the neurons preserve an important, task-specific feature of the cortical activity. This idea is very close to the notion of synergies advocated in this book. Are there multi-neuron synergies stabilizing a feature of the output of the whole population? This is an open question. It has not been investigated rigorously but the circumstantial evidence supporting this idea is impressive.

Most studies of neuronal populations follow a common scheme. First, a potentially important performance variable is identified. Commonly, this variable is an explicit component of the task such as direction of movement of a limb, force vector applied by the limb onto a stationary object (a handle), velocity vector of limb movement, and so on. Then, a large group of neurons are identified whose activity can be recorded reliably; typically, this has been done in animal studies using arrays of implanted electrodes. Further, the relations between the level of activity of each individual neuron within the group and the identified performance variable are computed based on a set of measurements at different values of the variable. Commonly, such relations look like cosine functions: A neuron has a *preferred direction* of the limb displacement or force vector, at which its activity shows maximal increase at action initiation (see Figure 7.1 earlier in this section). Deviations from the preferred direction are associated with a smaller increase in the activity of the neuron. Moving in the opposite direction may actually lead to a drop in the baseline neuron activity.

After defining preferred directions for many neurons within the population, the next step is to record their activity during an action and then process them in the following way. Each neuron is assigned a vector pointing in its preferred direction with the magnitude equal to the number of action potentials the neuron generates within a reasonably selected time window about the action initiation time. Note that this step reduces the amount of information defined at the previous step: The tuning curve for each neuron is substituted with only one number, the preferred direction. When all the individual vectors are summed up, the resultant vector is typically pointing in the direction of the target.

If the studies ended there, these results might not be as surprising and groundbreaking. Indeed, any set of elements that have their activity smoothly tuned with respect to a performance variable and whose preferred directions cover evenly the whole range of possible changes in the performance variable is expected to lead to this outcome. For example, a set of muscle spindle endings in all the muscles within a limb may be expected to show a similar result.

In a follow-up experiment, the task was modified (Georgopoulos et al. 1989). The monkeys were trained to move not in the direction of the target but at a certain angle to this direction. These experiments have been addressed as those involving "mental rotation." Indeed, if a target lights up corresponding to a certain movement direction from the initial position, the monkey has to identify this direction and then compute a new direction that makes the required angle with the vector pointing at the presented target (Figure 7.16). Only then, a movement can be initiated.

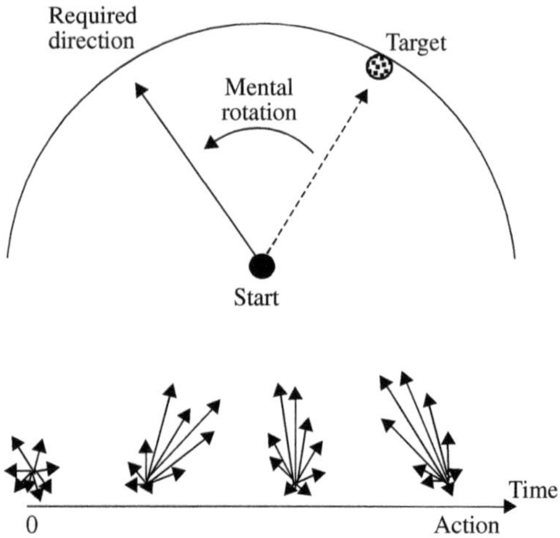

Figure 7.16. An illustration of an experiment with mental rotation. A target is lighted up. The task is to move not in the direction of the target but at a fixed angle. Neuronal population vector first points at the target and then rotates to the ultimate direction of the actual movement (as illustrated by the lower cartoon).

Experiments have shown that, after a target was presented, the neuronal population vector (computed in the same way as described earlier) initially pointed at the target and then rotated toward the required movement direction. After the vector pointed in the required direction (the dashed arrow in Figure 7.16), a movement was initiated. The rotation of the neuronal population vector is a very strong argument that the results of those experiments are indeed nontrivial and reflective of a task-specific change in the combined activity of a population of cortical neurons. In other words, they provide additional evidence for neuronal synergies existing in the cortical motor areas. Very much in line with the opening quotation from Bernstein, studies of cortical neuronal populations have revealed patterns of activity related to various performance variables such as the spatial trajectory of the effector's endpoint, or the force vector applied by an end-effector, or derivatives of these variables (Georgopoulos et al. 1982; Schwartz 1993; Coltz et al. 1999; Cisek and Kalaska 2005; Herter et al. 2007; Wang et al. 2007). A recent study has provided more support for this idea by showing that neurons in the primary motor cortex show activity patterns that can correlate with variables in both muscle activation space and hand-centered space (Morrow et al. 2007). Practice may decide what variables and in what spaces are encoded by neuronal populations in the primary motor cortex (Matsuzaka et al. 2007).

The ability of populations of cortical neurons to show patterns of activation that correlate with physical variables in the external world presents a tempting mechanism to improve impaired motor function in persons with both peripheral and neurological disorders. Indeed, if a population of neurons can reliably encode direction of movement, velocity, force, and other variables relevant to motor behavior (see Jackson et al. 2007), computer analysis of changes in the activity of a large set of neurons may be able to extract these variables and then generate them using a robotic device. Such approaches have been actively developed over the past years (Schwartz 2004; Lebedev and Nicolelis 2006; Wahnoun et al. 2006).

Typical experimental work involved monkeys as subjects (Wessberg and Nicolelis 2004; Wahnoun et al. 2006). For example, a monkey with an implanted set of numerous electrodes in the primary motor cortex could be trained to reach to an object in space (a piece of fruit) and bring it to its mouth. During the many repetitions, neuronal activity is recorded and processed to extract features related to direction of arm movement. Further, the monkey's arms are tied up to the chair while its neuronal activity is recorded and used to drive a robotic arm that moves the piece of fruit in directions encoded by the neuronal activity. Quickly, the monkey realizes that it does not have to activate the arm muscles but simply "think" about moving the fruit to its mouth, and the robot would obediently comply.

More recently, similar approaches have been tried to enable paralyzed persons to control a cursor on the computer screen. There are obvious limitations to such methods of compensation for a lost motor function involving, in particular, implantation of electrodes. However, with the current rate of technological advances, it seems realistic to expect in a near future robotic devices controlled by neuronal population activity of the human brain (for recent reviews see Birbaumer and Cohen 2007; Wolpaw 2007).

Studies of the cortical control of the human hand have resulted in a hypothesis that has direct relevance to the topic of synergies (Schieber 2001; Schieber and Santello 2004). These studies showed, in particular, that maps of cortical representations of the hand were rather disjointed and mosaic. They were characterized by both divergence and convergence. That is, one neuron could project on different muscles and induce motion of different digits, while neurons in different areas could project on one and the same digit. These projections have also demonstrated substantial plasticity with both long-term specialized training, such as playing musical instruments (Pascual-Leone 2001) or reading Braille (Pascual-Leone et al. 1995), and short-term practice (Classen et al. 1998; Latash et al. 2003). To complicate matters, many cortical areas seem to be involved in the control of the hand (Gardner et al. 2007; Suminski et al. 2007), and the neural activity patterns may differ depending on the context—for example in response to loading and unloading perturbations applied to the hand-held object (Ehrsson et al. 2007).

Marc Schieber synthesized these observations and suggested an idea of a *cortical piano* (Schieber 2001; Schieber and Rivlis 2007). This idea implies that individual cortical neurons are similar to piano keys, while functional movements involve "playing chords." In other words, neurons that can be separated anatomically may be united functionally in groups (*chords*) that produce desired motor effects. The cortical piano idea does not explicitly address the issue of motor (neuronal) redundancy. As such, its relevance to the issue of synergies is not obvious. It seems likely that *chords* may be organized in the course of everyday actions and specialized practice. However, are *chords* synergies or modes in our terminology? If the *chords* show high reproducibility across repetitive attempts at a task, they seem more like modes, that is stable groupings of variables (activity levels of individual neurons) that may simplify control but do not necessarily afford flexibility and adaptability as expected from synergies. On the other hand, if *chords* show trial-to-trial variability in their composition, they may indeed be viewed as cortical neuronal synergies.

I would like to mention one more recent study of changes in the excitability of projections from cortical neurons to spinal alpha-motoneurons induced by a change in position of a joint within a multi-joint limb (Ginanneschi et al. 2006). In those studies, excitability of projections to motoneurons innervating forearm muscles was studied. If was shown to be affected by changes in the shoulder joint position that did not affect the length of muscles of the forearm. The authors interpreted the observations as reflecting a global synergy uniting the proximal and distal forearm segments during reaching movements.

One of the major features of synergies is an ability to reach the same result with variable means, that is flexibility. In movement studies, this feature has been addressed as equifinality or motor equivalence (reviewed in Berkinblit et al. 1986a; Wing 2000; Wiesendanger and Serrien 2004). In recent studies, a group led by Michael Graziano applied long-lasting electrical stimulation to cortical areas (including the motor areas) of monkeys and observed rather striking behavioral responses (Graziano et al. 2002, 2005).

Unlike the earlier studies with brief pulses of electrical stimulation to the motor cortex that produced localized, brief motor responses, the studies of Graziano with long-lasting stimulation led to complex movement patterns that resembled those from the everyday repertoire of the monkey. They could look like elements of a defensive reaction (turning the head away and bringing a hand to the head in a defensive gesture) or of a feeding behavior (bringing a hand to the mouth and turning the head towards the hand). If the same stimulation was applied while the monkey was in different initial postures, different movements were observed leading to similar overall outcomes in the relative position of the hand and the head. This flexibility of motor actions leading to highly reproducible outcomes brings to memory the famous expression by Bernstein "repetition without repetition." Did the stimulation induce elements of functional motor synergies? The answer is not obvious.

The idea of applying long-lasting currents, rather than more commonly used brief pulses of stimulation, has been based on an intuitive consideration that during natural actions, changes in the activity of neurons are relatively slow and long-lasting. On the other hand, application of such long-lasting stimuli allows an involvement of virtually all the areas of the brain in the observed behavioral responses making neurophysiological interpretation of such observations difficult and ambiguous.

To summarize, it seems that many (if not all) major neurophysiological structures have an ability to form synergies. Philogenetically older structures, such as the spinal cord, seem to be more limited in their ability to learn new synergies. They seem to ensure that a set of absolutely crucial for survival synergies is always available and reliable; examples may involve the CPG for locomotion and complex reflex projections that may facilitate multi-joint synergies for accurate actions by the end-effector of the limb (Nichols 1994, 2002). Philogenetically younger, more rostral structures within the CNS seem to be more plastic and willing to rearrange neural projections with practice to develop new synergies or optimize existing ones. This seems to make sense in a hierarchically organized system. Such mechanisms as the tonic stretch reflex and recurrent inhibition, discussed as synergies earlier, should be reliable and not change by too much and too fast. Since higher levels of hypothetical control hierarchies rely on properties of lower levels, a gradient in the ability to change and typical rates of such changes may be expected with more flexible synergies seen higher up in the hierarchy.

Part Eight

Models and Beyond Motor Synergies

In recent years, the expression *mechanistic interpretation* has become some kind of a mantra for quite a few researchers in motor control (and other areas of system physiology and psychology). Requests to have a mechanistic interpretation are commonly seen in reviews of manuscripts and grant applications. Most frequently, this expression implies an explicit model, which can be rather simplistic or very sophisticated. For example, models can represent sets of boxes connected by arrows that show the direction of an assumed flow of signals within the scheme, or an explicit set of computational rules that produce a result given initial conditions.

Before expressing my personal opinion on the issue of *mechanistic interpretation*, let me quote the great scientist and philosopher Pavel Florensky. Pavel Alexandrovich Florensky (1882–1937) was trained as a mathematician and an engineer, but later he became best known as a theologian and a philosopher. He was regarded by those who were lucky to know him personally as Leonardo da Vinci of the twentieth century. Florensky was arrested and executed by the NKVD (later transformed into the infamous KGB, and even later—into FSB) in one of the most brutal Stalin campaigns. In one of his published lectures (Florensky 1999), he wrote: "It is clear that the so-called 'mechanistic modeling' is only a method of crude schematization of life, sometimes practically useful, but concealing the reality if it is regarded as something more than a metaphorical scheme" (p. 407). This opinion was shared by Israel Gelfand, who used "modeler" as a derogatory word to describe computationally inclined biologists and biologically inclined mathematicians who adored the equations they wrote but had little understanding of the systems they were trying to model.

There are models and models. Any research of a complex system starts with a model in a broad sense, that is, a simplified view of the system that allows concentrating on some of its features while ignoring others. This is true for research on human movements as well. Any movement study starts from selecting a level of analysis that ignores or downplays processes at other levels. For example, a study of joint kinematics may ignore how exactly muscles are activated to produce the kinematic patterns, while a study of muscle activation patterns may ignore how ions move through the membranes of the involved excitable cells and how brain structures interact to produce these patterns.

How does one select which features of the system to consider and which ones to ignore? There is no golden rule. Researchers use their knowledge and intuition to formulate research problems such that they address interesting questions and are studied at an appropriate level of analysis. Such simplifications come at a price, and sometimes researchers ignore very important contributors to the processes they study, based on poor understanding of the system. However, this is the price most researchers are willing to pay to be able to study complex systems. Qualitative jumps in understanding of such systems commonly happen when a scientist realizes that one of the previously ignored factors is indeed very important.

Formal models that are founded on a solid background of experimental material specific to the system of interest and a deep intuitive understanding of the system are indeed very important and useful. Even when a researcher thinks that the available data unambiguously point at a certain set of rules that define functioning of a system, such conclusions have to be tested formally. The system for movement production is very complex, and it is hard to predict how complex systems would behave, based on a set of rules derived from experimental data and intuitive thinking. A formal model allows testing such intuitively reached conclusions, and, quite frequently, results of model analysis point at missing elements that, for example, do not allow the model to show stable behavior. Then, the model becomes the source of new questions and leads to more thinking and new experimental studies. Such models help researchers find new ways to explore the system of interest.

Then, there are models of another type. Commonly, they are imported from other areas of science such as physics, engineering, and control theory that deal with less complex, inanimate systems. Such models are applied directly to problems of motor control and coordination, and their parameters are searched for in experimental data, for example, in characteristics of neurophysiological processes in certain structures of the body or in mechanical characteristics of movements. In a way, experimental studies of movements turn into means of finding parameters of models (equations) accepted *a priori*. This seems to be a rather distorted approach to movement (and other biological) studies: The main goal of a study should be to understand a system of interest, not to use this system to find parameters of equations imported from another area of science.

As mentioned in one of the earlier sections, human movements do not violate laws of mechanics. However, this is not what makes studies of human movements exciting. It would be naïve to expect equations of classical mechanics, by themselves, to provide insights into specific features of the neural control of movement. Along similar lines, it is naïve to expect equations of the control theory to help understand how the brain controls voluntary movements. The control theory was developed to deal with systems that are incomparably less complex and more predictable, for example, ballistic missiles. Imported models that are accepted *a priori*, without being based on a comprehensive analysis of the system of interest, are frequently useless and even harmful. They create an illusion of sophistication and depth that entices young researchers: The terminology may sound very impressive, and the equations may look formidable. Such models frequently lead to fashionable trends that quickly turn into religions: If you say the "right" words (write the "right" equations), you are a good researcher; if you do not say these words, your research is worthless.

I agree with Florensky and Gelfand wholeheartedly. A formal model is justified only when it is built on a deep understanding of the system of interest and when it is specific to this system. Otherwise, one quickly becomes a "modeler." So, my dear colleagues, next time you have a manuscript or a grant application to review, please think twice what you really want before asking for a *mechanistic interpretation.*

8.1 SYNERGIES AND THE CONTROL THEORY

8.1.1 Control: Basic Notions

The system for the production of voluntary movements is so complex, compared to typical man-made systems, that application of theories and computational methods developed for inanimate objects to biological movement will always be open to question. In the section on internal models (section 3.3), I have already discussed some of the obvious limitations of such approaches. Nevertheless, in this section, I will try to describe briefly elements of control theory and a recently introduced model of control of multi-element systems that shows some of the typical features of synergic behavior.

Consider a hierarchical structure consisting of at least two components, a controller and a controlled object (Figure 8.1). The controller receives an input from a smart center that comes up with decisions on what and when to do. The input specifies a desired state of the controlled object. Let us assume that state of the object can be described with a set of variables, $X_i(t)$, which can be represented as

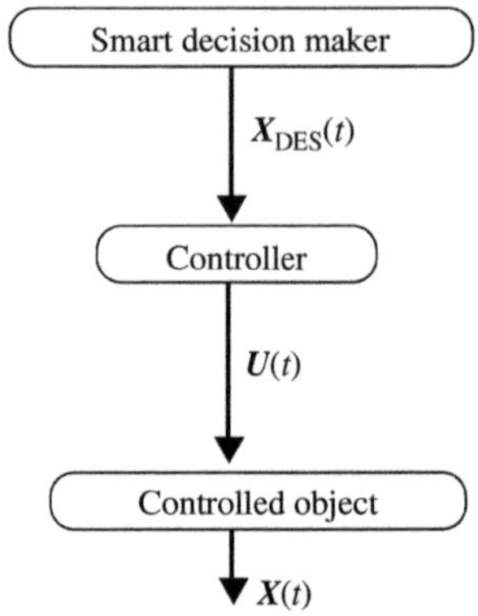

Figure 8.1. A smart decision-making system decides what a desired state ($\boldsymbol{X}_{DES}(t)$) of a controlled object should be. Based on this decision, a controller produces a vector of control variables $\boldsymbol{U}(t)$, which produces a change in the state of the object $\boldsymbol{X}(t)$.

a time-varying vector $\boldsymbol{X}(t)$. At any time t_0, the object can be described with a set of values for each of the variables, that is, with a vector $\boldsymbol{X}(t_0)$. The controller may also receive an input that informs it on the current (initial) state of the controlled object, but this is not always necessary.

The controller communicates with the controlled object using another set of time-varying variables, $U_i(t)$. We are going to address these as control variables. These variables can be specified by the controller independently of signals it may receive from the controlled object and the environment. This does not mean that the controller cannot change a control variable based on information from the controlled object and/or other sources. What is important is that it has a choice of reacting or not reacting to this information. Note that the controlled object does not have such freedom of choice and is expected to change its state obediently under the action of the vector of control variables, $\boldsymbol{U}(t)$, which we will address as the control vector. This latter feature justifies drawing the scheme in Figure 8.1 as a hierarchy.

In experimental studies of movements, researchers measure different sets of variables, defined partly by the formulation of their particular research problems and partly by the available tools. It is well known that if you have only a hammer, everything looks like a nail. It is not surprising, therefore, that the formulation of research problems follows closely the development of new research tools. We are going to address variables measured in experiments as performance variables, $\boldsymbol{P}(t)$, whether or not they are directly reflecting performance in a particular motor task. Performance variables can be measured at different levels of the neuromotor hierarchy. Consider, for example, variables recorded with such methods as electromyography, electroencephalography, single neuron recording, positron emission tomography, magnetic resonance imaging. Muscle activation patterns can be viewed as performance variables, since these patterns are direct precursors of muscle forces given muscle length, its changes, and a handful of parameters. But what about activation patterns within the central nervous system? Alpha-motoneurons are just one step upstream from the muscles they innervate, and their patterns of activation carry more or less the same information as electromyographic signals. So, they also belong to $\boldsymbol{P}(t)$. But what about patterns of activation of neuronal populations in the brain? Do they represent $\boldsymbol{P}(t)$, control variables $\boldsymbol{U}(t)$, or even the "smart input" into the controller? There is no answer to this question. Some studies assume (usually implicitly) that anything that happens "above the neck" is control, while anything "below the neck" is performance, but it is easy to present examples against such a simplistic view: Take a look at section 7.2 on synergies in the spinal cord.

8.1.2 Open-Loop and Closed-Loop (Feed-Forward and Feedback) Control

According to Figure 8.1, the controller receives an input corresponding to a desired state of the object or its desired change in time $\boldsymbol{X}_{DES}(t)$. However, it may or may not receive information about the current state of the object $\boldsymbol{X}(t)$. If the controller specifies a control vector $\boldsymbol{U}(t)$ independently of information on the current state of the controlled object, this type of control is called *open-loop* or *feed-forward* (these two terms are not perfect synonyms, but for the sake of the current discussion, I am not going to discuss differences between them). During open-loop control, a motor command is issued for a whole motor act before the outcome of the earliest change in the command is taken into account by the controller. A typical example of open-loop control is kicking a soccer ball. The central nervous system sends commands to the muscles involved in this motor act before the kick is initiated, and the brevity of this act does not allow the controller to adjust the commands during the kick.

If the controller changes the control vector based on its effects on the state of the object, it is called *closed-loop* or *feedback* (Figures 8.2A). An important component of a feedback control system is an error-detection mechanism, which uses signals about the current actual state of the controlled object. These signals may be supplied by sensory systems of different modalities such as vision, proprioception, vestibular system. They may lead to changes in muscle activation via relatively short-latency loops (cf. *reflexes* in Digression #6) or longer latency loops that are sometimes associated with a poorly defined notion of "voluntary reaction." Signals related to $\boldsymbol{X}_{act}(t)$ are compared to the desired progression of the state of the object $\boldsymbol{X}_{des}(t)$. Another structure called *comparator* computes an error, $\Delta\boldsymbol{X}(t)$ between $\boldsymbol{X}_{act}(t)$ and $\boldsymbol{X}_{des}(t)$ and uses it to adjust the vector of control variables $\boldsymbol{U}(t)$: $\mathrm{d}\boldsymbol{U}(t)/\mathrm{d}t = f[\Delta\boldsymbol{X}(t)]$.

For example, if you want to drive a car at a certain constant speed, say exactly 10 miles/hour above the speed limit, and the car is not equipped with a cruise control system, you have to use feedback control. The state variable of the controlled

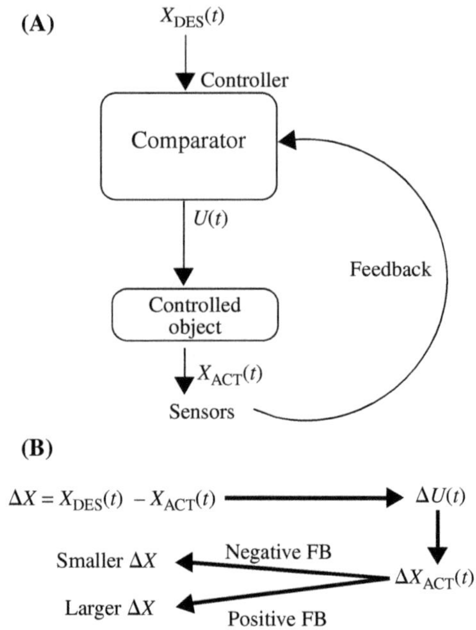

Figure 8.2. (A) A controller can change control variables, $\boldsymbol{U}(t)$, based on feedback signals from the controlled object. This is done by a comparator. (B) If a discrepancy exists between a desired and actual state of the object [$\Delta \boldsymbol{X}(t)$], the controller adjusts the control variable $\Delta \boldsymbol{U}(t)$. These adjustments can lead to a drop in $\Delta \boldsymbol{X}$ (negative feedback) or to its increase (positive feedback).

object is very simple—it is the velocity of the car. Your control vector is the force with which you press the gas pedal or the brake pedal. You can use visual information from the speedometer or from the moving environment to estimate the current value of the state variable (velocity), compare it with the desired value, and adjust the control variable, that is, press one of the pedals stronger or lighter. This type of control allows you to maintain a preferred speed, when the road goes uphill or downhill, when the wind changes, or when you spot a police car.

Two basic types of feedback control can be distinguished, negative and positive. The main purpose of the negative feedback control is to minimize deviations of a state variable from a certain desired value. Hence, it uses feedback signals on an error $\Delta \boldsymbol{X}(t)$ to adjust the control vector in such a way that the adjustment brings $\Delta \boldsymbol{X}(t)$ down closer to zero. Negative feedback control systems tend to counteract any deviations of the state variables of the controlled object from their desired values or their desired time evolution. The former example of driving a car at a constant speed is typical of negative feedback control. Another example from the area of motor control is stabilization of posture, in particular vertical posture during standing.

Positive feedback acts to amplify errors in state variables of the controlled object. In other words, an error $\Delta \boldsymbol{X}(t)$ leads to an adjustment of $\boldsymbol{U}(t)$, which leads to further increase in $\Delta \boldsymbol{X}(t)$ (Figure 8.2B). Positive feedback systems tend to get out of control leading to a qualitative change in the behavior of the system. At first glance, positive feedback control looks useless, since it destabilizes behavior. This is true with respect to most of the control schemes considered in the area of motor control. However, there are situations when positive feedback may be useful, for example, when one needs to quickly amplify a small signal and bring about a qualitative change in behavior.

Consider the following example. You stand quietly, and someone pushes you from behind. As long as the push is light, a negative feedback control system will help stabilize the vertical posture by generating forces against the push. However, if suddenly a strong push occurs, such negative feedback system may not be powerful enough to counteract the stronger push. Then, a positive feedback system may start to act, leading to further deviation of the body in the direction of the push, resulting in a protective step.

Imagine that at a certain time t_0 the state of a controlled object is $\boldsymbol{X}_{act}(t_0)$ different from its desired state $\boldsymbol{X}_{des}(t_0)$, leading to an error $\Delta \boldsymbol{X}(t_0)$. It takes the comparator time Δt_1 to detect the error signal $\Delta \boldsymbol{X}(t_0 + \Delta t_1)$. The controller takes time Δt_2 to adjust the control vector by a certain magnitude $\Delta \boldsymbol{U}(t_0 + \Delta t_1 + \Delta t_2)$, based on the signal from the comparator. This adjustment will lead after time Δt_3 to a corrective change in the state variables of the controlled object, $\Delta \boldsymbol{X}_{cor}(t_0 + \Delta t_1 + \Delta t_2 + \Delta t_3)$. If this is a negative feedback system, $\Delta \boldsymbol{X}_{cor}(t_0 + \Delta t_1 + \Delta t_2 + \Delta t_3)$ will be directed against $\Delta \boldsymbol{X}(t_0)$ and will lead to a reduction of the latter. If this is a positive feedback system, $\Delta \boldsymbol{X}_{cor}(t_0 + \Delta t_1 + \Delta t_2 + \Delta t_3)$ will amplify $\Delta \boldsymbol{X}(t_0)$.

Let me introduce two important parameters that characterize feedback loops, gain, and delay. Gain (G) can be defined as the ratio between the magnitude of a corrective vector $\Delta \boldsymbol{X}_{cor}$ and the magnitude of the original error vector $\Delta \boldsymbol{X}$ that gave rise to the correction $G = |\Delta X_{cor}|/|\Delta X|$. If $G = 1$ in a negative feedback system, a seemingly perfect correction mechanism issues a corrective signal that annihilates the error completely at a new steady-state. However, the perfection of such a system may only appear to be such, because of the unavoidable time delays involved in any feedback control systems. One should not forget that the correction takes place after a delay since the original error emerged. This delay is $\Delta T = (\Delta t_1 + \Delta t_2 + \Delta t_3)$. The controlled object is likely to change its state over this time interval. As a result, the correction may become suboptimal or detrimental, even if the gain of the negative feedback loop is unity. This is likely to happen if the original error $\Delta \boldsymbol{X}(t)$ changes its sign as a function of time over time intervals comparable to ΔT. Time delay can be measured in time units, for example, in seconds or milliseconds, or in relative timing units, for example, in percentage of a typical time interval characterizing the process (such as cycle duration for a cyclic action). Time delay may turn into an important drawback of feedback control, particularly if the magnitude of the delay is large and the motor process is fast. Therefore, if speed is vital, feed-ward control may be preferred while, if accuracy is important and speed is not, feedback control has an advantage. One possibility to adjust the relative contribution of the feed-forward and feedback components of a control system is through regulation of the gain in the feedback loop.

8.1.3 A Simple Feedback Scheme of Synergic Control of a Multi-Joint Movement

A simple computational model has been proposed to explain some features of a multi-joint synergy observed in studies of the wiping reflex by the spinal frog. This scheme (mentioned briefly earlier) is based on only kinematic variables. This means that the scheme does not address issues of physiological variables that may be used to control the actions. Rather, it assumes that the controller receives feedback on current limb configuration and is able to issue control variables directly related to individual joint rotations.

Recall that if a small piece of paper soaked in a weak acid solution is placed on the back or on a forelimb of a frog whose spinal cord has been surgically cut at a high level, the frog wipes the stimulus off with a coordinated multi-joint motion of the ipsilateral hindlimb (see section 3.3). The wiping movement remains accurate in conditions of different positions of the forelimb with the stimulus and in conditions of hindlimb loading and joint fixation.

Berkinblit, Gelfand, and Feldman (1986b) suggested a model, within which the vector of angular velocity in each joint is defined as a cross-product of two vectors, one from the joint to the target and the other from the joint to the limb endpoint (Figure 8.3). According to this rule, the joints will continue to move until the endpoint coincides with the target (assuming that the target is accessible). Then, the two vectors will be identical for each joint, in particular they will be parallel to each other, and their cross-product will become zero. Note that the model implies knowledge of the limb configuration at each moment of time and, as such, assumes using feedback signals. This model has an inherent ability to generate accurate movements to the target (stimulus) even if one of the joints gets off the track (e.g. as a result of an unexpected loading or joint fixation), because movement will continue until the endpoint reaches the target. In other words, it has a built-in error compensation among the joints. The model can also demonstrate a reproducible pattern of involvement of individual joints if the frog initiates the movement from the same initial configuration.

This model is certainly very simple and not very physiological. When some of its predictions were tested experimentally during multi-joint human movements (Karst and Hasan 1987, 1990), some of the results failed to follow model

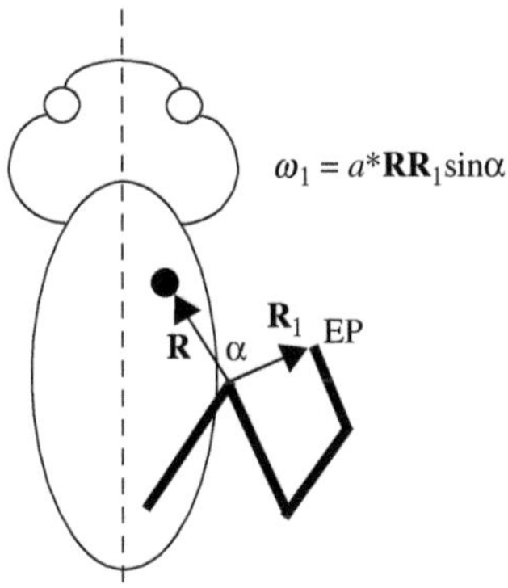

Figure 8.3. According to the model of Gelfand, Berkinblit, and Feldman, the vector of angular velocity (ω) in a joint is defined as a cross-product of two vectors, one from the joint to the target (**R**) and the other from the joint to the limb endpoint ($\mathbf{R}_1$).

predictions. In particular, during movement initiation, the direction of joint rotation was not always the same as one would predict based on Figure 8.3.

8.1.4 Optimal Control and Synergies

Optimal control theory is a branch of mathematics developed to find ways of controlling a system, which changes in time, such that certain criteria of optimality are satisfied. Let us assume that instantaneous rate of change in the state variables of a system is defined by their initial values (X_0 at time $t = 0$) and by the current values of the control variables:

$$\frac{\mathrm{d}X(t)}{\mathrm{d}t} = f[X(t),U(t),t] \quad \boldsymbol{X}(0) = \boldsymbol{X}_0. \qquad \text{Equation (8.1)}$$

If one knows the control function $U(t)$ over a time interval from 0 to T when the system is analyzed, the initial state of the system $\boldsymbol{X}_0$, and the form of the function f, Equation 8.1 can be integrated to find the trajectory of the state variables $\boldsymbol{X}(t)$. It is possible to choose a time profile $\boldsymbol{U}(t)$ such that the state and control variables optimize (minimize or maximize) an objective function over the same time interval:

$$I = \int_0^T F[X(t),U(t),t]\mathrm{d}t + S[X(T),T] \qquad \text{Equation (8.2)}$$

In this equation, F is a function reflecting the costs of the process associated with changes in both control and state variables. For example, this function may reflect a desire of the controller to minimize energy expenditure, fatigue, “effort” (a poorly defined notion related to changes in control variables), and other characteristics of the action. The *S*-function gives the so-called salvage value of the final state ***X***(*T*). Its purpose is to make sure that the control is not only optimal (in a sense of minimizing *I*) during the process, but it makes sense at the final state. For example, minimization of joint motion may be interpreted as making no motion at all, but the salvage function does not allow such a solution. Typically, there are constraints imposed on each component of *I*. For example, values of the control and state variables are usually limited to a certain range compatible with human anatomy and physiology, while the final state ***X***(*T*) is expected to comply with certain task-specific criteria.

For example, consider the (unsolvable) problem of finding an optimal swing of the bat in baseball. Obviously, ***U***(*t*) will be constrained by the abilities of the batter, there will be geometrical constraints on ***X***(*t*) to make sure that the bat does not fly away from the hand, and ***X***(*T*) will be constrained by the requirement to hit the ball. In this example, cost of the process is of little importance while the final outcome (the hit) is. Because of the complexity of the human motor control system, it is very hard to formulate Equations 8.1 and 8.2 sensibly. As a result, there are only a few examples of application of the optimal control theory to problems of motor control.

Recently, a method of stochastic optimal control has been applied by Todorov and Jordan (2002) to address the control of a redundant system (e.g. a kinematically redundant limb during a reaching movement; this problem has been considered in much detail in section 3.3). Todorov and Jordan formulated a cost function, similar to the function *I* in Equation 8.2, in such a way that it included a measure of internal effort spent on control and a measure of accuracy of performance of the whole system, namely, the model minimized the weighted sum where the first summand was the squared difference between a function of effector outputs and its required value, and the second summand was the effort defined as the variance of the control signals. In other words, the controller recomputes a new desired trajectory at every moment in time, making no effort to correct deviations away from the previously planned behavior, unless those deviations interfere with important performance characteristics and lead to an increase in the cost function. This idea is conceptually consistent with the principle of abundance and the computational method developed within the uncontrolled manifold (UCM) hypothesis. It suggests a particular method of producing flexible behaviors leading to a desired goal. The model has shown an ability to demonstrate co-variation patterns among elemental variables across repetitive trials similar to those observed in experiments. Note that the UCM hypothesis does not assume a particular mechanism that leads to the formation of the UCM and limits most of the variability to this subspace. This process may or may not be based on an optimization principle. A few alternatives are described later in this section.

As Todorov and Jordan admit, the computational part of their model introduces substantial complications (Todorov and Jordan 2002; Todorov 2004). There are other disadvantages of this scheme; for example, it assumes accurate knowledge of the current state of the effectors, based on sensory signals from proprioceptors. As such, its performance may suffer from such drawbacks inherent to systems with feedback loops as time delays and a chance of self-excitation for some values of system parameters. As mentioned in an earlier section (section 5.3.3), multi-finger synergies can emerge a few tens of milliseconds after the initiation of an action. This time delay is too short to allow the controller to use signals from proprioceptors for the synergy organization.

8.2 SYNERGIES AND NEURAL NETWORKS

The general idea of solving problems of motor redundancy by selecting each time a solution from a family of motor equivalent solutions can be realized in many ways. In particular, an artificial neural network can be organized and trained to perform such operations. Whether a family of solutions generated by a network satisfies the definition of synergy accepted in this book would have to be tested formally by analysis of the structure of variability.

Bullock and colleagues (1993) proposed a neural network that was able to handle the problem of inverse kinematics by learning transformations from end-effector coordinates to joint coordinates during multi-joint kinematic tasks. This network selected in each attempt one solution out of a set of motor-equivalent solutions. An analysis of variance was not part of that work, although extending the model to address the structure of variance seems feasible. Such an analysis would allow drawing links between the model of Bullock and his colleagues and the definition of synergies. For example, if one adds noise to all the joint variables, two components of joint variance can be quantified as in the UCM method (section 4.1). One could expect the *good variance* to be higher than the *bad variance* ($V_{UCM} > V_{ORT}$), which is a sign of a multi-joint synergy stabilizing the endpoint trajectory.

I would like to remind the reader one of the very first examples of inanimate objects that can show behaviors with some (but not all) of the main features of synergies, the rusty bucket (section 1.3). The bucket can show both sharing and flexibility/stability features of synergies, based on physical laws such as the law of conservation of matter and laws of hydrostatics. These laws are inapplicable to the human neuro-motor system. However, some of their salient features can be modeled using a simple neural network. Moreover, these features are compatible with some of the knowledge about the central nervous system.

An attempt to simulate the described features of the rusty bucket led to the development of a physiologically based model of synergic behavior (Latash et al. 2005). This model was developed for the task of multi-finger force production to allow direct comparison with the wealth of experimental data available for such tasks. As illustrated in Figure 8.4, the model starts with a task signal that comes from an undefined smart hierarchically higher structure. This command signal (level A) is being shared among a set of neural elements producing a particular sharing pattern. Four elements are illustrated in the figure corresponding to the four fingers of the hand. The resultant signals are corrupted with noise (otherwise, there would be no variability) leading to inputs to neurons at the next level (level B). The output of the level B neurons **m** is transformed by an enslaving matrix (see section 4.2.1, Digression #8 and Li et al. 1998; Zatsiorsky et al. 1998; Latash et al. 2001), resulting in a four-dimensional vector of finger forces: $\boldsymbol{f} = [\mathbf{E}]\mathbf{m}$.

The output of each of the level B elements also serves as an input into an interneuron (IN in Figure 8.4, level C), which projects back to all the level B neurons. These back-coupling loops are characterized by gains (g_{ij} comprising a 4×4 matrix $\boldsymbol{G}$), time delays, and thresholds. The controller is assumed to be able to vary the entries in the $\boldsymbol{G}$ matrix. This allows the controller to define both the required average behavior (by the TASK signal) and its stability properties (by varying $\boldsymbol{G}$).

For example, if all entries of the $\boldsymbol{G}$ matrix are negative, the model performs the task of accurate ramp force production and shows stabilization of the total

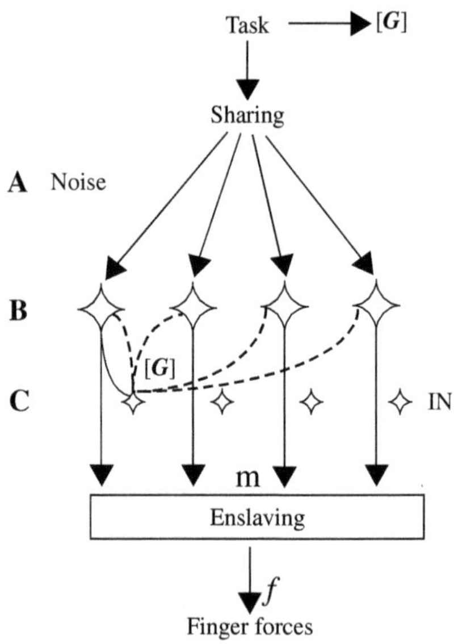

Figure 8.4. Two signals are generated by a hypothetical controller based on a task. One of them defines how the task is shared among the elements (Sharing). These signals are corrupted by noise and then projected on a set of "main" neurons (level B), whose outputs define modes (**m**) later transformed into finger forces, $\boldsymbol{f}$. The "main" neurons also project on small interneurons (level C) that have feedback projections on all four "main" neurons. The other control signal defines entries of a feedback matrix [$\boldsymbol{G}$], which reflects the strength and sign of the projections from the small interneurons at level C to the "main" neurons at level B.

force across trials. This was shown using the framework of the UCM hypothesis and the synergy index ΔV ($\Delta V > 0$, see section 5.6.1). Moreover, the model was able to replicate the finding of a time delay of the order of a few tens to 100 ms between the initiation of a slow ramp force production and the emergence of a force-stabilizing synergy (see critical time in section 5.3.3 and Shim et al. 2003b; Latash et al. 2004). The total moment produced by the finger forces with respect to an axis through the midpoint between the middle and ring fingers was not stabilized ($\Delta V < 0$ when computed with respect to the total moment of force). However, when 2 of the 16 entries of the ***G*** matrix were turned positive, the model performed the task of force production and showed stabilization of both the total force and the total moment of force as described in several experiments (Latash et al. 2001; Scholz et al. 2002).

Another attractive feature of this *central back-coupling* (CBC) model is that it allows the controller to modify required behavior and its stability properties independently (Figure 8.5). In particular, it allows changing entries of the ***G*** matrix without changing the TASK signal in anticipation of a quick action. This may result in anticipatory changes in the ΔV index of the force-stabilizing multi-finger synergy—the phenomenon of anticipatory synergy adjustments (ASAs) (see section 5.3.4).

Note that short-latency negative feedback loops, also addressed as lateral inhibition and surround suppression, are very common in the central nervous system. They have been described and/or postulated for sensory system of different modalities (Lund et al. 2003; Schoppa and Urban 2003; Wehr and Zador 2003; Ozeki et al. 2004), as well as for brain circuits traditionally associated with the production of movement (Fukai 1999). The well-known system of Renshaw cells (recurrent inhibition) may be viewed as a particular instantiation of this scheme. Recall that Renshaw cells are excited by alpha-motoneurons and make inhibitory

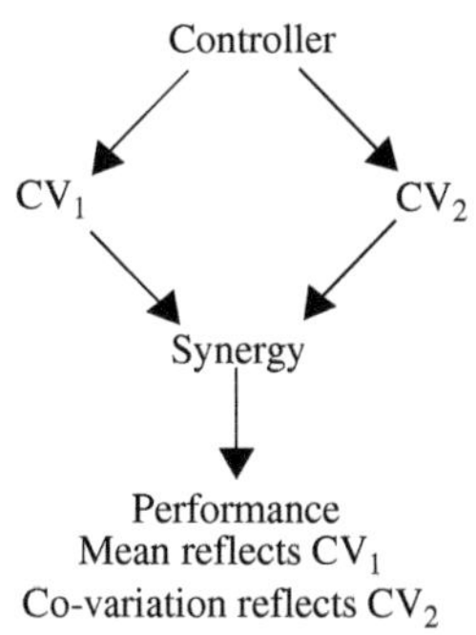

Figure 8.5. Controller generates two groups of control variables, CV_1 and CV_2. They project onto the synergy level and define the mean performance (CV_1) and co-variation patterns among elemental variables (CV_2). In Figure 8.4, CV_1 corresponds to the input to the "main" neurons after sharing, while CV_2 corresponds to entries of the [**G**] matrix.

projections on alpha-motoneurons of the same pool (and of pools controlling synergistic muscles, Figure 7.8). Renshaw cells have recently become incorporated into several hypotheses on the control of movement (van Heijst et al. 1998; Uchiyama et al. 2003). There is substantial variability in the organization and the strength of inhibitory projections mediated by Renshaw cells in different muscles (Katz et al. 1993). These projections can be modulated pharmacologically and by descending projections (Mattei et al. 2003; Hultborn et al. 2004), which may be the basis of how Renshaw cells stabilize the output of a motoneuronal pool in a way that could be muscle and task specific (cf. Hultborn et al. 2004).

The idea of feedback loops with very short latencies (*back-coupling*) was used in another model of synergies in redundant effector systems (Martin et al. 2004; Martin 2005). To solve the problem of motor redundancy, these researchers applied a Jacobian augmentation technique suggested earlier in robotics (Baillieul 1985). The idea of this method is to augment the Jacobian matrix with additional constraints thus making this matrix invertible and allowing to find a solution for the originally redundant problem. The authors achieved this goal by organizing an additional matrix combined from vectors spanning the null-space of the Jacobian. Further, they coupled this computational method with a biomechanical model of the effectors equipped with physiologically feasible muscle models following a simplified version of the equilibrium-point hypothesis (Gribble et al. 1998). The model has been able to account for the structure of variance observed in experiments that involved a pointing task performed by a kinematically redundant limb. In that model, three factors contributed to the observed variance. First, the neuronal computations of the equilibrium-point trajectories were assumed to be noisy. Second, motions of equilibrium points within the two subspaces (UCM and orthogonal to the UCM) of the joint space were assumed to be independent of each other. Third, the realized joint configuration was used as an input into the unit that performed the neural computation of the equilibrium trajectory, keeping the performance variable unchanged. This back-coupling played an essential role in the ability of the model to demonstrate different amounts of *good* and *bad* variance.

8.3 SYNERGIES WITHOUT FEEDBACK

8.3.1 Do Synergies Improve Accuracy?

Does one have to use a feedback control scheme to produce synergic behavior? At first glance, this is unavoidable: Indeed, how can one show error correction without getting information on errors? But maybe "error correction" is a misleading term. It implies that the presence of a synergy reduces errors (deviations from perfect performance) and is expected to lead to more accurate behavior. But is this so? Let me ask a very basic and simple question: Does having a synergy stabilizing a performance variable (in the meaning introduced earlier, $\Delta V > 0$) lead to a decrease in variability of that variable? Surprisingly, there is no clear answer to this basic question.

Indeed, it is very hard to compare indices of variability across different systems and tasks. Imagine that a person is asked to produce a certain level of constant force with one finger and then with two fingers. A two-finger force-stabilizing synergy, commonly seen in such tasks (Latash et al. 2002b; Gorniak et al. 2007a), leads to much higher values of *good* variability compared to *bad* variability. So, does it lead to more accurate performance? It seems that this question should have a direct answer: Simply compare the variability in the total force over a set of trials between one-finger and two-finger tasks. But here come a few complicating factors. First, one finger is weaker than two fingers. Should the subject of this mental experiment be asked to produce the same absolute level of force or the same percentage of their maximal force-generating capability? Should the subject be allowed to use visual feedback? What about proprioceptive feedback?

The last two issues are important because of the classical Weber–Fechner law of perception. This law states that the ability of a person to distinguish between two sensory signals depends on their magnitude (see Kunimoto et al. 2001; Johnson et al. 2002). Indeed, the smallest difference between the magnitudes of two signals that can be reliably perceived by a person increases nearly proportionally with the average magnitude of the signals. So, if a person is given visual feedback and asked to produce force such that the cursor on the screen follows a target line, accuracy of performance will depend first and foremost on the visual resolution of the feedback signal, not on the number of effectors (fingers) that participate in the task. What if a person is asked to remember a certain value of force and reproduce it with closed eyes? Then, subjects in such an experiment would rely on proprioceptive feedback, and the Weber–Fechner law might be expected to result in a linear increase in variability (standard deviation of force), with an increase in the force level. Indeed, an increase in the level of force-related feedback is expected to lead to an increase in deviations from this level that are perceived by the person and, therefore, corrected. So, measures of force variability in any experiment that allows the subject to use feedback system are likely to reflect the Weber–Fechner law—the perceptual component of the task—rather than the motor component such as co-variation of forces produced by the effectors.

What if a person is asked to produce a very fast action to a target (a very quick force pulse), an action that is too quick for any use of sensory feedback signals? This is indeed possible. However, such actions are associated with synergy indices (ΔV) that are close to zero or even negative, implying positive co-variation of finger forces that destabilizes the total force (Goodman et al. 2005).

This seems to be a catch! Strong synergies are typically associated with relatively slow changes in the performance variable (see section 5.3). However, such tasks offer the subjects plenty of time to use sensory signals to correct the performance variable. As a result, indices of variability of that variable are likely to reflect sensory processes governed by the Weber–Fechner law, not motor synergies.

There is another potential problem (this analysis was suggested by Dr. Simon Goodman). Several studies have shown an increase in the standard deviation of force with the force magnitude that is rather similar to the Weber–Fechner law and may even be its consequence (e.g. Carlton et al. 1993; Newell and Carlton 1993). Imagine now that a person produces the total force of 20 N with one finger, and the standard deviation across a series of trials is 2 N (Figure 8.6). Now let us ask this person to perform this task with two fingers. Each finger is expected to produce less force; for simplicity, assume that they share the force equally, 10 N each. If there is no co-variation and standard deviation of force is proportional to the force magnitude, each finger is expected to show a standard deviation of 1 N. Now recall that variance is standard deviation squared. This means that variance of the total force in the one-finger task will be 4 N^2. In the two-finger task, *in the absence of co-variation*, the variance of the total force will be the sum of the variances of individual finger forces: $1 + 1 = 2$ N^2. The variance in the two-finger task dropped. But we assumed no co-variation, that is, no force-stabilizing synergy. So, the dependence of variability of performance variables on the magnitude of these variables can by itself lead to an apparent improvement in accuracy in multi-effector tasks. This effect has to be taken into consideration when indices of variability are compared across different sets of effectors.

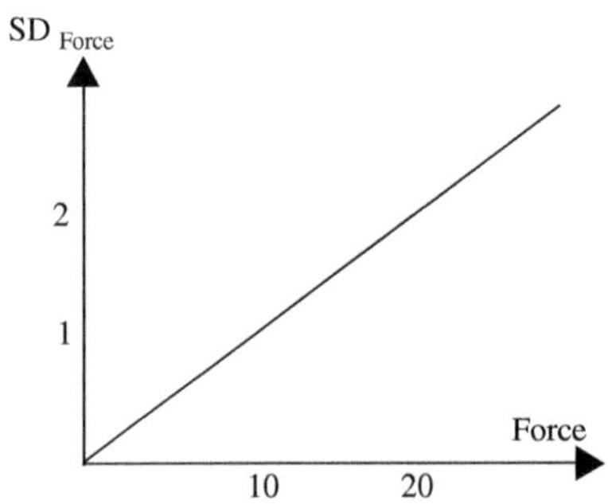

Task	Force-1	Force-2	Variance
1 finger	20	0	4
2 fingers	10	10	2

Figure 8.6. Standard deviation of force is typically proportional to the force level (the upper graph). So, if one finger produces 20 N, variance of total force is expected to be 4 N^2; if two fingers share the same task, variance of total force is expected to be 2 N^2. See the text for more.

To conclude, surprisingly, there is no unambiguous answer to the question formulated in the title of this subsection. This has allowed developing a model, which views two component of variance, *good* and *bad*, as specified independently.

8.3.2 A Feed-Forward Model with Separate Specification of Good and Bad Variability

In all the mentioned models, some kind of explicit feedback was used to ensure that deviations of important performance variables from their desired values (trajectories) are kept low. Even if the time delays of such feedback loops are small, as in the CBC schemes, they are still prone to all the disadvantages inherent to feedback control systems such as possible loss of stability. This is potentially a very high price to pay and, at least in some behavioral situations, such schemes do not seem evolutionarily feasible. Can interaction of elemental variables be organized to produce all the characteristics of synergies without using feedback loops?

A model of feed-forward control of a redundant motor system has been offered recently (Goodman and Latash 2006). Actually, the model contains an element of feedback, but these feedback signals are not used to correct the control process (as in closed-loop or feedback control, see section 8.1.2) but to update the knowledge of the controller on the relations between changes in elemental variables and in important performance variables (the Jacobian of the system). The structure of the model is illustrated in Figure 8.7.

The model is expected to generate a required time profile of a task variable, $g_{TASK}(t)$. There are several elemental variables that can be represented by a vector,

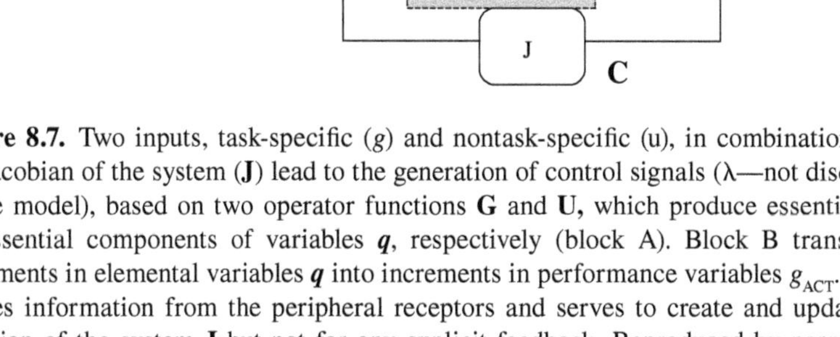

Figure 8.7. Two inputs, task-specific (g) and nontask-specific (u), in combination with the Jacobian of the system (**J**) lead to the generation of control signals (λ—not discussed in the model), based on two operator functions **G** and **U,** which produce essential and nonessential components of variables $\boldsymbol{q}$, respectively (block A). Block B transforms increments in elemental variables $\boldsymbol{q}$ into increments in performance variables g_{ACT}. Block C uses information from the peripheral receptors and serves to create and update the Jacobian of the system **J** but not for any explicit feedback. Reproduced by permission from Goodman SR, Latash ML (2006) Feed-forward control of a redundant motor system. *Biological Cybernetics* 95: 271–280. © Springer.

$\boldsymbol{q}(t)$. The number of elemental variables is larger than the dimensionality of g_{TASK}, that is, the system is redundant.

There are two explicitly task-related inputs to the model. One of them is related to an ideal, perfect execution of the task: It is $g_{TASK}(t)$. The other input is described by a function $g_i(t)$ that is produced in a particular trial *i*. To illustrate this idea, imagine that a person is asked to press on a set of force sensors with fingertips and produce a force profile shown on the screen (thick line in Figure 8.8). However, in individual trials, the person produces time profiles $g_i(t)$ (shown with thin lines in Figure 8.8) that deviate somewhat from $g_{TASK}(t)$. When many $g_i(t)$ are averaged across trials for each moment of time, the result is expected to match $g_{TASK}(t)$.

One more input into the model represents a function $\boldsymbol{u}(t)$, which has nothing to do with $g_{TASK}(t)$. For example, this function may reflect an optimization principle or even be selected arbitrarily. The purpose of this function is to create *good* variance which can be modified by the controller leading to a stronger or a weaker synergy stabilizing $g(t)$.

The next step involves the generation of elemental variables $\boldsymbol{q}(t)$. In this step, the authors of the model used the Moore–Penrose pseudo-inverse procedure, which had been used in both robotics and motor control (Whitney 1969; Mussa-Ivaldi et al. 1989; Feldman et al. 1990). This procedure, applied to a multi-link serial kinematic chain, results in a solution that minimizes the norm of a vector of joint angular velocities (Zatsiorsky 1998). Applying this procedure requires knowledge of the Jacobian matrix. The controller is assumed to get information from the periphery on the current state of the structures that produce elemental variables. This information is used only to create and update the Jacobian, not for any explicit feedback corrections.

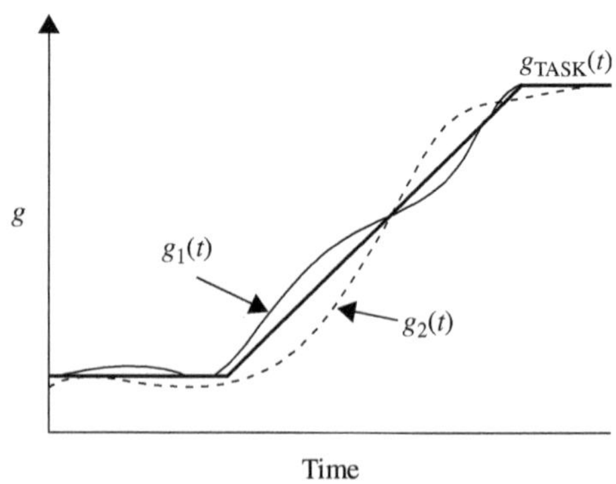

Figure 8.8. A task may be represented with a time function $g_{TASK}(t)$. In each particular trial, subjects may be expected to produce different time profiles $g_i(t)$. Reproduced by permission from Goodman SR, Latash ML (2006) Feed-forward control of a redundant motor system. *Biological Cybernetics* 95: 271–280. © Springer.

This model was able to describe a number of phenomena observed in multi-finger synergies (reviewed in section 5.3). These included possibilities of different shapes of data distribution corresponding to presence and absence of force-stabilizing synergies in two-finger tasks, changes in synergies with practice, and changes in synergy indices in preparation for a fast action. The biggest advantage of the model is its potential immunity to feedback-induced instabilities.

A recent model by Liebermann and colleagues (Liebermann et al. 2006) has also used a feed-forward scheme to address the problem of motor redundancy. These authors addressed this problem at the level of kinematics during multi-joint movements. In their model, a planned path is determined entirely before the action and is reduced to a certain plane called Listing's plane. Two laws, Donders' law and Listing's law, were formulated in the nineteenth century to address kinematic constraints on natural eye rotations. Donders observed that during saccadic eye movements (very fast eye rotations observed when a person shifts gaze from one object to another) performed with the head fixed, the eye always achieved the same orientation at any position in space, regardless of the path taken to reach it or the direction of the saccade. The realization of this rule could be achieved by constraining three-dimensional rotation vectors of the eye to a two-dimensional map called Listing's plane that maintains a constant direction throughout motion. This is known as Listing's law (Westheimer 1957), which allows only those postures that can be attained by direct, fixed-axis rotations relative to a primary reference vector. There is substantial evidence in support of a control scheme based on Listing's law in studies of eye movements (Tweed and Vilis 1987; Van Opstal 1993; Crawford and Vilis 1995; Tweed 1999), as well as in studies of limb movements (Miller et al. 1992; Gielen et al. 1997; Admiraal et al. 2001; Marotta et al. 2003). One may view Listing's law as a particular outcome of a centrally produced function analogous to function $\boldsymbol{u}$ in the described feed-forward model.

8.4 SYNERGIES AND THE EQUILIBRIUM-POINT HYPOTHESIS

All the aforementioned models operate with relations between performance variables and elemental variables. There is no qualitative distinction between the two groups—a performance variable stabilized by a synergy at one level of analysis may turn into an elemental variable at another level. Examples are synergies based on muscle modes (sections 4.2.2 and 5.5). Modes were viewed during that analysis as elemental variables. However, a mode may be considered as a performance variable stabilized by changes in the activation levels of the muscles comprising the mode, which are produced by co-varied changes of commands to the muscles (time profiles of the thresholds of their tonic stretch reflexes—see section 3.4 on the equilibrium-point hypothesis). As mentioned in one of the early sections (section 3.2), an input into a structural unit comes from another structural unit that may ensure a synergy stabilizing that input variable.

In virtually all the cited experimental and modeling studies, both elemental and performance variables have been associated with mechanical (or electromyographic) outputs of the elements of the system and of the system as a whole. This is not surprising, because the models are expected to match the available experimental material, and the only material available represents relations between elemental and performance variables both measured in physical units as the outputs of the corresponding elements. However, in sections 3.3 and 3.4, I have argued vehemently that control variables in the central nervous system cannot be associated with performance variables produced by the body. They have to be associated with *parameters* of equations (sets of rules) that describe the interactions among different parts of the body and between the body and the environment that are governed by natural laws.

If one wants to understand the central, neural mechanisms of synergies, another qualitative leap is necessary to approach the problem of how parametrization of the elements of a system brings about the basic features of a synergy. In my opinion, only one hypothesis in the area of motor control defines control variables in a noncontradictory, biologically specific way, and this is the equilibrium-point hypothesis (λ-model, Feldman 1966, 1986; section 3.3). Within the equilibrium-point hypothesis, control variables are associated with spatial coordinates of muscle activation thresholds (λs). Muscle activations, forces, joint torques, and displacements emerge as a result of an interaction between control variables, external forces, and reflex loops.

Unfortunately, at the current level of experimental sophistication, analysis of motor synergies at the level of control variables (λs or reference configurations) has been elusive. One can try to make an argument that analyzing experiments with isometric force production in predictable conditions allows the researchers to expect regularities seen at the level of mechanical elemental variables to be adequate reflections of some hypothetical regularities that exist at the level of control variables. This is, however, far from obvious. Even if no overt motion happens during an isometric force production action, two factors may lead to changes in the reflex contribution to the observed mechanical and electromyographic variables. First, muscle activation leads to shortening of the muscle fibers and lengthening of the tendon such that the total length of the muscle + tendon complex remains unchanged. Second, muscle activation is accompanied by activation of the gamma-system described in Digression #3. To remind, gamma-motoneurons send signals to muscle fibers inside the main length- and velocity-sensitive receptors muscle spindles and change the sensitivity of the spindle sensory endings to both muscle length and velocity. Since spindle endings are a major source of reflex effects, the reflex contribution to muscle activation is expected to change as well. So, even in perfectly controlled isometric conditions, changes in muscle activation and, consequently, in muscle force reflect effects of changes in both control variables and activity in the reflex loops.

There have been several attempts to reconstruct the time changes of λs (or of the {r,c} pairs) during natural actions (Latash and Gottlieb 1991c, 1992; Latash 1992a,b, 1994; Gomi and Kawato 1996). All these studies, however, used simplified mechanical models of the moving segments and, therefore, produced questionable results (see Gribble et al. 1998). In addition, these studies required many repetitions of a task to get a single estimate of the control pattern. To study synergies quantitatively, many estimates of the control pattern are needed, which may be impossible to obtain for practical reasons—too many trials are required.

Another approach to identifying reference configurations is based on an analysis of electromyographic patterns of large groups of muscles (Feldman et al. 1998a; Lestienne et al. 2000). This approach is based on the following idea. Because of the gravity field, which provides a nonzero external load during most actions, actual joint configurations during natural movements rarely coincide with reference configurations, thus leading to nonzero activation of certain muscle groups. However, during a quick movement with a reversal, one may expect that, at some point in time, the moving actual configuration will coincide with the moving reference configuration. If this happens, all the muscles participating in the action may be expected to show a minimum of their activation level at the same time. Such global minima of muscle activity were indeed observed in the cited studies. However, even this method offers only an indirect method of identifying a reference configuration at only one or few points during a movement. Recall that analysis of synergies has to be based on numerous observations along a trajectory (see Part 4).

There is, however, another link between the idea of synergies and the equilibrium-point hypothesis. With respect to single muscle control, the equilibrium point hypothesis may be viewed as based on a particular example of a multi-element synergy with motor units playing the role of elements and the overall muscle behavior (e.g. the level of activation, or the active muscle force, or the joint position) playing the role of task variables. The tonic stretch reflex mechanism (Matthews 1959; Feldman 1966) is an example of a feedback system that produces co-variation among the rates of the action potential generation by individual motor units stabilizing the total level of muscle activation (as illustrated in Figure 7.7). So, a set of motor units may show both preference for particular sharing patterns and the stability/flexibility feature. The former can be based on the well-established size principle (Henneman et al. 1965; Digression #1), while the latter can be a feature of the tonic stretch reflex.

The equilibrium-point hypothesis of single-muscle control is an example of how a large set of elements (motor units) can be united by a physiological mechanism (the tonic stretch reflex) to stabilize an important feature of performance—the equilibrium point characterized by values of muscle force and length. In a sense, this hypothesis unites problems of control and coordination. Can this hypothesis be generalized to the control of multi-effector systems, like the ones discussed in the numerous examples of motor synergies? Such an attempt has been made by Anatol Feldman and his colleagues, who introduced the notion of *reference configuration* as a control variable at a high level of a control hierarchy involved in the production of natural, multi-joint movements (Feldman and Levin 1995; Feldman et al. 2007; Pilon et al. 2007). Reference configuration defines, in the external space, a configuration, in which all the muscles would attain a minimal level of activity—a set of threshold values for muscle activation. External forces may not allow the body to reach a current reference configuration; then, the difference between the reference configuration and an actual configuration will result in the nonzero muscle activations, leading to force production against the external forces. As such, reference configuration is a natural extension of the notion of the threshold of the tonic stretch reflex to multi-muscle systems (see section 3.4).

The general idea of control using reference configurations may be described as following a *principle of minimal end-state action*: The body tries to achieve an end-state, compatible with the external force field, where its muscles show minimal activation levels. This principle is a natural extension of the principle of minimal interaction (section 3.2), which was illustrated by Michael Tsetlin as a desire of each element of the system to do minimal work.

If one understands the notion of *reference configuration* not as a detailed configuration of all the segments of the body (all its kinematic degrees-of-freedom) but as a combination of threshold positions of only important points (Pilon et al. 2007), it offers a very attractive framework to look at motor synergies. This framework assumes a hierarchical control system where, at each level of the hierarchy, the system is redundant, that is, it produces more output variables than the number of constraints specified by input variables. If the controller cares only about certain characteristics of a motor action, other characteristics (those within the UCM) may be allowed to vary either arbitrarily or based on some secondary considerations, possibly reflecting optimization of certain features of performance.

For example, during a multi-joint movement, frequently trajectory of the endpoint is viewed as an important performance variable. In particular, this assumption was used by Paul Gribble and David Ostry (2000) in an elegant model of movement adaptation to velocity-dependent external force fields based on the equilibrium-point hypothesis. In that model, control at the endpoint trajectory level was mapped on signals at the hierarchically lower level (six muscles crossing two joints). Let us follow this idea and assume that the central nervous system organizes feedback on the endpoint position and uses this feedback to produce a time profile of an output variable, which will be used as the input into the next, hierarchically lower control level. At each point in time, the controller specifies threshold values related to the endpoint location, and the discrepancy between the resulting reference endpoint location and its actual location drives the output of that highest level of the control hierarchy. These signals serve as the input into the next level of control, which will drive reference trajectories at a joint level.

Because the system is redundant, a reference trajectory at a higher hierarchical level does not specify unambiguously all the reference trajectories at a lower level. Emergence of particular lower level reference trajectories may be based on a feedback mechanism (e.g. like the CBC-model in section 8.2) or on a feed-forward mechanism (similar to the scheme in section 8.3). Hence, a hierarchy of control levels, where each level functions based on the equilibrium-point control principle, seems like a plausible control structure leading to motor synergies.

The problem gets a little more complicated if one wants to incorporate the possibility of muscle co-contraction without changing limb configuration. Co-contraction is an important mechanism of motor control, reflected at the single-joint level by a special command, *c*-command. Humans can easily co-contract muscles without changing joint position. This observation suggests that there is one more important input that may define a state of muscles different from a minimal end-state activation. This hypothetical input can be incorporated into the described hierarchical scheme of control as follows.

Assume that state of the endpoint of a multi-joint limb can be described with two commands, one defining its equilibrium position (R) and the other defining its stability about the equilibrium position (C)—equivalent to the $\{r,c\}$ pair of commands introduced within the equilibrium-point hypothesis (section 3.4). An {R, C} pair maps on commands sent to individual joints (Figure 8.9); the control of a relatively simple joint with one kinematic degree-of-freedom can be described with a pair of variables, $\{r,c\}$. Given an external torque, these variables define an equilibrium joint position and the slope of the dependence between small changes in the external torque and joint position changes, that is, the apparent joint stiffness (section 3.4). If a large number of joints define state of the

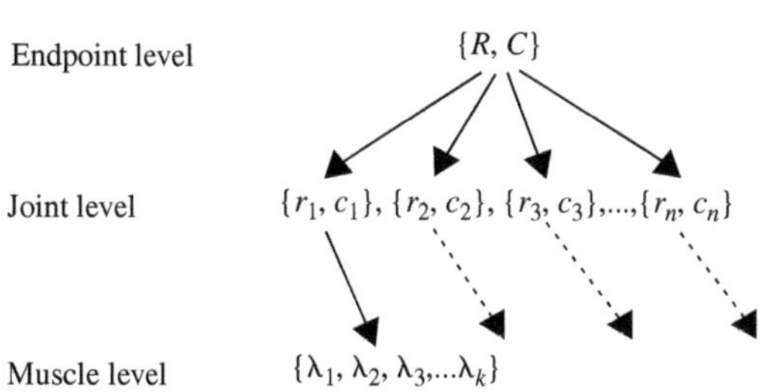

Figure 8.9. State of the endpoint may be described with a pair of commands, $\{R, C\}$. A multi-$\{r,c\}$ synergy may be expected to stabilize required values of R and C. To define an $\{r,c\}$ pair, the controller has to arrange a set of λ values for all the muscles, which can be done by arranging a multi-λ synergy that stabilizes required values of each $\{r,c\}$ pair.

endpoint, a multi-{r,c} synergy may be expected to stabilize required values of R and C. Most human joints, including relatively simple ones (such as the elbow joint), are crossed by more than two muscles. Hence, to define an {r,c} pair, the controller has to arrange a set of λ values for all the muscles. This is another typical problem of redundancy, and it makes sense to assume that the controller arranges a multi-λ synergy that stabilizes required values of the {r,c} pair.

How can such a synergy be organized? One option is to use a central feedback-based scheme, for example, like the one described in section 8.2. Note that the general structure of that scheme (see Figure 8.4) is not very different from the structure of the tonic stretch reflex loop. It is based on shorter loops and does not use peripheral sensory receptors, but otherwise the idea remains the same: Feedback to neural structures that define elemental variables is organized to stabilize a feature of the combined output of all the variables. Such CBC schemes may be used to stabilize any variable, based on a set of elemental variables, as long as feedback is arranged appropriately. Maybe, learning synergies consists of arranging new systems of such short-latency feedback loops that use appropriate feedback and project on appropriate output neural structures.

The notion of synergy has been introduced in the first sections of this book without any specificity to the motor function. Indeed, elemental and performance variables may be selected for any function of any biological object or group of objects. In the following sections, I would like to focus on two potential groups of synergies that go beyond the production of voluntary limb or whole-body movements.

8.5 SENSORY SYNERGIES

The best method to start discussing a complicated issue is to agree on terms and definitions. What is a *sensory synergy*? This expression can be understood in at least two different ways. According to the main definition of synergy, this is a neural organization of a set of elemental variables (in this case, sensory variables) with the purpose of stabilizing a certain performance variable. So, a sensory synergy may be viewed as an organization of perception, based on sensory signals from different sources and possibly of different modalities to ensure stable performance of a motor action. But one can also view sensory synergies separately from their potential role in the organization of action. Then, they can be viewed as neural organizations stabilizing percepts.

Sensory signals of different modalities are in most situations redundant. They create a single coherent picture of the world and enrich this picture with characteristics that are complementary, not contradictory. What we see usually fits well with what we hear, smell, touch, etc. Within this section, both types of sensory synergies will be discussed. I am going to start, however, with a few cases of neurological disorders associated with an impairment to one of the main senses participating in the control of movements—the feeling of one's own body (somatosensation). In all these cases, sensory signals of another modality (frequently, vision) help to compensate for the lack of somatosensory information, and one can witness a reorganization of the neural processing of relatively spared sensory signals. These are all signs of multi-modal sensory synergies stabilizing motor actions.

8.5.1 Sensory Synergies in Neurological Disorders

Sensory information is crucial for the production of movements. Indeed, people who lose one of the major sources of sensory signals show a significant impairment of their movements, if they cannot substitute for the lost source of information with information from a sensory system of another modality. Let me mention a few examples of disorders associated with a profound or even complete loss of information from peripheral sensory receptors, whose signals are used to create an image of one's own body (somatosensory receptors). These include proprioceptors in muscles, tendons, and joints (see Digression #3), as well as cutaneous and subcutaneous receptors that are sensitive to skin pressure.

In the second half of the nineteenth century and the first half of the twentieth century, syphilis was a relatively widespread and poorly treatable disease. Patients with advanced stages of syphilis could show degeneration of the dorsal columns of the spinal cord that contain neural fibers carrying sensory information to the brain. In many of the patients, the descending pathways from the brain to the spinal structures and the rest of the machinery participating in voluntary muscle activation were not affected to a similar degree. Despite sufficient muscle strength, those patients could stand only with their eyes open. Closing the eyes led to a major increase in postural sway and the possibility of a fall (Mauritz and Dietz 1980; Lanska 2002). Since healthy persons have little trouble standing with eyes closed, these observations indicate the importance of the lost sensory information for the control of vertical posture. But the fact that these patients could indeed stand with eyes open suggests that, at least to a degree, visual information could substitute for the lost somatosensory information. This conclusion comes very close to the idea of redundancy and *error compensation.*

Similar observations have been made in patients who suffer from advanced stages of diabetes (Lord et al. 1993; Van Deursen et al. 1998; Van Deursen and Simoneau 1999). Although by itself diabetes is not a neurological disorder (rather a disorder of the ability to metabolize glucose), it can lead to serious metabolic problems in the peripheral parts of the body, including the feet. These problems may lead to a loss of sensory signals from the feet and lower legs. Patients with such consequences of diabetes are impaired in their ability to stand with eyes closed, although other sources of information such as vision and even light finger touch (Dickstein et al. 2001) help improve stability of standing. Sensory signals from more proximal (closer to the trunk) segments of the body are relatively unimpaired and should help the postural control system. Indeed, experimental studies have suggested that the postural control system in such patients starts to pay more attention to sensory signals from more proximal parts of the legs compared to sensory signals from the feet and the lower leg (Van Deursen et al. 1998). These studies used muscle vibration to produce postural illusions. To explain them, another digression is needed.

Digression 12: Sensory and Motor Effects of Muscle Vibration

High-frequency, low-amplitude vibration (with a frequency of about 50–100 Hz and an amplitude of about 1 mm) applied to a muscle or to its tendon causes a variety of motor and sensory effects. This stimulus leads to very quick changes in the length of the muscle fibers and causes an unusually high level of activity of primary muscle spindle endings (Biancony and van der Meulen 1963; Brown et al. 1967) that are sensitive to muscle velocity (see Digression #3). Other receptors can also show substantial response to vibration, including secondary endings of the muscle spindles and skin receptors, but most of the central effects of vibration have typically been discussed as consequences of the very high activity of primary spindle endings. Indeed, vibration has been shown to be able to force primary spindle endings generate several action potentials for each cycle of vibration up to a frequency of 100 Hz (Roll and Vedel 1982; Cordo et al. 1998), much higher than typical frequencies of firing of these endings observed during natural actions (Vallbo 1970).

Muscle vibration can cause a reflex muscle contraction, the tonic vibration reflex (TVR, Eklund and Hagbarth 1966; Herman and Mecomber 1971) that has a number of unusual features (Figure 8.10). First, humans can easily suppress

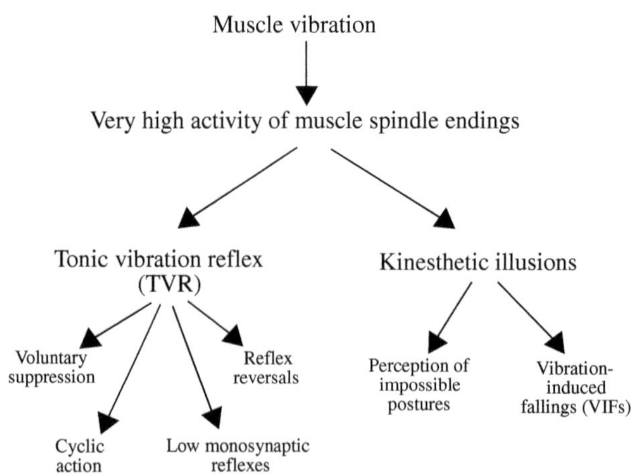

Figure 8.10. A schematic summary of unusual sensory-motor effects of muscle vibration mediated by the unusually high level of activity of the spindle sensory endings.

this reflex or allow it to develop. So, it is not clear whether the word *reflex* is at all appropriate in this case (see Digression #6).

Second, TVR contractions can be seen not only in muscles that are subjected to the vibration but also in other muscles, including muscles with an opposing action at the same joint (antagonists) and muscles crossing other joints of the limb. For example, when vibration is applied to the Achilles tendon, modification of such factors as the relative configuration of the segments of the leg and pressure on the sole of the foot can lead to TVR not only in the calf muscle group but also in other muscles acting at all three major joints of the leg (Latash and Gurfinkel 1976; Gurfinkel and Latash 1978). The distribution of the reflex activation patterns in different postures has been interpreted as indicating an important role of spinal neural structures that participate in the production of locomotor movements. In a more recent study, this guess has received impressive support: The vibration applied to the Achilles tendon was able to induce cyclic, locomotor-like movements in a subject lying on a side with the leg suspended in the air with a system of belts (Gurfinkel et al. 1998).

Third, activation of a muscle by TVR is accompanied by suppression of monosynaptic reflexes in that muscle (DeGail et al. 1966; Desmedt and Godaux 1978; Agarwal and Gottlieb 1980). Note that when a muscle is activated voluntarily, monosynaptic reflexes increase, reflecting the general increase in the excitability of the alpha-motoneuronal pool. The disparity between the apparently excited alpha-motoneuronal pool (TVR) and the decrease in its responsiveness to a standard peripheral stimulus has resulted in a hypothesis that vibration leads to a substantial increase in the presynaptic inhibition of afferent fibers from muscle spindles (Delwaide 1969; Gillies et al. 1970). Presynaptic inhibition is a mechanism of selective inhibition of particular inputs into a motoneuronal pool, which can leave other inputs unaffected. Suppression of monosynaptic reflexes by vibration has been used as an index of disorders associated with decreased spinal inhibition, such as spasticity (Hagbarth and Eklund 1968; Burke et al. 1972).

The dramatic change in the activity of muscle spindle endings produced by vibration has also been blamed for a range of perceptual effects such as vibration-induced illusions and vibration-induced fallings (VIFs). In particular, vibration of a muscle has been reported to cause a sensation of a joint motion corresponding to an increase in the muscle length (Goodwin et al. 1972; Lackner and Levine 1979). Sometimes, even anatomically impossible joint configurations have been reported by subjects (Craske 1977).

If muscle vibration is applied to the Achilles tendon while the person is standing, a major displacement of the body backward is observed, and sometimes the person has to take a step to avoid a fall (Eklund and Hagbarth 1967; Hayashi et al. 1981). Such VIFs have been interpreted as consequences of two processes. First, the unusually high level of spindle activity from the ankle extensor muscle group (triceps surae) leads to a central over-estimation of the muscle length. Second, the system for stabilization of the vertical posture interprets these signals as corresponding to a body motion forward and corrects this illusory motion with an actual sway backward. The VIFs are particularly strong when a subject in such an experiment stands with eyes closed. Opening the eyes can lead to alleviation of the effects of vibration, and VIFs can even disappear. This is another example when sensory signals of one modality compensate the disrupted or unreliable signals of another modality—a sign of a multi-sensory synergy.

Muscle vibration also shows effects on locomotor patterns. In particular, when a person steps in place, turning the vibration on leads to walking forward or backward depending on what muscle group is subjected to vibration (Ivanenko et al. 2000). Such vibration-induced effects depend strongly on visual information and light touch to an external object—another sign of a multi-sensory synergy. A recent study compared the effects of vibration applied to various muscles on posture and locomotion (Courtine et al. 2007). The authors of this study did not find correlations between the effects of vibration on posture and on locomotion. They have concluded that the vibration-induced changes in the sensory signals are processed within a general sensory-motor plan, which fits well the idea that sensory signals are combined into task-specific synergies.

End of Digression #12

In healthy persons, effects of vibration of the Achilles tendon are typically much larger than VIFs through vibration of more proximal muscles, for example, when the vibration is applied to the patellar tendon or to the hamstrings. However, in persons with diabetes, the effects of vibration on posture are reversed: They are decreased when the vibration is applied to the Achilles tendon and are increased when it is applied to the more proximal muscles (Van Deursen et al. 1998). These observations suggest that the postural control system pays less attention (gives lower weight) to signals from affected areas of the body and compensates the lack of reliable information from the lower leg by paying more attention (giving higher weight) to the relatively spared sensory signals from the upper leg.

There is a rare disorder—large fiber peripheral neuropathy—associated with a nearly complete loss of somatosensory information, in the absence of major problems with voluntary muscle activation. Patients who suffer from this disorder can stand, walk, and perform accurate limb movements but only under the continuous control of vision. When they try to produce rapid multi-joint arm movements, a major disruption of joint coordination is seen, suggesting that signals to muscles acting at different joints are not coupled properly to take into consideration the effects of interaction torques, that is, torques in one joint due to motion of other joints of the limb (Sainburg et al. 1995). This leads to nonstraight trajectories of the endpoint, which may interfere with everyday movements. In fact, in the cited study, the participants were asked to produce a horizontal hand movement simulating the slicing of a loaf of bread. The hand trajectory in a person with large fiber peripheral neuropathy was such that the knife did not follow a straight line (as in healthy persons) but a curve. These observations suggest that sensory signals may be crucial for the creation of multi-joint and multi-muscle synergies.

Whether their role is to provide information that is used in a feedback-based control process (as in the mentioned model by Todorov and Jordan 2002, section 8.1.4) or they are necessary to create an adequate mapping between deviations in joint motion and in endpoint motion (more in line with another model, Goodman and Latash 2006, section 8.3.2) remains unclear.

In a recent study (Tunik et al. 2003), two persons with large fiber peripheral neuropathy were required to reach to remembered targets without visual control. In some trials, the trunk motion was mechanically blocked. Healthy subjects showed nearly invariant hand trajectories in the blocked trials. The two "deafferented" persons showed a degree of compensation for the lack of trunk motion, which was not as good as in the healthy subjects. This study leads to two conclusions. First, the lack of proprioception does indeed impair multi-joint synergies. Second, another source of information, possibly from the vestibular receptors, helped introduce corrections of hand movements in blocked trials. This is an illustration of a multi-sensory synergy, which substituted for the unavailable somatosensory and visual information with signals of yet another sensory modality.

Note also that large fiber peripheral neuropathy eliminates major sensory inputs for the tonic stretch reflex. Since this reflex is assumed to play a central role in the equilibrium-point control of voluntary movements, patients who suffer from this disorder have to learn a new way to control their muscles. This may by itself disrupt basic synergies that have been developed during a lifetime, based on the undisturbed functioning of the tonic stretch reflex loop.

All the mentioned observations suggest that sensory signals of different modalities play an important role in the creation of motor synergies. Moreover, if sensory signals of one of the modalities become unavailable or unreliable, the other modality (commonly, vision) can take over and ensure reasonably accurate performance.

8.5.2 Sensory–Motor Interactions

We are now getting close to an important issue in movement studies—that of intimate links between perception and action. Until now, I have mostly mentioned how perception can play a role in action organization. However, there is also an opposite effect: Perception depends on action. This has been known since at least the nineteenth century. As mentioned in the earlier section on the history of movement studies (section 2.3), a great Russian physiologist, Ivan Mikhailovich Sechenov (1829–1905), wrote that humans did not see but looked, they did not hear but listened, and so on, emphasizing the active aspect of perceptual processes. With respect to perceiving variables that are directly related to movements, such as positions, velocities, and forces, a groundbreaking observation was made by Hermann von Helmholtz (1821–1894). He noticed that when a person moved the eyes (and/or the head) voluntarily, there was an adequate perception of self-motion in the motionless environment. In contrast, (as the reader can check on himself or herself), if one presses slightly on the eyeball with a finger and displaces the eyeball, there is a strong perception of a shifting environment. Von Helmholtz drew a conclusion that the difference was in commands sent to muscles that move the eye. When eye rotation is associated with typical changes in commands to the muscles, the person interprets a shift of the images over the retina as self-motion in a motionless environment. If the eye moves in an artificial way, a similar shift of the images over the retina is misinterpreted as motion of the environment.

Further, this view was generalized with an introduction of the notion of an *efferent copy* (sometimes called *efference copy* or *corollary discharges*) as a copy of control signals, which are sent to produce a movement of a body part, that participates in the perception of the relative positions (and forces) produced by the segments of that body part (von Holst 1954; Laszlo 1966; Feldman and Latash 1982a). Violations of perception have been reported in several studies, where one of the signals (afferent or efferent) was artificially changed (McCloskey 1978; Rymer and D'Almeida 1980; Feldman and Latash 1982c).

Studies of the relations between commands to a muscle and perception of that muscle's length and force have been complicated by the lack of agreement on the physiological nature and physical meaning of such commands. The equilibrium-point hypothesis once again offers an attractive framework to deal with this issue (Feldman and Latash 1982a,b). Indeed, if muscle states can be viewed as points on a force–length plane (for simplicity, let us consider only steady-states), selecting a motor command is associated with selecting a curve on that plane, that is, a characteristic of the tonic stretch reflex for the muscle (Figure 8.11). This means

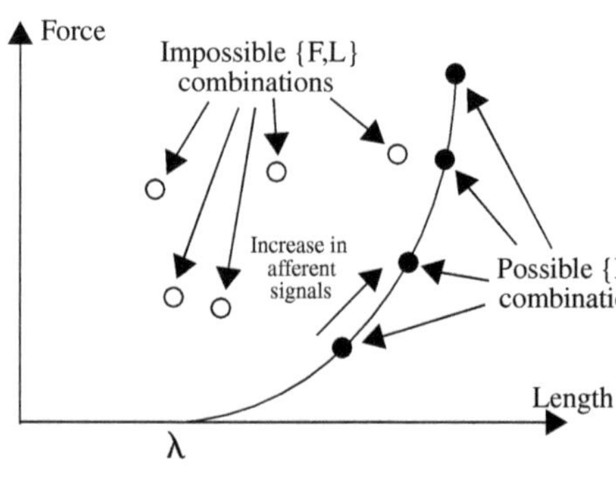

Figure 8.11. Within the equilibrium-point hypothesis, a central command λ defines a relation between active muscle force and muscle length. In static conditions, only points on that characteristic are possible, while points off the characteristic are not. Since afferent signals from most major proprioceptors increase along the line, they form an abundant set of sources of sensory information to define muscle force and length.

that when a person sends a control signal to a muscle, this by itself already solves a big part of the problem of perception, because only points on that curve are allowed as steady-states (filled circles in Figure 8.11), while points off the curve are impossible (empty circles).

In other words, selecting a motor command to a muscle reduces the problem of identifying its state in a two-dimensional space to a problem of finding its state along a one-dimensional line. Afferent signals are necessary to define a point on that line. Note, however, that the level of activity of all major proprioceptive signals increases along the line (in the direction of the arrow in Figure 8.11) due to an increase in both muscle length and muscle force. Hence, the set of signals from different sources becomes redundant (or abundant) and may be organized into a sensory synergy stabilizing a point on the tonic stretch reflex characteristic (an invariant characteristic) that corresponds to the muscle's state. Within this scheme, the efferent copy has a physical meaning of a relation between muscle force and length, and the abundance of sensory receptors allows expecting stable, adequate perception even when one of the sources generates unreliable signals (see Digression #3), of course, within limits.

A major step forward in understanding the relations between perception and action was made by James Gibson (1904–1979), the father of an area called *ecological psychology*. In particular, Gibson (1979) suggested that sensory signals could affect motor command directly (he called this "direct perception") without an elaborate central processing required for updating the perceived picture of the world and one's own body. The idea of action–perception coupling has been developed by Michael Turvey and his colleagues (reviewed in Turvey 1990, 1998). Some of the ideas expressed by Turvey's group are very close to the notion of synergies. Studies of this group focused mostly on patterns of natural coordination during simultaneous motion of two effectors when, typically, no other instruction was given to the subjects beyond "move naturally." It is not easy to guess what performance variables could be stabilized by sensory-motor synergies uniting the two moving limbs. However, stabilization of the relative phase has been shown in many studies (reviewed in Kugler and Turvey 1987; Kelso 1995). In experiments with artificial distortions of sensory information, stability of relative phase has been shown to depend more on the sensory signals compared to the actions performed by the subjects (reviewed in Hommel et al. 2001). Does a central neural organization responsible for stabilization of relative phase qualify as the basis of a synergy? This sounds possible, particularly because relative phase may be viewed as an important performance variable for a variety of rhythmic actions such as locomotion and breathing. If one considers relative phase in a more general sense as relative timing of actions by two effectors, not necessarily during cyclic actions (Schöner 1990), this variable becomes obviously important across a variety of actions, from a multi-joint action of throwing a baseball to playing piano. Such *timing synergies* have, however, been elusive, and a few experiments have failed so far to support their existence (section 4.5).

8.5.3 Sensory Synergies in Vertical Posture

Maintaining vertical posture in the field of gravity is not an easy task (Figure 8.12). The center of mass of the human body is located high over the ground, its projection has to fall within a relatively small support area, and there are several (potentially collapsible) joints connecting the center of mass to the feet. It is absolutely imperative for bipedal animals to know where "up" and "down" are. So, how do we know that our bodies are vertical?

There are three major sensory systems that provide information relevant to vertical posture (reviewed in Allison and Jeka 2004; Allison et al. 2006). One of them is the vestibular system that is sensitive to acceleration, both linear and rotational. The second one is vision that can extract the "up–down" information from the orientation of objects in the environment. The third one is the system of receptors sensitive to deformation produced by the field of gravity. These are sometimes united under the name of graviceptors. This group includes pressure receptors in the feet and proprioceptive receptors that signal such relevant variables as the length and force of muscles and the position and torque of joints along the vertical body axis. Information provided by these (and, sometimes, other) sensory systems in most everyday situations is noncontradictory and redundant. Can we view these three sources of information as elemental variables forming a synergy with the purpose of stabilizing perception of body orientation? Some of the earlier examples suggest that this is true; for example, the decrease or even disappearance of VIFs to vibration of the Achilles tendons, when the person opens the eyes (see the previous subsection).

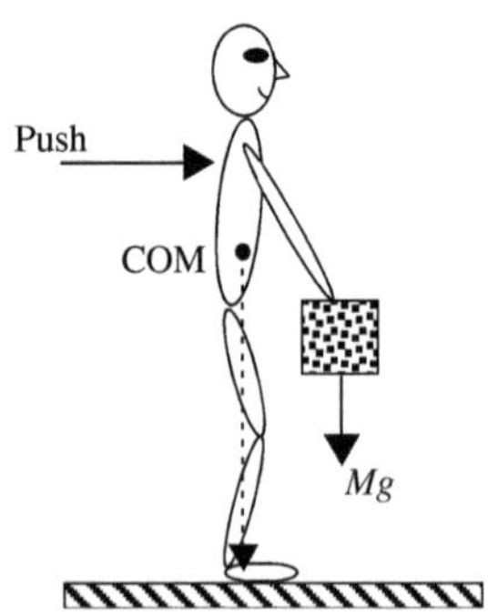

Figure 8.12. Vertical posture is a complex motor task. One has to make sure that the projection of the center of mass (COM, the black dot) falls into the small support area. Manipulation of objects (the weight *Mg*) and external forces (push) add to the complexity of the task.

One of the less well-known attractions of California is a place called *The Mystery Spot*, not far from Santa Cruz. I would recommend all those who are interested in multi-sensory integration to visit this place. *The Mystery Spot* is a park located on a rather steep slope of a hill. Due to the location of the park and the sunlight patterns, trees on the hill grow not vertically but at an angle. In addition, those who designed the park built a few buildings with the walls that are also inclined with respect to both the hillside and the direction of gravity. Visitors are also spending most of their time standing on inclined surfaces, which has been shown to lead to postural reorganization (Kluzik et al. 2007). This combination of factors creates a very confusing visual environment leading to strong illusions. In particular, a ball placed on a nearly horizontal surface, seems to roll up, not down. It is easier to stand on a wall of a house than on its floor, and so forth. The guides add to the confusion with clever pseudo-scientific comments about a "local gravitational anomaly" on that particular hill.

Note that the other two sources of information, the vestibular system and the graviceptors, are unaffected in *The Mystery Spot*. The presence of strong illusions demonstrates that, in the presence of vision, we, first and foremost, tend to believe our eyes, not other conflicting sources of information. The strong dependence of the vertical posture on visual information has also been demonstrated in experiments with subjects facing a screen, on which a pattern of points or stripes was projected (Dichgans and Brandt 1973; Van Asten et al. 1988; Dijkstra et al. 1994; Ravaioli et al. 2005). When the projection was changed to create a feeling that the points move toward the subject, the body of the subject swayed backward. When the pattern "moved away" from the subject, the body swayed forward.

Can other sensory signals, beyond the three mentioned ones, have strong effects on vertical posture? The answer is "yes." If a person stands and touches lightly an external object with a fingertip, information from the fingertip can be used to stabilize vertical posture (Holden et al. 1994; Jeka and Lackner 1994). Motion of the point of contact can lead to body sway similar to what is observed during motion of the visual scene (Jeka et al. 1998). Light touch may serve not only as an additional sensory input but also as an additional task (keeping the finger contact with the touched surface), imposing constraints that have an effect on postural mechanisms (Riley et al. 1999; Johannsen et al. 2007). Finger touch is not the only example of using pressure on another body part to help stabilize vertical posture. For example, an artificial plantar-based, tongue-placed biofeedback has been shown to decrease postural sway in healthy persons (Vuillerme et al. 2007).

Manipulations of auditory information can lead to illusory perception of body motion (Lackner 1977), and this information has been shown to help maintain balance when signals of other sensory modalities are unavailable or unreliable (Dozza et al. 2007). Sensory and orientation illusions are experienced by persons who are subjected to inertial forces (Clark and Graybiel 1968; DiZio et al. 2001).

So, perception of the vertical is not trivial and likely uses all the available sources of relevant information. Signals from these sources are likely to be weighed and united into a synergy as demonstrated by a number of studies showing an interaction among sensory signals from systems of different modality (McIntyre et al. 2001; reviewed in Soto-Faraco and Kingstone 2004; Lackner and DiZio 2004; Day and Guerraz 2007). A distortion of perception of verticality may lead to balance problems, for example those that happen with advanced age (Manckoundia et al. 2007).

I would like to mention here a phenomenon that everyone has probably observed and even experienced, at least a few times. When a person walks toward a moving walkway (an escalator), typically he or she can enter the escalator without any hesitation, any conscious effort, and any visible disruption of the walking pattern. This is not a trivial task: Stepping on a moving surface is associated with substantial forces at contact that may be expected to produce perturbations of the vertical posture. This means that seeing an escalator leads to preparation of the postural (and locomotion) control mechanisms for the upcoming perturbation. Adjustments in the motor patterns are done in anticipation of the perturbation, rather than in response to it. Now take a look at people (or recollect your experience of) stepping on a motionless escalator. The same persons hesitate, step on the escalator cautiously, grab the handrail, and show major disruptions of the vertical posture. Why? Apparently, the adjustments to the locomotor pattern and vertical posture associated with seeing an escalator are very hard to suppress, even if all the sources of sensory information suggest that this is simply a motionless continuation of the walkway with a smooth transition to motionless stairs. Here, the pre-existent knowledge of what to expect from an escalator dominates and overruns the multi-modal sensory synergy, all the components of which tell the brain that no adjustments in the posture/locomotion are necessary.

8.5.4 Multi-Sensory Mechanisms

Sensory signals of different modalities interact to create a coherent and noncontradictory image of the world and of one's own place in this world. Adequate perception is obviously crucial for the organization of actions that allow animals to survive and succeed. A bright illustration was given in one of the James Bond movies, "The Man with the Golden Gun," where a lot of action happens inside a house equipped with many mirrors and odd objects that completely distort perception of one's own location and interactions with objects. James Bond barely survives the experience (as he has been successfully doing for the last 40 years or so), but his evil opponent fails.

Animals have to have stable and adequate perception of the world, but these percepts should also be flexible to allow for unexpected objects and events to be quickly spotted and identified. In other words, sensory synergies are a must for survival. Sensory synergies should be studied as other synergies, that is, as a mapping between a higher dimensional space of elemental variables and a lower dimensional space of performance variables, percepts. There have been many studies of interactions among different senses and neurophysiological mechanisms that may form the basis of such interactions (for a comprehensive review see Calvert et al. 2004). However, there is no adequate language for perception that would be analogous to the language of mechanics for action. The notion of *qualia* introduced to describe properties of sensory experience and qualities of feelings (e.g. "redness" of a colored object) has been viewed as suspicious and controversial by many researchers (Churchland 1985; Lewis 1995). As a result, it is hard to identify adequate elemental variables and performance variables, and this seems to be a crucial step for any quantitative analysis of sensory synergies that would be analogous to the UCM approach (see section 4.1).

Most of our percepts are stable. If we talk to a person, we expect this person to remain the same over the entire time of our interaction, and all our senses confirm that this is indeed true. A stable percept of a dog is that of a hairy, four-legged, tail wagging, and barking animal that smells like a dog. What if we see a six-legged, hairy, barking animal with all other typical features of a dog? Likely, we will try to rub the eyes and reevaluate the image. In other words, this percept will not be stable. However, if we watch a science fiction movie with weird music, a six-legged dog could be alright—a reasonable percept that does not require a reevaluation. So, stability of sensory experiences is a complex issue that may depend not only on the sensory signals but also on the context in a general meaning.

Interactions among sensory systems of different modalities are many and varied. They are flexible and can show plasticity, particularly in persons who have impairment in one of the senses. Such cross-modal plasticity has been demonstrated in deaf persons (Campbell and MacSweeney 2004) and also in blind persons whose visual cortex may become engaged in the tactile function (Fridman et al. 2004). These adjustments are facilitated by the documented neuronal convergence of different modalities (Rolls 2004), in particular during the control of defensive movements in experiments on monkeys mentioned earlier (Graziano et al. 2004, section 7.5) and during audiovisual integration in humans (Fort and Giard 2004).

The earlier motionless escalator example suggests that sometimes one of the senses overrides all others and drives them. In a recent study, subjects were asked to hold a load on one of the palms and then lift the load with the other hand (Diedrichsen et al. 2007). Computer simulations were used to create a disparity between the unloading action and the sensation experienced by the first (postural) hand. The subjects perceived an illusionary increase in force on the postural hand, likely produced by a conflict between sensorimotor predictions and visual object information.

Another example is the ventriloquist effect, when auditory signals are being fooled by vision (Howard and Templeton 1966; reviewed in Woods and Recanzone 2004). Ventriloquists produce speech sounds without moving the lips, and these sounds are perceived as being produced by a doll whose lips move in synchrony. Human reliance on vision is evidently able to create inadequate percepts and override hypothetical multi-sensory synergies.

An example of a possible sensory synergy is synesthesia, an enigmatic and curious phenomenon. This term implies coupling of two (or more) sensory modalities, resulting in attributing characteristics typical of one of the modalities to sensations provided by the other one. One example is the interaction between the senses of smell and taste resulting in expressions like "sweet smell" that are used by many (reviewed in Stevenson and Boakes 2004). Fewer people—about one in every few hundred (Ramachandran et al. 2004)—describe interactions between sense of pain and color, voice and color, and letters/numbers and color (reviewed in Cytowic 1989, 1997). The latter example of synesthesia, that is, associating letters, digits, and words with colors is probably most common (Mattingley and Rich 2004). Synesthetic associations are stable over months or even years of reexamining the same persons. They are associated with neurophysiological phenomena such as activation in the visual cortex during presentation of auditory stimuli, phonemes, or words, which have been shown in several studies (Paulesu et al. 1995; Aleman et al. 2001).

Synesthesia seems to be a perceptual rather than a cognitive phenomenon. For example, some persons with synesthesia see Arabic numerals as colors but not Roman numerals. On the other hand, when the same ambiguous visual stimulus can mean different characters, depending on the words it is embedded in, persons with synesthesia report seeing different colors although the physical stimulus remains the same. So, this phenomenon shows top–down influences that underlie many well-known visual illusions. Synesthesia is more common in artists and poets (Domino 1989). A great Russian composer, Alexander Skriabin (1872–1915), said that he saw music as movement of colors. However, his attempts to introduce "colored music" have been largely unsuccessful, because most people could not appreciate it.

Synesthesia may be viewed as a sensory synergy, because it helps stabilize percepts in certain situations when two or more modalities supply redundant information about an object. For example, to identify a word "five" (or a written numeral 5), people without synesthesia can rely only on auditory information (or on black-and-white vision). A person who associates this word (numeral) with a color, has additional, redundant sensory feelings that may help identify the word quicker and/or with fewer mistakes. However, this is all getting rather philosophical and moving away from the main quest of this book: To make the word "synergy" exact, operationally defined, and measurable.

8.6 LANGUAGE AS A SYNERGY

Language and movement are closely related and even intertwined. This should not be surprising because language is produced by movement. For example, to pronounce a phrase, a person has to generate coordinated motor actions by articulators. To write a phrase—a coordinated action by the hand and arm is required. In the sign language, movements obviously play the central role. There is evidence from studies of Paleolithic times that watching actions by another person could be an earlier language form than listening to another person (Ruben 2005). Recently, a close link between action words and neural commands associated with action has been suggested, based on experiments with transcranial magnetic stimulation (Devlin and Watkins 2007).

The notion of *synergy* with respect to language may be explored at different levels of analysis. For example, one may wish to analyze motor synergies associated with speech actions or sensory synergies associated with perceiving sounds or written characters. On the other hand, language as a set of grammatical rules applied to a lexicon may also be viewed as a hierarchy of synergies.

With respect to the coordination of articulators, it has traditionally been accepted that the main, and maybe the only, goal of speech movements is to produce a certain sound (Johnson et al. 1993; Nieto-Castanon et al. 2005). For example, a perturbation applied to one of the articulators has been shown to induce very quick (at a delay of about 20 ms) responses in other articulators (Kelso et al. 1982, 1984; Abbs and Gracco 1984). These quick responses have been interpreted as signs of synergies among articulators that try to maintain certain salient features of the produced sound. A more recent study has suggested that the motor cortex mediates such responses (Ito et al. 2005). Similar conclusions have been reached in studies of co-variation of the lower and upper lip motion during speech movements in natural conditions and in unusual conditions, in particular, with a bite block (Folkins and Linville 1983).

Recently, this view has been challenged by results of a cleverly designed experimental study performed by David Ostry and his group (Tremblay et al. 2003). In those experiments, slight mechanical perturbations to articulators produced corrective actions, although the perturbations did not lead to any detectable changes in the acoustical signal. These observations suggest that speech actions are likely to have a somatosensory goal in addition to an acoustical one, which may be achieved by neural control of the resistance of articulators to perturbation (Nasir and Ostry 2006).

In a recent study by Shaiman (2002), articulatory kinematics were analyzed during two types of manipulation, of the phonetic context and of the speaking rate. Localized kinematic adjustments to manipulations of phonetic context were seen within an articulator, while changes in the speaking rate affected globally the whole utterance. Across the articulators, the upper lip and jaw kinematics co-varied over localized manipulations in phonetic context, while this *synergy* was reconfigured across the utterance for manipulations of speaking rate.

Such analyses of speech (or writing) fit well the introduced definition of motor synergies and may be expected to benefit from application of the UCM analysis. Several studies of speech used measures reflecting variability among elemental variables and tried to link it to stability of performance. In particular, there have been attempts to link the extent of within-a-person articulatory variability along any given articulatory direction to a measure of acoustic stability (Nieto-Castanon et al. 2005). Correlation analysis of the trajectories produced by articulators has suggested a possibility of multiple motor equivalent solutions for the production of the same acoustical effect (Perkell et al. 1993). Task-related differences in spatial variability of some articulators, such as the lower lip and tongue, have been reported (Tasko and McClean 2004). Taken together, these findings suggest that all three components of synergies are demonstrated by speech movements.

One of the phenomena that attracted much attention recently is the so-called *anticipatory coarticulation* (Perkell and Matthies 1992; Farnetani and Recasens 1993; Matthies et al. 2001). This term refers to adjustments in the patterns of trajectories of several individual articulators that are involved in the production of a sound in preparation to producing the next sound. These patterns differ, depending on what the next sound is going to be. Anticipatory coarticulation resembles another phenomenon, ASAs (see section 5.3.4), that describes changes in patterns of co-variation among elemental variables in preparation to an action.

To apply the UCM method of analysis one has to define and measure elemental variables and map (theoretically or experimentally) their changes on changes in performance variables, whether these are coordinates or sounds—see Part 4. These are nontrivial steps, but they do not seem to pose conceptual problems beyond those solved in the described applications of the UCM toolbox to analysis of multi-muscle synergies (section 5.5). For example, flexibility of muscle activation patterns that can be used to produce speech sounds has been emphasized (Gentil 1992). Articulator muscle groups (*modes* in our parlance) have been identified in both speech and nonspeech actions and shown to differ across tasks (Moore et al. 1988), even in 2-year-olds (Ruark and Moore 1997). Other aspects of language synergies seem much less trivial to analyze quantitatively. These include sensory synergies that may stabilize important features of a message and synergies at the level of linguistic variables such as phonemes and words.

Information exchange is a crucial element in animal life. Animal communication uses a variety of sensory modalities including visual, auditory, chemical, and tactile (Holldobler 1999; Partan and Marler 1999). Sometimes, the different channels are redundant to ensure that messages have higher chances of being transmitted successfully, even if the environment is noisy (Wilson 1975; Wiley 1983). On the other hand, using several modalities to convey a message does not mean that they are completely redundant. A message an animal wants to convey may be complex, and different sensory signals may complement each other in delivering different aspects of the message. So, animal communication may be viewed as based on multi-modal sensory synergies. Such synergies are also typical of human communication, when facial expressions and body language commonly emphasize messages of spoken phrases. Humans can also use touch as a complementary source of sensory signals (Lash 1980; Chomsky 1986), but they do not seem to use chemicals for this purpose, at least not as much as cats and dogs do.

So, perception of speech can be viewed as a special case of multi-sensory integration (Massaro 2004). Visual and auditory sources of information typically reinforce each other but sometimes they come into conflict. An example is ventriloquism, when the lack of motion of the lips of a speaker and the presence of motion of the lips of a doll create an impression that the doll produces the sounds. Another example is the so-called McGurk effect (McGurk and MacDonald 1976), when a discrepancy between sound and lip motion results in an intermediate percept, for example listening to "ba" and viewing a face saying "ga" may induce a perception of "da."

Human language is a very exciting and enigmatic phenomenon. This was appreciated by Pavel Florensky (1999), as illustrated by the following quotation: "The everpresent readiness of the words to serve our needs at any moment in no way can be a function of only memory. No memory could possibly satisfy the continuous and various requirement of the mind for its expression, if the speaker would not have a key to forming words in his/her instinctive feeling. Even mastering a foreign language can be achieved only by learning the mystery of its formation" (p. 147).

This quotation emphasizes important features of language: its flexibility and variability that brings it close to the notion of synergy as discussed in this book. I suggest that language may be viewed as a synergy built on utterances (words) as elemental variables. What is the performance variable that this synergy tries to stabilize? Probably, it is safe to say that when we speak, more commonly than not, we try to make sure that the listener gets the message. The message may be informative or emotional. If the goal is to induce an emotion, a lot of equivalent solutions can be offered with different utterances that may not even be words. One can also speak simply to attract attention—then, it does not matter what one says, rather how loudly the utterances are made. I am going to use a poorly defined word *meaning* to imply what a person tries to achieve by speaking. So, by definition, *meaning* is a performance variable that is supposedly stabilized by a multi-utterance synergy.

I wonder how many of my colleagues experienced the following feeling while presenting a lecture: If the lecture is on a topic that is relatively simple and routine, you sometimes start thinking about other things while continuing to speak without listening to your own voice. At some point, you "wake-up" because you hear something completely ridiculous. In a fraction of a second, you realize that you are the one who had actually said those words. In such situations, it is very important to introduce a quick on-line adjustment before the audience has realized what happened. Such an adjustment should make the ridiculous statement less ridiculous (maybe turn it into a joke) and/or compatible with the message you would like to deliver. These episodes look to me as synergic adjustments made at a very high level of the hierarchy stabilizing the meaning one wants to express.

As a nonnative English speaker, I have also frequently experienced episodes when the right word did not come to mind. Then, the rate of speech might have to be adjusted (decreased), and the whole phrase might have to be restructured to build the same message on a different set of words. This experience is similar to the one with a ridiculous statement in a lecture, but the adjustment is done in anticipation of an episode where the forgotten word should have been used rather than *post factum*, after a wrong word has already been spoken.

Is grammar a set of synergic rules that, when applied to a vocabulary, allows to build synergies stabilizing *meanings*? If so, this may be a unique case of a nearly explicit set of synergic rules (for Esperanto—the set is indeed explicit) produced by the human brain. Then, studying grammar may help decipher synergic rules for other areas such as perception and movement, rules that are currently unknown and have to be guessed.

As with any synergy, elemental variables (words) that are used to build performance variables (phrases, meanings) may themselves be viewed as synergies. In particular, words may be considered performance variables stabilized by synergies in space of phonemes as elemental variables. For example, the same word pronounced by different persons, with different accents or speech peculiarities, may sound differently, but it may still be perceived as the same word.

It seems that the areas of movement studies and linguistics are natural partners for cross-fertilization. Unfortunately, only a handful of researchers have worked in both these areas. In this context, I would like to emphasize the broad spectrum of activities of Noam Chomsky (2006, 2007), who has tried to cross-fertilize linguistics, mental processes, and politics—not a bad idea since all three are products of biological objects and, as such, are likely to obey similar synergic rules.

8.7 CONCLUDING COMMENTS: WHAT NEXT?

Smart readers start a book from the last page trying to guess whether it is worth reading. So, I feel that this is probably the most important subsection. If it happens to be boring, nine out of ten potential readers will remain potential, and I will have only myself to blame.

I would like to summarize the contents of this book using a set of axioms that have been accepted explicitly in the first sections.

Axiom-1 Not everything is a synergy (in particular, an atom is NOT a synergy of elementary particles, and a table is NOT a synergy of its legs).

Axiom-2 Synergies are trademarks of living beings (maybe, in future there will be robots endowed with synergies, but I have not seen or heard about one yet).

Axiom-3 Synergies are essential; animals with bad synergies are eaten before they pass their faulty genetic material to the next generation (alternatively they die of hunger and then are eaten posthumously anyway).

Axiom-4 Synergies are organized in hierarchies.

Axiom-5 At each level of a hierarchy, synergies ensure desired stability properties of functionally important performance variables.

Based on this set of axioms, which are closely related to the *principle of minimal interaction* and *principle of abundance*, several computational approaches have been offered to identify and quantify synergies. Personally, I like the computational apparatus of the *UCM hypothesis* very much. This apparatus has proven to be able to quantify a variety of synergies in a variety of spaces (e.g. kinetic, kinematic, and electromyographic), for a variety of tasks (standing, stepping, standing-up and sitting-down, reaching, throwing, pointing, pressing, manipulating objects, etc.), and in a variety of subpopulations (college students, persons with Down syndrome, healthy elderly, and neurological patients). This method has also allowed to track changes in synergies that happen with practice.

Despite this impressive (albeit incomplete) list of applications, the method is still in its infancy, and every new experimental or theoretical study leads to unexpected results and poses new questions. This, to me, is the most optimistic sign, suggesting that the approach is inherently rich. Such issues as *timing synergies*, *hierarchies of synergies*, and *changes in synergies with practice* have only had their surface scratched.

The lessons learned so far can be summarized as follows:

1. Synergies can be defined operationally and studied quantitatively.
2. Analysis of synergies is very rich; it leads to new methods of approaching the problem of motor control and coordinarion.
3. Synergies can emerge, change, recover, etc. They are living beings within living beings.

What seems to be urgently needed is development of biologically specific methods of quantitative research. Biology cannot rely on physics of inanimate nature for providing means of research but should take care of those itself. Limitations of the toolbox associated with the UCM hypothesis have to be overcome and alternative (complementary) methods for quantitative study of synergies have to be developed. But the problem is deeper. There is an even more general need to be able to measure biologically relevant variables such as control variables that are used to produce voluntary movements. Within the only reasonable hypothesis of motor control, the *equilibrium-point hypothesis*, these are time profiles of λs, {r,c} combinations, and reference configurations. There have been a few marginally successful studies with attempts at reconstructing such variables, but they have all been based on crude models of the body mechanics and neurophysiology.

With respect to synergies, the biggest challenge seems to be not identifying and quantifying them—progress in this area seems to be satisfactory—but rather mapping salient features of synergies onto neurophysiological mechanisms. Borrowing terms from Bernstein and his colleagues (Bassin et al. 1966; Lev Latash et al. 1999, 2000), what are the neurophysiological structures that might be responsible for various operators shared by different synergies? The current knowledge in this area is all but nonexistent. Another challenge is to generalize the method for nonmotor domains. In the last two sections, I tried to sketch a few potential approaches to synergies in perception and in language. However, this notion and the associated apparatus may be applicable at both much smaller and much larger scales compared to those described in this book. On one side of the spectrum, the notion of synergies seems to be directly relevant to processes at a molecular level, for example those involved in the regulation of compliance of the cardiac muscle (Fuchs and Martyn 2005; Fukuda et al. 2005). On the other side of the spectrum, this notion may be successfully applied to interpersonal interactions (see Schmidt et al. 1990; Tognoli et al. 2007) and interactions of large groups of people (sociology, economy, and politics).

To end on a more philosophical note: Are we getting closer to formulating an adequate language for biology, in particular for movement studies? I believe that we are on the right track. At least we feel discontent using imported sets of notions from other sciences, even such well-developed sciences as physics. And dissatisfaction with *status quo* (without being disgruntled) is a key to progress in science.

REFERENCES

Abbs JH, Gracco VL (1984) Control of complex motor gestures: Orofacial muscle responses to load perturbations of the lip during speech. *Journal of Neurophysiology* 51: 705–723.

Abend W, Bizzi E, Morasso P (1982) Human arm trajectory formation. *Brain* 105: 331–348.

Adamovich SV, Feldman AG (1984) A model of central regulation of movement parameters. *Biofizika* 29: 306–309.

Admiraal MA, Medendorp WP, Gielen CCAM (2001) Three dimensional head and upper arm orientations during kinematically redundant movements and at rest. *Experimental Brain Research* 142: 181–192.

Agarwal GC, Gottlieb GL (1980) Effect of vibration of the ankle stretch reflex in man. *Electroencephalography and Clinical Neurophysiology* 49: 81–92.

Albus JS (1971) A theory of cerebellar function. *Mathematical Bioscience* 10: 25–61.

Aleman A, Rutten GM, Sitskoorn MM, Dautzenberg G, Ramsey NF (2001) Activation of striate cortex in the absence of visual stimulation: An fMRI study of synaethesia. *Neuroreport* 12: 2827–2930.

Alexandrov A, Frolov AA, Massion J (1998) Axial synergies during human upper trunk bending. *Experimental Brain Research* 118: 210–220.

Alexandrov AV, Frolov AA, Horak FB, Carlson-Kuhta P, Park S (2005) Feedback equilibrium control during human standing. *Biological Cybernetics* 93: 309–322.

Alexandrov AV, Frolov AA, Massion J (2001a) Biomechanical analysis of movement strategies in human forward trunk bending. I. Modeling. *Biological Cybernetics* 84: 425–434.

Alexandrov AV, Frolov AA, Massion J (2001b) Biomechanical analysis of movement strategies in human forward trunk bending. II. Experimental study. *Biological Cybernetics* 84: 435–443.

Allard T, Clark SA, Jenkins WM, Merzenich MM (1991) Reorganization of somatosensory area 3b representations in adult owl monkeys after digital syndactyly. *Journal of Neurophysiology* 66: 1048–1058.

Allison L, Jeka JJ (2004) Multi-sensory integration: resolving ambiguities for human postural control. In: Calvert G, Spence C, Stein BE (Eds.) *The Handbook of Multi-Sensory Processes*, pp. 785–798, MIT Press: Cambridge, MA.

Allison LK, Kiemel T, Jeka JJ (2006) Multi-sensory reweighting of vision and touch is intact in healthy and fall-prone older adults. *Experimental Brain Research* 175: 342–352.

Almeida GL, Corcos DM, Latash ML (1994) Practice and transfer effects during fast single joint elbow movements in individuals with Down syndrome. *Physical Therapy* 74: 1000–1016.

Amirikian B, Georgopoulos AP (2003) Motor cortex: coding and decoding of directional operations. In: Arbib MA (Ed.) *The Handbook of Brain Theory and Neural Networks*, Second Edition, pp. 690–701, MIT Press: Cambridge, MA.

An CH, Atkeson CG, Hollerbach JM (1988) *Model-Based Control of a Robot Manipulator.* MIT Press: Cambridge, MA.

An KN, Chao EY, Cooney WP, 3rd, Linscheid RL (1979) Normative model of human hand for biomechanical analysis. *Journal of Biomechanics* 12: 775–788.

An KN, Chao EY, Cooney WP, 3rd, Linscheid RL (1985) Forces in the normal and abnormal hand. *Journal of Orthopedic Research* 3: 202–211.

Anson JG (1992) Neuromotor control and Down syndrome. In: Summers JJ (Ed.) *Approaches to the Study of Motor Control and Learning*, pp. 387–412, North-Holland: Amsterdam.

Aoki T, Latash ML, Zatsiorsky VM (2007) Adjustments to local friction in multi-finger prehension. *Journal of Motor Behavior* 39: 276–290.

Aoki T, Niu X, Latash ML, Zatsiorsky VM (2006) Effects of friction at the digit-object interface on the digit forces in multi-finger prehension. *Experimental Brain Research* 172: 425–438.

Arbib MA, Iberall T, Lyons D (1985) Coordinated control programs for movements of the hand. In: Goodwin AW, Darian-Smith I (Eds.) *Hand Function and the Neocortex*, pp. 111–129, Springer Verlag: Berlin.

Arimoto S, Nguyen PTA, Han HY, Doulgeri Z (2000) Dynamics and control of a set of dual fingers with soft tips. *Robotica* 18: 71–80.

Arimoto S, Tahara K, Yamaguchi M, Nguyen PTA, Han HY (2001) Principles of superposition for controlling pinch motions by means of robot fingers with soft tips. *Robotica* 19: 21–28.

Aruin AS, Almeida GL, Latash ML (1996) Organization of a simple two-joint synergy in individuals with Down syndrome. *American Journal of Mental Retardation* 101: 256–268.

Aruin AS, Latash ML (1995) Directional specificity of postural muscles in feed-forward postural reactions during fast voluntary arm movements. *Experimental Brain Research* 103: 323–332.

Asanuma H (1973) Cerebral cortical control of movements. *Physiologist* 16: 143–166.

Asatryan DG, Feldman AG (1965) Functional tuning of the nervous system with control of movements or maintenance of a steady posture. I. Mechanographic analysis of the work of the limb on execution of a postural task. *Biophysics* 10: 925–935.

Atkeson CG (1989) Learning arm kinematics and dynamics. *Annual Review of Neuroscience* 12: 157–183.
Atkeson CG, Hollerbach JM (1985) Kinematic features of unrestrianed vertical arm movements. *Journal of Neuroscience* 5: 2318–2320.
Atsuta Y, Garcia-Rill E, Skinner RD (1991) Control of locomotion *in vitro.* I. Deafferentation. *Somatosensory and Motor Research* 8: 45–53.
Babinski F (1899) De l'asynergie cerebelleuse. *Revue Neurologique* 7: 806–816.
Bagesteiro LB, Sainburg RL (2005) Interlimb transfer of load compensation during rapid elbow joint movements. *Experimental Brain Research* 161: 155–165.
Baillieul J (1985) Kinematic programming alternatives for redundant manipulators. In: *Proceedings of the IEEE International Conference of Robotic Automation,* St. Louis, 722–728.
Barto AG, Fagg AH, Sitkoff N, Houk JC (1999) A cerebellar model of timing and prediction in the control of reaching. *Neural Computing* 11: 565–594.
Bassin PV, Bernstein NA, Latash LP (1966) On the problem of the relation between structure and function in the brain from a contemporary point of view. In: Grastschenkov NI (Ed.) *Physiology in Clinical Practice*, pp. 38–71, Nauka: Moscow (in Russian).
Bastian AJ, Martin TA, Keating JG, Thach WT (1996) Cerebellar ataxia: abnormal control of interaction torques across multiple joints. *Journal of Neurophysiology* 76: 492–509.
Bastian AJ, Mugnaini E, Thach WT (1999) Cerebellum. In: Zigmond MJ, Bloom FE, Landis SC, Roberts JL, Squire LR (Eds.) *Fundamental Neuroscience*, pp. 973–992, Academic Press: San Diego.
Baud-Bovy G, Soechting JF (2001) Two virtual fingers in the control of the tripod grasp. *Journal of Neurophysiology* 86: 604–615.
Beer RF, Dewald JP, Dawson ML, Rymer WZ (2004) Target-dependent differences between free and constrained arm movements in chronic hemiparesis. *Experimental Brain Research* 156: 458–470.
Beer RF, Dewald JP, Rymer WZ (2000) Deficits in the coordination of multi-joint arm movements in patients with hemiparesis: evidence for disturbed control of limb dynamics. *Experimental Brain Research* 131: 305–319.
Bellugi U, Bihrle A, Jernigan T, Trauner D, Doherty S (1990) Neuropsychological, neurological, and neuroanatomical profile of Williams syndrome. *American Journal of Medical Genetics* 6 (suppl.): 115–125.
Bemben MG (1998) Age-related alterations in muscular endurance. *Sports Medicine* 25: 259–269.
Bennett DJ, Hollerbach JM, Xu Y, Hunter IW (1992) Time-varying stiffness of human elbow joint during cyclic voluntary movement. *Experimental Brain Research* 88: 433–442.
Beppu H, Suda M, Tanaka R (1984) Analysis of cerebellar motor disorders by visually guided elbow tracking movement. *Brain* 107: 787–809.
Berkinblit MB, Feldman AG, Fukson OI (1986a) Adaptability of innate motor patterns and motor control mechanisms. *Behavioral and Brain Sciences* 9: 585–638.
Berkinblit MB, Gelfand IM, Feldman AG (1986b) A model for the control of multi-joint movements. *Biofizika* 31: 128–138.
Bernstein NA (1930) A new method of mirror cyclographie and its application towards the study of labor movements during work on a workbench. *Hygiene, Safety and Pathology of Labor,* 5: 3–9 and 6: 3–11 (in Russian).
Bernstein NA (1935) The problem of interrelation between coordination and localization. *Archives of Biological Science* 38: 1–35 (in Russian).

Bernstein NA (1947) *On the Construction of Movements.* Medgiz: Moscow (in Russian).
Bernstein NA (1967) *The Co-ordination and Regulation of Movements.* Pergamon Press: Oxford.
Bernstein NA (1996) On dexterity and its development. In: Latash ML, Turvey MT (Eds.) *Dexterity and Its Development*, pp. 1–244, Erlbaum: Mahwah, NJ.
Bernstein NA (2003) *Contemporary Studies on the Physiology of the Neural Process.* Smysl: Moscow, Russia.
Bernstein NA, Popova LN (1930) Investigations on the biomechanics of hitting the keys during piano playing. In: *Proceedings of the Piano-methodological Section of the State Institute of Music Science,* Muzgiz, Moscow, vol. 1, pp. 5–47.
Berntson GG, Torello MW (1982) The paleocerebellum and the integration of behavioral function. *Physiological Psychology* 10: 2–12.
Bhushan N, Shadmehr R (1999) Computational nature of human adaptive control during learning of reaching movements in force fields. *Biological Cybernetics* 81: 39–60.
Biancony R, van der Meulen J (1963) The response to vibration of the end organs of mammalian muscle spindles. *Journal of Neurophysiology* 26: 177–190.
Birbaumer N, Cohen LG (2007) Brain-computer interfaces: communication and restoration of movement in paralysis. *The Journal of Physiology* 579: 621–636.
Bizzi E, Accornero N, Chapple W, Hogan N (1982) Arm trajectory formation in monkeys. *Experimental Brain Research* 46: 139–143.
Bizzi E, Giszter SF, Loeb E, Mussa-Ivaldi FA, Saltiel P (1995) Modular organization of motor behavior in the frog's spinal cord. *Trends in Neurosciences* 18: 442–446.
Bizzi E, Hogan N, Mussa-Ivaldi FA, Giszter S (1992) Does the nervous system use equilibrium-point control to guide single and multiple joint movements? *Behavioral and Brain Science* 15: 603–613.
Bizzi E, Mussa-Ivaldi FA, Giszter S (1991) Computations underlying the execution of movement: a biological perspective. *Science* 253: 287–291.
Blanpied P, Smidt GL (1992) Human plantarflexor stiffness to multiple single-stretch trials. *Journal of Biomechanics* 25: 29–39.
Blazquez PM, Fujii N, Kojima J, Graybiel AM (2002) A network representation of response probability in the striatum. *Neuron* 33: 973–982.
Bloedel JR (1992) Functional heterogeneity with structural homogeneity: how does the cerebellum operate? *Behavioral and Brain Science* 15: 666–678.
Bobath B (1978) *Adult Hemiplegia: Evaluation and Treatment,* William Heinemann: London.
Bongaardt R (2001) How Bernstein conquered movement. In: Latash ML, Zatsiorsky VM (Eds.) *Classics in Movement Science*, pp. 59–84, Human Kinetics: Urbana, IL.
Bongard MM (1970) *Pattern Recognition*, Spartan: New York.
Bosco G, Poppele RE (2002) Encoding of hindlimb kinematics by spinocerebellar circuitry. *Archives Italiennes de Biologie* 140: 185–192.
Bourbonnais D, Vanden Noven S (1989) Weakness in patients with hemiparesis. *American Journal of Occupational Therapy* 43: 313–319.
Braitenberg V (1967) Is the cerebellar cortex a biological clock in the millisecond range? *Progress in Brain Research* 25: 2334–2346.
Brazier MAB (1959) The historical development of neurophysiology. In: Field J, Magoun JHW, Hall VE (Eds.) *Handbook of Physiology, Neurophysiology, vol. I*, pp. 1–58, American Physiological Society: Washington, DC.
Brown MC, Engberg I, Matthews PB (1967) The relative sensitivity to vibration of muscle receptors of the cat. *The Journal of Physiology* 192: 773–800.

Bruwer M, Cruse H (1990) A network model for the control of the movement of a redundant manipulator. *Biological Cybernetics* 62: 549–555.
Bullock D, Grossberg S, Guenther F (1993) A self-organizing neural model of motor equivalence reaching and tool use by a multi-joint arm. *Journal of Cognitive Neuroscience* 5: 408–435.
Burgess PR, Clark FJ (1969) Characteristics of knee joint receptors in the cat. *The Journal of Physiology* 203: 317–335.
Burke D, Andrews CJ, Lance JW (1972) Tonic vibration reflex in spasticity, Parkinson's disease, and normal subjects. *Journal of Neurology, Neurosurgery and Psychiatry* 35: 477–486.
Burke RE, Rudomin P, Zajac FE (1970) Catch property in single mammalian motor units. *Science* 168: 122–124.
Burke RE, Rudomin P, Zajac FE (1976) The effect of activation history on tension production by individual muscle units. *Brain Research* 109: 515–529.
Burnett RA, Laidlaw DH, Enoka RM (2000) Coactivation of the antagonist muscle does not covary with steadiness in old adults. *Journal of Applied Physiology* 89: 61–71.
Burstedt MK, Flanagan JR, Johansson RS (1999) Control of grasp stability in humans under different frictional conditions during multi-digit manipulation. *Journal of Neurophysiology* 82: 2393–2405.
Bursztyn LL, Ganesh G, Imamizu H, Kawato M, Flanagan JR (2006) Neural correlates of internal-model loading. *Current Biology* 16: 2440–2445.
Burtet L, Raptis H, Tunik E, Latash ML, Forget R, Feldman AG (2006) Threshold control of wrist movements revealed by transcranial magnetic stimulation of the motor cortex. In: *Annual Meeting of the Society for Neuroscience*, Atlanta, GA, October 14–18.
Cadoret G, Smith AM (1996) Friction, not texture, dictates grip forces used during object manipulation. *Journal of Neurophysiology* 75: 1963–1969.
Calancie B, Needham-Shropshire B, Jacobs P, Willer K, Zyck G, Green BA (1994) Involuntary stepping after chronic spinal cord injury. Evidence for a central rhythm generator for locomotion in man. *Brain* 117: 1143–1159.
Calvert G, Spence C, Stein BE (Eds.) (2004) *The Handbook of Multi-Sensory Processes*, MIT Press: Cambridge, MA.
Campbell R, MacSweeney M (2004) Neuroimaging studies of cross-modal plasticity and language processing in deaf people. In: Calvert G, Spence C, Stein BE (Eds.) *The Handbook of Multi-Sensory Processes*, pp. 773–783, MIT Press: Cambridge, MA.
Capozzo A, Marchetti M, Tosi V (Eds.) (1992) *Biolocomotion: a Century of Research Using Moving Pictures*, Promograph: Roma, Italy.
Carlton LG, Kim KH, Liu YT, Newell KM (1993) Impulse variability in isometric tasks. *Journal of Motor Behavior* 25: 33–43.
Castiello, U (1997) Arm and mouth coordination during the eating action in humans: a kinematic analysis. *Experimental Brain Research* 115: 552–556.
Cauraugh JH (2004) Coupled rehabilitation protocols and neural plasticity: upper extremity improvements in chronic hemiparesis. *Restorative Neurology and Neuroscience* 22: 337–347.
Celnik PA, Cohen LG (2004) Modulation of motor function and cortical plasticity in health and disease. *Restorative Neurology and Neuroscience* 22: 261–268.
Chan CWY, Kearney RE (1982) Is the functional stretch reflex servo controlled or preprogrammed? *Electroencephalography and Clinical Neurophysiology* 53: 310–324.
Chao EY, Opgrande JD, Axmear FE (1976) Three-dimensional force analysis of finger joints in selected isometric hand functions. *Journal of Biomechanics* 9: 387–396.

Cheruel F, Dormont JF, Amalric M, Schmied A, Farin D (1994) The role of putamen and pallidum in motor initiation in the cat. I. Timing of movement-related single-unit activity. *Experimental Brain Research* 100: 250–266.

Chomsky C (1986) Analytic study of the Tadoma method: language abilities of three deaf-blind subjects. *Journal of Speech and Hearing Research* 29: 332–347.

Chomsky N (2006) *Language and Mind*, Third Edition, Cambridge University Press: Cambridge, MA.

Chomsky N (2007) *Failed States: The Abuse of Power and the Assault on Democracy*, Holt Paperbacks: New York.

Christou EA, Poston B, Enoka JA, Enoka RM (2007) Different neural adjustments improve endpoint accuracy with practice in young and old adults. *Journal of Neurophysiology* 97: 3340–3350.

Churchland P (1985) Reduction, qualia, and the direct introspection of brain states. *Journal of Philosophy* 82: 8–28.

Cioni M, Cocilovo A, Di Pasquale F, Araujo MB, Siqueira CR, Bianco M (1994) Strength deficit of knee extensor muscles of individuals with Down syndrome from childhood to adolescence. *American Journal of Mental Retardation* 99: 166–174.

Cirstea MC, Mitnitski AB, Feldman AG, Levin MF (2003) Interjoint coordination dynamics during reaching in stroke. *Experimental Brain Research* 151: 289–300.

Cisek P, Kalaska JF (2005) Neural correlates of reaching decisions in dorsal premotor cortex: specification of multiple direction choices and final selection of action. *Neuron* 45: 801–814.

Clark B, Graybiel A (1968) Influence of contact cues on the perception of the oculogravic illusion. *Acta Otolaryngology* 65: 373–380.

Clark FJ, Burgess PR (1975) Slowly adapting receptors in cat knee joint: can they signal joint angle? *Journal of Neurophysiology* 38: 1448–1463.

Classen J, Liepert J, Wise SP, Hallett M, Cohen LG (1998) Rapid plasticity of human cortical movement representation induced by practice. *Journal of Neurophysiology* 79: 1117–1123.

Cohen LG, Bandinelli S, Findley TW, Hallett M (1991a) Motor reorganization after upper limb amputation in humans: a study with focal magnetic stimulation. *Brain* 114: 615–627.

Cohen LG, Bandinelli S, Topka HR, Fuhr P, Roth BJ, Hallett M (1991b) Topographic maps of human motor cortex in normal and pathological conditions: mirror movements, amputations and spinal cord injuries. *Electroencephalography and Clinical Neurophysiology* 43 (suppl.): 36–50.

Cohen LG, Celnik P, Pascual-Leone A, Corwell B, Falz L, Dambrosia J, Honda M, Sadato N, Gerloff C, Catala MD, Hallett M (1997) Functional relevance of cross-modal plasticity in blind humans. *Nature* 389: 180–183.

Cole KJ (1991) Grasp force control in older adults. *Journal of Motor Behavior* 23: 251–258.

Cole KJ (2006) Age-related directional bias of fingertip force. *Experimental Brain Research* 175: 285–291.

Cole KJ, Abbs JH (1987) Kinematic and electromyographic responses to perturbation of a rapid grasp. *Journal of Neurophysiology* 57: 1498–1510.

Cole KJ, Abbs JH, Turner GS (1988) Deficits in the production of grip force in Down syndrome. *Developmental Medicine and Child Neurology* 30: 752–758.

Cole KJ, Johansson RS (1993) Friction at the digit-object interface scales the sensorimotor transformation for grip responses to pulling loads. *Experimental Brain Research* 95: 523–532.

Cole KJ, Rotella DL, Harper JG (1998) Tactile impairements cannot explain the effect of age on a grasp and lift. *Experimental Brain Research* 121: 263–269.

Cole KJ, Rotella DL, Harper JG (1999) Mechanisms for age-related changes of fingertip forces during precision gripping and lifting in adults. *Journal of Neuroscience* 19: 3238–3247.

Colebatch JG, Gandevia SC (1989) The distribution of muscular weakness in upper motor neuron lesions affecting the arm. *Brain* 112: 749–763.

Coltz JD, Johnson MTV, Ebner TJ (1999) Cerebellar Purkinje cell simple spike discharge encodes movement velocity in primates during visuomotor arm tracking. *Journal of Neuroscience* 19: 1782–1803.

Connolly BH (2001) Aging in individuals with lifelong disabilities. *Physical and Occupational Therapy in Pediatrics* 21: 23–47.

Contreras-Vidal JL, Grossberg S, Bullock D (1997) A neural model of cerebellar learning for arm movement control: cortico-spino-cerebellar dynamics. *Learning and Memory* 3: 475–502.

Cooke JD, Brown S, Forget R, Lamarre Y (1985) Initial agonist burst duration changes with movement amplitude in a deafferented patient. *Experimental Brain Research* 60: 184–187.

Corcos DM, Gottlieb GL, Agarwal GC (1988) Accuracy constraints upon rapid elbow movements. *Journal of Motor Behavior* 20: 255–272.

Corcos DM, Gottlieb GL, Latash ML, Almeida GL, Agarwal GC (1992) Electromechanical delay: an experimental artifact. *Journal of Electromyography and Kinesiology* 2: 59–68.

Cordo PJ, Burke D, Gandevia SC, Hales JP (1998) Mechanical, neural and perceptual effects of tendon vibration. In: Latash ML (Ed.) *Progress in Motor Control, Vol. 1,* pp. 151–171, Human Kinetics: Champaign, IL.

Cordo PJ, Nashner LM (1982) Properties of postural adjustments associated with rapid arm movements. *Journal of Neurophysiology* 47: 287–302.

Corning PA (2003) *Nature's Magic: Synergy in Evolution and the Fate of Humankind*, Cambridge University Press: Cambridge, UK.

Corning PA (2005) *Holistic Darwinism: Synergy, Cybernetics and the Bioeconomics of Evolution*, University of Chicago Press: Chicago, IL.

Couillandre A, Maton B, Brenière Y (2002) Voluntary toe-walking gait initiation: electromyographical and biomechanical aspects. *Experimental Brain Research* 147: 313–321.

Courchesne E (1997) Brainstem, cerebellar and limbic neuroanatomical abnormalities in autism. *Current Opinion in Neurobiology* 7: 269–278.

Courtine G, De Nunzio AM, Schmid M, Beretta MB, Marco Schieppati M (2007) Stance- and locomotion-dependent processing of vibration-induced proprioceptive inflow from multiple muscles in humans. *Journal of Neurophysiology* 97: 772–779.

Cramer SC (1999) Stroke recovery. Lessons from functional MR imaging and other methods of human brain mapping. *Physical Medicine and Rehabilitation Clinics of North America* 10: 875–886.

Cramer SC, Nelles G, Benson RR, Kaplan JD, Parker RA, Kwong KK, Kennedy DN, Finklestein SP, Rosen BR (1997) A functional MRI study of subjects recovered from hemiparetic stroke. *Stroke* 28: 2518–2527.

Craske B (1977) Perception of impossible limb positions induced by tendon vibration. *Science* 196: 71–73.

Crawford JD, Vilis T (1995) How do motor systems deal with the problems of controlling three-dimensional rotations? *Journal of Motor Behavior* 27: 89–99.

Crome LC, Stern J (1967) *Pathology of mental retardation,* Churchill: London.

Crossman ERFW, Goodeve PJ (1983) Feedback control of hand-movement and Fitts' law. *Quarterly Journal of Experimental Psychology* 35A: 251–278.

Crowninshield RD, Brand RA (1981) A physiologically based criterion of muscle force prediction in locomotion. *Journal of Biomechanics* 14: 793–801.

Cruse H, Bruwer M (1987) The human arm as a redundant manipulator: the control of path and joint angles. *Biological Cybernetics* 57: 137–144.

Cusumano JP, Cesari P (2006) Body-goal variability mapping in an aiming task. *Biological Cybernetics* 94: 367–379.

Cytowic RE (1989) *Synaesthesia: A Union of the Senses*, Springer-Verlag: New York.

Cytowic RE (1997) Synaesthesia: phenomenology and neuropsychology. A review of current knowledge. In: Baron-Cohen S, Harrison JE (Eds.) *Synaesthesia: Classic and Contemporary Readings*, pp. 17–39, Blackwell: Cambridge, UK.

Dailey A, Martindale C, Borkum J (1997) Creativity, synaesthesia and physiognomic perception. *Creativity Research Journal* 10: 1–8.

Danion F (2007) The contribution of non-digital afferent signals to grip force adjustments evoked by brisk unloading of the arm or the held object. *Clinical Neurophysiology* 118: 146–154.

Danion F, Latash ML, Li S (2003) Finger interactions studied with transcranial magnetic stimulation during multi-finger force production tasks. *Clinical Neurophysiology* 114: 1445–1455.

Danion F, Schöner G, Latash ML, Li S, Scholz JP, Zatsiorsky VM (2003) A force mode hypothesis for finger interaction during multi-finger force production tasks. *Biological Cybernetics* 88: 91–98.

Danna-Dos-Santos A, Slomka K, Zatsiorsky VM, Latash ML (2007) Muscle modes and synergies during voluntary body sway. *Experimental Brain Research* 179: 539–550.

d'Avella A, Bizzi E (2005) Shared and specific muscle synergies in natural motor behaviors. *Proceedings of the National Academy of Sciences USA* 102: 3076–3081.

d'Avella A, Saltiel P, Bizzi E (2003) Combinations of muscle synergies in the construction of a natural motor behavior. *Nature Neuroscience* 6: 300–308.

Davids K, Bennett S, Newell KM (Eds.) (2005) *Movement System Variability*, Human Kinetics: Urbana, IL.

Day BL, Guerraz M (2007) Feedforward versus feedback modulation of human vestibular-evoked balance responses by visual self-motion information. *The Journal of Physiology* 582: 153–161.

De Gail P, Lance JW, Neilson PD (1966) Differential effects on tonic and phasic reflex mechanisms produced by vibration of muscles in man. *Journal of Neurology, Neurosurgery and Psychiatry* 29: 1–11.

Delwaide PJ (1969) Approach to the physiopathology of spasticity: the Hoffmann reflex and vibrations applied to the Achilles tendon. *Revue Neurologique (Paris)* 121: 72–74.

Demieville HN, Partridge LD (1980) Probability of peripheral interaction between motor units and implications for motor control. *American Journal of Physiology* 238: R119–R137.

De Montaigne M (2003) *The Complete Works*. Everyman's Library, Alfred A. Knopf: New York.

De Schutter E, Maex R (1996) The cerebellum: cortical processing and theory. *Current Opinion in Neurobiology* 6: 759–764.

Desmedt JE, Godaux E (1978) Mechanism of the vibration paradox: excitatory and inhibitory effects of tendon vibration on single soleus muscle motor units in man. *The Journal of Physiology* 285: 197–207.

Desmurget M, Prablanc C, Rossetti Y, Arzi M, Paulignan Y, Urquizar C, Mignot JC (1995) Postural and synergic control for three-dimensional movements of reaching and grasping. *Journal of Neurophysiology* 74: 905–910.

Desrosiers J, Bourbonnais D, Bravo G, Roy P-M, Guay M (1996) Performance of the 'unaffected' upper extremity of elderly stroke patients. *Stroke* 27: 1564–1570.

Devlin JT, Watkins KE (2007) Stimulating language: insights from TMS. *Brain* 130: 610–622.

DeWald JP, Pope PS, Given JD, Buchanan TS, Rymer WZ (1995) Abnormal muscle coactivation patterns during isometric torque generation at the elbow and shoulder in hemiparetic subjects. *Brain* 118: 495–510.

De Wolf S, Slijper H, Latash ML (1998) Anticipatory postural adjustments during self-paced and reaction-time movements. *Experimental Brain Research* 121: 7–19.

Dichgans J, Brandt T (1973) Optokinetic motion sickness and pseudo-Coriolis effects induced by moving visual stimuli. *Acta Otolaryngology* 76: 339–348.

Dickstein R, Shupert CL, Horak FB (2001) Fingertip touch improves postural stability in patients with peripheral neuropathy. *Gait and Posture* 14: 238–247.

Diedrichsen J, Verstynen T, Hon A, Zhang Y, Ivry RB (2007) Illusions of force perception: the role of sensori-motor predictions, visual information, and motor errors. *Journal of Neurophysiology* 97: 3305–3313.

Diener HC, Bacher M, Guschlbauer B, Thomas C, Dichgans J (1993) The coordination of posture and voluntary movement in patients with hemiparesis. *Journal of Neurology* 240: 161–167.

Dietz V, Berger W (1984) Interlimb coordination of posture in patients with spastic paresis. Impaired function of spinal reflexes. *Brain* 107: 965–978.

Dietz V, Quintern J, Berger W (1984) Corrective reactions to stumbling in man: functional significance of spinal and transcortical reflexes. *Neuroscience Letters* 44: 131–135.

Dijkstra TM, Schöner G, Giese MA, Gielen CCAM (1994) Frequency dependence of the action-perception cycle for postural control in a moving visual environment: relative phase dynamics. *Biological Cybernetics* 71: 489–501.

DiZio P, Lackner JR (1995) Motor adaptation to Coriolis force perturbations of reaching movements: endpoint but not trajectory adaptation transfers to the nonexposed arm. *Journal of Neurophysiology* 74: 1787–1792.

DiZio R, Lackner JR, Held RM, Shinn-Cunningham B, Durlach NI (2001) Gravitoinertial force magnitude and direction influence heaf-centric auditory localization. *Journal of Neurophysiology* 85: 2455–2460.

Domino G (1989) Synaesthesia and creativity in fine arts students: an empirical look. *Creativity Research Journal* 2: 17–29.

Domkin D, Laczko J, Djupsjöbacka M, Jaric S, Latash ML (2005) Joint angle variability in 3D bimanual pointing: uncontrolled manifold analysis. *Experimental Brain Research* 163: 44–57.

Domkin D, Laczko J, Jaric S, Johansson H, Latash ML (2002) Structure of joint variability in bimanual pointing tasks. *Experimental Brain Research* 143: 11–23.

Donker SF, Roerdink M, Greven AJ, Peter J, Beek PJ (2007) Regularity of center-of-pressure trajectories depends on the amount of attention invested in postural control. *Experimental Brain Research* 181: 1–11.

Dounskaia N (2007) Kinematic invariants during cyclical arm movements. *Biological Cybernetics* 96: 147–163.

Dozza M, Fay B, Horak FB, Chiari L (2007) Auditory biofeedback substitutes for loss of sensory information in maintaining stance. *Experimental Brain Research* 178: 37–48.

Duarte M, Latash ML (2007) Effects of postural task requirements on the speed-accuracy trade-off. *Experimental Brain Research* 180: 457–467.

Dufosse M, Hugon M, Massion J (1985) Postural forearm changes induced by predictable in time or voluntary triggered unloading in man. *Experimental Brain Research* 60: 330–334.

Dum RP, Strick PL (2003) An unfolded map of the cerebellar dentate nucleus and its projections to the cerebral cortex. *Journal of Neurophysiology* 89: 634–639.

Dumont CE, Popovic MR, Keller T, Sheikh R (2006) Dynamic force-sharing in multi-digit task. *Clinical Biomechanics* 21: 138–146.

Duque J, Hummel F, Celnik P, Murase N, Mazzocchio R, Cohen LG (2005) Transcallosal inhibition in chronic subcortical stroke. *Neuroimage* 28: 940–946.

Duysens J, Loeb GE (1980) Modulation of ipsi- and contralateral reflex responses in unrestrained walking cats. *Journal of Neurophysiology* 44: 1024–1037.

Duysens J, Loeb GE, Weston BJ (1980) Crossed flexor reflex responses and their reversal in freely walking cats. *Brain Research* 197: 538–542.

Duysens J, Pearson KG (1976) The role of cutaneous afferents from the distal hindlimb in the regulation of the stepcycle in thalamic cats. *Experimental Brain Research* 24: 245–255.

Edin BB, Westling G, Johansson RS (1992) Independent control of human fingertip forces at individual digits during precision lifting. *The Journal of Physiology* 450: 547–564.

Edwards JM, Elliott D (1989) Asymmetries in intermanual transfer of training and motor overflow in adults with Down's syndrome and nonhandicapped children. *Journal of Clinical and Experimental Neuropsychology* 11: 959–966.

Edwards JM, Elliott D, Lee TD (1986) Contextual interference effects during skill acquisition and transfer in Down's syndrome adolescents. *Adapted Physical Education Quarterly* 3: 250–258.

Ehrsson HH, Fagergren A, Ehrsson GO, Forssberg H (2007) Holding an object: neural activity associated with fingertip force adjustments to external perturbations. *Journal of Neurophysiology* 97: 1342–1352.

Eklund G, Hagbarth KE (1966) Normal variability of tonic vibration reflexes in man. *Experimental Neurology* 16: 80–92.

Eklund G, Hagbarth KE (1967) Vibratory induced motor effects in normal man and in patients with spastic paralysis. *Electroencephalography and Clinical Neurophysiology* 23: 393.

Elble RJ, Moody C, Leffler K, Sinha R (2004) The initiation of normal walking. *Movement Disorders* 2: 139–146.

Ellaway PH, Davey NJ, Ljubisavljevic M (1999) Magnetic stimulation of the nervous system. In: Windhorst U, Johansson H (Eds.) *Modern Techniques in Neuroscience Research*, pp. 869–892, Springer Verlag: Berlin.

Ellis MD, Acosta AM, Yao J, Dewald JP (2007) Position-dependent torque coupling and associated muscle activation in the hemiparetic upper extremity. *Experimental Brain Research* 176: 594–602.

Ellis MD, Holubar BG, Acosta AM, Beer RF, Dewald JP (2005) Modifiability of abnormal isometric elbow and shoulder joint torque coupling after stroke. *Muscle and Nerve* 32: 170–178.

Emken JL, Benitez R, Sideris A, Bobrow JE, Reinkensmeyer DJ (2007) Motor adaptation as a greedy optimization of error and effort. *Journal of Neurophysiology* 97: 3997–4006.

Enoka RM, Christou EA, Hunter SK, Kornatz KW, Semmler JG, Taylor AM, Tracy BL (2003) Mechanisms that contribute to differences in motor performance between young and old adults. *Journal of Electromyography and Kinesiology* 13: 1–12.

Evarts EV (1968) Relation of pyramidal tract activity to force exerted during voluntary movement. *Journal of Neurophysiology* 31: 14–27.

Evarts EV, Teravainen H, Calne DB (1981) Reaction time in Parkinson's disease. *Brain* 104: 167–186.

Fahn S, Greene PE, Ford B, Bressman SB (1998) *Handbook of Movement Disorders*, Blackwell Science: Philadelphia, PA.

Farina D, Merletti R, Enoka RM (2004) The extraction of neural strategies from the surface EMG. *Journal of Applied Physiology* 96: 1486–1495.

Farnetani E, Recasens D (1993) Anticipatory consonant-to-vowel coarticulation in the production of VCV sequences in Italian. *Language and Speech* 36: 279–302.

Feldman AG (1966) Functional tuning of the nervous system with control of movement or maintenance of a steady posture. II. Controllable parameters of the muscle. *Biophysics* 11: 565–578.

Feldman AG (1979) *Central and Reflex Mechanisms of Motor Control*, Nauka: Moscow (in Russian).

Feldman AG (1980) Superposition of motor programs. I. Rhythmic forearm movements in man. *Neuroscience* 5: 81–90.

Feldman AG (1986) Once more on the equilibrium-point hypothesis (λ-model) for motor control. *Journal of Motor Behavior* 18: 17–54.

Feldman AG, Adamovich SV, Ostry Dl, Flanagan IR (1990) The origin of electromyograms—explanations based on the equilibrium point hypothesis. In: Winters IM, Woo SL-Y (Eds.) *Multiple Muscle Systems*, pp. 195–312, Springer: Berlin, Heidelberg, New York.

Feldman AG, Goussev V, Sangole A, Levin MF (2007) Threshold position control and the principle of minimal interaction in motor actions. *Progress in Brain Research* 165: 267–281.

Feldman AG, Latash ML (1982a) Interaction of afferent and efferent signals underlying joint position sense: empirical and theoretical approaches. *Journal of Motor Behavior* 14: 174–193.

Feldman AG, Latash ML (1982b) Afferent and efferent components of joint position sense: interpretation of kinaesthetic illusions. *Biological Cybernetics* 42: 205–214.

Feldman AG, Latash ML (1982c) Inversions of vibration-induced senso-motor events caused by supraspinal influences in man. *Neuroscience Letters* 31: 147–151.

Feldman AG, Latash ML (2005) Testing hypotheses and the advancement of science: recent attempts to falsify the equilibrium-point hypothesis. *Experimental Brain Research* 161: 91–103.

Feldman AG, Levin MF (1995) Positional frames of reference in motor control: their origin and use. *Behavioral and Brain Sciences* 18: 723–806.

Feldman AG, Levin MF, Mitnitski AM, Archambault P (1998) 1998 ISEK Congress Keynote Lecture: multi-muscle control in human movements. *Journal of Electromyography and Kinesiology* 8: 383–390.

Feldman AG, Orlovsky GN (1972) The influence of different descending systems on the tonic stretch reflex in the cat. *Experimental Neurology* 37: 481–494.

Feldman AG, Ostry DJ, Levin MF, Gribble PL, Mitnitski AB (1998) Recent tests of the equilibrium-point hypothesis (λ model). *Motor Control* 2: 189–205.

Feynman R (1994) *The Character of Physical Law*, Random House: New York.

Field EC, Stein PS (1997) Spinal cord coordination of hindlimb movements in the turtle: intralimb temporal relationships during scratching and swimming. *Journal of Neurophysiology* 78: 1394–1403.

Field-Fote EC, Volker Dietz V (2007) Single joint perturbation during gait: preserved compensatory response pattern in spinal cord injured subjects. *Clinical Neurophysiology* 118: 1607–1616.

Fingelkurts AA, Fingelkurts AA, Ermolaev VA, Kaplan AY (2006) Stability, reliability and consistency of the compositions of brain oscillations. *International Journal of Psychophysiology* 59: 116–126.

Fitts PM (1954) The information capacity of the human motor system in controlling the amplitude of movement. *Journal of Experimental Psychology* 47: 381–391.

Fitts PM, Radford BK (1966) Information capacity of discrete motor responses under different cognitive sets. *Journal of Experimental Psychology* 71: 475–482.

Flanagan JR, Wing AM (1993) Modulation of grasp force with load force during point-to-point arm movements. *Experimental Brain Research* 95: 131–143.

Flanagan JR, Wing AM (1995) The stability of precision grasp forces during cyclic arm movements with a hand-held load. *Experimental Brain Research* 105: 455–464.

Flash T (1987) The control of hand equilibrium trajectories in multi-joint arm movements. *Biological Cybernetics* 57: 257–274.

Flash T, Handzel AA (2007) Affine differential geometry analysis of human arm movement. *Biological Cybernetics* 96: 577–601.

Flash T, Hogan N (1985) The coordination of arm movements: an experimentally confirmed mathematical model. *Journal of Neuroscience* 5: 1688–1703.

Florenskiy PA (1999) *Selected Works, Vol. 3,* Mysl': Moscow.

Flowers KA (1975) Ballistic and corrective movements on an aiming task. *Neurology* 25: 413–421.

Folkins JW, Linville RN (1983) The effects of varying lower-lip displacement on upper-lip movements: implications for the coordination of speech movements. *Journal of Speech and Hearing Research* 26: 209–217.

Forssberg H (1979) Stumbling corrective reaction: a phase dependent compensatory reaction during locomotion. *Journal of Neurophysiology* 42: 936–953.

Forssberg H, Eliasson AC, Redon-Zouitenn C, Mercuri E, Dubowitz L (1999) Impaired grip-lift synergy in children with unilateral brain lesions. *Brain* 122: 1157–1168.

Forssberg H, Grillner S, Rossignol S (1975) Phase dependent reflex reversal during walking in chronic spinal cat. *Brain Research* 85: 103–107.

Forssberg H, Grillner S, Rossignol S (1977) Phasic gain control of reflexes from the dorsum of the paw during spinal locomotion. *Brain Research* 132: 121–139.

Forssberg H, Grillner S, Rossignol S, Wallen P (1976) Phasic control of reflexes during locomotion in vertebrates. In: R Herman (Ed.) *Neural Control of Locomotion,* pp. 647–674, Plenum Press: New York, London.

Fort A, Giard M-H (2004) Mutiple electrophysiological mechanisms of audiovisual integration in human perception. In: Calvert G, Spence C, Stein BE (Eds.) *The Handbook of Multi-Sensory Processes*, pp. 503–514, MIT Press: Cambridge, MA.

Freitas SMSF, Duarte M, Latash ML (2006) Two kinematic synergies in voluntary whole-body movements during standing. *Journal of Neurophysiology* 95: 636–645.

Fridman EA, Celnik P, Cohen LG (2004) Visual cortex engagement in tactile function in the presence of blindness. In: Calvert G, Spence C, Stein BE (Eds.) *The Handbook of Multi-Sensory Processes*, pp. 711–718, MIT Press: Cambridge, MA.

Fuchs F, Martyn DA (2005) Length dependent Ca^{2+} activation in cardiac muscle: some remaining questions. *Journal of Muscle Research and Cell Motility* 26: 199–212.

Fuhr P, Cohen LG, Dang N, Findley TW, Haghighi S, Oro J, Hallett M (1992) Physiological analysis of motor reorganization following lower limb amputation. *Electroencephalography and Clinical Neurophysiology* 85: 53–60.

Fujita T, Nakamura S, Ohue M, Fujii Y, Miyauchi A, Takagi Y, Tsugeno H (2005) Effect of age on body sway assessed by computerized posturography. *Journal of Bone and Mineral Metabolism* 23: 152–156.

Fukai T (1999) Sequence generation in arbitrary temporal patterns from theta-nested gamma oscillations: a model of the basal ganglia-thalamo-cortical loops. *Neural Networks* 12: 975–987.

Fukson OI, Berkinblit MB, Feldman AG (1980) The spinal frog takes into account the scheme of its body during the wiping reflex. *Science* 209: 1261–1263.

Fukuda N, Wu Y, Nair P, Granzier HL (2005) Phosphorylation of titin modulates passive stiffness of cardiac muscle in a titin isoform-dependent manner. *Journal of General Physiology* 125: 257–271.

Gao F, Latash ML, Zatsiorsky VM (2005a) Control of finger force direction in the flexion-extension plane. *Experimental Brain Research* 161: 307–315.

Gao F, Latash ML, Zatsiorsky VM (2005b) Internal forces during object manipulation. *Experimental Brain Research* 165: 69–83.

Gao F, Li S, Li Z-M, Latash ML, Zatsiorsky VM (2003) Matrix analyses of interaction among fingers in static force production tasks. *Biological Cybernetics* 89: 407–414.

Gardner EP, Babu KS, Reitzen SD, Ghosh G, Brown AS, Chen J, Hall AL, Herzlinger MD, Kohlenstein JB, Ro JY (2007) Neurophysiology of prehension. I. Posterior parietal cortex and object-oriented hand behaviors. *Journal of Neurophysiology* 97: 387–406.

Gelfand IM (1991) Two archetypes in the psychology of man. *Nonlinear Science Today* 1: 11–16.

Gelfand IM, Latash ML (1998) On the problem of adequate language in movement science. *Motor Control* 2: 306–313.

Gelfand IM, Latash ML (2002) On the problem of adequate language in biology. In: Latash ML (Ed.) *Progress in Motor Control. vol. 2: Structure-Function Relations in Voluntary Movement*, pp. 209–228, Human Kinetics: Urbana, IL.

Gelfand IM, Tsetlin ML (1961) The non-local search principle in automatic optimization systems. *Doklady Akademii Nauk SSSR* (*Proceedings of the Academy of Sciences of the USSR*) 137: 295.

Gelfand IM, Tsetlin ML (1962) On certain methods of control of complex systems. *Advances in Mathematical Sciences* 17: 103 (in Russian).

Gelfand IM, Tsetlin ML (1966) On mathematical modeling of the mechanisms of the central nervous system. In: Gelfand IM, Gurfinkel VS, Fomin SV, Tsetlin ML (Eds.) *Models of the Structural-Functional Organization of Certain Biological Systems*, pp. 9–26, Nauka: Moscow, Russia (in Russian, a translation is available in 1971 edition by MIT Press: Cambridge MA).

Gentil M (1992) Variability of motor strategies. *Brain and Language* 42: 30–37.

Georgopoulos AP (1986) On reaching. *Annual Review of Neuroscience* 9: 147–170.

Georgopoulos AP, Kalaska JF, Caminiti R, Massey JT (1982) On the relations between the direction of two-dimensional arm movements and cell discharge in primate motor cortex. *Journal of Neuroscience* 2: 1527–1537.

Georgopoulos AP, Lurito JT, Petrides M, Schwartz AB, Massey JT (1989) Mental rotation of the neuronal population vector. *Science* 243: 234–236.

Georgopoulos AP, Schwartz AB, Kettner RE (1986) Neural population coding of movement direction. *Science* 233: 1416–1419.

Geuze RH (2005) Postural control in children with developmental coordination disorder. *Neural Plasticity* 12: 183–196.

Gibson JJ (1979) *The Ecological Approach to Visual Perception*, Houghton Mifflin: Boston, MA.

Gielen CCAM, Ramaekers L, van Zuylen EJ (1988) Long-latency stretch reflexes as co-ordinated functional responses in man. *The Journal of Physiology* 407: 275–292.

Gielen CCAM, van der Oosten K, Pull ter Gunne F (1985) Relations between EMG activation patterns and kinematic properties of aimed movements. *Journal of Motor Behavior* 17: 421–442.

Gielen CCAM, van Zuylen EJ (1986) Coordination of arm muscles during flexion and supination: application of the tensor analysis approach. *Neuroscience* 17: 527–539.

Gielen CCAM, Vrijenhoek EJ, Flash T, Neggers SF (1997) Arm position constraints during pointing and reaching in 3-D space. *Journal of Neurophysiology* 78: 660–673.

Giese MA, Poggio T (2003) Neural mechanisms for the recognition of biological movements. *Nature Reviews in Neuroscience* 4: 179–192.

Gillberg C, Kadesjo B (2003) Why bother about clumsiness? The implications of having developmental coordination disorder (DCD). *Neural Plasticity* 10: 59–68.

Gillies JD, Lance JW, Tassinari CA (1970) The mechanism of the suppression of the monosynaptic reflex by vibration. *Proceedings of the Australian Association of Neurologists* 7: 97–102.

Ginanneschi F, Dominici F, Biasella A, Gelli F, Rossi A (2006) Changes in corticomotor excitability of forearm muscles in relation to static shoulder positions. *Brain Research* 1073–1074: 332–338.

Giszter SF, Mussa-Ivaldi FA, Bizzi E (1993) Convergent force fields organized in the frog's spinal cord. *Journal of Neuroscience* 13: 467–491.

Glansdorf P, Prigogine I (1971) *Thermodynamic Theory of Structures, Stability and Fluctuations,* Wiley: New York.

Glazer VD, Gauzelman VE (1997) Linear and nonlinear properties of simple cells of the striate cortex of the cat: two types of nonlinearity. *Experimental Brain Research* 117: 281–291.

Goldberger ME (1977) Locomotor recovery after unilateral hindlimb deafferentation in cats. *Brain Research* 123: 59–74.

Gomi H, Kawato M (1996) Equilibrium-point hypothesis examined by measured arm stiffness during multi-joint movement. *Science* 272: 117–120.

Goodman SR, Latash ML (2006) Feedforward control of a redundant motor system. *Biological Cybernetics* 95: 271–280.

Goodman SR, Shim JK, Zatsiorsky VM, Latash ML (2005) Motor variability within a multi-effector system: experimental and analytical studies of multi-finger production of quick force pulses. *Experimental Brain Research* 163: 75–85.

Goodwin GM, McCloskey DI, Matthews PB (1972) The contribution of muscle afferents to kinaesthesia shown by vibration induced illusions of movement and by the effects of paralysing joint afferents. *Brain* 95: 705–748.

Gorniak S, Zatsiorsky VM, Latash ML (2007a) Hierarchies of synergies: an example of the two-hand, multi-finger tasks. *Experimental Brain Research* 179: 167–180.

Gorniak S, Zatsiorsky VM, Latash ML (2007b) Emerging and disappearing synergies in a hierarchically controlled system. *Experimental Brain Research* 183: 259–270.

Gottlieb GL, Corcos DM, Agarwal GC (1989) Strategies for the control of voluntary movements with one mechanical degree of freedom. *Behavioral and Brain Sciences* 12: 189–250.

Gottlieb GL, Song Q, Hong DA, Almeida GL, Corcos D (1996) Coordinating movement at two joints: a principle of linear covariance. *Journal of Neurophysiology* 75: 1760–1764.

Graybiel AM (1995) Building action repertoires: memory and learning functions of the basal ganglia. *Current Opinion in Neurobiology* 5: 733–741.

Graybiel AM (1997) The basal ganglia and cognitive pattern generators. *Schizophrenia Bulletin* 23: 459–469.

Graybiel AM (2004) Network-level neuroplasticity in cortico-basal ganglia pathways *Parkinsonism and Related Disorders* 10: 293–296.

Graybiel AM (2005) The basal ganglia: learning new tricks and loving it. *Current Opinion in Neurobiology* 15: 638–644.

Graziano MS, Aflalo TN, Cooke DF (2005) Arm movements evoked by electrical stimulation in the motor cortex of monkeys. *Journal of Neurophysiology* 94: 4209–4223.

Graziano MS, Taylor CSR, Moore T (2002) Complex movements evoked by microstimulation of precentral cortex. *Neuron* 34: 841–851.

Graziano MSA, Gross CG, Taylor CSR, Moore T (2004) Multi-sensory neurons for the control of defensive movements. In: Calvert G, Spence C, Stein BE (Eds.) *The Handbook of Multi-Sensory Processes*, pp. 443–452, MIT Press: Cambridge, MA.

Gribble PL, Ostry DJ (2000) Compensation for loads during arm movements using equilibrium-point control. *Experimental Brain Research* 135: 474–482.

Gribble PL, Ostry DJ, Sanguineti V, Laboissiere R (1998) Are complex control signals required for human arm movements? *Journal of Neurophysiology* 79: 1409–1424.

Grillner S (1975) Locomotion in vertebrates: central mechanisms and reflex interaction. *Physiological Reviews* 55: 247–304.

Grillner S (1979) Interaction between central and peripheral mechanisms in the control of locomotion. In: Granit R, Pompeiano O (Eds.) *Reflex Control of Posture and Movement*, pp. 227–235, Elsevier: Amsterdam, New York, Oxford.

Grillner S, Wallen P (1985) Central pattern generators for locomotion, with special reference to vertebrates. *Annual Reviews in Neuroscience* 8: 233–261.

Grossman ED, Blake R, Kim CY (2004) Learning to see biological motion: brain activity parallels behavior. *Journal of Cognitive Neuroscience* 16: 1669–1679.

Grush R (2004) The emulation theory of representation: motor control, imagery, and perception. *Behavioral and Brain Sciences* 27: 377–396.

Guigon E, Baraduc P, Desmurget M (2007) Computational motor control: redundancy and invariance. *Journal of Neurophysiology* 97: 331–347.

Gurfinkel VS, Latash ML (1978) Reflex reversals in shin muscles. *Human Physiology* 4: 30–35.

Gurfinkel VS, Levik YS, Kazennikov OV, Selionov VA (1998) Locomotor-like movements evoked by leg muscle vibration in humans. *European Journal of Neuroscience* 10: 1608–1612.

Gupta R, Ashe J (2007) Lack of adaptation to random conflicting force fields of variable magnitude. *Journal of Neurophysiology* 97: 738–745.

Gutman A (1994) Gelfand-Tsetlin principle of minimal afferentation and bistability of dendrites. *International Journal of Neural Systems* 5: 83–86.

Gutman SR, Gottlieb GL (1992) Analysis of kinematic invariances of multi-joint reaching movement. *Biological Cybernetics* 73: 311–322.

Gutman SR, Latash ML, Gottlieb GL, Almeida GL (1993) Kinematic description of variability of fast movements: analytical and experimental approaches. *Biological Cybernetics* 69: 485–492.

Hagbarth KE, Eklund G (1968) The effects of muscle vibration in spasticity, rigidity, and cerebellar disorders. *Journal of Neurology, Neurosurgery and Psychiatry* 31: 207–213.

Halbertsma JM, Miller S, van der Meche FGA (1976) Basic programs for the phasing of flexion and extension movements of the limbs during locomotion. In: Herman R (Ed.) *Neural Control of Locomotion,* pp. 489–517, Plenum Press: New York, London.

Hallett M (2001) Plasticity of the human motor cortex and recovery from stroke. *Brain Research Reviews* 36: 169–174.

Hallett M (2007) Volitional control of movement: the physiology of free will. *Clinical Neurophysiology* 118: 1179–1192.

Harding DC, Brandt KD, Hillberry BM (1993) Finger joint force minimization in pianists using optimization techniques. *Journal of Biomechanics* 26: 1403–1412.

Hasan Z (1986) Optimized movement trajectories and joint stiffness in unperturbed, inertially loaded movements. *Biological Cybernetics* 53: 373–382.

Hasan Z (2005) The human motor control system's response to mechanical perturbation: should it, can it, and does it ensure stability? *Journal of Motor Behavior* 37: 484–493.

Hayashi R, Miyake A, Jijiwa H, Watanabe S (1981) Postural readjustment to body sway induced by vibration in man. *Experimental Brain Research* 43: 217–225.

Henneman E, Somjen G, Carpenter DO (1965) Excitability and inhibitability of motoneurones of different sizes. *Journal of Neurophysiology* 28: 599–620.

Herman R, Mecomber SA (1971) Vibration-elicited reflexes in normal and spastic muscle in man. *American Journal of Physical Medicine* 50: 169–183.

Herter TM, Kurtzer I, Cabel DW, Haunts KA, Scott SH (2007) Characterization of torque-related activity in primary motor cortex during a multi-joint postural task. *Journal of Neurophysiology* 97: 2887–2899.

Hess CW, Mills KR, Murray NM (1987) Responses in small hand muscles from magnetic stimulation of the human brain. *The Journal of Physiology* 388: 397–419.

Hill AV (1938) The heat of shortening and the dynamic constants of muscle. *Proceedings of the Royal Society of London B* 126: 136–195.

Hill AV (1953) The mechanics of active muscle. *Proceedings of the Royal Society of London B* 141: 104–117.

Hinder MR, Milner TE (2003) The case for an internal dynamics model versus equilibrium point control in human movement. *The Journal of Physiology* 549: 953–963.

Hogan N (1984) An organizational principle for a class of voluntary movements. *Journal of Neuroscience* 4: 2745–2754.

Hogan N (1985) The mechanics of multi-joint posture and movement control. *Biological Cybernetics* 52: 315–331.

Holdefer RN, Miller LE (2002) Primary motor cortical neurons encode functional muscle synergies. *Experimental Brain Research* 146: 233–243.

Holden M, Ventura J, Lackner JR (1994) Stabilization of posture by precision contact of the index finger. *Journal of Vestibular Research* 4: 285–301.

Holldobler B (1999) Multi-modal signals in ant communication. *Journal of Comparative Physiology, A: Sensory, Neural and Behavioral Physiology* 184: 129–141.

Hollerbach JM, Atkeson CG (1987) Deducing planning variables from experimental arm trajectories pitfalls and possibilities. *Biological Cybernetics* 56: 279–292.

Hollerbach JM, Flash T (1982) Dynamic interaction between limb segments during planar arm movements. *Biological Cybernetics* 44: 67–77.

Hommel B, Musseler J, Aschersleben G, Prinz W (2001) The theory of event coding (TEC): a framework for perception and action planning. *Behavioral and Brain Sciences* 24: 849–878.

Horak FB, Nashner LM (1986) Central programming of postural movements: adaptation to altered support-surface configurations. *Journal of Neurophysiology* 55: 1369–1381.

Horak FB, Shupert CL, Mirka A (1989) Components of postural dyscontrol in the elderly: a review. *Neurobiology of Aging* 10: 727–738.

Houk JC (2005) Agents of the mind. *Biological Cybernetics* 92: 427–437.

Houk JC, Buckingham JT, Barto AG (1996) Models of the cerebellum and motor learning. *The Behavioral and Brain Sciences* 19: 368–383.

Houk JC, Gibson AR (1987) Sensorimotor processing through the cerebellum. In: King JS (Ed.) *New Concepts in Cerebellar Neurobiology*, pp. 387–416, Alan R. Liss: New York.

Howard IP, Templeton WB (1966) *Human Spatial Orientation*, Wiley: New York.

Hsu W-L, Scholz JP, Schöner G, Jeka JJ, Kiemel T (2007) Control and estimation of posture during quiet stance depends on multi-joint coordination. *Journal of Neurophysiology* 97: 3024–3035.

Hughlings Jackson J (1889) On the comparative study of disease of the nervous system. *British Medical Journal* 2: 355–362.

Hultborn H, Brownstone RB, Toth TI, Gossard JP (2004) Key mechanisms for setting the input-output gain across the motoneuron pool. *Progress in Brain Research* 143: 77–95.

Imamizu H, Kuroda T, Miyauchi S, Yoshioka T, Kawato M (2003) Modular organization of internal models of tools in the human cerebellum. *Proceedings of the National Academy of Sciences USA* 100: 5461–5466.

Imamizu H, Kuroda T, Yoshioka T, Kawato M (2004) Functional magnetic resonance imaging examination of two modular architectures for switching multiple internal models. *Journal of Neuroscience* 24: 1173–1181.

Imamizu H, Miyauchi S, Tamada T, Sasaki Y, Takino R, Putz B, Yoshioka T, Kawato M (2000) Human cerebellar activity reflecting an acquired internal model of a new tool. *Nature* 403: 192–195.

Imamizu H, Uno Y, Kawato M (1995) Internal representation of the motor apparatus: implications from generalization in visuomotor learning. *Journal of Experimental Psychology: Human Perception and Performance* 21: 1174–1198.

Inglin B, Woollacott MH (1988) Anticipatory postural adjustments associated with reaction time arm movements: a comparison between young and old. *Journal of Gerontology* 43: M105–M113.

Ito M (1989) Long-term depression. *Annual Reviews of Neuroscience* 12: 85–102.

Ito M (2005) Bases and implications of learning in the cerebellum—adaptive control and internal model mechanism. *Progress in Brain Research* 148: 95–109.

Ito T, Azuma T, Yamashita N (2003) Anticipatory control in the initiation of a single step under biomechanical constraints in humans. *Neuroscience Letters* 352: 207–210.

Ito T, Kimura T, Gomi H (2005) The motor cortex is involved in reflexive compensatory adjustment of speech articulation. *Neuroreport* 16: 1791–1794.

Ivanenko YP, Cappellini G, Dominici N, Poppele RE, Lacquaniti F (2005) Coordination of locomotion with voluntary movements in humans. *Journal of Neuroscience* 25: 7238–7253.

Ivanenko YP, Grasso R, Lacquaniti F (2000) Influence of leg muscle vibration on human walking. *Journal of Neurophysiology* 84: 1737–1747.

Ivanenko YP, Poppele RE, Lacquaniti F (2004) Five basic muscle activation patterns account for muscle activity during human locomotion. *The Journal of Physiology* 556: 267–282.

Ivanenko YP, Wright WG, Gurfinkel VS, Horak F, Cordo P (2006) Interaction of involuntary post-contraction activity with locomotor movements. *Experimental Brain Research* 169: 255–260.

Ivry RB (2003) Cerebellar involvement in clumsiness and other developmental disorders. *Neural Plasticity* 10: 141–153.

Ivry RB, Keele SW, Diener HC (1988) Dissociation of the lateral and medial cerebellum in movement timing and movement execution. *Experimental Brain Research* 73: 167–180.

Ivry RB, Spencer RM (2004) The neural representation of time. *Current Opinion in Neurobiology* 14: 225–232.

Jackson A, Mavoori J, Fetz EE (2007) Correlations between the same motor cortex cells and arm muscles during a trained task, free behavior, and natural sleep in the macaque monkey. *Journal of Neurophysiology* 97: 360–374.

Jaeger D, Gilman S, Aldridge JW (1995) Neuronal activity in the striatum and pallidum of primates related to the execution of externally cued reaching movements. *Brain Research* 694: 111–127.

Jaric S, Latash ML (1999) Learning a pointing task with a kinematically redundant limb: emerging synergies and patterns of final position variability. *Human Movement Science* 18: 819–838.

Jaric S, Milanovic S, Blezic S, Latash ML (1999) Changes in movement kinematics during single-joint movements against expectedly and unexpectedly changed inertial loads. *Human Movement Science* 18: 49–66.

Jeka JJ, Lackner JR (1994) Fingertip contact influences human postural control. *Experimental Brain Research* 100: 495–502.

Jeka JJ, Oie K, Schoner G, Dijkstra T, Henson E (1998) Position and velocity coupling of postural sway to somatosensory drive. *Journal of Neurophysiology* 79: 1661–1674.

Jenkins WM, Merzenich MM (1987) Reorganization of neocortical representations after brain injury: a neurophysiological model of the bases of recovery from stroke. *Progress in Brain Research* 71: 249–266.

Jobin A, Levin MF (2000) Regulation of stretch reflex threshold in elbow flexors in children with cerebral palsy: a new measure of spasticity. *Developmental Medicine and Child Neurology* 42: 531–540.

Johansson BB (2000) Brain plasticity and stroke rehabilitation: the Willis lecture. *Stroke* 31: 223–230.

Johansson RS, Westling G (1984) Roles of glabrous skin receptors and sensorimotor memory in automatic control of precision grip when lifting rougher or more slippery objects. *Experimental Brain Research* 56: 550–564.

Johannsen L, Wing AM, Hatzitaki V (2007) Effects of maintaining touch contact on predictive and reactive balance. *Journal of Neurophysiology* 97: 2686–2695.

Johnson K, Ladefoged P, Lindau M (1993) Individual differences in vowel production. *The Journal of the Acoustical Society of America* 94: 701–714.

Johnson KO, Hsiao SS, Yoshioka T (2002) Neural coding and the basic law of psychophysics. *Neuroscientist* 8: 111–121.

Jones RD, Donaldson IM, Parkin PJ (1989) Impairment and recovery of ipsilateral sensory-motor function following unilateral cerebral infarction. *Brain* 112: 113–132.

Kandel ER, Schwartz JH, Jessell TM (Eds.) (1999) *Principles of Neural Science*, Fourth Edition, McGraw-Hill: New York.

Kang N, Shinohara M, Zatsiorsky VM, Latash ML (2004) Learning multi-finger synergies: an uncontrolled manifold analysis. *Experimental Brain Research* 157: 336–350.

Karst GM, Hasan Z (1987) Antagonist muscle activity during human forearm movements under varying kinematic and loading conditions. *Experimental Brain Research* 67: 391–401.

Karst GM, Hasan Z (1990) Direction-dependent strategy for control of multi-joint arm movements. In: Winters JM, Woo SL-Y (Eds.) *Multiple Muscle Systems. Biomechanics and Movement Organization*, pp. 268–281, Springer-Verlag: New York.

Katz R, Mazzocchio R, Penicaud A, Rossi A (1993) Distribution of recurrent inhibition in the human upper limb. *Acta Physiologica Scandinavica* 149: 183–198.

Kautz SA, Duncan PW, Perera S, Neptune RR, Studenski SA (2005) Coordination of hemiparetic locomotion after stroke rehabilitation. *Neurorehabilitation and Neural Repair* 19: 250–258.

Kawato M (1999) Internal models for motor control and trajectory planning. *Current Opinion in Neurobiology* 9: 718–727.

Kawato M, Gomi H (1992) The cerebellum and VOR/OKR learning models. *Trends in Neurosciences* 15: 445–453.

Kay BA, Turvey MT, Meijer OG (2003) An early oscillator model: studies on the biodynamics of the piano strike (Bernstein & Popova, 1930). *Motor Control* 7: 1–45.

Keele SW, Ivry R (1990) Does the cerebellum provide a common computation for diverse tasks? A timing hypothesis. *Annals of the New York Academy of Sciences* 608: 179–211.

Kelso JAS (1995). *Dynamic Patterns: The Self-Organization of Brain and Behavior,* MIT Press: Cambridge.

Kelso, JAS, Tuller B, Fowler CA (1982) On the functional specificity of articulatory control and coordination. *Journal of the Acoustical Society of America* 72: S103.

Kelso JAS, Tuller B, Vatikiotis-Bateson E, Fowler CA (1984) Functionally specific articulatory cooperation following jaw perturbations during speech: evidence for coordinative structures. *Journal of Experimental Psychology: Human Perception and Performance* 10: 812–832.

Keshner EA, Woollacott MH, Debu B (1988) Neck, trunk and limb muscle responses during postural perturbations in humans. *Experimental Brain Research* 71: 455–466.

Kiemel T, Oie KS, Jeka JJ (2002) Multi-sensory fusion and the stochastic structure of postural sway. *Biological Cybernetics* 87: 262–277.

Kiers L, Clouston P, Chiappa KH, Cros D (1995) Assessment of cortical motor output: compound muscle action potential versus twitch force recording. *Electroencephalography and Clinical Neurophysiology* 97: 131–139.

Kilbreath SL, Gandevia SC (1994) Limited independent flexion of the thumb and fingers in human subjects. *The Journal of Physiology* 479: 487–497.

Kim SW, Shim JK, Zatsiorsky VM, Latash ML (2006) Anticipatory adjustments of multi-finger synergies in preparation for self-triggered perturbations. *Experimental Brain Research* 174: 604–612.

Kinoshita H, Backstrom L, Flanagan JR, Johansson RS (1997) Tangential torque effects on the control of grip forces when holding objects with a precision grip. *Journal of Neurophysiology* 78: 1619–1630.

Kinoshita H, Francis PR (1996) A comparison of prehension force control in young and elderly individuals. *European Journal of Applied Physiology* 74: 450–460.

Kinoshita H, Kawai S, Ikuta K (1995) Contributions and co-ordination of individual fingers in multiple finger prehension. *Ergonomics* 38: 1212–1230.

Kinoshita H, Murase T, Bandou T (1996) Grip posture and forces during holding cylindrical objects with circular grips. *Ergonomics* 39: 1163–1176.

Kirkendall DT, Garrett WE Jr (1998) The effects of aging and training on skeletal muscle. *American Journal of Sports Medicine* 26: 598–602.

Kline TL, Schmit BD, Kamper DG (2007) Exaggerated interlimb neural coupling following stroke. *Brain* 130: 159–169.

Kluzik J, Peterka RJ, Horak FB (2007) Adaptation of postural orientation to changes in surface inclination. *Experimental Brain Research* 178: 1–17.

Knight AA, Dagnall PR (1967) Precision in movements. *Ergonomics* 10: 321–330.

Koshland GF, Gerilovsky L, Hasan Z (1991) Activity of wrist muscles elicited during imposed or voluntary movements about the elbow joint. *Journal of Motor Behavior* 23: 91–100.

Kowalski N, Depireux SA, Shamma SA (1996) Analysis of dynamic spectra in ferret primary auditory cortex. II. Prediction of unit responses to arbitrary dynamic spectra. *Journal of Neurophysiology* 76: 3524–3534.

Krakauer J, Ghez C (2000) Voluntary movement. In: Kandel ER, Schwartz JH, Jessell TM (Eds.) *Principles of Neural Science*, Fourth Edition, pp. 756–780, McGraw-Hill: New York.

Krishnamoorthy V, Goodman SR, Latash ML, Zatsiorsky VM (2003a) Muscle synergies during shifts of the center of pressure by standing persons: identification of muscle modes. *Biological Cybernetics* 89: 152–161.

Krishnamoorthy V, Latash ML, Scholz JP, Zatsiorsky VM (2003b) Muscle synergies during shifts of the center of pressure by standing persons. *Experimental Brain Research* 152: 281–292.

Krishnamoorthy V, Latash ML, Scholz JP, Zatsiorsky VM (2004) Muscle modes during shifts of the center of pressure by standing persons: effects of instability and additional support. *Experimental Brain Research* 157: 18–31.

Krishnamoorthy V, Scholz JP, Latash ML (2007) The use of flexible arm muscle synergies to perform an isometric stabilization task. *Clinical Neurophysiology* 118: 525–537.

Krishnamoorthy V, Yang JF, Scholz JP (2005) Joint coordination during quiet stance: effects of vision. *Experimental Brain Research* 164: 1–17.

Kudo K, Tsutsui S, Ishikura T, Ito T, Yamamoto Y (2000) Compensatory coordination of release parameters in a throwing task. *Journal of Motor Behavior* 32: 337–345.

Kugler PN, Turvey MT (1987) *Information, natural law, and the self-assembly of rhythmic movement*, Erlbaum: Hillsdale, NJ.

Kunimoto C, Miller J, Pashler H (2001) Confidence and accuracy of near-threshold discrimination responses. *Conscious Cognition* 10: 294–340.

Kutch JJ, Buchanan TS (2001) Human elbow joint torque is linearly encoded in electromyographic signals from multiple muscles. *Neuroscience Letters* 311: 97–100.

Lackner JR (1977) Induction of illusory self-rotation and nystagmus by a rotating soundfield. *Aviation, Space and Environmental Medicine* 48: 129–131.

Lackner JR, DiZio P (1994) Rapid adaptation to Coriolis force perturbations of arm trajectory. *Journal of Neurophysiology* 72: 1–15.

Lackner JR, DiZio P (2004) Multi-sensory influences on orientation and movement control. In: Calvert G, Spence C, Stein BE (Eds.) *The Handbook of Multi-Sensory Processes*, pp. 409–424, MIT Press: Cambridge, MA.

Lackner JR, Levine MS (1979) Changes in apparent body orientation and sensory localization, induced by vibration of postural muscles; vibratory myesthetic illusions. *Aviation and Space Environmental Medicine* 50: 346–354.

Landsmeer JMF, Long C (1965) The mechanism of finger control, based on electromyograms and location analysis. *Acta Anatomica* 60: 330–347.

Lang CE, Schieber MH (2003) Differential impairment of individuated finger movements in humans after damage to the motor cortex or the corticospinal tract. *Journal of Neurophysiology* 90: 1160–1170.

Langolf GD, Chaffin DB, Foulke JA (1976) An investigation of Fitts' law using a wide range of movement amplitudes. *Journal of Motor Behavior* 8: 113–128.

Lanska DJ (2002) The Romberg sign and early instruments for measuring postural sway. *Seminars in Neurology* 22: 409–418.

Lash J (1980) *Helen and Teacher*, Addison-Wesley: Reading, MA.

Laszlo JI (1966) The performance of a simple motor task with kineasthetic sense loss. *Quarterly Journal of Experimental Psychology* 18: 1–8.

Latash LP (1979) Trace changes in the spinal cord and some basic problems of the neurophysiology of memory. In: Oniani TN (Ed.) *Seventh Gagra Talks: The Neurophysiological Basis of Memory*, pp. 118–130, Metsniereba: Tbilisi.

Latash LP, Latash ML, Mejier OG (1999) Thirty years later: on the problem of the relation between structure and function in the brain from a contemporary viewpoint (1966). Part I. *Motor Control* 3: 329–345.

Latash LP, Latash ML, Mejier OG (2000) Thirty years later: on the problem of the relation between structure and function in the brain from a contemporary viewpoint (1966). Part II. *Motor Control* 4: 125–149.

Latash ML (1992a) Virtual trajectories, joint stiffness, and changes in natural frequency during single-joint oscillatory movements. *Neuroscience* 49: 209–220.

Latash ML (1992b) Independent control of joint stiffness in the framework of the equilibrium-point hypothesis. *Biological Cybernetics* 67: 377–384.

Latash ML (1993) *Control of Human Movement*, Human Kinetics: Urbana, IL.

Latash ML (1994) Reconstruction of equilibrium trajectories and joint stiffness patterns during single-joint voluntary movements under different instructions. *Biological Cybernetics* 71: 441–450.

Latash ML (1996) How does our brain make its choices? In: Latash ML, Turvey MT (Eds.) *Dexterity and Its Development*, pp. 277–304, Erlbaum: Mahwah, NJ.

Latash ML (1997) The answer may be 42. So, what is the question? *Motor Control* 1: 205–207.

Latash ML (2000) The organization of quick corrections within a two-joint synergy in conditions of unexpected blocking and release of a fast movement. *Clinical Neurophysiology* 111: 975–987.

Latash ML (2006) A new book by Nikolai Bernstein: contemporary studies in the physiology of the neural process. *Motor Control* 10: 1–6.

Latash ML, Almeida GL, Corcos DM (1993) Pre-programmed reactions in individuals with Down syndrome: the effects of instruction and predictability of the perturbation. *Archives of Physical Medicine and Rehabilitation* 73: 391–399.

Latash ML, Anson JG (1996) What are normal movements in atypical populations? *Behavioral and Brain Sciences* 19: 55–106.

Latash ML, Anson JG (2006) Synergies in health and disease: relations to adaptive changes in motor coordination. *Physical Therapy* 86: 1151–1160.

Latash ML, Aruin AS, Shapiro MB (1995) The relation between posture and movement: a study of a simple synergy in a two-joint task. *Human Movement Science* 14: 79–107.

Latash ML, Aruin AS, Zatsiorsky VM (1999) The basis of a simple synergy: reconstruction of joint equilibrium trajectories during unrestrained arm movements. *Human Movement Science* 18: 3–30.

Latash ML, Corcos DM (1991) Kinematic and electromyographic characteristics of single-joint movements of individuals with Down syndrome. *American Journal of Mental Retardation* 96: 189–201.

Latash ML, Danion F, Scholz JF, Schöner G (2003) Coordination of multi-element motor systems based on motor abundance. In: Latash ML, Levin MF (Eds.) *Progress in Motor Control, Vol. 3: Effects of Age, Disorder, and Rehabilitation*, pp. 97–124, Human Kinetics: Urbana, IL.

Latash ML, Ferreira SS, Wieczorek SA, Duarte M (2003) Movement sway: changes in postural sway during voluntary shifts of the center of pressure. *Experimental Brain Research* 150: 314–324.

Latash ML, Gottlieb GL (1990) Equilibrium-point hypothesis and variability of the amplitude, speed, and time of sinle-joint movements. *Biofizika* 35: 870–874.

Latash ML, Gottlieb GL (1991a) An equilibrium-point model of dynamic regulation for fast single-joint movements: I. Emergence of strategy-dependent EMG patterns. *Journal of Motor Behavior* 23: 163–177.

Latash ML, Gottlieb GL (1991b) An equilibrium-point model of dynamic regulation for fast single-joint movements: II. Similarity of isometric and isotonic programs. *Journal of Motor Behavior* 23: 179–191.

Latash ML, Gottlieb GL (1991c) Reconstruction of elbow joint compliant characteristics during fast and slow voluntary movements. *Neuroscience* 43: 697–712.

Latash ML, Gottlieb GL (1992) Virtual trajectories of single-joint movements performed under two basic strategies. *Neuroscience* 47: 357–365.

Latash ML, Gurfinkel VS (1976) Tonic vibration reflex and position of the body. *Physiologiya Cheloveka (Human Physiology)* 2: 593–598.

Latash ML, Gutman SR (1993) Variability of fast single-joint movements and the equilibrium-point hypothesis. In: Newell KM, Corcos DM (Eds.) *Variability in Motor Control*, pp. 157–182, Human Kinetics: Urbana, IL.

Latash ML, Gutman SR (1994) Abnormal motor patterns in the framework of the equilibrium-point hypothesis: a cause for dystonic movements? *Biological Cybernetics* 71: 87–94.

Latash ML, Jaric S (2002) The organization of drinking: postural characteristics of the arm-head coordination. *Journal of Motor Behavior* 34: 139–150.

Latash ML, Kang N, Patterson D (2002a) Finger coordination in persons with Down syndrome: atypical patterns of coordination and the effects of practice. *Experimental Brain Research* 146: 345–355.

Latash ML, Li S, Danion F, Zatsiorsky VM (2002b) Central mechanisms of finger interaction during one- and two-hand force production at distal and proximal phalanges. *Brain Research* 924: 198–208.

Latash ML, Penn RD (1996) Changes in voluntary motor control induced by intrathecal baclofen. *Physiotherapy Research International* 1: 229–246.

Latash ML, Scholz JF, Danion F, Schöner G (2001) Structure of motor variability in marginally redundant multi-finger force production tasks. *Experimental Brain Research* 141: 153–165.

Latash ML, Scholz JF, Danion F, Schöner G (2002c) Finger coordination during discrete and oscillatory force production tasks. *Experimental Brain Research* 146: 412–432.

Latash ML, Scholz JP, Schöner G (2002d) Motor control strategies revealed in the structure of motor variability. *Exercise and Sport Science Reviews* 30: 26–31.

Latash ML, Scholz JP, Schöner G (2007) Toward a new theory of motor synergies. *Motor Control* 11: 275–307.

Latash ML, Shim JK, Smilga AV, Zatsiorsky V (2005) A central back-coupling hypothesis on the organization of motor synergies: a physical metaphor and a neural model. *Biological Cybernetics* 92: 186–191.

Latash ML, Shim JK, Zatsiorsky VM (2004) Is there a timing synergy during multi-finger production of quick force pulses? *Experimental Brain Research* 159: 65–71.

Latash ML, Yarrow K, Rothwell JC (2003) Changes in finger coordination and responses to single pulse TMS of motor cortex during practice of a multi-finger force production task. *Experimental Brain Research* 151: 60–71.

Latash ML, Zatsiorsky VM (1993) Joint stiffness: myth or reality? *Human Movement Science* 12: 653–692.

Latash ML, Zatsiorsky VM (Eds.) (2001) *Classics in Movement Science*, Human Kinetics: Urbana, IL.

Latash ML, Zatsiorsky VM (2006) Principle of superposition in human prehension. In: Kawamura S, Swinin M (Eds.) *Advances in Robot Control: From Everyday Physics to Human-Like Movements*, pp. 249–261, Springer: New York.

Latash ML, Zatsiorsky VM (in press) Multi-finger prehension: control of a redundant mechanical system. In: Sternad D (Ed.) *Progress in Motor Control-V*, Springer: New York.

Latt LD, Sparto PJ, Furman JM, Redfern MS (2003) The steady-state postural response to continuous sinusoidal galvanic vestibular stimulation. *Gait and Posture* 18: 64–72.

Lebedev MA, Nicolelis MA (2006) Brain-machine interfaces: past, present and future. *Trends in Neurosciences* 29: 536–546.

Lee G, Fradet L, Ketcham CJ, Dounskaia N (2007) Efficient control of arm movements in advanced age. *Experimental Brain Research* 177: 78–94.

Lee WA, Buchanan TS, Rogers MW (1987) Effects of arm acceleration and behavioral conditions on the organization of postural adjustments during arm flexion. *Experimental Brain Research* 66: 257–270.

Leijnse JN, Snijders CJ, Bonte JE, Landsmeer JM, Kalker JJ, Van Der Meulen JC, Sonneveld GJ, Hovius SE (1993) The hand of the musician: the kinematics of the bidigital finger system with anatomical restrictions. *Journal of Biomechanics* 10: 1169–1179.

Leijnse JN, Walbeehm ET, Sonneveld GJ, Hovius SE, Kauer JM (1997) Connections between the tendons of the musculus flexor digitorum profundus involving the synovial sheaths in the carpal tunnel. *Acta Anatomica (Basel)* 160: 112–122.

Lemon RN, Baker SN, Davis JA, Kirkwood PA, Maier MA, Yang HS (1998) The importance of the cortico-motoneuronal system for control of grasp. *Novartis Foundation Symposium* 218: 202–215.

Lepers R, Brenière Y (1995) The role of anticipatory postural adjustments and gravity in gait initiation. *Experimental Brain Research* 107: 118–124.

Lestienne FG, Thullier F, Archambault P, Feldman AG, Levin MF (2000) Multi-muscle control of head movements in monkeys: the referent configuration hypothesis. *Neuroscience Letters* 283: 65–68.

Levin MF (1996) Interjoint coordination during pointing movements is disrupted in spastic hemiparesis. *Brain* 119: 281–293.

Levin MF, Michaelsen SM, Cirstea CM, Roby-Brami A (2002) Use of the trunk for reaching targets placed within and beyond the reach in adult hemiparesis. *Experimental Brain Research* 143: 171–180.

Levin MF, Selles RW, Verheul MH, Meijer OG (2000) Deficits in the coordination of agonist and antagonist muscles in stroke patients: implications for normal motor control. *Brain Research* 853: 352–369.

Lewis D (1995) Should a materialist believe in qualia? *Australasian Journal of Philosophy* 73: 140–144.

Li S, Danion F, Latash ML, Li Z-M, Zatsiorsky VM (2000) Characteristics offinger force production during one- and two-hand tasks. *Human Movement Science* 19: 897–924.

Li ZM (2006) Functional degrees of freedom. *Motor Control* 10: 301–310.

Li ZM, Dun S, Harkness DA, Brininger TL (2004) Motion enslaving among multiple fingers of the human hand. *Motor Control* 8: 1–15.

Li ZM, Latash, ML, Zatsiorsky VM (1998) Force sharing among fingers as a model of the redundancy problem. *Experimental Brain Research* 119: 276–286.

Li ZM, Zatsiorsky VM, Latash ML (2000) Contribution of the extrinsic and intrinsic hand muscles to the moments in finger joints. *Clinical Biomechanics* 15: 203–211.

Li ZM, Zatsiorsky VM, Latash ML (2001) The effects of finger extensor mechanism on the flexor force during isometric tasks. *Journal of Biomechanics* 34: 1097–1102.

Liddell EGT, Sherrington CS (1924) Reflexes in response to stretch (myotatic reflexes) *Proceedings of the Royal Society of London B* 96: 212–242.

Liebermann DG, Biess A, Gielen CC, Flash T (2006) Intrinsic joint kinematic planning. II: hand-path predictions based on a Listing's plane constraint. *Experimental Brain Research* 171: 155–173.

Liepert J, Classen J, Cohen LG, Hallett M (1998) Task-dependent changes of intracortical inhibition. *Experimental Brain Research* 118: 421–426.

Linden DJ (1996) Cerebellar long-term depression as investigated in a cell culture preparation. *Behavioral and Brain Sciences* 19: 339–346.

Lisin VV, Frankstein SI, Rechtman MB (1973) The influence of locomotion on flexor reflex of the hind limb in cat and man. *Experimental Neurology* 38: 180–183.

Llinas R (1985) Functional significance of the basic cerebellar circuit in motor coordination. In: Bloedel JR, Dichgans J, Precht W (Eds.) *Cerebellar Functions*, Springer-Verlag: Berlin.

Long C (1965) Intrinsic-extrinsic muscle control of the fingers. Electromyographic studies. *The Journal of Bone and Joint Surgery* 50A: 973–984.

Lord SR, Caplan GA, Colagiuri R, Colagiuri S, Ward JA (1993) Sensori-motor function in older persons with diabetes. *Diabetic Medicine* 10: 614–618.

Lund, JS, Angelucci A, Bressloff PC (2003) Anatomical substrates for functional columns in macaque monkey primary visual cortex. *Cerebral Cortex* 13: 15–24.

Ma S-P, Zahalak GI (1985) The mechanical response of the active human triceps brachii muscle to very rapid stretch and shortening. *Journal of Biomechanics* 18: 585–598.

McAlonan GM, Cheung V, Cheung C, Suckling J, Lam GY, Tai KS, Yip L, Murphy DG, Chua SE (2005) Mapping the brain in autism. A voxel-based MRI study of volumetric differences and intercorrelations in autism. *Brain* 128: 268–276.

McIsaac TL, Fuglevand AJ (2007) Motor-unit synchrony within and across compartments of the human flexor digitorum superficialis. *Journal of Neurophysiology* 97: 550–556.

MacKay WA, Crammond DJ, Kwan HC, Murphy JT (1986) Measurements of human forearm viscoelasticity. *Journal of Biomechanics* 19: 231–238.

MacKenzie CL, Iberall T (1994) *The Grasping Hand,* North-Holland: Amsterdam, New York.

MacKenzie CL, Marteniuk RG, Dugas C, Liske D, Bickmeier B (1987) Three-dimensional movement trajectories in Fitts' task: implications for control. *Quarterly Journal of Experimental Psychology* 39A: 629–647.

Macpherson JM, Rushmer DS, Dunbar DC (1986) Postural responses in the cat to unexpected rotations of the supporting surface: evidence for a centrally generated synergic organization. *Experimental Brain Research* 62: 152–160.

Maier MA, Hepp-Reymond MC (1995) EMG activation patterns during force production in precision grip. II. Muscular synergies in the spatial and temporal domain. *Experimental Brain Research* 103: 123–136.

Maki BE, Holliday PJ, Fernie GR (1990) Aging and postural control. A comparison of spontaneous- and induced-sway balance tests. *Journal of American Geriatrics Society* 38: 1–9.

Malfait N, Gribble PL, Ostry DJ (2005) Generalization of motor learning based on multiple field exposures and local adaptation. *Journal of Neurophysiology* 93: 3327–3338.

Malfait N, Ostry DJ (2004) Is interlimb transfer of force-field adaptation a cognitive response to the sudden introduction of load? *Journal of Neuroscience* 24: 8084–8089.

Manckoundia P, Mourey F, Pfitzenmeyer P, Van Hoecke J, Pérennou D (2007) Is backward disequilibrium in the elderly caused by an abnormal perception of verticality? A pilot study. *Clinical Neurophysiology* 118: 786–793.

Marotta JJ, Medendorp WP, Crawford JD (2003) Kinematic rules for upper and lower arm contributions to grasp orientation. *Journal of Neurophysiology* 90: 3816–3827.

Marr D (1969) A theory of cerebellar cortex. *The Journal of Physiology* 202: 437–470.

Martin TA, Greger BE, Norris SA, Thach WT (2001) Throwing accuracy in the vertical direction during prism adaptation: not simply timing of ball release. *Journal of Neurophysiology* 85: 2298–2302.

Martin TA, Norris SA, Greger BE, Thach WT (2002) Dynamic coordination of body parts during prism adaptation. *Journal of Neurophysiology* 88: 1685–1694.

Martin V (2005) A dynamical systems account of the uncontrolled manifold and motor equivalence in human pointing movements. PhD Thesis, Ruhr-Universiteit, Bochum, Germany.

Martin V, Scholz JP, Schöner G (2004) Theory of the uncontrolled manifold: variance, self-motion, and neuronal noise. Program No. 871.17. *Abstract Viewer and Itinerary Planner.* Society for Neuroscience: Washington, DC.

Marzke MW (1992) Evolutionary development of the human thumb. *Hand Clinics* 8: 1–8.

Massaro DW (2004) From multi-sensory integration to talking heads and language learning. In: Calvert G, Spence C, Stein BE (Eds.) *The Handbook of Multi-Sensory Processes*, pp. 153–176, MIT Press: Cambridge, MA.

Massion J (1992) Movement, posture and equilibrium—interaction and coordination. *Progress in Neurobiology* 38: 35–56.

Mattei B, Schmied A, Mazzocchio R, Decchi B, Rossi A, Vedel JP (2003) Pharmacologically induced enhancement of recurrent inhibition in humans: effects on motoneurone discharge patterns. *The Journal of Physiology* 548: 615–629.

Mathewson RC, Nava PB (1985) Effects of age on Meissner corpuscles: a study of silver-impregnated neurites in mouse digital pads. *Journal of Comparative Neurology* 231: 250–259.

Matsuzaka Y, Picard N, Strick PL (2007) Skill representation in the primary motor cortex after long-term practice. *Journal of Neurophysiology* 97: 1819–1832.

Matthews PBC (1959) The dependence of tension upon extension in the stretch reflex of the soleus of the decerebrate cat. *The Journal of Physiology* 47: 521–546.

Matthies M, Perrir P, Perkell JS, Zandipour M (2001) Variation in anticipatory coarticulation with changes in clarity and rate. *Journal of Speech, Language and Hearing Research* 44: 340–353.

Mattingley JB, Rick AN (2004) Behavioral and brain correlates of multi-sensory experience in synesthesia. In: Calvert G, Spence C, Stein BE (Eds.) *The Handbook of Multi-Sensory Processes*, pp. 851–865, MIT Press: Cambridge, MA.

Mauritz KH, Dietz V (1980) Characteristics of postural instability induced by ischemic blocking of leg afferents. *Experimental Brain Research* 38: 117–119.

McCloskey DI (1978) Kineasthetic sensibility. *Physiological Reviews* 58: 763–820.

McGurk H, MacDonald J (1976) Hearing lips and seeing voices. *Nature* 264: 746–748.

McIntyre J, Lipshits M, Zaoui M, Berthoz A, Gurfinkel V (2001) Internal reference frames for representation and storage of visual information: the role of gravity. *Acta Astronautica* 49: 111–121.

Meijer O (2002) Bernstein versus Pavlovianism: an interpretation. In: Latash ML (Ed.) *Progress in Motor Control, vol. 2: Structure-Function Relations in Voluntary Movement*, pp. 229–250, Human Kinetics: Urbana, IL.

Melzer I, Benjuya N, Kaplanski J (2004) Postural stability in the elderly: a comparison between fallers and non-fallers. *Age and Ageing* 33: 602–607.

Mercier C, Bourbonnais D (2004) Relative shoulder flexor and handgrip strength is related to upper limb function after stroke. *Clinical Rehabilitation* 18: 215–221.

Merton PA (1953) Speculations on the servo-control of movements. In: Malcolm JL, Gray JAB, Wolstenholm GEW (Eds.) *The Spinal Cord*, pp. 183–198, Little, Brown: Boston, MA.

Merzenich MM, Nelson RJ, Stryker MS, Cynader MS, Schoppman A, Zook JM (1984) Somatosensory cortical map changes following digit amputation in adult monkeys. *Journal of Comparative Neurology* 224: 591–605.

Messier J, Adamovich S, Berkinblit M, Tunik E, Poizner H (2003) Influence of movement speed on accuracy and coordination of reaching movements to memorized targets in three-dimensional space in a deafferented subject. *Experimental Brain Research* 150: 399–416.

Meunier S, Kwon J, Russmann H, Ravindran S, Mazzocchio R, Cohen L (2007) Spinal use-dependent plasticity of synaptic transmission in humans after a single cycling session. *The Journal of Physiology* 579: 375–388.

Meyendorff J (1964) *A Study of Gregory Palamas*. The Faith Press: London.

Meyendorff J (1974) *St. Gregory Palamas and Orthodox Spirituality*, St. Vladimir's Seminary: USA.

Meyer DE, Abrams RA, Kornblum S, Wright CE, Smith JE (1988a) Optimality in human motor performance: ideal control of rapid aimed movements. *Psychological Reviews* 95: 340–370.

Meyer DE, Irwin DE, Osman AM, Kounios J (1988b) The dynamics of cognition and action: mental processes inferred from speed-accuracy decomposition. *Psychological Reviews* 95: 183–237.

Miall RC (1998) The cerebellum, predictive control and motor coordination. *Novartis Foundation Symposium* 218: 272–284.

Micera S, Carpaneto J, Posteraro F, Cenciotti L, Popovic M, Dario P (2005) Characterization of upper arm synergies during reaching tasks in able-bodied and hemiparetic subjects. *Clinical Biomechanics* 20: 939–946.

Michaelsen SM, Jacobs S, Roby-Brami A, Levin MF (2004) Compensation for distal impairments of grasping in adults with hemiparesis. *Experimental Brain Research* 157: 162–173.

Miller LE, Holdefer RN, Houk JC (2002) The role of the cerebellum in modulating voluntary limb movement commands. *Archives Italiennes de Biologie* 140: 175–183.

Miller LE, Houk JC (1995) Motor co-ordinates in primate red nucleus: preferential relation to muscle activation versus kinematic variables. *The Journal of Physiology* 488: 533–548.

Miller LE, Theeuwen M, Gielen CCAM (1992) The control of arm pointing movements in three dimensions. *Experimental Brain Research* 90: 415–426.

Miller S, Ruit JB, van der Meche FGA (1977) Reversal of sign of long spinal reflexes dependent on the phase of the step cycle in the high decerebrate cat. *Brain Research* 128: 447–459.

Miller S, van der Meche FGA (1976) Coordinated stepping of all four limbs in the high spinal cat. *Brain Research* 109: 395–398.

Mochizuki L, Duarte M, Amadio AC, Zatsiorsky VM, Latash ML (2006) Changes in postural sway and its fractions in conditions of postural instability. *Journal of Applied Biomechanics* 22: 51–66.

Molnar GE (1978) Analysis of motor disorder in retarded infants and young children. *American Journal of Mental Deficiency* 83: 213–222.

Monzee J, Lamarre Y, Smith AM (2003) The effects of digital anesthesia on force control using a precision grip. *Journal of Neurophysiology* 89: 672–683.

Moore CA, Smith A, Ringel RA (1988) Task-specific organization of activity in human jaw muscles. *Journal of Speech and Hearing Research* 31: 670–680.

Morasso P (1981) Spatial control of arm movements. *Experimental Brain Research* 42: 223–227.

Morasso P (1983) Three-dimensional arm trajectories. *Biological Cybernetics* 48: 187–194.

Morasso P, Mussa Ivaldi FA (1982) Trajectory formation and handwriting: a computational model. *Biological Cybernetics* 45: 131–142.

Mori S (1987) Integration of posture and locomotion in acute decerebrate cats and in awake, freely moving cats. *Progress in Neurobiology* 28: 161–195.

Moritz CT, Lucas TH, Perlmutter SI, Fetz EE (2007) Forelimb movements and muscle responses evoked by microstimulation of cervical spinal cord in sedated monkeys. *Journal of Neurophysiology* 97: 110–120.

Morris AF, Vaughan SE, Vaccaro P (1982) Measurements of neuromuscular tone and strength in Down's syndrome children. *Journal of Mental Deficiency Research* 26: 41–46.

Morrow MM, Jordan LR, Miller LE (2007) Direct comparison of the task-dependent discharge of M1 in hand space and muscle space. *Journal of Neurophysiology* 97: 1786–1798.

Mouchnino L, Aurenty R, Massion J, Pedotti A (1992) Coordination between equilibrium and head-trunk orientation during leg movement: a new strategy build up by training. *Journal of Neurophysiology* 67: 1587–1598.

Mouchnino L, Mille ML, Cincera M, Bardot A, Delarque A, Pedotti A, Massion J (1998) Postural reorganization of weight-shifting in below-knee amputees during leg raising. *Experimental Brain Research* 121: 205–214.

Muellbacher W, Facchini S, Boroojerdi B, Hallett M (2000) Changes in motor cortex excitability during ipsilateral hand muscle activation in humans. *Clinical Neurophysiology* 111: 344–349.

Müller H, Sternad D (2003) A randomization method for the calculation of covariation in multiple nonlinear relations: illustrated with the example of goal-directed movements. *Biological Cybernetics* 89: 22–33.

Müller H, Sternad D (2004) Decomposition of variability in the execution of goal-oriented tasks: three components of skill improvement. *Journal of Experimental Psychology: Human Perception and Performance* 30: 212–233.

Mussa-Ivaldi FA, Bizzi E (2000) Motor learning through the combination of primitives. *Philosophical Transactions of the Royal Society of London B Biological Sciences* 355: 1755–1769.

Mussa-Ivaldi FA, Giszter SF (1992) Vector field approximation: a computational paradigm for motor control and learning. *Biological Cybernetics* 67: 491–500.

Mussa-Ivaldi FA, Giszter SF, Bizzi E (1994) Linear combinations of primitives in vertebrate motor control. *Proceedings of the National Academy of Sciences USA* 91: 7534–7538.

Mussa-Ivaldi FA, Morasso P, Zaccaria R (1989) Kinematic networks. A distributed model for representing and regularizing motor redundancy. *Biological Cybernetics* 60: 1–16.

Myklebust BM, Gottlieb GL (1993) Development of the stretch reflex in the newborn: reciprocal excitation and reflex irradiation. *Child Development* 64: 1036–1045.

Myklebust BM, Gottlieb GL, Agarwal GC, Penn RD (1982) Reciprocal excitation of antagonist muscles as a differentiating feature in spasticity. *Annals of Neurology* 12: 367–374.

Nakao M, Inoue Y, Murkami H (1989) Aging process of leg muscle endurance in males and females. *European Journal of Applied Physiology* 59: 209–214.

Narici MV, Bordini M, Cerretelli P (1991) Effect of aging on human adductor pollicis muscle function. *Journal of Applied Physiology* 71: 1277–1281.

Nashner LM (1976) Adapting reflexes controlling human posture. *Experimental Brain Research* 26: 59–72.

Nashner LM (1979) Organization and programming of motor activity during posture control. *Progress in Brain Research* 50: 177–184.

Nashner LM, Cordo PJ (1981) Relation of automatic postural responses and reaction-time voluntary movements of human leg muscles. *Experimental Brain Research* 43: 395–405.

Nashner LM, Woollacott M (1979) The organization of rapid postural adjustments of standing humans: an experimental-conceptual model. In: Talbott RE, Humphrey DR (Eds.) *Posture and Movement*, pp. 243–257, Raven: New York.

Nasir SM, Ostry DJ (2006) Somatosensory precision in speech production. *Current Biology* 16: 1918–1923.

Neckel N, Pelliccio M, Nichols D, Hidler J (2006) Quantification of functional weakness and abnormal synergy patterns in the lower limb of individuals with chronic stroke. *Journal of Neuroengineering and Rehabilitation* 3: 17.

Nelson W (1983) Physical principles for economies of skilled movements. *Biological Cybernetics* 46: 135–147.

Newell KM (1991) Motor skill acquisition. *Annual Reviews in Psychology* 42: 213–237.

Newell KM, Broderick MP, Deutsch KM, Slifkin AB (2003) Task goals and change in dynamical degrees of freedom with motor learning. *Journal of Experimental Psychology: Human Perception and Performance* 29: 379–387.

Newell KM, Carlton LG (1993) Force variability in isometric responses. *Journal of Experimental Psychology: Human Perception and Performance* 14: 37–44.

Newell KM, Corcos DM (Eds.) (1993) *Variability in Motor Control*, Human Kinetics: Urbana, IL.

Newell KM, Vaillancourt DE (2001) Woodworth (1889) Movement variability and theories of motor control. In: Latash ML, Zatsiorsky VM (Eds.) *Classics in Movement Studies*, pp. 409–435, Human Kinetics: Champaign, IL.

Nichols TR (1989) The organization of heterogenic reflexes among muscles crossing the ankle joint in the decerebrate cat. *The Journal of Physiology* 410: 463–477.

Nichols TR (1994) A biomechanical perspective on spinal mechanisms of coordinated muscular action: an architecture principle. *Acta Anatomica* 151: 1–13.

Nichols TR (2002) Musculoskeletal mechanics: a foundation of motor physiology. *Advances in Experimental and Medical Biology* 508: 473–479.

Nichols TR, Steeves JD (1986) Resetting of resultant stiffness in ankle flexor and extensor muscles in the decerebrate cat. *Experimental Brain Research* 62: 401–410.

Nieto-Castanon A, Guenther FH, Perkell JS, Curtin HD (2005) A modeling investigation of articulatory variability and acoustic stability during American English /r/ production. *The Journal of the Acoustical Society of America* 117: 3196–3212.

Nudo RJ (2003) Adaptive plasticity in motor cortex: implications for rehabilitation after brain injury. *Journal of Rehabilitation Medicine* 41 (suppl.): 7–10.

Nudo RJ, Plautz EJ, Frost SB (2001) Role of adaptive plasticity in recovery of function after damage to motor cortex. *Muscle and Nerve* 24: 1000–1019.

Nudo RJ, Wise BM, SiFuentes F, Milliken GW (1996) Neural substrates for the effects of rehabilitative training on motor recovery after ischemic infarct. *Science* 272: 1791–1794.

O'Hare A, Khalid S (2002) The association of abnormal cerebellar function in children with developmental coordination disorder and reading difficulties. *Dyslexia* 8: 234–248.

Ohtsuki T (1981) Inhibition of individual fingers during grip strength exertion. *Ergonomics* 24: 21–36.

Ojakangas CL, Ebner TJ (1992) Purkinje cell complex and simple spike changes during a voluntary arm movement learning task in the monkey. *Journal of Neurophysiology* 68: 2222–2236.

Olafsdottir H, Yoshida N, Zatsiorsky VM, Latash ML (2005) Anticipatory covariation of finger forces during self-paced and reaction time force production. *Neuroscience Letters* 381: 92–96.

Olafsdottir H, Yoshida N, Zatsiorsky VM, Latash ML (2007a) Elderly show decreased adjustments of motor synergies in preparation to action. *Clinical Biomechanics* 22: 44–51.

Olafsdottir H, Zatsiorsky VM, Latash ML (2005) Is the thumb a fifth finger? A study of digit interaction during force production tasks. *Experimental Brain Research* 160: 203–213.

Olafsdottir H, Zhang W, Zatsiorsky VM, Latash ML (2007b) Age related changes in multi-finger synergies in accurate moment of force production tasks. *Journal of Applied Physiology* 102: 1490–1501.

Orlovsky GN, Deliagina TG, Grillner S (1999) *Neuronal Control of Locomotion. From Mollusc to Man*, Oxford University Press: New York.

Ostry DJ, Feldman AG (2003) A critical evaluation of the force control hypothesis in motor control. *Experimental Brain Research* 153: 275–288.

Ostry DJ, Gribble PL, Gracco VL (1996) Coarticulation of jaw movements in speech production: is context sensitivity in speech kinematics centrally planned? *Journal of Neuroscience* 16: 1570–1579.

Ozeki H, Sadakane O, Akasaki T, Naito T, Shimegi S, Sato H (2004) Relationship between excitation and inhibition underlying size tuning and contextual response modulation in the cat primary visual cortex. *Journal of Neuroscience* 24: 1428–1438.

Palmen SJ, van Engeland H, Hof PR, Schmitz C (2004) Neuropathological findings in autism. *Brain* 127: 2572–2583.

Partan S, Marler P (1999) Communication goes multi-modal. *Science* 283: 1272–1273.

Pascual-Leone A (2001) The brain that plays music and is changed by it. *Annals of the New York Academy of Sciences* 930: 315–329.

Pascual-Leone A, Dang N, Cohen LG, Brasil-Neto JP, Cammarota A, Hallett M (1995) Modulation of muscle responses evoked by transcranial magnetic stimulation during the acquisition of new fine motor skills. *Journal of Neurophysiology* 74: 1037–1045.

Pataky TC, Latash ML, Zatsiorsky VM (2004a) Prehension synergies during non-vertical grasping. I. Experimental observations. *Biological Cybernetics* 91: 148–158.

Pataky TC, Latash ML, Zatsiorsky VM (2004b) Prehension synergies during non-vertical grasping. II. Modeling and optimization. *Biological Cybernetics* 91: 231–242.

Pataky TC, Latash ML, Zatsiorsky VM (2007) Finger interaction during radial and ulnar deviation: experimental data and neural network modeling. *Experimental Brain Research* 179: 301–312.

Paulesu E, Harrison J, Baron-Cohen S, Watson JDG, Goldstein L, Heather J, Frackowiak AS, Frith CD (1995) The physiology of coloured hearing: a PET activation study of colour-word synaethesia. *Brain* 118: 661–676.

Paulignan Y, Dufosse M, Hugon M, Massion J (1989) Acquisition of co-ordination between posture and movement in a bimanual task. *Experimental Brain Research* 77: 337–348.

Paulson HL, Stern MB (2004) Clinical manifestations of Parkinson's disease. In: Watts RL, Koller WC (Eds.) *Movement Disorders. Neurological Principles and Practice*, pp. 233–246, McGraw-Hill: New York.

Paus T, Jech R, Thompson CJ, Comeau R, Peters T, Evans AC (1997) Transcranial magnetic stimulation during positron emission tomography: a new method for studying connectivity of the human cerebral cortex. *Journal of Neuroscience* 17: 3178–3184.

Pelz J, Hayhoe M, Loeber R (2001) The coordination of eye, head, and hand movements in a natural task. *Experimental Brain Research* 139: 266–277.

Penfield W, Rasmussen T (1950) *The Cerebral Cortex of Man. A Clinical Study of Localization of Function*. MacMillan: New York.

Pereira HS, Landgren M, Gillberg C, Forssberg H (2001) Parametric control of fingertip forces during precision grip lifts in children with DCD (developmental coordination disorder) and DAMP (deficits in attention motor control and perception). *Neuropsychologia* 39: 478–488.

Perkell JS, Matthies ML (1992) Temporal measures of anticipatory labial coarticulation for the vowel/u/: within- and cross-subject variability. *The Journal of the Acoustical Society of America* 91: 2911–2925.

Perkell JS, Matthies ML, Svirsky MA, Jordan MI (1993) Trading relations between tongue-body raising and lip rounding in production of the vowel /u/: a pilot "motor equivalence" study. *The Journal of the Acoustical Society of America* 93: 2948–2961.

Pigeon A, Yehia LH, Mitnitski AB, Feldman AG (2000) Superposition of independent units of coordination during pointing movements involving the trunk with and without visual feedback. *Experimental Brain Research* 131: 336–349.

Pilon J-F, De Serres SJ, Feldman AG (2007) Threshold position control of arm movement with anticipatory increase in grip force. *Experimental Brain Research* 181: 49–67.

Plamondon R, Alimi AM (1997) Speed/accuracy trade-offs in target-directed movements. *Behavioral and Brain Sciences* 20: 279–349.

Polit A, Bizzi E (1978) Processes controlling arm movements in monkey. *Science* 201: 1235–1237.

Polit A, Bizzi E (1979) Characteristics of motor programs underlying arm movement in monkey. *Journal of Neurophysiology* 42: 183–194.

Pollick FE, Kay JW, Heim K, Stringer R (2005) Gender recognition from point-light walkers. *Journal of Experimental Psychology: Human Perception and Performance* 31: 1247–1265.

Popescu FC, Hidler JM, Rymer WZ (2003) Elbow impedance during goal-directed movements. *Experimental Brain Research* 152: 17–28.

Popescu FC, Rymer WZ (2000) End points of planar reaching movements are disrupted by small force pulses: an evaluation of the hypothesis of equifinality. *Journal of Neurophysiology* 84: 2670–2679.

Popescu FC, Rymer WZ (2003) Implications of low mechanical impedance in upper limb reaching movements. *Motor Control* 7: 323–327.

Prilutsky BI (2000) Coordination of two- and one-joint muscles: functional consequences and implications for motor control. *Motor Control* 4: 1–44.

Prochazka A, Clarac F, Loeb GE, Rothwell JC, Wolpaw JR (2000) What do reflex and voluntary mean? Modern views on an ancient debate. *Experimental Brain Research* 130: 417–432.

Quaney BM, Cole KJ (2004) Distributing vertical forces between the digits during gripping and lifting: the effects of rotating the hand versus rotating the object. *Experimental Brain Research* 155: 145–155.

Raibert MH (1977) Motor control and learning by the state space model. Doctoral dissertation, MIT, Cambridge, MA.

Ramachandran VS, Hubbard EM, Butcher PA (2004) Synesthesia, cross-activation, and the foundations of neuroepistemology. In: Calvert G, Spence C, Stein BE (Eds.) *The Handbook of Multi-Sensory Processes*, pp. 867–883, MIT Press: Cambridge, MA.

Rarick GL, Dobbins DA, Broadhead GG (1976) *The Motor Domain and Its Correlates in Educated Handicapped Children*, Prentice Hall: Englewood Cliffs, NJ.

Ravaioli E, Oie KS, Kiemel T, Chiari L, Jeka JJ (2005) Nonlinear postural control in response to visual translation. *Experimental Brain Research* 160: 450–459.

Ravnborg M, Blinkenberg M, Dahl K (1991) Standardization of facilitation of compound muscle action potentials evoked by magnetic stimulation of the cortex. *Electroencephalography and Clinical Neurophysiology* 81: 195–201.

Raynor AJ (2001) Strength, power, and coactivation in children with developmental coordination disorder. *Developmental Medicine and Child Neurology* 43: 676–684.

Recanzone GH, Merzenich MM, Jenkins WM, Grajski KA, Dinse HR (1992) Topographic reorganization of the hand representation in cortical area 3b of owl monkeys trained in a frequency-discrimination task. *Journal of Neurophysiology* 67: 1031–1056.

Reisman D, Scholz JP (2003) Aspects of joint coordination are preserved during pointing in persons with post-stroke hemiparesis. *Brain* 126: 2510–2527.

Reisman DS, Scholz JP (2006) Workspace location influences joint coordination during reaching in post-stroke hemiparesis. *Experimental Brain Research* 170: 265–276.

Reisman DS, Scholz JP, Schöner G (2002) Differential joint coordination in the tasks of standing up and sitting down. *Journal of Electromyography and Kinesiology* 12: 493–505.

Riccio GE (1993) Information in movement variability about the qualitative dynamics of posture and orientation. In: Newell KM, Corcos DM (Eds.) *Variability and Motor Control*, pp. 317–358, Human Kinetics: Champaign, IL.

Ridding MC, Brouwer B, Nordstrom MA (2000) Reduced interhemispheric inhibition in musicians. *Experimental Brain Research* 133: 249–253.

Riley MA, Stoffregen TA, Grocki MJ, Turvey MT (1999) Postural stabilization for the control of touching. *Human Movement Science* 18: 795–817.

Riley MA, Wong S, Mitra S, Turvey MT (1997) Common effects of touch and vision on postural parameters. *Experimental Brain Research* 117, 165–170.

Rispal-Padel L, Cicirata F, Pons C (1981) Contribution of the dentato-thalamo-cortical system to control of motor synergy. *Neuroscience Letters* 22: 137–144.

Roll JP, Vedel JP (1982) Kinaesthetic role of muscle afferents in man, studied by tendon vibration and microneurography. *Experimental Brain Research* 47: 177–190.

Rolls ET (2004) Multi-sensory neuronal convergence of taste, somatosensory, visual, olfactory, and auditory inputs. In: Calvert G, Spence C, Stein BE (Eds.) *The Handbook of Multi-Sensory Processes*, pp. 311–332, MIT Press: Cambridge, MA.

Rosenbaum DA, Engelbrecht SE, Busje MM, Loukopoulos LD (1993) Knowledge model for selecting and producing reaching movements. *Journal of Motor Behavior* 25: 217–227.

Rosenbaum DA, Meulenbroek RJ, Vaughan J, Jansen C (2001) Posture-based motion planning: applications to grasping. *Psychological Reviews* 108: 709–734.

Rossini PM, Barker AT, Berardelli A, Caramia MT, Caruso G, Gracco RQ, Dimitrijevic MR, Hallett M, Katayama Y, Lucking CH, Maertens-de Noordhout AL, Marsden CD, Murray NMF, Rothwell JC, Swash M, Tomberg C (1994) Non-invasive electrical and magnetic stimulation of the brain, spinal cord and roots. basic principles and procedures for routine clinical application. *Electroencephalography and Clinical Neurophysiology* 91: 79–92.

Rossini PM, Pauri F (2000) Neuromagnetic integrated methods tracking human brain mechanisms of sensorimotor areas 'plastic' reorganisation. *Brain Research Reviews* 33: 131–154.

Rothwell JC, Traub MM, Marsden CD (1982) Automatic and "voluntary" responses compensating for disturbances of human thumb movements. *Brain Research* 248: 33–41.

Ruark JL, Moore CA (1997) Coordination of lip muscle activity by 2-year-old children during speech and nonspeech tasks. *Journal of Speech, Language and Hearing Research* 40: 1373–1385.

Ruben RJ (2005) Sign language: its history and contribution to the understanding of the biological nature of language. *Acta Otolaryngologica* 125: 464–467.

Ruegg DG, Bongioanni F (1989) Superposition of ballistic on steady contractions in man. *Experimental Brain Research* 77: 412–420.

Rymer WZ, D'Almeida A (1980) Joint position sense. The effects of muscle contraction. *Brain* 103: 1–22.

Sabatini AM (2002) Identification of neuromuscular synergies in natural upper-arm movements. *Biological Cybernetics* 86: 253–262.

Sainburg RL, Ghilardi MF, Poizner H, Ghez C (1995) Control of limb dynamics in normal subjects and patients without proprioception. *Journal of Neurophysiology* 73: 820–835.

Saltiel P, Wyler-Duda K, D'Avella A, Tresch MC, Bizzi E (2001) Muscle synergies encoded within the spinal cord: evidence from focal intraspinal NMDA iontophoresis in the frog. *Journal of Neurophysiology* 5: 605–619.

Saltzmann EL, Kelso JAS (1987) Skilled actions: a task-dynamic approach. *Psychological Reviews* 94: 84–106.

Sanes JN (1985) Information processing deficits in Parkinson's disease during movement. *Neuropsychology* 23: 381–392.

Santello M, Soechting JF (2000) Force synergies for multi-fingered grasping. *Experimental Brain Research* 133: 457–467.

Schieber MH (1999) Somatotopic gradients in the distributed organization of the human primary motor cortex hand area: evidence from small infarcts. *Experimental Brain Research* 128: 139–148.

Schieber MH (2001) Constraints on somatotopic organization in the primary motor cortex. *Journal of Neurophysiology* 86: 2125–2143.

Schieber MH, Hibbard LS (1993) How somatotopic is the motor cortex hand area? *Science* 261: 489–492.

Schieber MH, Rivlis G (2007) Partial reconstruction of muscle activity from a pruned network of diverse motor cortex neurons. *Journal of Neurophysiology* 97: 70–82.

Schieber MH, Santello M (2004) Hand function: peripheral and central constraints on performance. *Journal of Applied Physiology* 96: 2293–2300.

Schmahmann JD (2004) Disorders of the cerebellum: ataxia, dysmetria of thought, and the cerebellar cognitive affective syndrome. *Journal of Neuropsychiatry and Clinical Neuroscience* 16: 367–378.

Schmahmann JD, Sherman JC (1998) The cerebellar cognitive affective syndrome. *Brain* 121: 561–579.

Schmidt RA (1975) A schema theory of discrete motor skill learning. *Psychological Reviews* 82: 225–260.

Schmidt RA (1980) Past and future issues in motor programming. *Research Quarterly of Exercise and Sport* 51: 122–140.

Schmidt RA, McGown C (1980) Terminal accuracy of unexpected loaded rapid movements: evidence for a mass-spring mechanism in programming. *Journal of Motor Behavior* 12: 149–161.

Schmidt RC, Carello C, Turvey MT (1990) Phase transitions and critical fluctuations in the visual coordination of rhythmic movements between people. *Journal of Experimental Psychology: Human Perception and Performance* 16: 227–247.

Scholz JP, Danion F, Latash ML, Schöner G (2002) Understanding finger coordination through analysis of the structure of force variability. *Biological Cybernetics* 86: 29–39.

Scholz JP, Kang N, Patterson D, Latash ML (2003) Uncontrolled manifold analysis of single trials during multi-finger force production by persons with and without Down syndrome. *Experimental Brain Research* 153: 45–58.

Scholz JP, Latash ML (1998) A study of a bimanual synergy associated with holding an object. *Human Movement Science* 17: 753–779.

Scholz JP, Reisman D, Schoner G (2001) Effects of varying task constraints on solutions to joint coordination in a sit-to-stand task. *Experimental Brain Research* 141: 485–500.

Scholz JP, Schöner G (1999) The uncontrolled manifold concept: identifying control variables for a functional task. *Experimental Brain Research* 126: 289–306.

Scholz JP, Schöner G, Latash ML (2000) Identifying the control structure of multi-joint coordination during pistol shooting. *Experimental Brain Research* 135: 382–404.

Schöner G (1990) A dynamic theory of coordination of discrete movement. *Biological Cybernetics* 63: 257–270.

Schöner G (1995) Recent developments and problems in human movement science and their conceptual implications. *Ecological Psychology* 8: 291–314.

Schöner G (2002) Timing, clocks, and dynamical systems. *Brain and Cognition* 48: 31–51.

Schoppa NE, Urban NN (2003) Dendritic processing within olfactory bulb circuits. *Trends in Neurosciences* 26: 501–506.

Schotland JL, Lee WA, Rymer WZ (1989) Wiping reflex and flexion withdrawal reflexes display different EMG patterns prior to movement onset in the spinalized frog. *Experimental Brain Research* 78: 649–653.

Schwartz AB (1993) Motor cortical activity during drawing movements: population representation during sinusoid tracing. *Journal of Neurophysiology* 70: 28–36.

Schwartz AB (2004) Cortical neural prosthetics. *Annual Reviews in Neuroscience* 27: 487–507.

Schweighofer N, Arbib MA, Kawato M (1998) Role of the cerebellum in reaching movements in humans. I. Distributed inverse dynamics control. *European Journal of Neuroscience* 10: 86–94.

Shadmehr R, Mussa-Ivaldi FA (1994) Adaptive representation of dynamics during learning of a motor task. *Journal of Neuroscience* 14: 3208–3224.

Shadmehr R, Wise SP (2005) *The computational neurobiology of reaching and pointing*, MIT Press: Cambridge, MA.

Shaiman S (2002) Articulatory control of vowel length for contiguous jaw cycles: the effects of speaking rate and phonetic context. *Journal of Speech, Language and Hearing Research* 45: 663–675.

Shapkov YuT, Shapkova EYu, Mushkin AYu (1995) Spinal generators of human locomotor movements. In: *4-th IBRO World Congress in Neuroscience*, Kyoto, Japan. Abstracts, pp. 349.

Shapkova EYu (2004) Spinal locomotor capability revealed by electrical stimulation of the lumbar enlargement in paraplegic patients. In: Latash ML, Levin MF (Eds.) *Progress in Motor Control-3*, pp. 253–290, Human Kinetics: Champaign, IL.

Shelton FNAP, Reding MJ (2001) Effect of lesion location on upper limb motor recovery after stroke. *Stroke* 32: 107–112.

Shemmell J, Hasan Z, Gottlieb GL, Corcos DM (2007) The effect of movement direction on joint torque covariation. *Experimental Brain Research* 176: 150–158.

Sherrington CS (1910) Flexion reflex of the limb, crossed extension reflex, and reflex stepping and standing. *The Journal of Physiology* 40: 28–121.

Shiffrar M, Lichtey L, Heptulla Chatterjee S (1997) The perception of biological motion across apertures. *Perception and Psychophysics* 59: 51–59.

Shik ML, Orlovskii GN, Severin FV (1966) Organization of locomotor synergism. *Biofizika* 11: 879–886 (in Russian).

Shik ML, Severin FV, Orlovskii GN (1967) Structures of the brain stem responsible for evoked locomotion. *Sechenov Physiological Journal of the USSR* 53: 1125–1132 (in Russian).

Shim JK (2005) Rotational equilibrium control in multi-digit human prehension. PhD Dissertation, Pennsylvania State University: University Park, PA.

Shim JK, Kim SW, Oh SJ, Kang N, Zatsiorsky VM, Latash ML (2005a) Plastic changes in interhemispheric inhibition with practice of a two-hand force production task: a transcranial magnetic stimulation study. *Neuroscience Letters* 374: 104–108.

Shim JK, Latash ML, Zatsiorsky VM (2003a) Prehension synergies: trial-to-trial variability and hierarchical organization of stable performance. *Experimental Brain Research* 152: 173–184.

Shim JK, Latash ML, Zatsiorsky VM (2003b) The central nervous system needs time to organize task-specific covariation of finger forces. *Neuroscience Letters* 353: 72–74.

Shim JK, Latash ML, Zatsiorsky VM (2004) Finger coordination during moment production on a mechanically fixed object. *Experimental Brain Research* 157: 457–467.

Shim JK, Latash ML, Zatsiorsky VM (2005a) Prehension synergies in three dimensions. *Journal of Neurophysiology* 93: 766–776.

Shim JK, Latash ML, Zatsiorsky VM (2005b) Prehension synergies: trial-to-trial variability and principle of superposition during static prehension in three dimensions. *Journal of Neurophysiology* 93: 3649–3658.

Shim JK, Lay B, Zatsiorsky VM, Latash ML (2004) Age-related changes in finger coordination in static prehension tasks. *Journal of Applied Physiology* 97: 213–224.

Shim JK, Olafsdottir H, Zatsiorsky VM, Latash ML (2005b) The emergence and disappearance of multi-digit synergies during force production tasks. *Experimental Brain Research* 164: 260–270.

Shim JK, Park J, Zatsiorsky VM, Latash ML (2006) Adjustments of prehension synergies in response to self-triggered and experimenter-triggered load and torque perturbations. *Experimental Brain Research* 175: 641–653.

Shinohara M, Latash ML, Zatsiorsky VM (2003a) Age effects on force production by the intrinsic and extrinsic hand muscles and finger interaction during maximal contraction tasks. *Journal of Applied Physiology* 95: 1361–1369.

Shinohara M, Li S, Kang N, Zatsiorsky VM, Latash ML (2003b) Effects of age and gender on finger coordination in maximal contractions and submaximal force matching tasks. *Journal of Applied Physiology* 94: 259–270.

Shinohara M, Scholz JP, Zatsiorsky VM, Latash ML (2004) Finger interaction during accurate multi-finger force production tasks in young and elderly persons. *Experimental Brain Research* 156: 282–292.

Sinkjaer T, Toft E, Andreassen S, Hornemann BC (1988) Muscle stiffness in human ankle dorsiflexors: intrinsic and reflex components. *Journal of Neurophysiology* 60: 1110–1121.

Slijper H, Latash ML (2000) The effects of instability and additional hand support on anticipatory postural adjustments in leg, trunk, and arm muscles during standing. *Experimental Brain Research* 135: 81–93.

Slobounov S, Johnston J, Chiang H, Ray WJ (2002a) Motor-related cortical potentials accompanying enslaving effect in single versus combination of fingers force production tasks. *Clinical Neurophysiology* 113: 1444–1453.

Slobounov S, Johnston J, Chiang H, Ray WJ (2002b) The role of submaximal force production in the enslaving phenomenon. *Brain Research* 954: 212–219.

Smith AM (1993) Babinski and movement synergism. *Revue Neurologique (Paris)* 149: 764–770.

Smith JL, Hoy MG, Koshland GF, Phillips DM, Zernicke RF (1985) Intralimb coordination of the paw-shake response: a novel mixed synergy. *Journal of Neurophysiology* 54: 1271–1281.

Smits-Engelsman BCM, Rameckers EAA, Duysens J (2007) Children with congenital spastic hemiplegia obey Fitts' Law in a visually guided tapping task. *Experimental Brain Research* 177: 431–439.

Soechting JF, Lacquaniti F (1981) Invariant characteristics of a pointing movement in man. *Journal of Neuroscience* 1: 710–720.

Soto-Faraco S, Kingstone A (2004) Multi-sensory integration in dynamic information. In: Calvert G, Spence C, Stein BE (Eds.) *The Handbook of Multi-Sensory Processes*, pp. 49–68, MIT Press: Cambridge, MA.

Spencer RM, Ivry RB, Cattaert D, Semjen A (2005) Bimanual coordination during rhythmic movements in the absence of somatosensory feedback. *Journal of Neurophysiology* 94: 2901–2910.

St. Gregory Palamas (1983) *The Triads.* Classics of Western Spirituality. Paulist Press: Mahwah, NJ.

St. Gregory Palamas (1988) *The One Hundred and Fifty Chapters.* Pontifical Institute of Mediaeval Studies: Toronto.

Stedman A, Davey NJ, Ellaway PH (1998) Facilitation of human first dorsal interosseous muscle responses to transcranial magnetic stimulation during voluntary contraction of the contralateral homonymous muscle. *Muscle and Nerve* 21: 1033–1039.

Stelmach GE, Worringham CJ, Strand EA (1986) Movement preparation in Parkinson's disease: the use of advance information. *Brain* 109: 1179–1194.

Sterr A, Muller MM, Elbert T, Rockstroh B, Pantev C, Taub E (1998) Perceptual correlates of changes in cortical representation of fingers in blind multi-finger Braille readers. *Journal of Neuroscience* 18: 4417–4423.

Stevenson RJ, Boakes RA (2004) Sweet and sour smells: learned synesthesia between the senses of taste and smell. In: Calvert G, Spence C, Stein BE (Eds.) *The Handbook of Multi-Sensory Processes*, pp. 69–84, MIT Press: Cambridge, MA.

Stuart DG, Pierce PA, Callister RJ, Brichta AM, McDonagh JC (2001) Sir Charles S. Sherrington: humanist, mentor, and movement neuroscientist. In: Latash ML, Zatsiorsky VM (Eds.) *Classics in Movement Science*, pp. 317–374, Human Kinetics: Urbana, IL.

Sugiuchi Y, Kakei S, Izawa Y, Shinoda Y (2003) Functional synergies among neck muscles revaled by branching patterns of single long descending motor-tract axons. In:

Mori S, Stuart DG, Wisendanger M (Eds.) *Brain Mechanisms for the Integration of Posture and Movement,* pp. 353–368, Elsevier: Amsterdam.

Suminski AJ, Rao SM, Mosier KM, Scheidt RA (2007) Neural and electromyographic correlates of wrist posture control. *Journal of Neurophysiology* 97: 1527–1545.

Tang PF, Woollacott MH (1998) Inefficient postural responses to unexpected slips during walking in older adults. *Journal of Gerontology A Biological Sciences and Medical Sciences* 53: M471–M480.

Tasko SM, McClean MD (2004) Variations in articulatory movement with changes in speech task. *Journal of Speech, Language and Hearing Research* 47: 85–100.

Taylor JL, Allen GM, Butler JE, Gandevia SC (1997) Effect of contraction strength on responses in biceps brachii and adductor pollicis to transcranial magnetic stimulation. *Experimental Brain Research* 117: 472–478.

Thach WT (1998) A role for the cerebellum in learning movement coordination. *Neurobiology of Learning and Memory* 70: 177–188.

Thach WT, Bastian AJ (2003) Role of the cerebellum in the control and adaptation of gait in health and disease. In: Mori S, Stuart DG, Wisendanger M (Eds.) *Brain Mechanisms for the Integration of Posture and Movement*, pp. 411–422, Elsevier: Amsterdam.

Thach WT, Goodkin HG, Keating JG (1992a) Cerebellum and the adaptive coordination of movement. *Annual Reviews in Neuroscience* 15: 403–442.

Thach WT, Kane SA, Mink JW, Goodkin HP (1992b) Cerebellar output: multiple maps and motor modes in movement coordination. In: Llinas R, Sotelo C (Eds.) *The Cerebellum Revisited*, pp. 283–300, Springer-Verlag: New York.

Ting LH, Macpherson JM (2005) A limited set of muscle synergies for force control during a postural task. *Journal of Neurophysiology* 93: 609–613.

Todorov E (2004) Optimality principles in sensorimotor control. *Nature Neuroscience* 7: 907–915.

Todorov E, Jordan MI (2002) Optimal feedback control as a theory of motor coordination. *Nature Neuroscience* 5: 1226–1235.

Toffin D, McIntire J, Droulez J, Kemeny A, Berthoz A (2003) Perception and reproduction of force direction in the horizontal plane. *Journal of Neurophysiology* 90: 3040–3053.

Tognoli E, Lagarde J, DeGuzman GC, Kelso JA (2007) The phi complex as a neuromarker of human social coordination. *Proceedings of the National Academy of Sciences USA* 104: 8190–8195.

Tong C, Flanagan JR (2003) Task-specific internal models for kinematic transformations. *Journal of Neurophysiology* 90: 578–585.

Topka H, Konczak J, Dichgans J (1998) Coordination of multi-joint arm movements in cerebellar ataxia: analysis of hand and angular kinematics. *Experimental Brain Research* 119: 483–492.

Tremblay S, Shiller DM, Ostry DJ (2003) Somatosensory basis of speech production. *Nature* 423: 866–869.

Tresch MC, Cheung VC, d'Avella A (2006) Matrix factorization algorithms for the identification of muscle synergies: evaluation on simulated and experimental data sets. *Journal of Neurophysiology* 95: 2199–2212.

Tseng Y, Scholz JP, Schöner G (2002) Goal-equivalent joint coordination in pointing: affect of vision and arm dominance. *Motor Control* 6: 183–207.

Tseng Y, Scholz, JP, Schöner G, Hotchkiss L (2003) Effect of accuracy constraint on the underlying joint coordination of pointing movements. *Experimental Brain Research* 149: 276–288.

Tunik E, Poizner H, Levin MF, Adamovich SV, Messier J, Lamarre Y, Feldman AG (2003) Arm-trunk coordination in the absence of proprioception. *Experimental Brain Research* 153: 343–355.

Turvey MT (1990) Coordination. *American Psychologist* 45: 938–953.

Turvey MT (1998) Dynamics of effortful touch and interlimb coordination. *Journal of Biomechanics* 31: 873–882.

Tweed D (1999) Three-dimensional model of the human eye–head saccadic system. *Journal of Neurophysiology* 77: 654–666.

Tweed D, Vilis T (1987) Implications of rotational kinematics for the oculomotor systems in three dimensions. *Journal of Neurophysiology* 58: 832–849.

Twitchell TE (1951) The restoration of motor function following hemiplegia in man. *Brain* 74: 443–480.

Uchiyama T, Johansson H, Windhorst U (2003) Static and dynamic input-output relations of the feline medial gastrocnemius motoneuron-muscle system subjected to recurrent inhibition: a model study. *Biological Cybernetics* 89: 264–273.

Ugawa Y, Terao Y, Hanajima R, Sakai K, Kanazawa I (1995) Facilitatory effect of tonic voluntary contraction on responses to motor cortex stimulation. *Electroencephalography and Clinical Neurophysiology* 97: 451–454.

Vallbo A (1970) Discharge patterns in human muscle spindle afferents during isometric voluntary contractions. *Acta Physiologica Scandinavica* 80: 552–566.

Vallbo A (1974) Human muscle spindle discharge during isometric voluntary contractions. Amplitude relations between spindle frequency and torque. *Acta Physiological Scandinavica* 90: 310–336.

Van Asten WN, Gielen CC, Denier van der Gon JJ (1988) Postural adjustments induced by simulated motion of differently structured environments. *Experimental Brain Research* 73: 371–383.

Van der Kamp J, Steenbergen B (1999) The kinematics of eating with the spoon: bringing the food to the mouth or the mouth to the food? *Experimental Brain Research* 129: 68–76.

van der Linden MH, Marigold DS, Gabreëls FJM, Duysens J (2007) Muscle reflexes and synergies triggered by an unexpected support surface height during walking. *Journal of Neurophysiology* 97: 3639–3650.

Vandervoort AA, Hayes KC (1989) Plantarflexor muscle function in young and elderly women. *European Journal of Applied Physiology and Occupational Physiology* 58: 389–394.

Van Deursen RW, Sanchez MM, Ulbrecht JS, Cavanagh PR (1998) The role of muscle spindles in ankle movement perception in human subjects with diabetic neuropathy. *Experimental Brain Research* 120: 1–8.

Van Deursen RW, Simoneau GG (1999) Foot and ankle sensory neuropathy, proprioception, and postural stability. *Journal of Orthopedics, Sports, and Physical Therapy* 29: 718–726.

Van Heijst JJ, Vos JE, Bullock D (1998) Development in a biologically inspired spinal neural network for movement control. *Neural Networks* 11: 1305–1316.

Van Hemmen JL (2007) Biology and mathematics: a fruitful merger of two cultures. *Biological Cybernetics* 97: 1–3.

Van Opstal J (1993) Representation of eye position in three dimensions. In: Berthoz A (Ed.) *Multi-Sensory Control of Movement*, pp. 27–41, Oxford University Press: Oxford.

Vereijken B, van Emmerick REA, Whiting HTA, Newell KM (1992) Free(z)ing degrees of freedom in skill acquisition. *Journal of Motor Behavior* 24: 133–142.

Vernazza-Martin S, Martin N, Massion J (1999) Kinematic synergies and equilibrium control during trunk movement under loaded and unloaded conditions. *Experimental Brain Research* 128: 517–526.

Vilensky JA, Moore AM, Eidelberg E, Walden JG (1992) Recovery of locomotion in monkeys with spinal cord lesions. *Journal of Motor Behavior* 24: 288–296.

Viviani P (1985) Segmentation and coupling in complex movements. *Journal of Experimental Psychology* 11: 828–845.

Viviani P, McCollum G (1983) The relation between linear extend and velocity in drawing movements. *Neuroscience* 10: 211–218.

Viviani P, Terzuolo C (1980) Space-time invariance in learned motor skills. In: Stelmach GE, Requin J (Eds.) *Tutorials in Motor Behavior*, pp. 525–533, Amsterdam: North-Holland

Von Holst E (1954) Relations between the central nervous system and the peripheral organs. *British Journal of Animal Behaviour* 2: 89–94.

Von Holst E, Mittelstaedt H (1950/1973) Daz reafferezprincip. Wechselwirkungen zwischen Zentralnerven-system und Peripherie, Naturwiss. 37: 467–476, 1950. The reafference principle. In: *The Behavioral Physiology of Animals and Man. The Collected Papers of Erich von Holst Vol. 1.* Martin R (translator) University of Miami Press: Coral Gables, FL, pp. 139–173.

Vuillerme N, Chenu O, Demongeot J, Payan Y (2007) Controlling posture using a plantar pressure-based, tongue-placed tactile biofeedback system. *Experimental Brain Research* 179: 409–414.

Wahnoun R, He J, Helms Tillery SI (2006) Selection and parameterization of cortical neurons for neuroprosthetic control. *Journal of Neural Engineering* 3: 162–171.

Walmsley A, Williams L, Rosenbaum D, Latash ML (2001) Equilibrium-point hypothesis and equifinality of voluntary movements under transient perturbations. In: Gantchev N (Ed.) *From Basic Motor Control to Functional Recovery—II*, pp. 309–316, Academic Publ. House "Prof. M. Drinov": Sofia.

Wang J, Sainburg RL (2006) Interlimb transfer of visuomotor rotations depends on handedness. *Experimental Brain Research* 175: 223–230.

Wang J, Stelmach GE (1998) Coordination among the body segments during reach-to-grasp action involving the trunk. *Experimental Brain Research* 123: 346–350.

Wang W, Chan SS, Heldman DA, Moran DW (2007) Motor cortical representation of position and velocity during reaching. *Journal of Neurophysiology* 97: 4258–4270.

Wang Y, Asaka T, Zatsiorsky VM, Latash ML (2006) Muscle synergies during voluntary body sway: combining across-trials and within-a-trial analyses. *Experimental Brain Research* 174: 679–693.

Wang Y, Zatsiorsky VM, Latash ML (2005) Muscle synergies involved in shifting center of pressure during making a first step. *Experimental Brain Research* 167: 196–210.

Wang Y, Zatsiorsky VM, Latash ML (2006) Muscle synergies in preparation to a step made under self-paced and reaction-time instructions. *Clinical Neurophysiology* 117: 41–56.

Wannier T, Liu J, Morel A, Jouffrais C, Rouiller EM (2002) Neuronal activity in primate striatum and pallidum related to bimanual motor actions. *Neuroreport* 13: 143–147.

Ward NS (2005) Neural plasticity and recovery of function. *Progress in Brain Research* 150: 527–535.

Wasserman EM, Kimbrell TA, George MS, Danielson AL, Herscovitch P, Hallett M, Cohen LG (1997) Local and distant changes in cerebral glucose metabolism during repetitive transcranial magnetic stimulation (rTMS). *Neurology* 48: A107.

Wehr M, Zador AM (2003) Balanced inhibition underlies tuning and sharpens spike timing in auditory cortex. *Nature* 426: 442–446.

Weiss AC, Weiller C, Liepert J (2003) Pre-movement motor excitability is reduced ipsilateral to low force pinch grips. *Journal of Neural Transmission* 110: 201–208.

Weiss EJ, Flanders M (2004) Muscular and postural synergies of the human hand. *Journal of Neurophysiology* 92: 523–535.

Welford AT, Norris AH, Schock NW (1969) Speed and accuracy of movement and their changes with age. In: WG Koster (Ed.) *Attention and Performance II*, pp. 3–15, North-Holland: Amsterdam.

Welsh JP, Llinas R (1997) Some organizing principles for the control of movement based on olivocerebellar physiology. *Progress in Brain Research* 114: 449–461.

Wessberg J, Nicolelis MA (2004) Optimizing a linear algorithm for real-time robotic control using chronic cortical ensemble recordings in monkeys. *Journal of Cognitive Neuroscience* 16: 1022–1035.

Westheimer G (1957) Kinematics of the eye. *Journal of the Optical Society of America* 47: 967–974.

Westling G, Johansson RS (1984) Factors influencing the force control during precision grip. *Experimental Brain Research* 53: 277–284.

Whitney DE (1969) Resolved motion rate control of manipulators and human prostheses. *IEEE Transactions on Man Machine Systems* 10: 47–53.

Wiesendanger M, Serrien DJ (2004) The quest to understand bimanual coordination. *Progress in Brain Research* 143: 491–505.

Wiley RH (1983) The evolution of communication. In: Halliday TR, Slater PJB (Eds.) *Animal Behavior 2: Communication*, pp. 156–189, Freeman: New York.

Wilson EO (1975) *Sociobiology,* Harvard University Press: Cambridge, MA.

Winegard KJ, Hicks AL, Vandervoort AA (1997) An evaluation of the length-tension relationship in elderly human plantarflexor muscles. *Journal of Gerontology Series A Biological Sciences and Medical Sciences* 52: B337–B343.

Wing AM (2000) Motor control: mechanisms of motor equivalence in handwriting. *Current Biology* 10: R245–R248.

Winstein CJ, Abbs JH, Petashnick D (1991) Influences of object weight and instruction on grip force adjustments. *Experimental Brain Research* 87: 465–469.

Winter DA, Prince F, Frank JS, Powell C, Zabjek KF (1996) Unified theory regarding A/P and M/L balance in quiet stance. *Journal of Neurophysiology* 75: 2334–2343.

Wolpaw JR (1987) Operant conditioning of primate spinal reflexes: the H-reflex. *Journal of Neurophysiology* 57: 443–459.

Wolpaw JR (2007) Brain-computer interfaces as new brain output pathways. *The Journal of Physiology* 579: 613–619.

Wolpaw JR, Carp JS (1993) Adaptive plasticity in spinal cord. *Advances in Neurology* 59: 163–174.

Wolpaw JR, Tennissen AM (2001) Activity-dependent spinal cord plasticity in health and disease. *Annual Reviews in Neuroscience* 24: 807–843.

Wolpert DM, Miall RC, Kawato M (1998) Internal models in the cerebellum. *Trends in Cognitive Science* 2: 338–347.

Woods TM, Recanzone GH (2004) Cross-modal interactions evidenced by the ventriloquism effects in humans and monkeys. In: Calvert G, Spence C, Stein BE (Eds.) *The Handbook of Multi-Sensory Processes*, pp. 35–48, MIT Press: Cambridge, MA.

Woollacott M, Inglin B, Manchester D (1988) Response preparation and posture control. Neuromuscular changes in the older adult. *Annals of the New York Academy of Sciences* 515: 42–53.

Woollacott MH, Shumway-Cook A (1990) Changes in posture control across the life span—a systems approach. *Physical Therapy* 70: 799–807.

Xerri C, Merzenich MM, Peterson BE, Jenkins W (1998) Plasticity of primary somatosensory cortex paralleling sensorimotor skill recovery from stroke in adult monkeys. *Journal of Neurophysiology* 79: 2119–2148.

Yang J-F, Scholz JP (2005) Learning a throwing task is associated with differential changes in the use of motor abundance. *Experimental Brain Research* 163: 137–158.

Yang J-F, Scholz JP, Latash ML (2007) The role of kinematic redundancy in adaptation of reaching. *Experimental Brain Research* 176: 54–69.

Zaal FT, Daigle K, Gottlieb GL, Thelen E (1999) An unlearned principle for controlling natural movements. *Journal of Neurophysiology* 82: 255–259.

Zahariev MA, MacKenzie CL (2007) Grasping at 'thin air': multi-modal contact cues for reaching and grasping. *Experimental Brain Research* 180: 69–84.

Zatsiorsky VM (1997) On muscle and joint viscosity. *Motor Control* 1: 299–309.

Zatsiorsky VM (1998) *Kinematics of Human Motion*, Human Kinetics: Champaign, IL.

Zatsiorsky VM (2002) *Kinetics of Human Motion,* Human Kinetics: Champaign, IL.

Zatsiorsky VM, Duarte M (2000) Rambling and trembling in quiet standing. *Motor Control* 4: 185–200.

Zatsiorsky VM, Gao F, Latash ML (2003) Prehension synergies: effects of object geometry and prescribed torques. *Experimental Brain Research* 148: 77–87.

Zatsiorsky VM, Gao F, Latash ML (2005) Motor control goes beyond physics: differential effects of gravity and inertia on finger forces during manipulation of hand-held objects. *Experimental Brain Research* 162: 300–308.

Zatsiorsky VM, Gregory RW, Latash ML (2002a) Force and torque production in static multi-finger prehension: biomechanics and control. Part I. Biomechanics. *Biological Cybernetics* 87: 50–57.

Zatsiorsky VM, Gregory RW, Latash ML (2002b) Force and torque production in static multi-finger prehension: biomechanics and control. Part II. Control. *Biological Cybernetics* 87: 40–49.

Zatsiorsky VM, King DL (1998) An algorithm for determining gravity line location from posturographic recordings. *Journal of Biomechanics* 31: 161–164.

Zatsiorsky VM, Latash ML (2004) Prehension synergies. *Exercise and Sport Science Reviews* 32: 75–80.

Zatsiorsky VM, Latash ML, Gao F, Shim JK (2004) The principle of superposition in human prehension. *Robotica* 22: 231–234.

Zatsiorsky VM, Li ZM, Latash ML (1998) Coordinated force production in multi-finger tasks: finger interaction and neural network modeling. *Biological Cybernetics* 79: 139–150.

Zatsiorsky VM, Li ZM, Latash ML (2000) Enslaving effects in multi-finger force production. *Experimental Brain Research* 131: 187–195.

Zehr EP, Balter JE, Ferris DP, Hundza SR, Loadman PM, Stoloff RH (2007) Neural regulation of rhythmic arm and leg movement is conserved across human locomotor tasks. *The Journal of Physiology* 582: 209–227.

Zhang W, Zatsiorsky VM, Latash ML (2006) Accurate production of time-varying patterns of the moment of force in multi-finger tasks. *Experimental Brain Research* 175: 68–82.

Zhang W, Zatsiorsky VM, Latash ML (2007) Finger synergies during multi-finger cyclic production of moment of force. *Experimental Brain Research* 177: 243–254.

Zigmond MJ, Bloom FE, Landis SC, Roberts JL, Squire LR (Eds.) (1999) *Fundamental Neuroscience*, Academic Press: San Diego, CA.

Zimmerman SD, McCormick RJ, Vadlamudi RK, Thomas DP (1993) Age and training alter collagen characteristics in fast- and slow-twitch rat limb muscle. *Journal of Applied Physiology* 75: 1670–1674.

Zoia S, Castiello U, Blason L, Scabar A (2005) Reaching in children with and without developmental coordination disorder under normal and perturbed vision. *Developmental Neuropsychology* 27: 257–273.

INDEX

Abnormal movement synergy, 228, 230–231, 264
Abundance principle, 59, 60, 227, 330
Action–perception coupling, 25, 349–352
Action potentials, 36–37, 64, 66, 69, 70, 71–72, 73, 74, 77, 107
Active inhibition, within CNS, 27
Aging effects, 66
  on motor coordination, 242–248
  on muscles and neurons, 238–242
Agonist–antagonist muscles, 260
Alpha–gamma coactivations, 74
Alpha-motoneurons, 64, 65, 69, 241, 301
Altenburger, Hans, 29
Alzheimer's disease, 239, 249
American Society of Biomechanics, 21–22
Angular velocity, in multi-joint movement, 328, 329
Animate objects, movements of, 18–19, 48
Ankle strategy, 173
Antagonist muscle, 21, 93
Anticipatory coarticulation, 358
Anticipatory co-variation. *See* Anticipatory synergy adjustment (ASA)
Anticipatory postural adjustment (APA), 170, 173–174, 201, 251
Anticipatory synergy adjustment (ASA), 199–201, 246
APA. *See* Anticipatory postural adjustment (APA)
Apology for Raymond Sebond, 21
Aristotle, 20
Articular receptors, 75
Artificial neural network, 331–334
ASA. *See* Anticipatory synergy adjustment (ASA)
Astruc, Jean, 21
Atomic theory of universe, 19
Atomism, 287
Atypical synergy, 39, 231
  and cerebellum changes, 261–263
Autism, 262
Autogenic (homonymous) reflexes, 92

Babinski, Joseph Felix Francois, 35
Barlaam, 5
Basal ganglia, 307–309
Beritashvili, Nicolas, 27
Bernstein, Alexander Nikolaevich, 31
Bernstein, Alexandra Karlovna, 31
Bernstein, Nikolai Alexandrovich, 1–2, 27, 29, 54, 56–57, 82, 85, 86, 168

Bernstein, Nikolai Alexandrovich (*contd.*)
and movement science, 30–35
Bernstein, Sergei Natanovich, 31
Bernstein problem, 33
Biological movement, 19, 22, 228
problems with studying, 45–50
Biological synergies, 12–13
Biomechanics. *See* Movement studies
Body–soul relationship, 21, 22, 24
BOLD (blood-oxygenation-level-dependent), 83
Borelli, Giovanni Alfonso, 21–22
Brain imaging techniques, 81–83
Braune, Christian Wilhelm, 24
Brown, Thomas Graham, 28

Catch property, 233
Cattell, James McKenn, 25
CBC. *See* Central back-coupling model (CBC)
Center of mass, 46, 145, 158, 171, 174–176, 185, 221, 222, 352
Center of pressure, 148, 170, 220–224
Center-out motor task, 79
Central back-coupling model (CBC), 333–334, 344
Central nervous system (CNS), 18, 201
active inhibition in, 27
motor redundancy and, 38, 227–228
plasticity in, 47, 233–238
Central pattern generators (CPGs), 295, 297
"Cerebellar asynergies", 35
Cerebellar nuclei, 302
Cerebellum, 302–307
atypical synergy and, 261–263
climbing fibers and, 302–303
mossy fibers and, 302
multi-joint actions and, 306
Chain effects, 212–213
Chromosome-21, 248
Chronic spinal animals, 295
Climbing fibers, 302–303
Closed-loop control. *See* Feedback control
CNS. *See* Central nervous system (CNS)
Coactivation command (*c*), 103, 104, 210, 253, 254
Co-contraction, 252–253, 254, 343–344
Conditioned reflexes, 26–27, 92
Control. *See* EP control; Feedback control; Feed-forward control; Motor control; Negative feedback control; Stochastic optimal control; Sensory feedback-based control; Positive feedback control
Control hypotheses, 129, 175
Control theory, 323–325
Control trajectory, 104, 105
Convergence, 133, 234
Coriolis force, 110, 111
Corollary discharges. *See* Efferent copy
CPGs. *See* Central pattern generators (CPGs)
Cross-bridges, 66
Cybernetics, 33–34

Damping, 46, 232
DCD. *See* Developmental coordination disorder (DCD)
Deafferented patients, 189
Democritus, 19
Descartes, René, 21
Developmental coordination disorder (DCD), 262
Diabetes, 345
Digit interaction and indices, 134–139
Direct model, 78
Divergence, 133, 234
Doctrine of synergy, 5
Donders' law, 25, 339
Dopamine, 240
Down syndrome persons, 248
atypical synergy and cerebellum changes in, 261–263
movements in, 249–254
multi-finger coordination in, 254–257
practice of movements in, 257–261
Drug, 93
DuBois-Reymond, Etienne, 22
Dynamic systems, 20, 56, 113, 117, 149–150

Economy principle, 59, 60
EEG. *See* Electroencephalography (EEG)
Efference copy. *See* Efferent copy
Efferent copy, 77, 350
Eigen-movements, 173
Elasticity, 22, 46, 89, 90, 96
Electroencephalography (EEG), 82
Electromyography (EMG), 140–142
alignment, 142
averaging, 142
filtering, 141–142
intramuscular, 140
rectification, 141
surface, 141
Elemental variable, 120
joints, 131–132
modes as, 131
force modes, 132–139
Jacobian, experimental identification of, 147–148
muscle modes, 139–147

EMG. *See* Electromyography (EMG)
Engrams, 57
Enslaving, 132–134, 192
EP control:
  of multi-muscle systems, 105–109
  of simple systems, 100–104
Equifinality, 85, 110
Equilibrium-point (EP) hypothesis, 62, 88
  EP control of multi-muscle systems, 105–109
  EP control of simple systems, 100–104
  experimental foundations of, 89–100
  mass–spring analogy, 109–117
  motor synergies and, 339–344
    single-muscle control, 341–342
  TMS and, 310–314
  trajectories within, 104–105
Equilibrium trajectory, 105
Error compensation, 10, 14, 41, 61, 126, 151
  See also Flexibility/stability
Error correction, 334
Extrafusal fibers, 73
Extrinsic finger flexor muscles, 135, 136, 239

FDP. *See* Flexor digitorum profundis (FDP)
FDS. *See* Flexor digitorum superficialis (FDS)
Feedback control, 325–328
  gain, 327
  of multi-joint synergy, 328–329
  negative, 326–327
  positive, 327
  time delay, 327–328
Feed-forward control, 325–328
  with good and bad variability specification, 337–339
Feigenberg, Josef, 33
Fenn, Wallace Osgood, 28
Fenn effect, 28
Feynman, Richard, 46
Fiber peripheral neuropathy, 241, 348, 349
Finger interaction:
  characteristics, 135
  enslaving effects, matrix of, 137
  See also Thumb
Fischer, Otto, 24
Fitts' law, 258, 259
Flexibility/stability, 13
  See also Error compensation
Flexion center, 23
Flexor digitorum profundis (FDP), 135, 136
Flexor digitorum superficialis (FDS), 135
Florensky, Pavel Alexandrovich, 321
fMRI. *See* Functional magnetic resonance imaging (fMRI)
Fookson, Olga, 54
Force deficit, 135
Force modes, 132–139
Force-stabilizing synergy, 260–261
Fork strategy, 229, 257, 260
Foster, Sir Michael, 25
Freezing degrees-of-freedom, 266, 273
Friction, 243
Functional magnetic resonance imaging (fMRI), 290

Galen, 21
Galvani, Luigi, 22
Gamma-motoneurons, 74
Gastev, Alexei, 31
GDR. *See* Generalized displacement reflex (GDR)
Gelfand, Israel M., 34, 49, 51–56, 322
"Gelfand seminar", 34
Generalized displacement reflex (GDR), 106, 179
Generalized motor program, 86
Gibson, James, 350
Golgi tendon organs, 74–75
Granule cells, 302
Grasping, 204, 207
Grip force, 205, 243
Gurfinkel, Victor, 53

H-reflex, 93
Hand force production, 224–226
Haptic perception, 72
Healthy aging, 239
Henneman principle, 38, 70
Heracleitus, of Ephesus, 19
Heteronymous (heterogenic) reflexes, 92
Hill, A.V., 68
Hip strategy, 242
Homonymous (autogenic) reflexes, 92
Human body:
  adaptation to force fields, effects of, 79–81
  brain imaging techniques, 81–83
  muscle, slow and visco-elastic, 65–70
  neural pathways, long and slow, 70–72
  reflexes and nonreflexes, 91–96
  sensors, 72–77
Human brain, 46, 140, 218, 234
  imaging techniques, 81–83
  stroke and, 263–264
Hypotonia, 250

Inanimate objects, movement of, 17–18, 48, 49
Inanimate synergies, 7–11

Inborn reflexes, 92, 93
Indeterminicity principle, 48, 232–233
Information exchange, 358–359
Institute of Labor, 31
Integration, 291
Interaction, 135, 137
  digit interaction and indices, 134–139
  maximal interaction principle, 62
  minimal interaction principle, 61–63
  sensory-motor interactions, 349–352
Interoceptive information, 72
Intrafusal fibers, 73
Intramuscular EMG, 140
Intrinsic flexor muscles, 134–135, 239
Invariant characteristic (IC) of muscle, 100, 351
Inverse dynamics problem, 64
Inverse kinematics problem, 63, 176, 331
Inverse model, 65
Isometric contraction, 76

Jackson, J. Hughlings, 35
Jacobian, 147, 224
Jacobian augmentation technique, 334
Joint displacement, 171–174
Joint Session of the Academy of Sciences and the Academy of Medical Sciences, 30
Joints, 131–132, 151, 171

Kalman filter, 78
Kimocyclography, 29, 31–32
Kinematic synergy, 167, 265
  multi-joint pointing, 182–184
  postural synergy, in standing, 170–174
  principal component analysis, 168
  quick-draw pistol shooting, 184–188
  reaching, 176–180
    in changing force field, 180–182
  sit-to-stand task, 174–176
Kinematics, 31–33, 259
Kinesthetic illusions, 77, 114
Kinetic synergy, 188–192
Knowledge hypothesis, 106

λ-model, 101, 115
Language, as synergy, 357–360
Latash, Lev Pavlovich, 26, 33
Law all-or-none, 66
Learning movement synergies:
  kinematic tasks, practice, 269–273
  kinetic tasks, practice, 274–279
  motor learning, traditional views on, 266–268
  plastic neural changes, 279–283
  with practice, 268–269
Leonardo da Vinci, 21
Leontjev, 52
Listing's law, 339
Ljapunov, A.A., 53
Locomotion, 28, 295
Lombard, Warren Plimpton, 29
Lombard's paradox, 29
Lysenko, Trofim, 30

Magnetic resonance imaging (MRI), 83
Magnetoencephalography (MEG), 83
Marey, Etienne-Jules, 24
Mass–spring analogy, 109–117
Matteucci, Carlo, 22
Matthews, Peter, 101
Maximal interaction principle, 62
McGurk effect, 359
Mechanistic modeling, 321–322
MEG. *See* Magnetoencephalography (MEG)
Mental rotation, 315
Meyerhold, Vsevolod, 52
Minimal end-state action principle, 89, 342
Minimal interaction principle, 61–63
Minimum-jerk criterion, 167
Mongolism, 248
Monosynaptic reflexes, 92–93
Montaigne, Michel Eyguem de, 21, 25
Moore–Penrose pseudo-inverse procedure, 338–339
Mosaic Down syndrome. *See* Down syndrome persons
Mossy fibers, 302
Motor control, 51
  EP-hypothesis and, 88
  movement studies and, 18, 22
  principle of minimal interaction and, 61–63
  programs and internal models, 63
  structural units and, 56–60
Motor coordination, 242–248
Motor learning, 257–258
  traditional views on, 266–268
Motor primitives, 293, 294
Motor redundancy, 33, 89, 227, 339
  CNS and, 227–228
  feed-forward control model of, 337–339
  neural networks and, 331–334
  synergy history and, 35–45
Motor synergies, 339
  formal models, 322
  zoo of, 167
Motor units, 37, 69–70, 116, 140

Motor variability:
analysis of, 148–155
computational tools, 155
surrogate data sets, analysis of, 159–162
modes, 131
timing synergies, 162–165
UCM hypothesis, 120–131
and principal component analysis, 155–159
Movement kinematics, 19, 24, 168
Movement studies, 17
in ancient Greece, 19–20
biological movement, problems with studying, 45–50
branches, 17–18
in Down syndrome persons 249–254
in frog, 23
indeterminicity principle, 232–233
motor control and, 18, 22
movement science in Soviet Union 30–35
photography and, 24
renaissance, 20–22
in Rome, 20
synergy history and motor redundancy, 35–45
in twentieth century, 25–30
MRI. *See* Magnetic resonance imaging (MRI)
Multi-digit synergy, 192–204
anticipatory synergy adjustment 199–201
emergence and disappearance of 197–199
multi-finger pressing
force and moment stabilization during, 192–195
timing errors, 195–197
Multi-finger coordination:
in Down syndrome persons, 254–257
Multi-finger pressing, 192–195
Multi-joint movement:
end-point trajectory and, 342–343
synergic control of, 328–329
Multi-joint pointing, 182–184
Multi-joint reaching, 180–182
Multi-muscle synergy, 217–226
anticipatory postural adjustment 220–222
in hand force production, 224–226
postural adjustments, to stepping 222–224
Multi-sensory mechanisms, 354–356
Muscle activation, 105, 140, 259, 325, 340
during fast voluntary movements, 29
reciprocal patterns of, 250
Muscle contractions, 65–70
Muscle elasticity, 90
Muscle fibers, 140, 239, 240–241
Muscle mode, 139–140
electromyography, 140–142
push-back, 145, 221, 223
push-forward, 145, 221, 223
Muscle physiology, 28
Muscle reflexes, 91
Muscle spindles, 72, *73*, 75
Muscle vibration:
sensory and motor effects of, 346–348
Muscles, 37, 134–135
activation, 29, 41, 301, 342
aging effects on, 238–242
contraction, 64, 66
current understanding of, 28
effects of age on, 238–242
extrinsic finger flexor muscles, 135, 136
intrinsic flexor muscles, 134–135, 136
modes, 139–140, 145
slow and visco-elastic, 65–70
synergies, 35, 217–220, 224–226
vibration, 348
See also individual entries
Muscular structure, 239
Muybridge, Eadweard J., 24
Myelinated fibers, 71
The Mystery Spot, 353

National Institutes of Health, 235
Negative feedback control, 326–327
Nerve spirits, 22
Neural pathways, 70–72
Neuromuscular synapses, 66
Neurons, aging effects on, 238 242
Neurophysiological mechanism, of synergy, 285
and basal ganglia, 307–309
brain structure and function
and cerebellum, 302–307
cortex of large hemispheres, 309
neuronal populations, 314–319
TMS and equilibrium-point hypothesis, 310–314
in spinal cord, 290–302
Newton, Isaac, 22
Nonindividualized control principle, 57
Nonreflexes, 91
Non-synergies, examples, 1–5
Normal synergy, 227–232
Normal tone, 250

Oligosynaptic reflexes, 93
On Dexterity and Its Development, 1, 266
On the Construction of Movements, 33
Open-loop control. *See* Feed-forward control
Optimal control theory, 329–331
Optimization, 176–179
Orbeli, Leon, 27
Oxygen debt, 28

Palamas concept, of synergy, 5–7
Parallel fibers, 302
Parkinson's disease, 239, 240
Pavlov, Ivan Petrovich, 23, 26, 27, 92
PCA. *See* Principal component analysis (PCA)
Perception:
  and action relationship, 349–352
  of muscle length and force, 114
Performance variable, 120
Pflüger, Eduard Friedrich Wilhelm, 23
Phasic polysynaptic reflexes, 92, 93–95
Physics, 17–18, 24–25
Physics of Living Systems, 46
Plasticity, 287, 355
  in CNS, 47, 233–238
Plato, 6, 20
Point-to-point reaching, 176–180
Polysynaptic reflexes
  phasic polysynaptic reflexes, 93–95
  tonic stretch reflex, 95, 96–100
Positive feedback control, 327
Postural synergy, in standing, 170–174
Posture-based planning hypothesis, 178
Posture-stabilizing mechanism, 98
Preferred direction, 315
Prehensile synergy, 204
  chain effects, 212–213
  hierarchical control of, 207–209
  hierarchies, of synergies, 213–217
  object, 204
  superposition principle, 209–212
Presynaptic inhibition, 347
Principal component analysis (PCA), 172, 219
  kinematic synergy and, 168
  UCM hypothesis and, 155–159
Prismatic grasp, 206
Proprioception, 72
Proprioceptive feedback, 335
Purkinje cell, 302, 305–306
Push-back muscle mode, 145, 221, 223
Push-forward muscle mode, 145, 221, 223
Push-side muscle mode, 145, 223
Pythagoras, 19

Qualia, 355
Quick-draw pistol shooting, 184–188

Raikin, Arkady, 4
Ramon Y Cajal, Santiago, 25
Reaction time. *See* Simple reaction time
Reactive forces, 185
Re-afference principle, 99
Reciprocal command (*r*), 103, 104, 210, 253, 254
Reciprocal inhibition, 93
Reciprocal muscle activation patterns 250, 251, 253
Reductionism, 287
Redundancy. *See* Motor redundancy
Reference configuration, 106, 179, 341, 342
Reflex-like actions, 242
Reflex reversal, 299
Reflexes, 21, 72, 242
  autogenic (homonymous) reflexes, 92
  conditioned reflexes, 26–27, 92
  generalized displacement reflex (GDR), 106, 179
  H-reflex, 93
  heteronymous (heterogenic) reflexes, 92
  homonymous (autogenic) reflexes, 92
  human body, 91–96
  inborn reflexes, 92, 93
  monosynaptic reflexes, 92–93
  muscle reflexes, 91
  oligosynaptic reflexes, 93
  phasic polysynaptic reflexes, 92, 93–95
  tonic stretch reflex, 95, 96–100
Renshaw cells, 301–302
Robotic arm, 272
Russian (Soviet) school of motor control, 34

Sarcopenia, 239
Saw-tooth tetanus, 67
Sechenov, Ivan Mikhailovich, 23, 25, 76–77, 349
Self-motion, 20
Sensors, 72–77
Sensory feedback-based control:
  and central pattern generation, 28
Sensory-motor interactions:
  action and perception relationship, 349–352

Sensory synergy, 115, 344–356
multi-sensory mechanisms, 354–356
in neurological disorders, 345–349
sensory-motor interactions 349–352
in vertical posture, 352–354
Series elastic element, 28
Servo-hypothesis, 30
Sharing pattern, 9, 10, 14, 42, 135
Sherrington, Sir Charles, 23, 25, 27, 96
Shtern, Lina Solomonovna, 26
Simple reaction time paradigm, 84
Single-trial UCM analysis, 151–155, 257
Sit-to-stand task, 174–176
Size principle, 70
Skriabin, Alexander, 356
Smooth tetanus, 67
Somatosensory cortex, 235
Soul and movement studies, 19–20, 21
Soviet school of movement science, 31
Speed–accuracy trade-off, 258–259
Spinal cord, 238, 290–302
Spinal frog, 42–43, *45*
Spring-like behavior, 109, 110, 113
St. Augustine, 20
St. Palamas, Gregory:
biography, 5
philosophical views of, 5–7
Stalin, 27
State variables, 113
Stochastic optimal control, 330–331
Stroke, 237, 239, 263
clinical features of, 264
Structural units, 56–60
axioms, 58–59
Superposition principle, 59, 209–212
Surface EMG, 141
Synapse, 25, 27
Synapsis. *See* Synapse
Synergies:
after stroke, 263–266
components of, 13–15
examples, 1–5
motor redundancy problem and 35–45
Synesthesia, 356
Syphilis, 345

Task dependence, 14–15
Tendon tap reflex (T-reflex), 93
Thumb, 137–139
Timing errors, in multi-digit synergy, 195–197
Timing synergies, 162–165
TMS. *See* Transcranial magnetic stimulation (TMS)
Tonic stretch reflex, 92, 95, 96–100, 107, 108, 109, 301, 341
Tonic vibration reflex (TVR) mechanism, 346–347
Transcranial magnetic stimulation (TMS), 235–237, 274, 275, 280–282
and equilibrium-point hypothesis 310–314
Tri-phasic EMG pattern, 250, 251
Trisomy-21, 248
Tsetlin, Michael, 34, 53–54
TVR. *See* Tonic vibration reflex (TVR) mechanism
Twitch contraction, 66, 67
Two-third power law, 167
Typical reciprocal pattern, 253
Typical synergy, 231

UCM. *See* Uncontrolled manifold (UCM) hypothesis
Uncontrolled manifold (UCM) hypothesis, 120–131, 256
control hypotheses, 129
finger force, 122–127
correlation analysis, 126–127
variance, 123
variance per dimension, 124–126
linearized approximation, 129–130
Unzer, J.A., 23

Variability, across systems, 335
See also Motor variability
Ventroloquist effect, 356
Vertical posture, sensory synergies in, 352–354
graviceptors, 352, 353
vestibular system, 352, 353
vision, 352, 353
Virtual finger, 207–208, 214
Visual feedback, 335
von Helmholtz, Hermann Ludwig Ferdinand, 24–25, 76, 349–350

Wachholder, Kurt, 29, 90
Weber, Eduard Friedrick Wilhelm 23–24
Weber, Emst Heinrich, 23

Weber, Wilhelm Eduard, 23
Weber–Fechner law, 335–336
Wholism, 287
Wiener, Norbert, 33
Willis, Thomas, 21
Woodworth, Robert Sessions, 25
Wrist–elbow synergy, 41–42, 189, 190–191

www.ingramcontent.com/pod-product-compliance
Ingram Content Group UK Ltd.
Pitfield, Milton Keynes, MK11 3LW, UK
UKHW020850150926
475595UK00003B/106